MASSEY-FERGUSON

SHOP MANUAL MF-202

Models ■ MF 175 ■ MF 180

Models ■ MF 205 ■ MF 210 ■ MF 220

Models ■ MF 2675 ■ MF 2705

Models ■ MF 2745 ■ MF 2775 ■ MF 2805

I&T

SHOP MANUAL

Information and Instructions

This shop manual contains several sections each covering a specific group of wheel type tractors. The Tab Index on the preceding page can be used to locate the section pertaining to each group of tractors. Each section contains the necessary specifications and the brief but terse procedural data needed by a mechanic when repairing a tractor on which he has had no previous actual experience.

Within each section, the material is arranged in a systematic order beginning with an index which is followed immediately by a Table of Condensed Service Specifications. These specifications include dimensions, fits, clearances and timing instructions. Next in order of arrangement is the procedures paragraphs.

In the procedures paragraphs, the order of presentation starts with the front axle system and steering and proceeding toward the rear axle. The last paragraphs are devoted to the power take-off and power lift systems. Interspersed where needed are additional tabular specifications pertaining to wear limits, torquing, etc.

HOW TO USE THE INDEX

Suppose you want to know the procedure for R&R (remove and reinstall) of the engine camshaft. Your first step is to look in the index under the main heading of ENGINE until you find the entry "Camshaft." Now read to the right where under the column covering the tractor you are repairing, you will find a number which indicates the beginning paragraph pertaining to the camshaft. To locate this wanted paragraph in the manual, turn the pages until the running index appearing on the top outside corner of each page contains the number you are seeking. In this paragraph you will find the information concerning the removal of the camshaft.

More information available at haynes.com
Phone: 805-498-6703

Haynes UK
Sparkford Nr Yeovil
Somerset BA22 7JJ England

Haynes North America, Inc
859 Lawrence Drive
Newbury Park
California 91320 USA

ISBN-10: 0-87288-362-0
ISBN-13: 978-0-87288-362-8

MASSEY-FERGUSON

Models ■ MF 175 ■ MF 180

Previously contained in I&T Shop Manual No. MF-29

SHOP MANUAL

MASSEY-FERGUSON

MODELS
MF175   MF180

Tractor serial number stamped on instrument panel name plate.

Engine serial number stamped on side of engine.

INDEX (By Starting Paragraph)

	MF175	MF180		MF175	MF180
BELT PULLEY	202	202	**NON-DIESEL ENGINE**		
			Assembly R&R	31, 55	31, 55
BRAKES			Camshaft	43, 70	43, 70
Adjustment	200	200	Connecting rods and bearings	47, 75	47, 75
R&R and Overhaul	201	201	Crankshaft and bearings	48, 76	48, 76
			Cylinder head	32, 56	32, 56
CARBURETOR (Gasoline)	116	116	Cylinder sleeves	45, 73	45, 73
(LP-Gas)	119	119	Engine balancer	49, 77	49, 77
			Flywheel	52, 81	52, 81
CLUTCH (Engine)	150	150	Front oil seal	41, 62	41, 62
			Ignition timing	144	144
COOLING SYSTEM			Main bearings	48, 76	48, 76
Pump and fan	143	143	Oil Pump	53, 83	53, 83
Radiator	141	141	Pistons	45, 73	45, 73
Thermostat	142	142	Piston and rod removal	44, 72	44, 72
			Piston rings	45, 73	45, 73
DIESEL FUEL SYSTEM			Relief valve	54, 84	54, 84
Bleeding system	124	124	Rear oil seal	51, 80	51, 80
Cold weather starting aid	137	137	Rocker arms	40, 60	40, 60
Fuel Filters	123	123	Tappets	39, 59	39, 59
Injection nozzles			Timing gear cover	41, 62	41, 62
Overhaul	133	133	Timing gears	42, 63	42, 63
Remove and reinstall	127	127	Valve guides	36, 58	36, 58
Testing	128	128	Valves and seats	34, 57	34, 57
Injection pump			Valve rotators	38	38
Remove and reinstall	135	135	Valve springs	37, 58A	37, 58A
Timing	136	136			
Trouble shooting	126	126	**DIESEL ENGINE**		
DIFFERENTIAL	189	189	Assembly R&R	85	85
			Camshaft	102	102
DIFFERENTIAL LOCK			Connecting rods and bearings	106	106
Adjustment	192	192	Crankshaft and bearings	107	107
Operation	191	191	Cylinder head	86	86
Remove and reinstall	193	193	Cylinder sleeves	104	104
			Engine balancer	108	108
ELECTRICAL			Flywheel	112	112
Alternator & regulator	147	147	Front oil seal	94	94
Distributor	144	144	Main bearings	107	107
Generator	146	146	Oil pan	113	113
Generator regulator	146	146	Oil pump	114	114
Starting motor	149	149	Pistons	104	104
			Piston and rod removal	103	103
			Piston rings	104	104
			Relief valve	115	115
			Rear oil seal	111	111
			Rocker arms	91	91

DIESEL ENGINE (Cont'd.)	MF175	MF180
Tappets	90	90
Timing gear cover	94	94
Timing gears	95	95
Timing gear housing	101	101
Valve guides	88	88
Valves and seats	87	87
Valve springs	89	89
FINAL DRIVE		
Axle shaft and housing	196	196
Bevel gears	194	194
Differential R&R	189	189
Differential overhaul	190	190
Planetary gear assembly	196	196
FRONT AXLE		
Axle main member	1	2
Axle pivot pin and bushing	1	2
Dual wheels and spindle	—	2
Steering spindles	3	3
Tie rods and/or drag links	4	4
GOVERNOR NON-DIESEL		
Adjustment	138	138
R&R and overhaul	139	139
HYDRAULIC SYSTEM		
Adjustment	216	216
Auxiliary control valve	231	231
Lift cover R&R	221	221
Operating pressure	215	215
Pump	225	225
Remote cylinder	233	233
Rockshaft	222	222
Work cylinder	222	222
INDEPENDENT POWER TAKE-OFF	206	206

	MF175	MF180
IGNITION SYSTEM	144	144
POWER TAKE-OFF	203, 206	203, 206
STEERING SYSTEM		
Cylinder	20	29
Gear unit	18	—
Filling and bleeding	6	8
Operating pressure	17	17
Pedestal	23	—
Pump	15	15
Relief valve	17	17
Trouble shooting	13	14
Hydramotor	—	25
Orbitrol	—	27
TRANSMISSION (Except Multi-Power)		
Countershaft	165	—
Drive shaft and gear	161	—
Main shaft	164	—
Planetary unit	167	—
Reverse idler	166	—
Shifter rails and forks	163	—
Remove and reinstall	158	—
R&R top cover	159	—
TRANSMISSION (Multi-Power)		
Clutch	184	184
Control valve	176	176
Countershaft	185	185
Input shaft	183	183
Mainshaft	182	182
Planetary unit	180	180
Pump	187	187
Remove and reinstall	171	172
Shifter rails and forks	179	179

CONDENSED SERVICE DATA

GENERAL	Diesel	Non-Diesel	Non-Diesel
Engine Make	Perkins	Cont'l.	Perkins
Engine Model	A4.236	G-206	AG4.236
Number of Cylinders	4	4	4
Bore—Inches	3⅞	3¹³⁄₁₆	3⅞
Stroke—Inches	5	4½	5
Displacement—Cu. In.	235.9	260	235.9
Compression Ratio	16:1	7.25:1	7.0:1
Main Bearings, No. of	5	3	5
Cylinder Sleeves	Dry	Dry	Dry
Forward Speeds	12*	12*	12*
Reverse Speeds	4*	4*	4*

*Except MF175 not equipped with "Multipower".

TUNE-UP	Diesel	Non-Diesel	Non-Diesel
Firing Order	1-3-4-2	1-3-4-2	1-3-4-2
Valve Tappet Gap (Hot)			
Intake	0.010	0.016	0.012
Exhaust	0.010	0.018	0.012
Compression Pressure			
@ Cranking Speed	......	145-160	......
Intake Valve Face Angle	44°	30°	45°
Intake Valve Seat Angle	45°	30°	45°
Exhaust Valve Face Angle	44°	44°	45°
Exhaust Valve Seat Angle	45°	45°	45°
Timing (Ign. or Inj.)			
Static	23° BTC	6° BTC	10° BTC
High Idle		26° BTC	32° BTC
Timing Location	Flywheel	Crank Pulley	Crank Pulley
Battery			
Volts	12	12	12
Capacity Amp/Hr.	96	41	95
Ground Polarity	Neg.	Neg.	Neg.
Distributor Contact			
Gap	......	0.022	0.022
Spark Plug Size	......	18 MM	14 MM
Electrode Gap	......	0.025	0.025

TUNE UP (Cont'd.)	Diesel	Non-Diesel	Non-Diesel
Injectors			
Opening Pressure	2500 psi	......	......
Spray Hole Dia.	0.009	......	......
Governed Speeds			
Engine			
Low Idle	750 rpm	475 rpm	750 rpm
High Idle	2160 rpm	2225 rpm	2250 rpm
Loaded	2000 rpm	2000 rpm	2000 rpm
Power Take-Off			
High Idle	686 rpm	705 rpm	706 rpm
Loaded	635 rpm	635 rpm	635 rpm
Horsepower @ PTO Shaft	63.7*	63.7*	60(1)

*According to Nebraska Tests. (1) Manufacturer's Rating.

Hydraulic System			
Maximum Pressure	————3000 psi————		
Rated Delivery	———— 12.8 gpm ————		

SIZES - CAPACITIES - CLEARANCES			
Crankshaft Journal Diameter	2.999	2.3745	2.999
Crankpin Diameter	2.499	2.062	2.499
Camshaft Journal Diameter:			
Front	1.997	1.8085	1.997
Center	1.987	1.746	1.987
Rear	1.967	1.6835	1.967
Crankshaft Bearing Clearance:			
Main Bearings	0.003	0.002	0.0035
Crankpin	0.0025	0.002	0.0025
Crankshaft End Play	0.009	0.006	0.009
Piston to Cylinder Clearance	0.007	......	......
Camshaft Bearing Clearance	0.004	0.0035	0.004
Camshaft End Play	0.010	0.005	0.010
Cooling System—Quarts	13.5	10.5	13.5
Crankcase—Quarts	9.5	6	8.5
Transmission, Differential and Hyd. Lift—Gal.	8	8	8
Power Steering—Quarts	——— Refer to para. 6 or 8———		
Steering Gear Housing—Quarts	——Refer to para. 5 or 7——		
Planetary Final Drive	1¾	1¾	1

FRONT SYSTEM

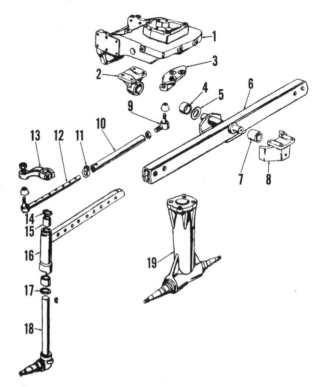

Model MF175 tractors are available only in standard clearance, adjustable axle type. Model MF180 units are high-clearance, row crop types with either a wide adjustable or dual-wheel tricycle front end assembly. Refer to the appropriate following paragraphs for service procedures.

AXLE ASSEMBLY
Model MF175

1. Refer to Fig. 1 for exploded view of front axle and associated parts. Pivot pin (12) is retained in front support casting (15) by the cone point set screw (14).

Axle center member (9) or the complete axle assembly can be lowered out of front support after removing pivot pin (12). Bushings (8 & 10) and/or pivot pin (12) may be renewed at this time. When reinstalling, make sure the $\frac{3}{32}$-inch thick thrust washer (11) is installed at rear of axle member (9), and vary the number and thickness of shims (7) at front to reduce end play to a minimum without binding. Shims (7) are available in thicknesses of 0.029, 0.035 and 0.040.

Model MF180

2. Refer to Fig. 2 for an exploded view of front axle and associated parts. The front axle pivot pin is a component part of axle center member (6).

Fig. 2—Exploded view of front support, front axle and associated parts of the type used on Model MF180. Tricycle pedestal (19) attaches in place of center steering arm (3).

1. Front support
2. Pivot bracket
3. Center steering arm
4. Bushing
5. Thrust washer
6. Axle center member
7. Bushing
8. Pivot bracket
9. Tie rod end
10. Adjusting sleeve
11. Clamp
12. Tie rod end
13. Steering arm
14. Dust seal
15. Bushing
16. Axle extension
17. Thrust washer
18. Spindle
19. Tricycle pedestal

To remove either the axle center member or complete axle assembly, remove the cap screws securing front and rear pivot brackets (2 & 8) to front support (1); then remove axle and pivot brackets as a unit. Bushings (4 & 7) may be renewed at this time. When reinstalling, make sure the $\frac{3}{32}$-inch thick thrust washer (5) is at rear of pivot pin as shown.

Tricycle pedestal (19) can be installed on tractors originally equipped with adjustable axle after removing axle assembly and center steering arm (3).

SPINDLE BUSHINGS
All Models

3. Each axle extension contains two renewable bushings (3—Fig. 1) or (15—Fig. 2) which require final sizing after installation to provide the recommended 0.0035-0.005 clearance for the spindle. Nominal spindle diameter is 1½-inches for all models.

TIE RODS AND TOE-IN
All Models

4. Automotive type tie rod and drag link ends are used. Recommended toe-in is 0-¼ inch. Adjust both drag links an equal amount by removing clamp bolt (B—Fig. 3) and turning adjusting sleeve (A) as required.

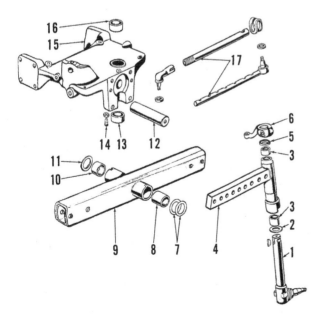

Fig. 1—Exploded view of front support, axle and associated parts of the type used on Model MF-175.

1. Spindle
2. Thrust washer
3. Bushing
4. Axle extension
5. Dust seal
6. Steering arm
7. Shims
8. Bushing
9. Axle center member
10. Bushing
11. Thrust washer
12. Pivot pin
13. Bushing
14. Lock screw
15. Front support
16. Bushing
17. Tie rod

Fig. 3—To adjust the toe-in, loosen clamp bolt (B) and turn adjusting sleeve (A) as required. Both tie rods should be adjusted an equal amount.

POWER STEERING SYSTEM

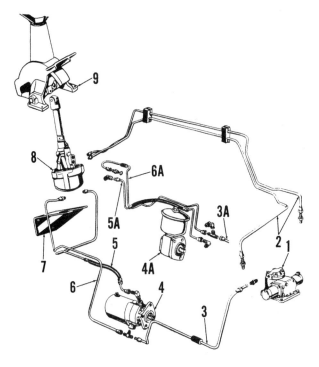

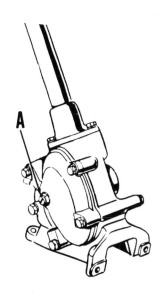

Fig. 4—Exploded view of hydrostatic power steering lines and components used on Model MF180 with Saginaw "Hydramotor."

1. Cylinder housing
2. Cylinder lines
3. Bleed line (Dsl)
3A. Bleed line (Gas)
4. Pump (Dsl)
4A. Pump (Gas)
5. Pressure line (Dsl)
5A. Pressure line (Gas)
6. Return line (Dsl)
6A. Return line (Gas)
7. Bracket
8. Hydramotor
9. Steering column

Fig. 5—On Model MF175, steering gear backlash is adjusted at screw (A) on gear housing side cover.

All Models are equipped with power steering. Model MF175 uses linkage-booster type power assist while Model MF180 is equipped with hydrostatic steering.

LUBRICATION AND BLEEDING
Model MF175

5. **STEERING GEAR.** The recommended lubricant for steering gear is SAE 90, mineral gear lubricant. Housing capacity is approximately one quart. Oil level and filler plug is located on left side of steering gear housing, and fluid should be maintained at level of filler plug hole.

6. **POWER STEERING SYSTEM.** Automatic Transmission Fluid, Type A or Massey Ferguson Spec. M-1110 oil is the recommended operating fluid. System capacity is approximately 1⅓ quarts for non-diesel models; and 1⅔ quarts for diesels. Power steering fluid reservoir is mounted on the gear-driven power steering pump, on left side of engine block on non-diesel models; right side on diesels. Fluid should be maintained at level of filler plug on diesel models and at "FLUID LEVEL" mark on side of reservoir on other models.

To bleed the system, fill reservoir and start the engine, then maintain fluid level at or near full mark by adding fluid, as steering wheel is turned to bleed air from the system.

Model MF180

7. All lubrication of the hydrostatic steering gear units is provided by the operating fluid for the system. Automatic Transmission Fluid, Type A, or Massey Ferguson Spec. M-1110 oil is the recommended operating fluid. System capacity is approximately 3 quarts, only a small portion of which is contained in the pump-mounted fluid reservoir. A much greater portion of the system fluid circulates through the lines, cylinders and manual pump and fills the cylinder and hydramotor housings where it serves as the lubricating fluid. It is important, therefore, that all leaks be eliminated, the reservoir kept at or near the full level and the filling and bleeding instructions be carefully followed.

8. **FILLING AND BLEEDING.** Only a small portion of the system capacity of approximately 3 quarts of fluid is contained in the reservoir. The remainder of the fluid circulates through cylinder housing and cylinders (1—Fig. 4), manual pump (8) and lines. Correct fluid level is even with filler plug on diesel power steering pump (4), or to "FLUID LEVEL" mark on pump (4A) for other models. If cylinder housing (1) has been disassembled, drained or renewed, make sure spindle and gear reservoir is filled as outlined in paragraph 30.

Fill the pump reservoir to the specified level and start and idle the engine, adding fluid as the level drops. Increase engine speed slightly and cycle the system a little bit at a time, keeping the reservoir full, until complete turns are made in both directions and spongy feeling disappears. Recheck and refill if necessary, after tractor has been run a few hours.

ADJUSTMENT
Model MF175

9. **STEERING GEAR.** Steering gear backlash is adjusted by means of the adjusting screw (A—Fig. 5) located on right side of steering gear housing. Turning the adjusting screw clockwise reduces the backlash. Adjustment is correct when a barely perceptible drag or resistance is felt when steering wheel is turned through the mid-position, and no backlash exists. Before attempting to adjust the backlash, first check the steering wheel shaft (camshaft) bearings as outlined in paragraph 10, and adjust the preload if required.

10. **CAMSHAFT BEARINGS.** To check the camshaft bearings, first loosen the backlash adjusting screw (A—Fig. 5) at least two full turns. With adjustment screw loosened, check camshaft bearings for end play by pulling up and pushing down on steering wheel. Camshaft should turn

freely with no perceptible looseness of bearings.

If end play exists, unbolt the steering column from gear housing and raise the column; then split and remove a sufficient quantity of shims (10—Fig. 14) to remove all end play. Shims are available in thicknesses of 0.002, 0.003 and 0.010. Adjust steering gear backlash as outlined in paragraph 9 after camshaft bearings have been adjusted.

11. **POWER STEERING LINKAGE.** Adjustments are provided for valve sensitivity and synchronization of the valve linkage. Two types of linkage have been used, refer to Figs. 6 and 8. To adjust the linkage, refer to the appropriate following paragraphs.

Early Models

To adjust the linkage on models before tractor serial No. 9A 18007, refer to Fig. 6 and proceed as follows:

Open the grille door and remove pin (P) which connects valve link (L) to actuating arm (A). With engine running at about ½-throttle and wheels in a straight-ahead position, loosen locknut (1) and turn adjusting pin (2) clockwise until it bottoms. Loosen the two locknuts (N) and turn adjusting sleeve (S) until pin (P) can be reinserted without moving control valve. Install and secure pin (P); then back out adjusting pin (2) seven full turns.

Check the adjustment by turning steering wheel. Sensitivity can be increased by backing out adjusting pin (2) or decreased by turning in the pin. If tractor steers easier when making a right-hand turn, lengthen valve link (L). If tractor steers more easily to left, shorten link (L).

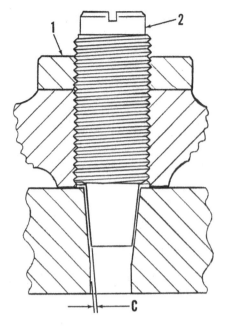

Fig. 7—Cross sectional view of sensitivity adjusting screw shown installed in Fig. 6. Clearance (C) should be approximately 0.040.

Tighten the locknuts (1 & N) when adjustment is correct, then recheck the steering response, making adjustments as required.

Late Models

On models after tractor serial number 9A 18007 (or earlier models which have been converted to late type), steering arm is not centered when adjusting screw is tightened and a different adjusting procedure is required. Refer to Fig. 9 for cross section.

To adjust the linkage, open grille door and remove pin (1—Fig. 8) con-

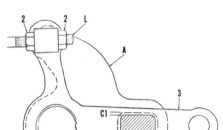

Fig. 9—Cross sectional view of actuating linkage used on late Model MF175. Neutral clearance (C1) is adjusted by shortening or lengthening link (L); neutral clearance (C2) by turning set screw (4). Clearances (C1 & C2) should each be 0.060. Refer to paragraph 11 for adjustment procdeure.

necting actuating link to valve spool.

Loosen locknut (5). Turn adjusting screw (4) clockwise until it stops, then back screw out ¾ turn.

Start and run engine at about ½ throttle. Hold lug of actuating arm (3) solidly against point of adjusting screw (4); loosen adjusting nuts (2) if necessary, and shorten or lengthen connecting link until pin (1) can be inserted without moving valve or actuating arm.

Tighten nuts (2), back adjusting screw (4) out an additional ¾ turn then tighten locknut (5).

NOTE: Valve operating clearance (C1—Fig. 9) for left hand turn is established at first loosening of adjusting screw (4); clearance (C2) for right hand turn at second loosening of screw. Recheck from the beginning and/or check for binding or linkage wear if steering action is erratic.

Model MF180

12. The only adjustment required is adjustment of the cylinder rack clearance. If clearance is excessive, wear is indicated and cylinder unit should be removed as outlined in paragraph 27 and overhauled as in paragraph 28.

TROUBLE SHOOTING

Model MF175

13. The power steering control valve is mounted on the cylinder and attached to the front steering pedestal located inside the radiator grille. External adjustments are provided for steering sensitivity and synchronization of the control valve with wheel movement.

Fig. 6—View of power steering cylinder and actuating linkage of the type used on early Model MF175. Valve is synchronized by shortening or lengthening the link (L); sensitivity adjusted by turning the tapered pin (2) in or out of arm (A).

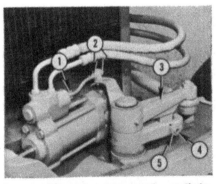

Fig. 8—View of power steering cylinder and actuating linkage used on late Model MF175. Refer also to Fig. 9 for cross section of actuating arms.

1. Link pin
2. Adjusting nuts
3. Actuating arm
4. Adjusting screw
5. Locknut

Loss of power assistance in one direction only, is most likely caused by improper adjustment of the control valve link on early models; and link or adjusting screw on late models.

Loss of power assistance in both directions is most likely caused by improper adjustment of the sensitivity adjusting screw on early models or linkage on late models. Malfunction could also be caused by lack of fluid, a sticking control valve, internal leakage within the cylinder, an improperly adjusted or malfunctioning relief valve or a malfunctioning pump.

Erratic action could be caused by air in the system, a sticking control valve, improper adjustment, wear or binding in steering linkage or the use of incorrect operating fluid.

Wheel shimmy could be caused by an improperly adjusted sensitivity screw or wear in steering linkage.

On early models, lengthening the valve link (L—Fig. 6) will cause the tractor to steer more easily (or self-steer) to the left. Shortening the link will cause tractor to steer more easily to the right. Fig. 7 shows a cross sectional view of the power steering adjusting screw. Clearance (C) of the tapered screw in its bore is approximately 0.040 when adjusted as outlined in paragraph 11. On late models, refer to Fig. 9 for cross section and to note following paragraph 11 for operation.

Model MF180

14. Some of the troubles which may be encountered in the operation of the hydrostatic steering system and their possible causes are as follows:

(Early Models)

1. Hard Steering
   a. Low pump pressure
   b. Insufficient fluid in reservoir
   c. Faulty pump relief valve
   d. Hydramotor control valve damaged or sticking
   e. Faulty steering cylinder

2. Power assistance in One Direction Only
   a. Hydramotor control valve sticking
   b. Damaged or restricted oil lines
   c. Faulty steering cylinder

3. Erratic Steering Control
   a. Vanes sticking in Hydramotor rotor
   b. Hydramotor metering unit scored or worn
   c. Broken Hydramotor control valve spring

4. Loss of Power Assistance
   a. Hydramotor torsion shaft broken
   b. Faulty supply pump
   c. Relief valve stuck open
   d. Manual control ball not seating

5. Unequal Turning Radius
   a. Improperly adjusted tie rods
   b. Rack & spindle improperly timed

6. Noisy Operation
   a. Insufficient fluid in reservoir
   b. Air in system
   c. Faulty relief valve
   d. Damaged or restricted oil lines

POWER STEERING PUMP

Cessna or Wooster gear type power steering pumps are used. Refer to Fig. 10, 11 or 12. All pumps are gear driven from the engine camshaft. Refer to the appropriate following paragraphs for overhaul procedure.

Cessna Pump

15. Refer to Fig. 10 for exploded view of pump. If pump body or gears are worn or scored, it is recommended that the complete pump be renewed.

When reassembling the pump, make sure that check valve (4) is properly installed in the correct hole. Install diaphragm seal (5) open side down and work the seal firmly in its groove using a blunt tool. Press protecter gasket (6) and plastic gasket (7) into relief in diaphragm seal; then install diaphragm (8) with bronze face toward the gears. Install shaft seal (2) with lip toward inside of pump. After assembly, pump should have a slight amount of drag, but should rotate evenly.

Bleed the system as outlined in paragraph 6 after pump is reinstalled.

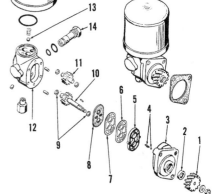

1. Gear
2. Oil seal
3. Mounting plate
4. Check valve
5. Diaphragm seal
6. Protector
7. Gasket
8. Diaphragm
9. Snap rings
10. Pump gear & shaft
11. Pump gear
12. Pump body
13. Check ball
14. Relief valve
15. Reservoir
16. Adapter
17. Filter
18. Spring
19. Cover

Fig. 10—Exploded view of Cessna power steering pump of the type used on Model MF175 gasoline tractors.

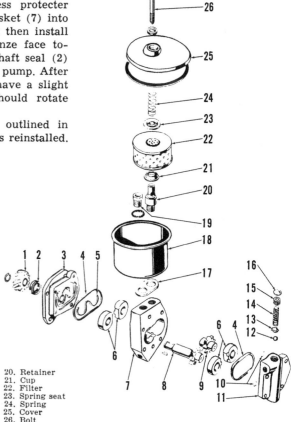

Fig. 11—Exploded view of Wooster gear type power steering pump of the type used on Model MF180 gasoline tractors.

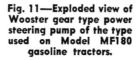

1. Gear
2. Oil seal
3. Mounting plate
4. Seal
5. Seal
6. Bearings
7. Pump body
8. Driven gear
9. Follow gear
10. "O" rings
11. Rear cover
12. Valve ball
13. Retainer
14. Spring
15. Adjusting screw
16. Expansion plug
17. Gasket
18. Reservoir
19. Special screw
20. Retainer
21. Cup
22. Filter
23. Spring seat
24. Spring
25. Cover
26. Bolt

OPERATING PRESSURE AND RELIEF VALVE

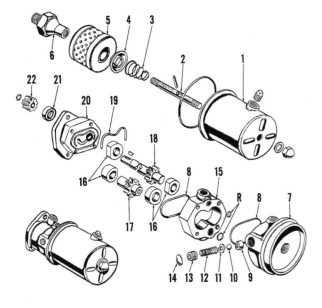

Fig. 12—Exploded view of Wooster power steering pump of the type used on diesel models.

1. Reservoir
2. Stud
3. Spring
4. Spring seat
5. Filter
6. Retainer
7. Rear cover
8. Seal
9. Valve seat
10. Valve ball
11. Retainer
12. Spring
13. Adjusting screw
14. Plug
15. Pump body
16. Bearings
17. Follow gear
18. Driven gear
19. Seal
20. Front cover
21. Seal
22. Drive gear

All Models

17. To check the relief pressure, connect a pressure test gage and shut-off valve in series with the pump pressure line as shown in Fig. 13. Open the shut-off valve and run engine until operating fluid is warm. Open the throttle and close the shut-off valve only long enough to obtain a reading. Gage reading should be 1100-1150 psi for all models. If it is not, either the pump or relief valve is malfunctioning.

On Model MF175 non-diesel with Cessna pump, a cartridge type relief valve (14—Fig. 10) is used, and service consists of renewing the cartridge. On all models with Wooster pump, relief pressure can be adjusted after removing the expansion plug (16—Fig. 11) or (14—Fig. 12) and turning the exposed Allen head adjusting plug.

Wooster Pump

16. Refer to Figs. 11 and 12 for exploded views. All parts are available individually; however, it is recommended that the internal pump gears and four bearings be renewed at the same time if they are damaged because of wear or scoring. If center housing (7—Fig. 11) or (15—Fig. 12) is also worn or scored, it is recommended that the pump be renewed.

To assemble the pump, use a suitable lubricant on all moving parts and tighten the retaining through-bolts evenly to a torque of 28-32 ft.-lbs.

Fill and bleed the system as outlined in paragraph 6 or 8 after pump is installed, and check and adjust relief pressure as outlined in paragraph 17.

Fig. 14—Exploded view of steering gear of the type used on Model MF175.

1. Steering arm
2. Oil seal
3. Bushing
4. Housing
5. Bushing
6. Lever shaft
7. Side cover
8. Side cover
9. Adjusting screw
10. Shims
11. Bearing cup
12. Bearing
13. Worm shaft
14. Bearing
15. Bearing cup
16. "O" ring
17. Steering column

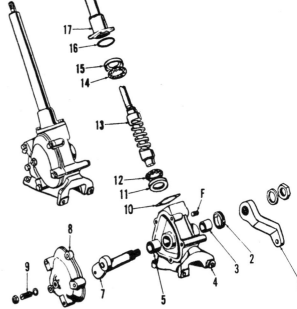

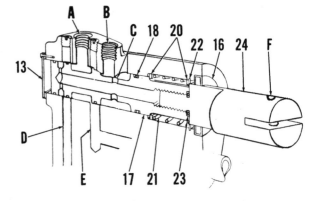

Fig. 13—View of typical power steering pump with pressure gage and shut-off valve installed for checking relief pressure. Refer to text.

Fig. 15—Cross sectional view of power steering control valve of the type used on Model MF175. Refer to Fig. 16 for parts identification except for the following.

A. Pressure port
B. Return port
C. Cross drilling
D. Pressure passage.
E. Pressure passage

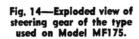

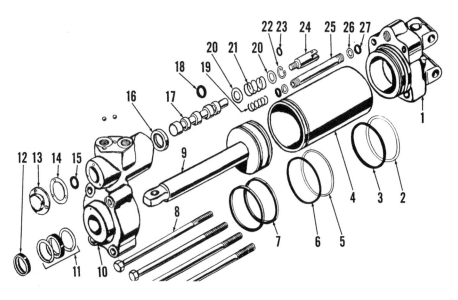

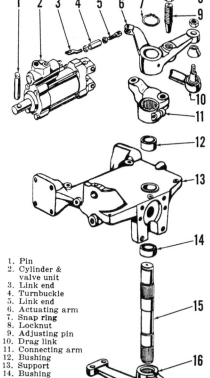

Fig. 16—Exploded view of power steering cylinder and control valve assembly of the type used on Model MF175. Refer to Fig. 15 for cross sectional view of valve.

1. End cap	10. Valve body	19. Spring
2. Backup ring	11. Rod seal	20. Centering washer
3. "O" ring	12. Wiper	21. Centering spring
4. Cylinder tube	13. Valve cap	22. Snap ring
5. Backup ring	14. Gasket	23. "O" ring
6. "O" ring	15. "O" ring	24. Rod end
7. Piston ring	16. Oil seal	25. Crossover tube
8. Through-bolts	17. Valve spool	26. Backup ring
9. Piston & rod assy.	18. "O" ring	27. "O" ring

Fig. 17—Exploded view of front support, steering pedestal, steering cylinder and associated parts used on early Model MF-175. Late models are similar but parts cannot be intermixed.

1. Pin
2. Cylinder & valve unit
3. Link end
4. Turnbuckle
5. Link end
6. Actuating arm
7. Snap ring
8. Locknut
9. Adjusting pin
10. Drag link
11. Connecting arm
12. Bushing
13. Support
14. Bushing
15. Shaft
16. Steering arm

STEERING GEAR UNIT
Model MF175

18. REMOVE AND REINSTALL. To remove the steering gear, remove hood, rear side panels, battery, steering wheel and instrument panel. Remove or block up fuel tank and remove rear support and support frame. Steering gear housing may now be removed; or overhauled without removal from transmission housing. Install by reversing the removal procedure.

19. OVERHAUL. Refer to Fig. 14 for an exploded view of steering gear. To overhaul the gear, remove pitman arm (1) and check the exposed end of pitman shaft (7) for paint, rust or burrs which might damage shaft bushing or seal. Remove side cover (8) and withdraw pitman shaft (7) from housing.

Remove steering column (17) and steering shaft (13), saving shim pack (10) for reinstallation when gear is reassembled.

Examine all parts and renew any which are questionable. The two bushings (3 & 5) for pitman shaft should be reamed after installation, to an inside diameter of 1.6235-1.625, to provide a diametral clearance of 0.0005-0.003 for pitman shaft.

Pitman shaft and arm are provided

with master splines for correct timing. When installing camshaft (13), vary the thickness of shim pack (10) to remove all end play and provide camshaft bearings with a very slight drag when shaft is turned. Shims are available in thicknesses of 0.002, 0.003 and 0.010. Adjust steering gear backlash as outlined in paragraph 9 after unit is reassembled.

CYLINDER AND VALVE UNIT
Model MF175

20. REMOVE AND REINSTALL. To remove the steering valve and cylinder unit, first remove grille door, remove the retaining wing nut and move oil cooler radiator to the side out of the way. Disconnect hydraulic pressure and return lines at valve housing and control link at actuating arm. Remove piston rod anchor pin and cylinder pivot pin and lift out the complete cylinder and valve unit.

Reinstall by reversing the removal procedure, bleed the system as outlined in paragraph 6 and adjust the valve as in paragraph 11, after unit is installed.

21. OVERHAUL VALVE. To disassemble the control valve, refer to Fig. 15 and proceed as follows:

Remove end cover (13). Turn valve spool (17) until cross-drilling (C) is

visible through return port (B) and carefully insert a small punch through port (B) and drilling (C). Insert a second punch through pin hole (F) in end of valve rod (24) and unscrew rod end (24) from valve (17) by turning rod end counter-clockwise. Remove seal (16) and snap ring (22); then push valve (17), centering washers (20) and centering spring (21) out of valve bore.

Examine valve bore and spool for wear, scoring or other damage. Spool and valve housing are available separately and do not require selective fitting. Refer to Fig. 16 for an exploded view of valve and cylinder unit.

When assembling the valve spool, install a new "O" ring (18) in annular groove on valve spool, lubricate the "O" ring with petroleum jelly and insert valve spool carefully from spring end of bore. Install end cap (13) using a new "O" ring (15) to prevent spool from sliding too far into bore. With spring end of valve bore up, install centering washers (20), spring (21) and snap ring (22). Install seal (16). Insert a small punch through return

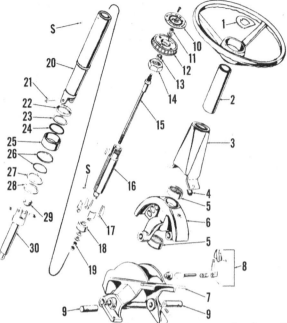

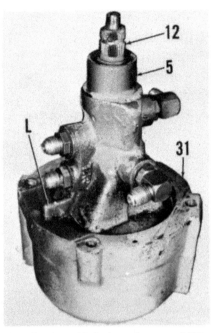

Fig. 18—Exploded view of steering column and associated parts used on Model MF180.

1. Tab washer
2. Sleeve
3. Shroud
4. Shim
5. Needle bearing
6. Pivot
7. Support
8. Latch
9. Pin
10. Cover
11. Nut
12. Cap
13. Spacer
14. Nut
15. Shaft
16. Shaft
17. Wedge
18. Nut
19. Nut
20. Tube
21. Roll pin
22. Shield
23. Thrust washer
24. Thrust washer
25. Retainer
26. Seal
27. Thrust washer
28. Snap ring
29. Ball
30. Coupling

Fig. 21—Assembled view of removed "Hydramotor" unit. Refer to Fig. 24 for parts identification except for anti-rotation lug (L).

port (B—Fig. 15) and cross drilling (C), make sure "O" ring (23) is properly positioned, and thread rod end (24) into control valve (17). Install cylinder and valve unit as outlined in paragraph 20.

22. OVERHAUL CYLINDER. To overhaul the power steering cylinder, first remove the cylinder and valve unit as outlined in paragraph 20. Refer to Fig. 16.

Clamp end cap (1) in a vise with piston rod end up. Remove the four through-bolts (8) and disassemble the cylinder, using Fig. 16 as a guide.

Clean all parts thoroughly and renew any which are scored, worn or damaged. Renew "O" rings, back-up rings and seals whenever cylinder is disassembled. Piston rod seals (11)

and wiper (12) can be removed with a sharp pointed tool after piston rod (9) is withdrawn. Be careful not to scratch the bore when removing or installing seals. Backup rings (2, 5 and 26) must be installed away from pressure side of their respective "O" rings; use Fig. 16 as a guide.

Assemble by reversing the disassembly procedure. Tighten the four through-bolts (8) evenly, to a torque of 15-20 ft.-lbs. Reinstall cylinder and valve unit as outlined in paragraph 20.

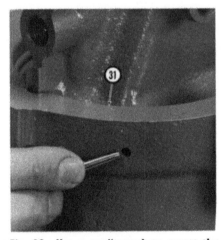

Fig. 22—Use a small punch to unseat the cover retaining ring.

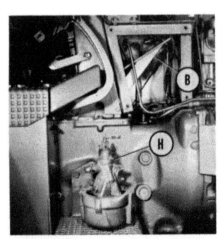

Fig. 19—Hydramotor (H) can be removed from right side after removing side panel as shown.

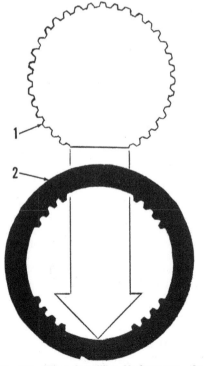

Fig. 20—When installing Hydramotor, the flat in shaft splines (1) must align with blank area in coupling as shown by arrow for maximum spline engagement.

Fig. 23—Lift out retaining ring (6), then remove the cover.

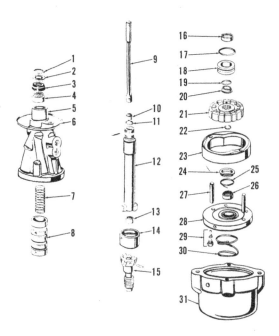

1. Snap ring
2. Dust seal
3. Oil seal
4. Needle bearing
5. Shaft housing
6. Snap ring
7. Spring
8. Valve spool
9. Torsion bar
10. Sleeve
11. "O" ring
12. Stub shaft
13. Needle bearing
14. Actuator
15. Drive shaft
16. Needle bearing
17. "O" ring
18. Bearing support
19. "O" ring
20. Seal
21. Rotor
22. Snap ring
23. Cam ring
24. Seal
25. "O" ring
26. Needle bearing
27. Dowel
28. Pressure plate
29. Check valve
30. Spring
31. Cover

Align ream the bushings after installation, to an inside diameter of 1.8755-1.8765.

Renew any damaged, worn or questionable parts, assemble by reversing the disassembly procedure, and adjust the valve linkage as outlined in paragraph 11.

STEERING COLUMN

Model MF180

24. Refer to Fig. 18 for an exploded view of the adjustable steering column. To remove the steering column pivot base (7), it is first necessary to disassemble the gearshift linkage as outlined in paragraph 174.

To remove the telescoping steering shaft (16) and associated parts for service on the locking mechanism, first remove the steering wheel, sliding sleeve (2) and shroud (3). Using a small punch, drive locking pin (S) into steering column tube (20); then withdraw items (15 through 19) as a unit.

HYDRAMOTOR

Model MF180 So Equipped

25. **REMOVE AND REINSTALL.** To remove the Hydramotor steering unit, first remove the hood and right steering support side panel as shown in Fig. 19. Disconnect the hydraulic lines, remove the retaining cap screws; then lower the Hydramotor unit out through side opening in steering support frame.

The Hydramotor stub shaft splines (1—Fig. 20) have a flat area and coupling yoke splines (2) have four blank areas as shown. When installing the hydramotor, make sure the flat is aligned with one of the blanks as indicated by arrow, to assure full

Fig. 25—Rotor, vanes and rotor ring can be removed from stub shaft after removing snap ring as shown.

POWER STEERING PEDESTAL

Model MF175

23. Refer to Fig. 17 for an exploded view of power steering pedestal and associated parts. Most service work can be performed without major disassembly of the unit.

Actuating pin (9) can be threaded upward out of actuating arm (6) if removal is indicated. Actuating arm (6) can be removed after disconnecting control valve link (5) and drag link (10), and removing snap ring (7). Power steering arm (11) is secured to steering shaft (15) by a clamping screw, and can be lifted from shaft after removing actuating arm and loosening the clamp screw.

Bushings (12 & 14) are contained in front supporting housing (13).

Fig. 26—Stub shaft may be removed from housing for seal renewal. Major parts are not available separately.

Fig. 27—Assembled view of rotor, rotor ring and vanes.

Fig. 28—Pump vanes (V) can be easily installed behind springs (S) in the six slots aligned with larger diameter of rotor ring as shown. Turn rotor (21) a quarter turn to install remainder of vanes.

spline engagement. Also make sure that full length of stub shaft splines enter the coupling yoke and that shaft and yoke are in perfect alignment when bolts are tightened. On some early models, it may be necessary to remove the Hydramotor mounting bracket (B—Fig. 19) and elongate the mounting holes to assure perfect alignment. Complete the installation by reversing the removal procedure.

26. **OVERHAUL.** To disassemble the removed Hydramotor, insert a small punch in hole in Hydramotor cover (31—Fig. 22) as shown; and unseat retaining ring (6—Fig. 23). Remove the ring using a blade screw driver. With ring removed, pressure plate spring (30—Fig. 24) should push shaft housing (5) and associated parts out of cover (31). If cover sticks, tap mounting ears with a plastic hammer to dislodge the cover. Lift off pressure plate (28) and alignment dowels (27).

Clamp stub shaft housing in a vise as shown in Fig. 26, unseat and remove snap ring (22—Fig. 25) and lift off the rotor, vanes and rotor ring as a unit. The rotor, vanes, ring and associated parts are only available as an assembly as shown in Fig. 27; therefore disassembly is not advised unless required for cleaning. If rotor unit is disassembled, reassembly can be facilitated by referring to Fig. 28 and proceeding as follows:

Install six vanes (V) in vane slots which align with largest diameter of cam ring (23), with rounded edge of vanes to outside and springs (S) positioned over inside edge. Turn rotor 90° until the six empty vane slots are aligned with larger ID of cam ring, then install the remaining vanes while pushing springs inward. Refer to Fig. 27 for correct assembled view.

Except for seals, component parts of stub shaft, valve, and shaft housing are not serviced. After rotor and ring assembly is removed from shaft, the shaft may be removed from housing as shown in Fig. 26. Refer to Figs. 29 and 30 for views of "Hydramotor" seals. When installing the cover, make sure anti-rotation lug fits in notch as shown in Fig. 31 and that cover is not cocked. Use an arbor press and sleeve and apply pressure at shaft housing, to overcome pressure plate spring tension. Insert one end of retaining ring under anti-rotation lug and make sure other end overlaps knockout hole in cover. Install the assembled "Hydramotor" as outlined in paragraph 25.

ORBITROL HAND PUMP
Model MF 180 So Equipped

Model MF 180 tractors are optionally equipped with a Char-Lynn ORBITROL hydrostatic hand pump. For tractors equipped with Saginaw HYDRAMOTOR, refer to paragraph 25.

27. **REMOVE AND REINSTALL.** To remove the ORBITROL hydrostatic hand pump, first remove left accessory panel, disconnect the four hydraulic lines; then unbolt and remove hand pump unit. Refer to Fig. 32.

Install by reversing the removal procedure. Fill and bleed steering system as outlined in paragraph 8 after unit is installed.

28. **OVERHAUL.** Refer to Fig. 33 for partially exploded view of the Char-Lynn ORBITROL hydrostatic hand pump. Rotary valve spool (13), sleeve (14) and housing (12) are available only as a matched set which also includes reaction springs (9),

cross pin (10) and component parts of check valve (5, 6, 7 & 8).

To disassemble the removed Orbitrol unit, first remove the cap screws retaining metering pump to bottom of housing and remove items (15) through (19). Remove the four screws retaining top cover (2) and lift off cover. Insert a suitably bent wire through port nearest check valve plug (5) and push out the plug as shown in Fig. 34. Remove check valve seat using an Allen wrench and remove check valve ball and spring (Fig. 35). Carefully push control valve spool and sleeve assembly out bottom of housing as shown in Fig. 36.

NOTE: Be careful valve unit does not bind. Parts are fit to extremely close tolerance and a twisting motion may be required to withdraw the spool.

Remove nylon cross pin discs (11—Fig. 33), and cross pin (10), then separate spool and sleeve as shown in Fig. 37. Thoroughly clean all parts in a suitable mineral solvent and blow dry with air. Inspect lapped surfaces for scoring or other damage. Mating surfaces of cover (19—Fig. 33), rotor assembly (17) and metering plate (16) can be hand lapped to remove burrs, slight scratches or other imperfections. Use 600 grit abrasive paper or lapping compound. Rinse in clean solvent and blow dry after lapping is complete. Rotor units are only available as a matched set.

Inspect centering springs (9) for fractures or distortion. Springs should have a minimum arch of 7/32-inch when measured at center. Inspect spool, sleeve and body for nicks, scoring or wear. If any part is damaged, renew the housing assembly which includes spool, sleeve and check valve.

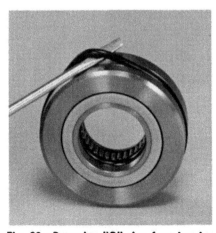
Fig. 29—Removing "O" ring from bearing support.

Fig. 30—Removing seal from pressure plate.

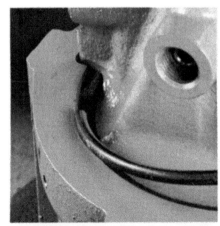
Fig. 31—When reassembling the "Hydramotor" insert end of retaining ring under anti-rotation lug as shown and make sure one end of ring overlaps the knockout hole in cover.

Fig. 32—Installed view of Char-Lynn OR-BITROL hydrostatic hand pump.

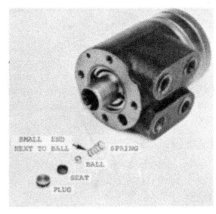

Fig. 35—Check Valve components removed from valve body.

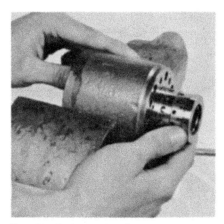

Fig. 36—Push spool and sleeve assembly out bottom of housing as shown.

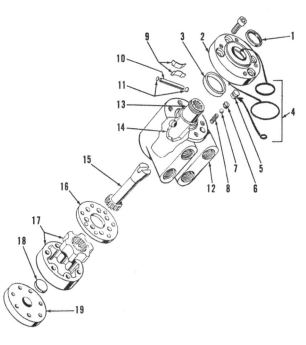

Fig. 33 — Exploded view of ORBITROL hydrostatic hand pump showing component parts.

1. Oil seal
2. Cover
3. Bushing
4. O-rings
5. Valve plug
6. Valve seat
7. Valve ball
8. Valve spring
9. Centering springs
10. Drive pin
11. Teflon disc
12. Valve body
13. Valve spool
14. Valve sleeve
15. Drive link
16. Metering plate
17. Rotor assembly
18. Disc
19. End plate

Renew O-rings (4) when assembling the unit.

Tighten check valve seat (6) to a torque of 150 inch-pounds. Install springs (9) with notched edge down, in sets of three as shown in Fig. 38.

Insert the assembled spool and sleeve unit from bottom of valve body ussing a twisting motion. Do not allow sleeve to move up beyond flush with upper machined surface of body. Be sure pin slot in drive shaft (15—Fig. 33) is aligned with valley of inner rotor as shown in Fig. 40. Rotor has six gear teeth and 12 splines; if drive slot is improperly aligned, steering unit will operate in reverse of hand pressure. Tighten the seven lower cap screws evenly to a torque of 250 inch-pounds and the four cap screws in top cover to a torque of 220 inch-pounds.

Install the assembled unit as outlined in paragraph 27 and bleed steering system as outlined in paragraph 8 after unit is installed.

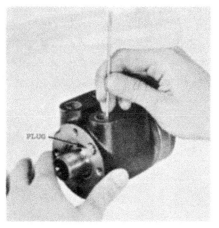

Fig. 34—Use a bent wire and work through valve port to push out check valve plug as shown.

Fig. 37 — Spool can be pushed from sleeve as shown after removing cross pin.

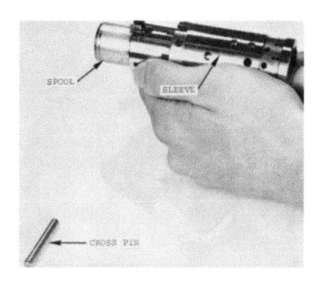

STEERING CYLINDER
Model MF180

29. REMOVE AND REINSTALL. To remove the steering cylinder housing, first remove grille door and the wing nut (W—Fig. 41) securing oil cooler (C) to grille. Pull out top of oil cooler

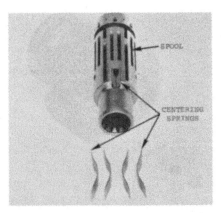

Fig. 38—View showing correct installation of centering springs (arch to arch in sets of three, notched edge down).

Fig. 39—Be sure O-rings, seal and bushing are properly installed when reassembling.

Fig. 40—When properly assembled, slot in drive link must align with valley of inner rotor as shown.

and unhook from guide pins (P); then move oil cooler radiator out of the way as shown in Fig. 42.

Disconnect cylinder lines (L—Fig. 43) and bleed line (B). Disconnect center steering arm (axle types) or remove lower pedestal (tricycle models). Remove the attaching cap screws and lift out the power rack housing and associated parts as a unit.

Install by reversing the removal procedure.

30. OVERHAUL. Refer to Fig. 44 for an exploded view of steering cylinder housing and associated parts.

NOTE: Cylinder housing is lubricated by the power steering fluid. Because housing capacity is greater than the capacity of power steering reservoir, it is extremely im-

Fig. 41—Model MF180 tractor with grille screen removed, showing "Multipower" oil cooler (C), pivot pins (P) and retaining wing nut (W).

Fig. 42—Multipower oil cooler can be easily removed for service on steering cylinder.

portant that reservoir and housing be full before tractor is released for service. Fill gear housing through opening for drain plug (20—Fig. 44) before housing is installed, or through either of the drain holes (using a funnel) before installing top cover (1).

When disassembling the steering gear housing, first remove top cover (1) and rack guide (21) and turn spindle (9) until timing marks (M—Fig. 45) are aligned; then mark the front flange hole in spindle (9—Fig. 44) for convenience in reassembly. Remove snap ring (3) and washer (4), and lower the spindle from housing (6) and gear (17). Remove cylinders (23) and one piston (14); withdraw rack (16), then remove gear (17) and thrust bearing (18 & 19) through front opening of housing (6).

Needle bearing (5) in top bore of housing (6) should be installed from top, with top end just below flush with housing top face. Lower needle bearing (7) should be installed from bottom with lower edge just above flush with counterbore for seal (8). Install seal (8) with spring-loaded lip to top, until it bottoms in housing counterbore.

Assemble by reversing the disassembly procedure, making sure timing marks (M—Fig. 45) on rack and pinion are aligned and that one of the attaching holes in spindle flange is pointing straight to the front with timing marks aligned. Vary the thickness of shim pack (22—Fig. 44) to establish a clearance of 0.001-0.008

Fig. 43—Installed view of Model MF180 steering cylinder housing, showing cylinder line (L) and bleed line (B). Pivot pin (P) is attaching point for "Multipower" oil cooler.

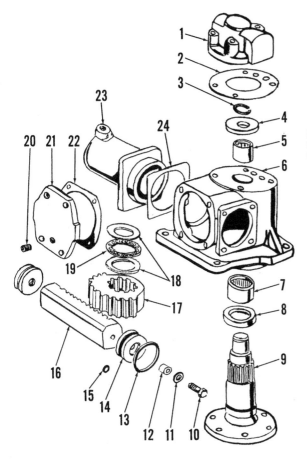

Fig. 44—Exploded view of hydrostatic steering cylinder housing and associated parts.

1. Cover
2. Gasket
3. Snap ring
4. Thrust washer
5. Needle bearing
6. Spindle housing
7. Needle bearing
8. Oil seal
9. Steering spindle
10. Cap screw
11. Washer
12. Spacer
13. Piston ring
14. Piston
15. Piston ring
16. Rack
17. Pinion
18. Bearing races
19. Thrust bearing
20. Drain plug
21. Rack guide
22. Shim
23. Cylinder
24. Gasket

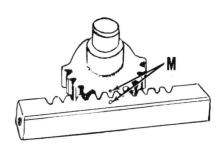

Fig. 45—Make sure match marks (M) on rack and pinion are aligned when cylinder unit is reassembled.

between rack (16) and rub block on guide (21). Shims are available in thicknesses of 0.006, 0.010, 0.014 and 0.020. Clearance can be measured with a feeler gage before installing one cylinder (23) and piston (14). Clearance should be measured with spindle (9) in a straight-ahead position, and the unit checked for binding by turning spindle through normal range of travel before unit is installed.

CONTINENTAL NON-DIESEL ENGINE AND COMPONENTS

Early non-diesel models are equipped with a Continental G-206, four cylinder engine having a bore of $3\frac{1}{16}$ inches, a stroke of $4\frac{1}{2}$ inches and a piston displacement of 206 cubic inches. Late models use a Perkins non-diesel engine which is covered separately beginning with paragraph 55.

R&R ENGINE WITH CLUTCH

All Models

31. To remove the engine and clutch, first drain cooling system and if engine is to be disassembled, drain oil pan. Remove hood. Disconnect drag link, air cleaner hose, radiator hoses and power steering hoses or lines. Support tractor under transmission housing and unbolt front support casting from engine; then roll front axle,

support and radiator as an assembly away from tractor. Shut off the fuel and remove fuel tank. On Model MF-180, remove side rails. Disconnect heat indicator sending unit, oil gage line, wiring harness, starter cable, tachometer cable and throttle linkage. Support engine in a hoist and unbolt engine from transmission case.

Install by reversing the removal procedure. Bleed the power steering as outlined in paragraph 6 or 8 and adjust governed speed as in paragraph 138 after tractor is assembled.

CYLINDER HEAD

All Models

32. **REMOVE AND REINSTALL.** To remove the cylinder head, drain cooling system, shut off the fuel and re-

move hood and fuel tank. Disconnect throttle and choke linkage from carburetor, remove air cleaner pipe and exhaust pipe; then unbolt and remove manifold and carburetor assembly. Disconnect heat indicator sending unit, bypass tube and upper radiator hose. Remove rocker arm cover, rocker arms assembly and push rods, then unbolt and remove cylinder head from tractor.

Install by reversing the removal procedure. Head gasket is pre-coated, and one side is marked "TOP" for proper assembly. Tighten the cylinder head cap screws to a torque of 70-75 ft.-lbs. using the sequence shown in Fig. 46. Reinstall push rods.

NOTE: Push rods can be dropped into cylinder block if misaligned, bumped, or improperly installed. Use care in installa-

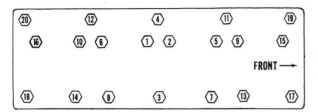

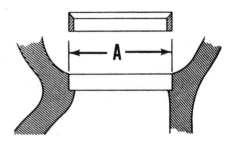

Fig. 46 — Cylinder head tightening sequence for non-diesel engines. Recommended tightening torque is 70-75 ft.-lbs.

Fig. 48—Valve seat counterbore (A) should be remachined to accept the 0.010 oversize insert if seats are renewed. Refer to text.

tion until rocker arms are in place and push rods safely secured.

Recommended tightening torque for rocker arm support screws is 35-40 ft.-lbs. When installing manifold, tighten end stud nuts to a torque of 35 ft.-lbs. and remaining stud nuts to 45-50 ft.-lbs. Adjust valve tappet gap using the procedure outlined in paragraph 35.

33. FUSE PLUG. Engines are equipped with a thermal type "Fuse" plug (19—Fig. 47) which serves as a warning device to indicate possible engine damage due to over-heating. The slotted head, 1/8-inch pipe plug threads into the water jacket of cylinder head. Inner end of plug contains a tin-lead alloy insert having a melting point of approximately 260° F.

After removing rocker arm cover and before engine is serviced, remove and inspect the plug. If insert has melted, engine has overheated; check for cracked block or head, warped or damaged gasket surfaces, or other heat damage to engine.

35. **VALVE TAPPET GAP.** The recommended cold tappet gap setting is 0.018 for intake valves and 0.020 for exhaust. Cold (static) setting of all valves can be made from just two crankshaft positions, using the procedure shown in Figs. 50 and 51, as follows:

Turn crankshaft until "TDC" mark on crankshaft pulley is aligned with timing pointer as shown in Fig. 49. Check the rocker arms for No. 1 and No. 4 cylinders. If rocker arms on No. 4 cylinder are tight, No. 1 piston is on compression stroke and the tappets indicated in Fig. 50 can be adjusted. If rocker arms on No. 1 cylinder are tight, No. 4 piston is on compression stroke and tappets indicated in Fig. 51 can be adjusted.

Make the indicated adjustments; then turn crankshaft one complete turn and adjust the remaining four tappets. Recheck the adjustment, if desired, with engine running at slow idle speed.

Recommended tappet clearance with engine at operating temperature is:

Intake—0.016; Exhaust—0.018. Clearance may be adjusted with engine running at slow idle speed or by following the procedure recommended for initial adjustment.

VALVE GUIDES
All Models

36. The pre-sized valve guides are interchangeable for intake and exhaust, and non-directional, capable of being installed either end up. Inside diameter of new guides is 0.3420-0.3435. Inner bore of guide has a fine, spiral groove (or rifling) which gives guide an unfinished appearance upon inspection, but guide must not be reamed.

When renewing guides, press old guides downward out of cylinder head

VALVES AND SEATS
All Models

34. Intake valves seat directly in cylinder head and valve stems are equipped with neoprene oil seals. Exhaust valves have renewable seat inserts. Gasoline models use positive type valve rotators (Rotocaps) on exhaust valves. LP Gas models use identical spring retainers on intake and exhaust valves and Rotocaps are not used.

Replacement exhaust valve seat inserts are supplied in 0.010 oversize only. When seats are to be renewed, re-machine the head counterbore (A—Fig. 48) to 1.447-1.448 to provide the recommended 0.003-0.005 interference fit for the service inserts.

Intake valve face and seat angle is 30°. Exhaust valves have a face angle of 44° and seat angle of 45° to provide the recommended 1° interference angle. Desired seat width is $\frac{1}{16}-\frac{3}{32}$ inch for all valves. Seats can be narrowed using 15 and 75 degree stones.

Fig. 47 — Exploded view of cylinder head and associated parts used on gasoline engines. Fuse plug (19) can be removed to check for evidence of overheating, refer to text. LP gas models use retainers (9) on intake valves instead of rotocaps (8).

1. Cylinder head
2. Intake valve
3. Valve guide
4. Push rod
5. Exhaust valve
6. Seat insert
7. Valve spring
8. Valve rotator
9. Retainer
10. Keepers
11. Ball socket
12. End plug
13. Spring
14. Rocker arm
15. Support
16. Rocker arm
17. Spring
18. Rocker shaft
19. Fuse plug

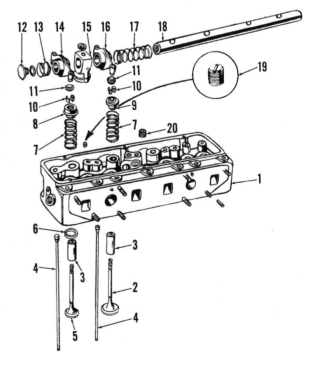

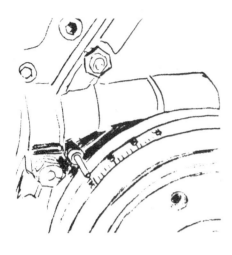

Fig. 49—View of timing gear cover and crankshaft pulley showing crankshaft timing marks. TDC mark must be aligned when adjusting tappets as shown in Fig. 50 and 51.

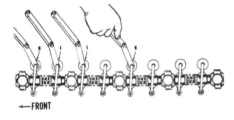

←FRONT

Fig. 50—With TDC timing mark align as shown in Fig. 49 and No. 1 piston on compression stroke, adjust the indicated intake valves to 0.018 and the indicated exhaust valves to 0.020.

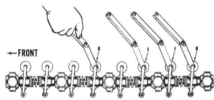

←FRONT

Fig. 51—With TDC timing mark aligns as shown in Fig. 49 and No. 4 piston on compression stroke, adjust the indicated intake valves to 0.018 and the indicated exhaust valves to 0.020.

using a piloted mandrel. Press new guide in from top until distance (A—Fig. 52) measured from rocker arm cover gasket surface to top of guide is approximately $\frac{1}{16}$-inch.

Valve stem diameters and clearance limits in guides are as follows:

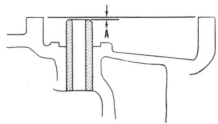

Fig. 52—When renewing valve guides, distance (A) from top of guide to rocker cover gasket surface should measure 1/16-inch.

Stem Diameter:

Intake 0.3406-0.3414
Exhaust 0.3382-0.3390

Desired Clearance:

Intake 0.0015
Exhaust 0.004

Maximum Clearance:

Intake 0.0049
Exhaust 0.0074

VALVE SPRINGS

All Models

37. Intake and exhaust valve springs are interchangeable. Renew any spring which is rusted, discolored, or fails to meet the test specifications which follow:

Free length (Approx) 2-5/64 inches
Lbs. Test @ $1\frac{21}{32}$ in. 41-47
Lbs. Test @ $1\frac{7}{32}$ in. 103-110

VALVE ROTATORS

Gasoline Models

38. Normal servicing of the positive type exhaust valve rotators ("Rotocaps") consists of renewing the units. It is important, however, to observe the valve action after engine is assembled. Rotator can be considered satisfactory if valve turns a slight amount each time valve opens.

VALVE TAPPETS

All Models

39. The mushroom type tappets (cam followers) operate directly in machined bores in cylinder block. The 0.5615-0.5620 diameter tappets are furnished in standard size only, and should have a clearance of 0.0005-0.002 in block bores. Tappets can be removed after removing camshaft as outlined in paragraph 43.

Refer to paragraph 35 for valve tappet gap adjustment procedure.

ROCKER ARMS

All Models

40. The rocker arm shaft (18—Fig. 47) is positively positioned by the support bolts, which pass through drilled holes in shaft and support brackets (15). Oil holes for rocker arms must be located on push rod side of engine.

Rocker arms should have 0.001-0.0021 clearance on the 0.9671-0.9677 diameter shaft. The stamped steel

Fig. 53 — Timing marks (T) are properly positioned on crankshaft when marked splines (A) are aligned. Install the short timing gear cover cap screw at (B).

rocker arms are right hand and left hand units which are installed in pairs. Bronze bushings in rocker arms are not serviced; if bushing is worn, renew the rocker arm. Refer to paragraph 35 for valve tappet gap adjustment procedure.

TIMING GEAR COVER

All Models

41. **REMOVE AND REINSTALL.** To remove the timing gear cover, drain cooling system and remove hood. Disconnect drag link on Model MF175, and radiator hoses and air cleaner hose on all models. On Model MF175, power steering pump can be removed from mounting pad on engine block without disconnecting hoses or draining system. On Model MF180, remove power steering cylinder lines and cylinder housing bleed line. On all models, remove the lines leading to Multipower filter and oil cooler.

Support tractor under transmission housing, unbolt front axle support from engine and roll axle, support and radiator as an assembly, away from tractor.

Remove fan blades, fan belt and the cap screw securing crankshaft pulley to shaft, then remove the pulley. Disconnect throttle and carburetor rods from governor lever. Remove the oil pan or remove the two front oil pan cap screws and loosen remainder of pan screws. Remove the rocker arm cover breather tube. Remove the cap screws securing timing gear cover to engine block, then remove cover from its doweled position on engine block.

Crankshaft front oil seal can be renewed at this time. Install the seal with lip to rear and rear edge flush with seal bore in cover. Refer to paragraph 139 for procedure for overhaul of governor shaft and associated parts.

Install by reversing the removal procedure, making sure that the short (⅞-inch) cover cap screw is installed in lower corner on tractor's left (B—Fig. 53). Tighten the cover retaining cap screws to a torque of 25-30 ft.-lbs. and adjust governor linkage as in paragraph 138.

When installing crankshaft pulley, make sure marks (A) on pulley and shaft splines are aligned as shown in Fig. 53 to properly position crankshaft timing marks (T). Tighten pulley retaining cap screw to a torque of 150 ft.-lbs.

TIMING GEARS

All Models

42. Timing gears can be renewed after removing timing gear cover as outlined in paragraph 41 and withdrawing governor weight unit from crankshaft.

Refer to Fig. 54 for a view of timing gear train with timing gear cover removed. Before attempting to remove either the camshaft gear (1) or crankshaft gear (4), first remove the retaining snap rings and withdraw balance shaft idler gears (2 and 6). Camshaft and crankshaft gears are a light press fit and can be removed using a suitable puller.

Recommended timing gear backlash is 0.002. Gears are available in standard size, plus undersizes or oversizes of 0.001 and 0.002. Gears are marked "S" (Standard), "U" (Undersize), or "O" (Oversize); the "U" or "O" enclosing the number marking "1" or "2", which indicates the exact size.

NOTE: For production reference, a timing gear size marking is stamped on block front face as shown at (S—Fig. 54). When

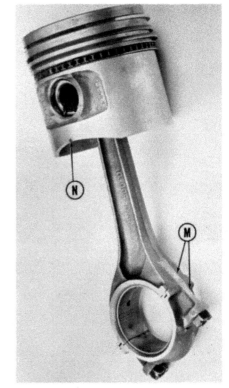

Fig. 55—Exploded view of balance shaft idler gears and associated parts showing bearing arrangement. Gears (2 & 6) are not interchangeable.

2. Idler gear	10. Bushing
6. Idler gear	11. Thrust washer
8. Snap ring	12. Shaft
9. Thrust washer	13. Set screw

Fig. 54—View of timing gear train with timing marks properly aligned. Code marks (S) are used for selection of gears to obtain the correct backlash as outlined in paragraph 42. Balancer idler gears (2 & 6) have odd number of teeth and will not align each time camshaft timing marks are aligned. Misalignment of marks on idler gears does not necessarily mean gears are mis-timed.

1. Camshaft gear
2. Idler gear
3. Balance shaft
4. Crankshaft gear
5. Oil pump gear
6. Idler gear
7. Balance shaft
S. Code marks
T. Timing marks

Fig. 56—Piston and connecting rod properly assembled. Notch (N) is installed to front and correlation marks (M) toward camshaft side.

renewing timing gears, any combination of standard, oversize or undersize gears in which the total equals the stamped reference number should give the correct backlash.

Balancer idler gears (2 and 6) are not interchangeable, the right hand gear (6) having 19 teeth while left hand gear (2) has 20. Bushings (10—Fig. 55), thrust washers (9 or 11), or shafts (12) are interchangeable for either gear. Shafts are retained in block and bearing cap by cup point set screws (13). Idler gear diametral clearance should be 0.0025-0.004 and end play should be 0.004-0.008. Renew bushings, shafts and/or thrust washers if clearances are incorrect.

During installation, mesh the single punch-marked tooth on crankshaft gear between the two marked teeth on camshaft gear as shown in Fig. 54. The single marking on camshaft gear and double marking on crankshaft gear are used for timing the balance gears as shown. Heating camshaft gear in oil or in an oven to approximately 300° F. will facilitate gear installation. Remove oil pan and support camshaft in a forward position while gear is being installed, to prevent loosening and leakage of camshaft rear plug.

Tighten the camshaft gear retaining nut to a torque of 125-130 ft.-lbs. and reinstall and time the balance shaft idler gears after crankshaft and camshaft gears are installed. Balance shaft driven gears are an integral part of shafts, and renewal requires removal of engine block. Refer to paragraph 49 for procedure.

CAMSHAFT

All Models

43. To remove the camshaft, first remove camshaft timing gear and balance shaft idler gears as outlined in paragraph 42. Remove fuel tank, rocker arm cover, rocker arms and push rods. Remove the ignition distributor and oil pan. Block up or suitably support the cam followers, remove the cap screws retaining camshaft thrust plate to cylinder block front face and withdraw camshaft from front of engine.

The front camshaft bore in engine block contains a renewable bushing and the outboard camshaft support in timing gear cover also contains a bushing. The two rear camshaft journals ride in unbushed bores in engine block. Normal diametral clearance of camshaft journals is 0.0025-0.0045 in block bores or 0.0025-0.0035 for out-

board (front) bearing. Renew camshaft front bushing and/or camshaft if clearance exceeds 0.006; or camshaft and/or cylinder block if clearance for other journals exceeds 0.007. Desired camshaft end play of 0.003-0.007 is controlled by the retaining thrust plate. Renew the plate if end play exceeds 0.008.

Camshaft journal diameters are as follows:

Outboard Journal 0.9965-0.997
Front Journal . . . . .1.808 -1.809
Center Journal 1.7455-1.7465
Rear Journal . . . . . . .1.683 -1.684

ROD AND PISTON UNITS

All Models

44. Connecting rod and piston units are removed from above after removing cylinder head and oil pan. Correlation marks (M—Fig. 56) should be installed toward camshaft side of block with identification notch (N) in piston skirt toward the front. Replacement rods are not marked, but should be stamped with cylinder number before installation. Tighten the connecting rod cap screws to a torque of 40-45 ft.-lbs.

PISTONS, SLEEVES AND RINGS

All Models

45. Pistons are available in standard size only and available only in a kit consisting of piston, pin, rings and sleeve for one cylinder. If piston and/or sleeve are scored, if piston ring grooves are worn or damaged; or if cylinder wall taper exceeds 0.008, renew the piston and sleeve assembly. Renew piston and sleeve assembly if top ring side clearance exceeds 0.008 or 2nd or 3rd ring side clearance exceeds 0.006, using a new ring. Recommended piston ring end gap is 0.010-0.020 for all rings.

The dry type cylinder sleeves should have 0.0005-0.0025 clearance in block bores and top of sleeve is shouldered to fit in counterbores in cylinder block. The installed sleeve should not be recessed more than 0.0015 below nor stand out more than 0.0025 above gasket surface of cylinder block. Interchange the kits in block bores to obtain the correct standout. Cutout notches in lower edge of cylinder sleeve must be installed at right angle to crankshaft centerline to provide connecting rod clearance.

PISTON PINS

All Models

46. The 1.250-1.252 diameter piston pins are full floating and retained in piston bosses by snap rings. Piston pins should have 0.0002-0.0006 clearance in both the piston bosses and connecting rod. Piston pins are available in standard size and oversizes of 0.003 and 0.005.

CONNECTING RODS AND BEARINGS

All Models

47. Connecting rod bearings are renewable from below after removing oil pan and oil pump intake screen and line. When installing new bearing shells, make sure the projection engages the milled slot in rod and cap and that correlation marks (M—Fig. 56) are in register. Replacement rods are not marked and should be stamped with cylinder number before rod is installed. Bearings are available in standard size and undersizes of 0.002, 0.010 and 0.020.

NOTE: Shop re-grinding of the hardened crankshaft is not recommended; however reground and treated crankshafts are available from the manufacturer on an exchange basis.

Standard specifications are as follows:

Crankpin diameter 2.0615-2.0625
Diametral clearance . . .0.0007-0.0027
Wear limit 0.0038
Rod side clearance . . .0.006 -0.010
Cap screw torque . . . .40-45 ft.-lbs.

CRANKSHAFT AND BEARINGS

All Models

48. The crankshaft is supported in three main bearings renewable from below without removing the crankshaft. To renew the main bearings, it is first necessary to remove timing gear cover and balance shaft idler gears as outlined in paragraph 40; and oil pump as outlined in paragraph 50.

Normal crankshaft end play of 0.004-0.008 is controlled by the flanged center main bearing inserts. Bearings are available in undersizes of 0.002, 0.010 and 0.020 as well as standard.

NOTE: Shop re-grinding of the hardened crankshaft is not recommended; however, reground and heat treated crankshafts are available from the manufacturer on an exchange basis.

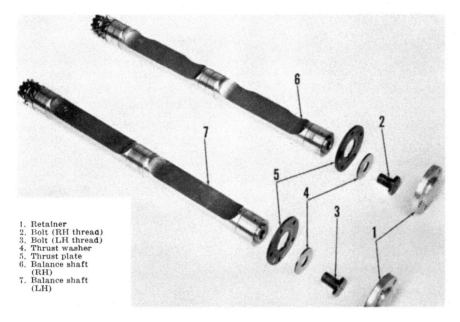

1. Retainer
2. Bolt (RH thread)
3. Bolt (LH thread)
4. Thrust washer
5. Thrust plate
6. Balance shaft (RH)
7. Balance shaft (LH)

Fig. 57—Exploded view of balance shafts and associated parts used on G-206 engine.

Fig. 58—Rear view of engine block with flywheel removed. To remove the balance shafts, unbolt and remove retainers (1).

Standard crankshaft specifications are as follows:

Main journal diameter 2.374 -2.375

Crankpin diameter ...2.0615-2.0625

Main bearing diametral
clearance 0.0005-0.0028

Wear limit 0.0038

Main bearing cap
screw torque ......85-95 ft.-lbs.

ENGINE BALANCE SHAFTS

All Models

49. **OPERATION.** The balancer assembly consists of two shafts (6 & 7—Fig. 57) lying parallel to crankshaft and driven by the timing gears in opposite directions at twice crankshaft speed. The shafts are weighted and timed to cancel out the natural tendency to vibrate inherent to a four cylinder in-line engine.

50. **OVERHAUL.** To remove the balance shafts, first remove the engine as outlined in paragraph 31; then remove clutch, flywheel, timing gear cover and oil pan. Unbolt and remove the two retainers (1—Fig. 58) as shown in Fig. 59, then carefully withdraw the balance shafts as shown in Fig. 60.

Balance shafts (6 & 7—Fig. 57) are not interchangeable, the right shaft (6) having relief notches to clear the connecting rods. Also, the left shaft (7) and cap screw (3) have left-hand threads because of the reverse rotation of the shaft.

Balance shaft bearing journals are heat-treated and should have 0.003-

0.0045 clearance in bushings. Journal diameters are 1.683-1.684. Bushings in cylinder block are renewable. Bushings have 0.003-0.0055 interference fit in block bores and should be installed with "LOCTITE" to prevent loosenings or turning during operation. Align ream bushings after installation to a finished diameter of 1.687-1.6875.

Examine thrust plates (5) and washers (4) for wear. If renewal is required, remember that left hand cap screw (3) has a left-hand thread. Tighten both cap screws (2 & 3) to a torque of 140-150 ft.-lbs.

Time the balance shafts as outlined in paragraph 42 when units are installed.

CRANKSHAFT REAR OIL SEAL

All Models

51. The crankshaft rear oil seal is contained in a one-piece retainer and serviced only as an assembly. To renew the seal, first separate engine from transmission case as outlined in paragraph 152 or 153 and remove the flywheel. Remove the oil pan and the two cap screws securing rear seal retainer to rear main bearing cap; remove the three remaining cap screws and lift off the seal and retainer unit.

Apply a light coating of oil to seal lip and gasket sealer to mounting gasket. Position seal retainer with the two threaded holes down. Reinstall cap screws and tighten evenly to a torque of 8-10 ft.-lbs. Complete the assembly by reversing the disassembly procedure.

Fig. 59—Removing balance shaft retainers.

Fig. 60 — Withdrawing balance shaft assembly.

FLYWHEEL

All Models

52. To remove the flywheel, separate engine from transmission case as outlined in paragraph 152 or 153 and remove the clutch. Flywheel is retained to crankshaft flange by four special cap screws, one of which is slightly offset to allow flywheel to be installed in one position only.

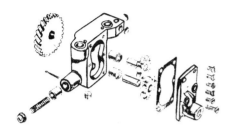

Fig. 61—Exploded view of engine oil pump.

Tighten flywheel retaining cap screws to a torque of 70-75 ft.-lbs. Flywheel face runout should not exceed 0.001 for each 1-inch distance from crankshaft centerline to point where measurement is taken.

The starter ring gear can be renewed after flywheel is removed. Heat ring gear evenly to approximately 450° F. and install on flywheel with beveled edge of teeth facing front of engine.

OIL PUMP

All Models

53. The gear type oil pump is mounted below the front main bearing cap and gear driven from the crankshaft timing gear. Pump is accesible after oil pan is removed.

The recommended backlash of 0.004-0.008 for oil pump drive gear is controlled by shims installed between oil pump body and main bearing cap.

Check pump parts for wear, scoring or other damage and renew parts or pump assembly as required. When installing pump cover, tighten the retaining screws to a torque of approximately 8-10 ft.-lbs. Refer to paragraph 54 for data on the pressure relief valve.

RELIEF VALVE

All Models

54. The plunger type pressure relief valve is located in pump body as shown in Fig. 61. Normal operating pressure is not less than 7 psi at slow idle speed and 20-30 psi at 1800 rpm.

PERKINS NON-DIESEL ENGINE AND COMPONENTS

Late Model MF175 and MF180 tractors are equipped with a Perkins AG4.236 four cylinder gasoline engine having a bore of 3⅞ inches, a stroke of 5 inches and a piston displacement of 235.9 cubic inches.

The engine is similar in design to the Perkins A4.236 diesel engine which is optionally used, and some parts are interchangeable.

Refer to the separate section beginning with paragraph 31 for service on the CONTINENTAL non-diesel engine used on early models and to section beginning with paragraph 85 for diesel engine service.

R&R ENGINE WITH CLUTCH

All Models

55. To remove the engine and clutch as a unit, first drain cooling system and if engine is to be disassembled, drain oil pan. Remove hood. Disconnect drag link on Model MF175. Disconnect air cleaner hose, radiator hose and power steering hoses or lines. Support tractor under transmission housing and unbolt front support casting from engine; then roll front axle, support and radiator as an assembly away from tractor. Shut off the fuel and remove fuel tank. On Model MF-180, remove side rails. Disconnect heat indicator sending unit, oil gauge line, wiring harness, starter cable, tachometer cable and throttle linkage. Support engine in a hoist and unbolt engine from transmission case.

Install by reversing the removal procedure. Bleed the power steering as outlined in paragraph 6 or 8 and adjust governed speed as in paragraph 138.

CYLINDER HEAD

All Models

56. **REMOVE AND REINSTALL.** To remove the cylinder head, drain cooling system, shut off the fuel and remove hood and fuel tank. Disconnect throttle and choke linkage from carburetor and remove air cleaner pipe and exhaust pipe. Remove manifold and carburetor assembly, rocker arm cover, rocker arms unit and push rods; then unbolt and remove the cylinder head.

The cylinder block, head and head gasket are internally ported to provide pressure lubrication to rocker arms. Make sure passages are open and aligned when cylinder head is reinstalled.

Head gasket is marked "TOP FRONT" for proper installation. Head gasket is treated and must be installed dry. Tighten cylinder head stud nuts to a torque of 85-90 ft.-lbs. using the sequence shown in Fig. 62. Adjust valve tappet gap as outlined in paragraph 61. Head should be retorqued and valves adjusted after engine is warm.

VALVES AND SEATS

All Models

57. Intake and exhaust valves seat directly in cylinder head and production valves and seat locations are numbered consecutively from front to rear. Production face and seat angle is 46 degrees; 45° may be used for service.

Fig. 62 — On Perkins Gasoline engines, tighten cylinder head stud nuts to a torque of 85-90 ft.-lbs. in the sequence shown.

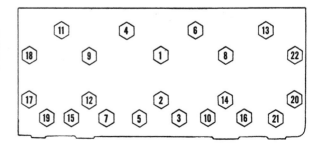

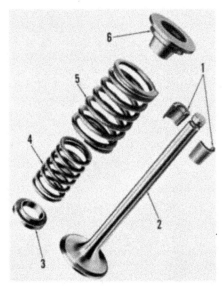

Fig. 63—Disassembled view of valve components showing correct positioning of springs and retainers.

1. Keepers
2. Valve
3. Spring seat
4. Inner spring
5. Outer spring
6. Retainer

Standard valve stem diameter is 0.3725-0.3735 for intake valves and 0.372-0.373 for exhaust valves. Service valves are available with stem oversizes of 0.003, 0.015 and 0.030 as well as standard. Intake valves are equipped with umbrella type rubber valve stem seals. Intake valves are recessed approximately 0.045 below gasket surface of cylinder head. Exhaust valves are not recessed.

Recommended valve tappet gap is 0.012 (hot) for intake valves and 0.015 (hot) for exhaust. Tappet gap should be adjusted using the procedure outlined in paragraph 61.

VALVE GUIDES
All Models

58. Intake and exhaust valve guides are cast into cylinder head and oversize valve stems are provided for service.

Standard guide bore diameter is 0.3745-0.3755 for both the intake and exhaust valves.

VALVE SPRINGS
All Models

58A. Inner and outer springs are used as shown in Fig. 63. Springs have close coils which should be installed next to cylinder head. The inner spring (4) has a shorter assembled length than outer spring (5) due to stepped seating washer (3) and retainer (6). Renew springs if they are distorted, discolored or fail to meet the test specifications which follow:

	Lbs. Test @ Inches
Inner spring	15 @ 1 9/16
Outer spring	40 @ 1 25/32

CAM FOLLOWERS
All Models

59. The mushroom type cam followers (tappets) operate directly in machined bores in engine block and can be removed after removing camshaft as outlined in paragraph 70. The 0.7475-0.7485 diameter cam followers are furnished in standard size only and should have a diametral clearance of 0.0015-0.0037 in block bores. Adjust tappet gap as outlined in paragraph 61.

ROCKER ARMS
All Models

60. The rocker arms and shaft assembly can be removed after removing hood, fuel tank and rocker arm cover. Rocker arms are right hand and left hand units which must be installed on shaft as shown in Fig. 64. Desired diametral clearance between new rocker arms and new shaft is 0.001-0.0035. Renew shaft and/or rocker arms if clearance is excessive. Adjust tappet gap as outlined in paragraph 61 after shaft is reinstalled.

Fig. 64 — View of rocker shaft showing rocker arms correctly installed. O-ring seal (shown) fits groove in oil supply tube.

Fig. 65 — Timing marks are located on crankshaft pulley as shown.

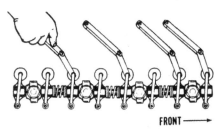

Fig. 66—With TDC timing marks aligned as shown in Fig. 65 and No. 1 piston on compression stroke, adjust the indicated valves. Suggested intial (cold) setting is 0.012 for intake valves and 0.015 for exhaust. Turn crankshaft one complete turn and refer to Fig. 67 for remainder of valves.

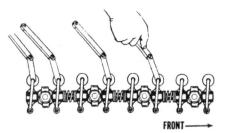

Fig. 67—With TDC timing marks aligned as shown in Fig. 65 and No. 1 piston on exhaust stroke, adjust the indicated valves as outlined for Fig. 66.

VALVE TAPPET GAP
All Models

61. The recommended hot tappet gap is 0.012 for both intake and exhaust valves. Cold (static) settings are 0.012 for intake valves and 0.015 for exhaust valves. All valves can be adjusted from just two crankshaft positions using the procedure outlined in this paragraph and illustrated in Figs. 65, 66 and 67.

Turn crankshaft until TDC timing marks on crankshaft pulley aligns

Fig. 68—Crankshaft pulley alignment marks (shown) insure proper positioning of engine timing marks.

with timing pointer as shown in Fig. 65 and check rocker arms for front and rear cylinders. If rear rocker arms are tight and front rocker arms loose, No. 1 piston is on compression stroke; adjust the valves indicated in Fig. 66, turn crankshaft one complete turn until timing marks again align and adjust remainder of valves. If No. 4 piston is on compression stroke (front rocker arms tight), adjust the four valves indicated in Fig. 67, turn crankshaft one complete turn and adjust remainder of valves.

TIMING GEAR COVER

All Models

62. To remove the timing gear cover, drain cooling system and remove hood. Disconnect drag link on Model MF175; and radiator hoses and air cleaner hose on all models. Disconnect power steering and Multipower oil cooler oil lines. Support tractor under

Fig. 69 — Timing gear train with cover removed. Governor drive gear timing marks align only occasionally but alignment is not necessary for proper engine timing. Refer to text for procedure.

transmission case, unbolt front support from engine block and roll front support, front axle and radiator as an assembly away from tractor.

Remove fan belt, fan blades and ignition coil. Remove the cap screw securing pulley to front of crankshaft and check to see that timing marks are legible (See Fig. 68), then remove pulley. Unbolt and remove timing gear cover.

Crankshaft front oil seal can be renewed at this time. Install seal with lip to rear, with front edge of seal recessed 0.380-0.390 into seal bore when measured from front of cover. A special tool (MFN 747B) and spacer (MFN 747C) are available to properly position the seal.

The timing gear cover is not doweled. Special tool MFN 747B can be used as a pilot when reinstalling. If tool is not available, reinstall crankshaft pulley to center the seal when reinstalling cover retaining screws. Assemble by reversing the disassembly procedure, making sure pulley timing marks are aligned as shown in Fig. 68. Tighten the pulley retaining cap screw to a torque of 280-300-ft.lbs.

TIMING GEARS
All Models

63. Refer to Fig. 69 for view of timing gear train with cover removed. Before attempting to remove any of the timing gears, first remove fuel tank, rocker arm cover and rocker arms to avoid the possibility of damage to pistons or valve train if camshaft or crankshaft should either one be turned independently of the other.

Because of the odd number of teeth in idler gear, all timing marks will align only once in many crankshaft revolutions. To accurately check the

Fig. 70—Idler gear end play should be 0.003-0.007 when measured as shown.

timing, turn crankshaft until the proper marks on crankshaft gear and camshaft gear mesh with idler gear; then remove and reinstall idler gear with marks aligned. Governor drive gear does not need to be timed, but alignment of marks is a convenience for engine assembly. Crankshaft timing gear also drives engine balancer. Balancer must be timed to crankshaft as outlined in paragraph 78.

Service gears are available in standard size only. Refer to the appropriate following paragraphs for renewal of gears, idler shaft or bushings if any of the parts are damaged or if noise is excessive.

64. IDLER GEAR AND HUB. Diametral clearance of idler gear on hub should be 0.0027-0.0047. Permissible end play is 0.0003-0.007.

Idler gear hub is a light press fit in timing gear housing bore. Due to uneven spacing of the three studs, hub can only be installed in one position. Tighten the three idler hub retaining cap screws to a torque of 20-25 ft.-lbs. Measure end play with a feeler gauge (Fig. 70) after idler gear is installed.

65. CAMSHAFT GEAR. Camshaft gear is keyed to shaft and retained by a cap screw. Camshaft gear is a transition fit (0.001 tight to 0.001 clearance) and threaded holes are provided for pulling gear. Make sure timing marks are aligned and tighten retaining cap screw to 45-50 ft.-lbs. when installing gear.

66. GOVERNOR GEAR. The governor gear also drives the distributor. Gears contain timing marks which are convenient for assembly but not essential for engine timing, as timing can be accomplished externally when distributor is installed.

Governor gear is a slip fit (0.0003-0.0012 clearance) on shaft. Shaft contains governor weight unit as well as distributor drive, and unit can be removed from rear as outlined in paragraph 71 after gear is off. When installing gear, tighten retaining nut to a torque of 25-35 ft.-lbs.

67. CRANKSHAFT GEAR. Crankshaft gear is keyed to shaft and is a transition fit (0.001 tight to 0.001 clearance) on shaft. It is usually possible to remove the gear using two small pry bars to move the gear forward. Engine balancer must be removed if a puller is required.

68. TIMING THE GEARS. To install and time the gears, first install camshaft and crankshaft gears as outlined in the appropriate preceding paragraphs with timing marks to front. Turn the shafts until the appropriate timing marks point toward idler gear hub, then install idler gear with marks aligned as shown in Fig. 69. The governor gear timing marks may be aligned for convenience when all gears are removed, but alignment is not necessary for proper ignition timing. Secure idler gear as outlined in paragraph 64.

TIMING GEAR HOUSING

All Models

69. To remove the timing gear housing, first remove timing gears as outlined in paragraphs 63 through 67 and the distributor and drive unit as in paragraph 71. Timing gear housing must be removed before camshaft can be withdrawn. Remove cap screws retaining timing gear housing to engine block and oil pan and lift off housing. Install by reversing the removal procedure.

Fig. 71—Camshaft is retained by thrust washer which can be removed only after removing timing gear housing.

CAMSHAFT

All Models

70. To remove the camshaft, first remove timing gears, distributor drive unit (governor) and timing gear housing as outlined in paragraphs 63 through 69. Secure cam followers (tappets) in uppermost position and lift off thrust washer (Fig. 71), then withdraw camshaft from block bores.

Thrust washer (Fig. 71) retains camshaft and controls end play. Thrust washer thickness is 0.216-0.218; check the washer for correct thickness and for wear or scoring. Recommended camshaft end play is 0.0004-0.016 and diametral clearance of camshaft journals in their bores is 0.0025-0.0053. Camshaft journal diameters are as follows:

Front 1.9965-1.9975
Center 1.9865-1.9875
Rear 1.9665-1.9675

DISTRIBUTOR DRIVE SHAFT

All Models

71. A disassembled view of distributor drive unit and governor assembly is shown in Fig. 72. To remove or disassemble the unit, first remove timing gear cover as outlined in paragraph 62 and governor gear as in paragraph 66. Remove the distributor as shown in Fig. 73 and governor shaft thrust plate as shown in Fig. 74. Disconnect governor and throttle linkage and unbolt and remove housings as shown in Fig. 75.

Unbolt and remove governor housing from drive body. Withdraw thrust bearing and distributor drive gear then remove drive shaft assembly and governor weight unit.

Distributor drive shaft end play is 0.004-0.008 and thrust is forward, because of governor action. Diametral clearance of shaft in housing bore is 0.001-0.0025.

Assemble by reversing the disassembly procedure. Turn crankshaft

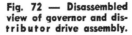
Fig. 72 — Disassembled view of governor and distributor drive assembly.

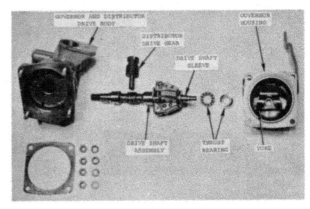

until No. 1 piston is coming up on compression stroke and 10° BTDC timing mark on crankshaft pulley is aligned, then install distributor gear with drive slot perpendicular to crankshaft and offset to rear as shown in Fig. 76. Install distributor and turn body until points just begin to open, then install and tighten clamp. Adjust governed speed as outlined in paragraph 138 and timing as in paragraph 144.

ROD AND PISTON UNITS

All Models

72. Connecting rod and piston units are removed from above after removing cylinder head, oil pan, engine balancer and rod bearing caps. Cylinder numbers are stamped on connecting rod and cap. Make certain correlation numbers are in register and face away from camshaft side of engine when reassembling. Tighten connecting rod nuts to a torque of 65-70 ft.-lbs.

PISTONS, SLEEVES AND RINGS

All Models

73. The aluminum alloy pistons have a combustion chamber cavity cast into piston crown and an "F" marking on top of piston to indicate front. Pistons are available in standard size only.

Each piston is fitted with a plain faced chrome top ring which may be installed either side up as may the plain faced second compression ring. The third ring is marked "BTM" (bottom) for correct installation. A chrome railed, segmented oil control ring is used in fourth grove.

Recommended side clearance for compression rings is 0.002-0.004. End gaps for compression rings should be as follows:

Top rings 0.016-0.021
Other rings 0.012-0.017

Production sleeves are a tight press fit in cylinder block and are finished

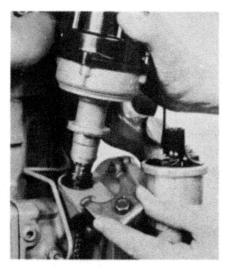

Fig. 73 — Remove distributor clamp and lift out distributor assembly.

Fig. 74 — Removing governor shaft thrust plate.

Fig. 75 — Removing governor housing assembly.

after installation. Service sleeves are a transition fit and are prefinished. Sleeves should not be bored and oversize pistons are not available. When installing new sleeves, make sure sleeves and bores are absolutely clean and dry, then chill the sleeves and press fully into place by hand. Top edge of sleeve should extend 0.030-0.035 above gasket face of cylinder block.

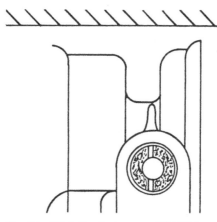

Fig. 76—With No. 1 piston at 12° BTDC on compression stroke, distributor drive slot should be perpendicular to crankshaft and offset to rear as shown. Insert a screwdriver in drive slot and turn counter-clockwise to remove the gear.

PISTON PINS
All Models

74. The full floating piston pins are retained in piston bosses by snap rings and are available in standard size only. The renewable connecting rod bushing must be final sized after installation to provide a diametral clearance of 0.0005-0.0015 for the pin. Be sure the pre-drilled oil hole in bushing is properly aligned with hole in top of connecting rod and install bushing from chamfered side of bore. Piston pin should be a thumb press fit in piston after piston is heated to 160° F. Piston pin diameter is 0.9998-1.0000.

CONNECTING RODS AND BEARINGS
All Models

75. Connecting rod bearings are precision type, renewable from below after removing oil pan, balancer unit and rod bearing caps. When renewing bearing shells, be sure the projection engages the milled slot in rod and cap and that correlation marks are in register and face away from camshaft side of engine.

Connecting rod bearings should have a diametral clearance of 0.0015-0.003 on the 2.499-2.4995 diameter crankpin. Recommended connecting rod side clearance is 0.010-0.015. Renew the self-locking connecting rod nuts and tighten to a torque of 65-70 ft.-lbs.

CRANKSHAFT AND BEARINGS
All Models

76. The crankshaft is supported in five precision type main bearings. To

remove the rear main bearing cap, it is first necessary to remove engine, clutch, flywheel and rear oil seal. All other main bearing caps can be removed after removing oil pan and engine balancer.

Upper and lower main bearing inserts are not interchangeable. The upper (block) half is slotted to provide pressure lubrication to crankshaft and connecting rods. Inserts are interchangeable in pairs for all journals except center main bearing. The center journal controls crankshaft end thrust and renewable thrust washers are installed at front and rear of cap and block bearing bore. Lower half of insert is anchored by a tab to bearing cap; upper half can be rolled out and renewed after cap is removed.

Bearing inserts are available in undersizes of 0.010, 0.020 and 0.030 as well as standard. Standard thrust washers are 0.089-0.091 in thickness and oversizes of 0.0075 are available. Oversizes may be installed in pairs at front or rear of journal or at both the front and rear. When renewing rear main bearing, refer to paragraph 80 for installation procedure of rear seal and oil pan bridge piece. Recommended tightening torque for main bearing cap screws is 145-150 ft.-lbs. Check crankshaft journals against the values which follow:

Main journal diameter.. 2.9985-2.999
Crankpin diameter .... 2.499 -2.4995
Diametral clearance:
　Main bearings 0.0025-0.0045
　Crankpin bearings ... 0.0015-0.003

Fig. 77—Renewable thrust washers (B & C) control crankshaft end play. Main bearing caps are positively located by ring dowels (A).

ENGINE BALANCER

All Models

77. OPERATION. The Lanchester type engine balancer consists of two unbalanced shafts which rotate in opposite directions at twice crankshaft speed. The inertia of the shaft weights is timed to cancel out natural engine vibration, thus producing a smoother running engine. The balancer is correctly timed when the balance weights are at their lowest point when pistons are at TDC and BDC of their stroke.

The balancer unit is driven by the crankshaft timing gear through an idler gear attached to balancer frame. The engine oil pump is mounted at rear of balancer frame and driven by the balancer shaft. Refer to Figs. 78 through 80.

78. REMOVE AND REINSTALL. The balancer assembly can be removed after removing the oil pan and mounting cap screws. Engine oil is pressure fed through balancer frame and cylinder block. Balancer frame bearings are also pressure fed. Refer to Fig. 78 for an installed view of balancer unit.

When installing balancer with engine in tractor, timing marks will be difficult to observe without removing gear cover. The balancer assembly can be safely installed as follows:

Turn crankshaft until No. 1 and No. 4 pistons are at the exact bottom of their stroke. Remove balancer idler gear (5—Fig. 80) if necessary, and reinstall with single punch-marked tooth of idler gear meshed between the two marked teeth on weight drive shaft as shown at (B—Fig. 78). Install balancer frame with balance weights hanging normally. If carefully installed, timing will be correct although timing marks at front of engine may not be aligned. (Marks align only occasionally when engine is running). If engine is mounted in stand or tractor front end is removed, timing marks can be observed by removing timing gear cover.

NOTE: Balancer can be safely installed with No. 1 & 4 pistons at either TDC or BDC. BDC is selected because interference between connecting rod and balance weights can give warning if unit is badly out of time. Also, alignment of timing marks is not essential but is a convenience for original engine assembly.

With balancer correctly installed, tighten the retaining cap screws to a torque of 32-36 ft.-lbs. and complete the assembly by reversing the removal procedure.

Fig. 78—Installed view of engine balancer with timing marks aligned. Refer to paragraph 78 for installation procedure.

79. OVERHAUL. Refer to Fig. 80 for an exploded view of balancer frame and associated parts. To disassemble the removed balancer unit, unbolt and remove oil pump housing (20) and associated parts; and idler gear (5) and associated parts. Set screws (S) retaining balance weights (15 & 16) are installed using Grade "A" (Red) LOCTITE. Loosen screws, then push balance shafts (11 & 12) forward out of frame and weights. NOTE: Use care when removing shafts, not to allow keys (13) to damage frame bushings as bushings are not available as a service item.

Recommended diametral clearance for shafts (11 & 12) in frame bushings is 0.002-0.0045 for front bushings or 0.002-0.0035 for rear. Shaft diameters are 1.2485-1.249 for front journal and 0.9987-1.0002 for rear. When assembling the balancer, used Grade "A" (Red) LOCTITE for installing the screws retaining gears (14) to balance weights (15 & 16) and the set screws (S) retaining balance weights to shafts. Also make sure flat surfaces of weights are aligned when installed, as shown in Fig. 79.

Recommended diametral clearance for idler gear (5—Fig. 80) on hub (2) is 0.001-0.0032. End play should be 0.008-0.014. Bushing for idler gear is not available as a service item.

Refer to paragraph 83 for overhaul of engine oil pump and to paragraph 78 for installation of balancer assembly.

Fig. 79—Assembled view of removed engine balancer and oil pump unit. Refer to Fig. 80 for exploded view.

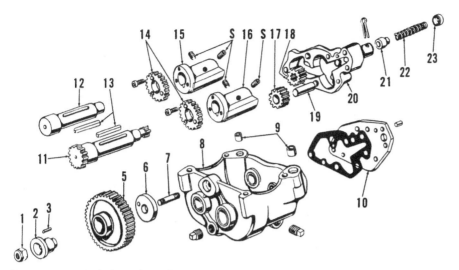

Fig. 80—Exploded view of Lanchester type engine balancer, engine oil pump and associated parts.

1. Locknut	10. Plate	18. Pump gear
2. Hub	11. Drive shaft	19. Shaft
3. Dowel	12. Driven shaft	20. Pump body
5. Idler gear	13. Key	21. Valve piston
6. Washer	14. Gear	22. Valve spring
7. Stud	15. Balance weight	23. Cap
8. Frame	16. Balance weight	S. Set screw
9. Ring dowels	17. Pump gear	

Fig. 83—Cylinder block bridge is equipped with end seals as shown.

CRANKSHAFT REAR OIL SEAL

All Models

80. The asbestos rope type rear oil seal is contained in a two-piece seal retainer attached to rear face of engine block as shown in Fig. 81. The seal retainer can be removed after removing flywheel.

The rope type crankshaft seal is precision cut to length, and must be installed in retainer halves with 0.010-0.020 of seal ends protruding from each end of retainer groove. Do not trim the seal. To install the seal, clamp each retainer half in a vise as shown in Fig. 82. Make sure seal groove is clean. Start each end in groove with the specified amount of seal protruding. Allow seal rope to buckle in the center until about an inch of each end is imbedded in groove, work center of seal into position, then roll with a round bar as shown. Repeat the process with other half of seal retainer.

When installing cylinder block bridge piece, insert end seals as shown in Fig. 83. Use a straight edge as shown in Fig. 84 to make sure bridge piece is flush with rear face of cylinder block.

Coat both sides of retainer gasket and end joints of retainer halves with a suitable gasket cement. Coat surface of rope seal with graphite grease. Install retainer halves and cap screws loosely and tighten clamp screws thoroughly before tightening the retaining cap screws.

FLYWHEEL

All Models

81. To remove the flywheel, first separate engine from transmission housing and remove the clutch. Flywheel is secured to crankshaft flange. To properly time the flywheel, align the seventh (unused) hole in flywheel with untapped hole in crankshaft flange. Tighten flywheel retaining cap screws to a torque of 75-80 ft.-lbs.

OIL PAN

All Models

82. The heavy cast iron oil pan serves as part of tractor frame and as attaching point for tractor front support. To remove the oil pan, first drain pan. Support tractor under

Fig. 81—Rear view of engine block showing oil seal retainer installed.

Fig. 82—Use a round bar to bed the asbestos rope seal in retainer half. Refer to text for details.

Fig. 84—Use a straight edge to align the cylinder block bridge.

transmission housing. On Model MF-180, loosen side rail rear cap screws. On all models, remove cap screws securing oil pan to cylinder block, front support and transmission housing and lower the oil pan from cylinder block.

Install by reversing the removal procedure.

OIL PUMP

All Models

83. The gear type oil pump is mounted on engine balancer frame and driven by balancer shaft as shown in Fig. 80. Oil pump can be removed after removing oil pan. The thickness of oil pump gears (17 & 18) should be

0.001 greater to 0.004 less than gear pocket depth in pump body (20). Radial clearance of gears in body bores should be 0.002-0.009. Examine body, gears and plate for wear or scoring, and renew any parts which are questionable.

RELIEF VALVE

All Models

84. The plunger type relief valve is located in oil pump body as shown in Fig. 80. Oil pressure should be 30-60 psi at full engine speed with engine at normal operating temperature. Valve spring is retained by a cap and cotter pin.

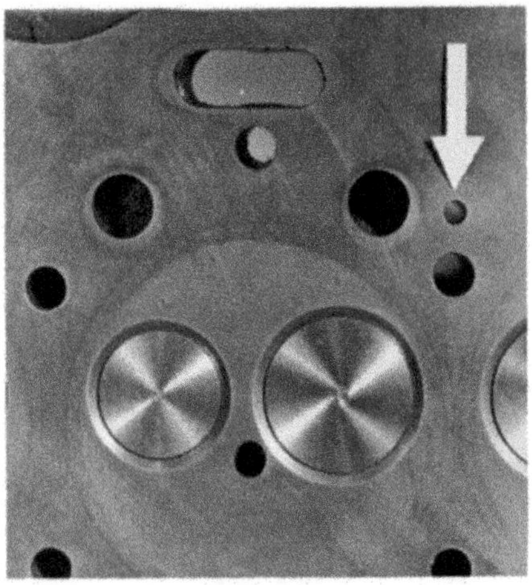

Fig. 85—View of diesel cylinder head showing valves and valve numbers. Arrow indicates oil pressure passage to rocker arms.

DIESEL ENGINE
AND COMPONENTS

Diesel models are equipped with a Perkins A4.236, four cylinder diesel engine having a bore of 3⅞-inches, a stroke of 5 inches and a displacement of 235.9 cubic inches.

R&R ENGINE WITH CLUTCH

All Models

85. To remove the engine and clutch as a unit, first drain the cooling system and if engine is to be disassembled, drain oil pan. Remove hood. Disconnect drag link, air cleaner hose, radiator hoses, oil cooler hoses and power steering hoses or lines. Support tractor under transmission housing and unbolt front support casting from engine; then roll front axle, support and radiator as an assembly away from tractor. Shut off the fuel and remove fuel tank. On Model MF180, remove side rails. Disconnect heat indicator sending unit, oil gage line, wiring harness, starter cable, tachometer cable and injection pump linkage. Support engine in a hoist and unbolt engine from transmission case.

Install by reversing the removal procedure. Bleed the power steering as outlined in paragraph 6 or 8 and the fuel system as in paragraph 124 after tractor is assembled.

CYLINDER HEAD

All Models

86. **REMOVE AND REINSTALL.** To remove the cylinder head, first remove hood and side panels. Shut off the fuel and remove fuel tank. Drain cooling system, disconnect upper radiator hose and remove heat indicator sending unit from side of head. Remove manifolds, injector lines and injectors. Remove rocker arm cover, rocker arms and push rods; then unbolt and remove the cylinder head.

The cylinder block, head and head gasket are internally ported to provide pressure lubrication to rocker arms; see Arrow—Fig. 85. Make sure passages are open and aligned when head is reinstalled. Cylinder head may be milled to remove surface defects

provided no more than 0.012 of metal is removed or total head thickness is not reduced below 4.0355. If head is resurfaced, valve height must be checked as outlined in paragraph 87.

Head gasket is marked "TOP FRONT" for proper installation. Gasket is treated and must be installed dry. Tighten cylinder head stud nuts to a torque of 85 ft.-lbs. using the sequence shown in Fig. 86. Adjust valve tappet gap as outlined in paragraph 92. Head should be re-torqued and tappet gap readjusted after engine has been warmed to operating temperature.

VALVES AND SEATS

All Models

87. Intake and exhaust valves seat directly in the cylinder head. Valve heads and seat locations are numbered consecutively from front to rear. Intake and exhaust valves have a face angle of 44 degrees, a seat angle of 45 degrees and a desired seat width of $\frac{1}{16}$-$\frac{3}{32}$ inch.

Valve heads should be recessed a specified amount into the cylinder head. Clearance can be measured

Fig. 86—On diesel models, tighten cylinder head stud nuts to a torque of 85 ft.-lbs. using the sequence shown.

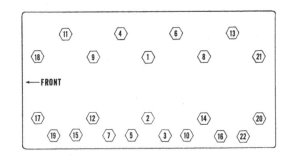

Fig. 87—Using a straight edge and feeler gage to check valve head height. Refer to paragraph 87.

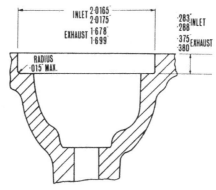

Fig. 88—Machining dimensions for installation of service valve seat inserts.

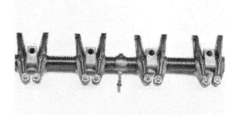

Fig. 89—Assembled view of rocker arm shaft.

using a straight edge and feeler gage as shown in Fig. 87. Production clearances are held within the limits of 0.029-0.039. A maximum clearance of 0.055 is permissible. Inlet and exhaust valve seats are available for service on those units where valve recess exceeds the recommended limits. Fig. 88 shows machining dimensions for the seats. After a new seat is installed, recess the valve by using a 30° stone. Inlet and exhaust valve tappet gap should be set to 0.010 hot.

CAM FOLLOWERS

All Models

90. The mushroom type cam followers (tappets) operate directly in machined bores in engine block and can only be removed after removing the camshaft as outlined in paragraph 102. The 0.7475-0.7485 diameter cam followers are furnished in standard size only and should have a diametral clearance of 0.0015-0.0037 in block bores. Adjust the tappet gap as outlined in paragraph 92.

ROCKER ARMS

All Models

91. The rocker arms and shaft assembly can be removed after removing the hood, fuel tank and rocker arm cover. Rocker arms are right hand and left hand units and should be installed on shaft as shown in Fig. 89. Desired diametral clearance between rocker arms and shaft is 0.001-0.0035. Renew shaft and/or rocker arms if clearance is excessive. When assembling the rocker arm shaft, oil feed holes in shaft must be installed toward valve stem side. Tighten rocker arm support stud nuts to a torque of 28-32 ft.-lbs.

VALVE GUIDES

All Models

88. Intake and exhaust valve guides are not renewable, and oversize valve stems are provided for service.

Standard guide bore diameter is 0.375-0.376 for both the intake and exhaust valves, with a desired stem to bore clearance of 0.0015-0.0035 for intake valves and 0.002-0.004 for exhaust valves. Oversizes of 0.003, 0.015 and 0.030 are provided for intake and exhaust valve stems.

Fig. 90 — With "TDC" flywheel timing marks aligned and No. 1 piston on compression stroke, adjust the indicated valves to 0.012 cold.

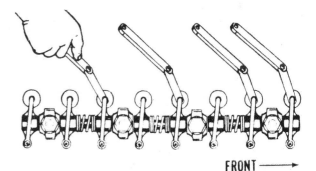

VALVE SPRINGS

All Models

89. Springs, retainers and locks are interchangeable for intake and exhaust valves. Umbrella type oil deflectors are used on intake valve stems only. Valve springs have close damper coils which should be installed next to cylinder head. Renew valve springs if they are rusted, discolored or distorted, or fail to meet the specifications which follow:

Lbs. test @ 1 25/32 inches ..38-42
Lbs. test @ 1 23/64 inches ..69-76

Fi. 91 — With "TDC" flywheel timing marks aligned and No. 4 piston on compression stroke, adjust the indicated valves to 0.012 cold.

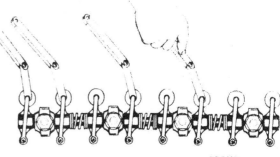

VALVE TAPPET GAP

All Models

92. The recommended cold tappet gap setting is 0.012 for both the intake and exhaust valves. Cold (static) setting of all valves can be made from just two crankshaft positions, using the procedure outlined in this paragraph and illustrated in Figs. 90 and 91.

Remove plug from left side of cylinder block and turn crankshaft until "TDC" timing mark on flywheel is aligned with timing pointer. Check the rocker arms for front and rear cylinders. If rear rocker arms are tight and front rocker arms have clearance, No. 1 piston is on compression stroke; adjust the four tappets shown in Fig. 90. If front rocker arms are tight and rear rocker arms have clearance, No. 4 piston is on compression stroke; adjust the four tappers shown in Fig. 91. After adjusting the indicated tappets, turn crankshaft one complete turn until "TDC" timing mark is again aligned and adjust the remaining tappets. Recheck the adjustment, if desired, with engine running at slow idle speed.

Recommended tappet clearance with engine at operating temperature is 0.010. Clearance may be adjusted with engine running at slow idle speed, or by following the procedure recommended for initial adjustment.

VALVE TIMING

All Models

93. The crankshaft gear and camshaft gear are both keyed to their shafts. Valve timing will be correct if timing marks are properly aligned as outlined in paragraph 100.

TIMING GEAR COVER

All Models

94. To remove the timing gear cover, drain cooling system and remove the hood. Disconnect drag link, air cleaner hose, radiator hoses, oil cooler hoses and power steering hoses or lines. Support tractor under transmission housing and unbolt front support casting from engine; then roll front axle, support and radiator as an assembly from tractor.

Remove fan belt, fan blades and crankshaft pulley; then unbolt and remove timing gear cover.

Crankshaft front oil seal can be renewed at this time. Install seal in cover with sealing lip to rear, with front edge of seal recessed approxi-

mately ¼-inch into seal bore when measured from front of cover.

Timing gear cover is not doweled; use Special Tool MFN-747B or the crankshaft pulley as a pilot to properly align the oil seal, when reinstalling the cover. Aluminum sealing washers must be installed on the four lower timing gear cover cap screws. Tighten the crankshaft pulley retaining cap screw to a torque of 250-300 ft.-lbs. and complete the assembly by reversing the disassembly procedure.

TIMING GEARS

All Models

95. Fig. 92 shows a view of timing gear train with cover removed. Before attempting to remove any of the timing gears, first remove fuel tank, rocker arm cover and rocker arms to avoid the possibility of damage to pistons or valve train if camshaft or crankshaft should either one be turned independently of the other.

Timing gear backlash should be 0.003-0.006 between idler gear (5) and any of its mating gears; or 0.006-0.009 between balancer idler gear (2) and either of its mating gears. Replacement gears are available in standard size only. If backlash is not within the specified limits, renew gears, idler shafts, bushings or other items concerned.

NOTE: Because of the odd number of teeth in idler gear (5), all timing marks will align only once in 18 crankshaft rev-

olutions. To check the timing, remove and reposition the idler gear or count the teeth as follows: With marked teeth on camshaft gear (4) and injection pump gear (6) meshed with idler gear, there should be nine (9) idler gear teeth between the nearest marked teeth on crankshaft and camshaft gears; and 23 idler gear teeth between marked tooth on injection pump drive gear and nearest marked tooth on camshaft gear.

To remove the timing gears or time the engine, refer to the appropriate following paragraphs:

96. **IDLER GEAR AND HUB.** The timing idler gear (5—Fig. 92) should have a diametral clearance of 0.0028-0.0047 and end play of 0.003-0.007 on idler gear hub. Hub is a light press fit in block front face, and can be loosened with a soft hammer if renewal is indicated. Due to uneven spacing of hub studs, hub can only be installed in one position. The two flanged bushings in gear are renewable. Bushings must be reamed after installation to an inside diameter of 1.9998-2.0007. Tighten idler gear retaining stud nuts to a torque of 21-24 ft.-lbs.

97. **CAMSHAFT GEAR.** The camshaft gear (4—Fig. 92) is pressed and keyed to shaft and retained by a special cap screw, tab washer and retaining plate. Use a suitable puller to remove the gear. Use the retaining plate and cap screw to draw gear into position when reinstalling, tighten cap screw to a torque of 45-50 ft.-lbs.

Fig. 92—Front view of engine with timing gear cover and oil pan removed, showing timing marks aligned. Because of the odd number of teeth in idler gears (2 & 5) all marks will not align with each camshaft revolution. Nonalignment of marks does not necessarily indicate improper engine timing.

T. Timing marks
1. Balancer gear
2. Idler gear
3. Crankshaft gear
4. Camshaft gear
5. Idler gear
6. Injection pump drive gear

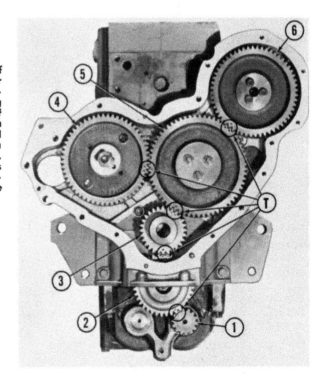

and lock in place by bending tab washer. Time the gears as outlined in paragraph 100.

98. INJECTION PUMP DRIVE GEAR. The injection pump drive gear (6—Fig. 92) is retained to pump adapter by three cap screws. When installing the gear, align dowel pin (2—Fig. 93) with slot (1) in adapter hub, then install the retaining cap screws. The injection pump drive gear and adapter are supported by the injection pump rotor bearings.

99. CRANKSHAFT GEAR. The crankshaft timing gear (3—Fig. 92) is keyed to the shaft and is a transition fit (0.001 tight to 0.001 loose) on shaft. It is usually possible to remove the gear using two small pry bars to move the gear forward. Remove timing gear housing and engine balancer, then use a suitable puller if gear cannot be removed with pry bars.

Fig. 93—Correct installation of injection pump drive gear is simplified by the dowel pin (2) which fits in machined notch (1) in pump drive shaft.

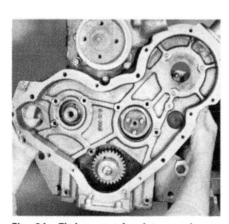

Fig. 94—Timing gear housing must be removed as shown before camshaft can be withdrawn.

100. TIMING THE GEARS. To install and time the gears, first install crankshaft, camshaft and injection pump gears as outlined in paragraphs 97, 98 and 99, with timing marks to front. Turn the shafts until timing marks (T—Fig. 92) point toward idler gear hub; then install idler gear with all marks aligned as shown. Secure idler gear as in paragraph 96.

TIMING GEAR HOUSING
All Models

101. To remove the timing gear housing, first remove timing gears as outlined in paragraphs 95 through 98 and the injection pump as in paragraph 135. Remove power steering pump if not previously removed. If oil pan has not been removed, remove the four front oil pan cap screws. Remove the cap screws securing timing gear housing to block front face and lift off the housing as shown in Fig. 94.

Timing gear housing must be removed before camshaft can be withdrawn. Install by reversing the removal procedure. Oil pan must be installed and tightened before tightening the cap screws securing timing gear housing to block front face.

CAMSHAFT
All Models

102. To remove the camshaft, first remove timing gear housing as outlined in paragraph 101. Secure cam followers (tappets) in their uppermost position, remove the fuel lift pump, then withdraw camshaft and front thrust washer.

The camshaft runs in three bearings. The front bearing bore contains

Fig. 95—Front face of engine block with camshaft removed, showing front camshaft bushing and thrust washer locating dowel. The other two camshaft bores are not bushed.

a pre-sized renewable bushing, while the two rear journals ride directly in machined bores in engine block. Camshaft bearing journals have a recommended diametral clearance of 0.0025-0.0053 in all three bearing bores. End play of 0.004-0.016 is controlled by the camshaft thrust washer which should be renewed during reassembly if end play is excessive. Camshaft journal diameters are as follows:

Front 1.9965-1.9975
Center 1.9865-1.9875
Rear 1.9665-1.9675

ROD AND PISTON UNITS
All Models

103. Connecting rod and piston units are removed from above after removing cylinder head, oil pan, engine balancer and rod bearing caps. Cylinder numbers are stamped on the connecting rod and cap. When reinstalling, make sure correlation numbers are in register and face away from camshaft side of engine. When installing connecting rod caps, use new self-locking nuts and tighten to a torque of 65-70 ft.-lbs.

PISTONS, SLEEVES AND RINGS
All Models

104. The aluminum alloy, cam ground pistons are supplied in standard size only and are available in a kit consisting of piston, pin and rings for one cylinder. The toroidal combustion chamber is offset in piston crown and piston is marked "FRONT" for proper assembly as shown at (F—Fig. 96).

Each piston is fitted with a plain faced chrome top ring which may be installed either side up. The internally stepped cast iron 2nd & 3rd compression rings must be installed with the marking "BTM" toward the bottom, away from piston crown. Oil rings may be installed either side up. Specifications for fitting piston rings are as follows:

End Gap
    Top ring 0.018-0.022
    Other rings 0.014-0.019
Side Clearance
    Compression rings 0.0019-0.0039
    Oil control rings 0.0025-0.0045
Piston Skirt Clearance 0.006-0.008

The production cylinder sleeves are 0.001-0.003 press fit in cylinder block bores. Service sleeves are a transition fit (0.001 tight to 0.001 clearance) in block bores. When installing new sleeves, make sure sleeves and bores are absolutely clean and dry, then chill the sleeves and press fully into place by hand. When properly installed, liner should extend 0.030-0.035 above gasket face of cylinder block as shown in Fig. 97.

PISTON PINS

All Models

105. The 1.3748-1.3750 diameter, floating type piston pins are retained in piston bosses by snap rings and are available in standard size only. The renewable connecting rod bushing must be final sized after installation to provide a diametral clearance of 0.00075-0.0017 for the pin. Be sure the predrilled oil hole in bushing is properly aligned with hole in top of connecting rod and install bushing from chamfered side of bore. Piston pin should be a thumb press fit in piston after piston is heated to 160° F.

CONNECTING RODS AND BEARINGS

All Models

106. Connecting rod bearings are precision type, renewable from below

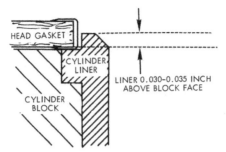

Fig. 97—Installed cylinder sleeve (liner) should extend 0.030 - 0.035 above gasket face of block as shown.

after removing oil pan, balancer unit and rod bearing caps. When renewing bearing shells, be sure that the projection engages milled slot in rod and cap and that the correlation marks are in register and face away from camshaft side of engine.

Connecting rods are graded as to weight and a code number is etched on side of rod as shown in Fig. 98. Three weight codes are used, 11, 12 and 13, the larger number being the heaviest rod. When renewing a connecting rod, the same weight code should be used as was on the removed rod. Replacement rods should be marked with the cylinder number in which they are installed. Bearings are available in standard size and undersizes of 0.010, 0.020 and 0.030.

Connecting rod bearings should have a diametral clearance of 0.0015-0.003 on the 2.499-2.4995 diameter

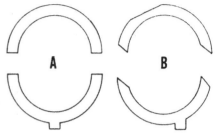

Fig. 99—Old type (A) and new type (B) crankshaft thrust washers are not interchangeable. Refer to paragraph 107.

crankpin. Recommended connecting rod side clearance is 0.0095-0.013. Renew the self-locking connecting rod nuts and tighten to a torque of 65-70 ft.-lbs.

CRANKSHAFT AND BEARINGS

All Models

107. The crankshaft is supported in five precision type main bearings. To remove the rear main bearing cap, it is first necessary to remove the engine, clutch, flywheel and rear oil seal. All other main bearing caps can be removed after removing oil pan and engine balancer.

Upper and lower inserts are not interchangeable, the upper (block) half being slotted to provide pressure lubrication to crankshaft and connecting rods. Inserts are interchangeable in pairs for all journals except the center main bearing. Crankshaft end play is controlled by thrust washers installed on front and rear of center main bearing. Two types of thrust washers have been used, which are not interchangeable; refer to Fig. 99. When installing the early type (A), make sure the steel back is positioned next to block and cap and the grooved bearing surface next to crankshaft thrust faces. The new (B) type cannot be installed backward.

Bearing inserts are available in undersizes of 0.010, 0.020 and 0.030 as well as standard; and thrust washers are available in standard thickness of 0.089-0.093, and oversizes of 0.007. Recommended main bearing diametral clearance is 0.0025-0.0045 and recommended crankshaft end play is 0.004-0.014. Tighten the main bearing retaining cap screws to a torque of 145-150 ft.-lbs. and secure by bending tabs on locking washers. When renewing rear main bearing, refer to paragraph 111 for installation of rear seal and oil pan bridge piece. Check

Fig. 96—Assembled rod and piston unit showing location of marks.

C. Correlation mark
F. "Front" mark
N. Cylinder number

Fig. 98—Connecting rods are graded by weight and a code number is etched on rod as shown. Refer to paragraph 106.

Fig. 100—Installed view of engine balancer with timing marks aligned. Refer to paragraph 109 for installation procedure.

the crankshaft journals against the values which follow:

Main journal diameter ..2.9985-2.999
Crankpin diameter .....2.499 -2.4995

ENGINE BALANCER
All Models

108. **OPERATION.** The Lanchester type engine balancer consists of two unbalanced shafts which rotate in opposite directions at twice crankshaft speed. The inertia of the shaft weights is timed to cancel out natural engine vibration, thus producing a smoother running engine. The balancer is correctly timed when the balance weights are at their lowest point when pistons are at TDC or BDC of their stroke.

The balancer unit is driven by the crankshaft timing gear through an idler gear attached to balancer frame. The engine oil pump is mounted at rear of balancer frame and driven by the balancer shaft. Refer to Figs. 100 through 102.

109. **REMOVE AND REINSTALL.** The balancer assembly can be removed after removing the oil pan and mounting cap screws. Engine oil is pressure fed through balancer frame and cylinder block. Balancer frame bearings are also pressure fed. Balancer and oil pump can be removed as a unit as shown in Fig. 101, after removing oil pan and the four retaining cap screws.

When installing balancer with en-

gine in tractor, timing marks will be difficult to observe without removing timing gear cover. The balancer assembly can be safely installed as follows: Turn crankshaft until No. 1 and No. 4 pistons are at top of their stroke and "TDC" flywheel timing marks are aligned. Remove balancer idler gear (5—Fig. 100) if necessary, and align the single punch-marked tooth between the two marked teeth on weight drive shaft (11) as shown at (B). Install balancer frame with balance weights hanging normally. If carefully installed, timing will be correct. If engine is mounted on a stand, timing marks can be observed by removing timing gear cover.

110. **OVERHAUL.** Refer to Fig. 102 for an exploded view of balancer frame and associated parts. To disassemble the removed balancer unit, unbolt and remove oil pump housing (20) and associated parts; and idler gear (5) and associated parts. Set screws (S) retaining balance weights (15 & 16) are installed using Grade "C" (Blue) LOCITE. Loosen the set screws, then push balance shafts (11 & 12) forward out of frame and weights.

Shaft bushings are renewable in balancer frame (8). Install new bushings with oil holes in bushing and frame aligned, making sure bushing does not extend beyond frame bore. Align ream bushings after installation to a finished diameter of 1.251-1.2526 for front bushings and 1.001-1.0022 for rear bushings. Recommended diametral clearance for shafts (11 & 12) in frame bushings is 0.002-0.004 for front bushings or 0.002-0.0035 for rear. When assembling the balancer use Grade "C" (Blue) LOCTITE for installing the screws retaining gears (14) to balance weights (15 & 16) and the set screws (S) retaining balance weights to shafts. Also make sure flat surfaces of weights are aligned when installed, as shown in Fig. 101.

Bushing (4—Fig. 102) is renewable in idler gear. Bushing is pre-sized to provide the recommended 0.001-0.0032 diametral clearance for hub (2). End play of installed idler gear should be 0.008-0.014.

Refer to paragraph 114 for overhaul of engine oil pump and to paragraph 109 for installation of balancer assembly.

CRANKSHAFT REAR OIL SEAL
All Models

111. The asbestos rope type rear oil seal is contained in a two-piece seal

Fig. 101 — Assembled view of removed engine balancer and oil pump unit. Refer to Fig. 102 for exploded view.

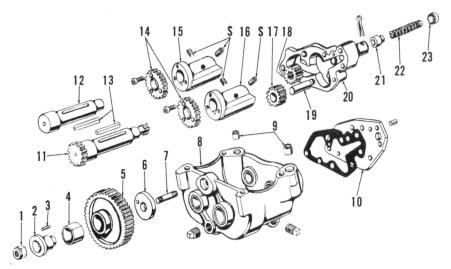

Fig. 102—Exploded view of Lanchester type engine balancer, engine oil pump and associated parts used on diesel engine.

1. Locknut
2. Hub
3. Dowel
4. Bushing
5. Idler gear
6. Washer
7. Stud
8. Frame
9. Ring dowels
10. Plate
11. Drive shaft
12. Driven shaft
13. Key
14. Gear
15. Balance weight
16. Balance weight
17. Pump gear
18. Pump gear
19. Shaft
20. Pump body
21. Valve piston
22. Valve spring
23. Cap

Fig. 105—Cylinder block bridge is equipped with end seals as shown.

Fig. 106—Use a straight edge to align the cylinder block bridge.

retainer attached to rear face of engine block as shown in Fig. 103. The seal retainer can be removed after removing flywheel.

The rope type crankshaft seal is precision cut to length, and must be installed in retainer halves with 0.010-0.020 of seal ends projecting from each end of retainer. Do not trim the seal. To install the seal, clamp each half of retainer in a vise as shown in Fig. 104. Make sure seal groove is clean. Start each end in groove with the specified amount of seal protruding. Allow seal rope to buckle in the center until about an inch of each end is bedded in groove, work center of seal into position, then roll with a round bar as shown. Repeat the process with the other half of seal.

When installing the cylinder block bridge piece (Fig. 105), insert end seals as shown, and use a straight edge as shown in Fig. 106 to make sure bridge piece is flush with rear face of cylinder block.

Sealing surface of crankshaft contains a machined spiral groove 0.004-0.008 deep.

FLYWHEEL

All Models

112. To remove the flywheel, first separate engine from transmission housing and remove the clutch. Flywheel is secured to crankshaft by six evenly spaced cap screws. To properly time flywheel to engine during

installation, be sure that unused hole in flywheel aligns with untapped hole in crankshaft flange.

CAUTION: Flywheel is only lightly piloted to crankshaft. Use caution when unbolting flywheel, to prevent flywheel from falling and causing possible injury.

Fig. 103—Rear view of engine block showing oil seal retainer.

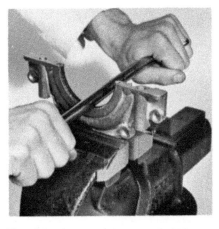

Fig. 104—Use a round bar to bed the asbestos rope seal in retainer half. Refer to text for details.

Fig. 107—Checking flywheel runout using a dial indicator, refer to text for details.

The starter ring gear can be renewed after flywheel is removed. Heat ring gear evenly to approximately 475° F. and install on flywheel with beveled end of teeth facing front of engine.

Check flywheel runout with a dial indicator as shown in Fig. 107 after flywheel is installed. Maximum allowable flywheel runout is 0.001 for each inch from flywheel centerline to point of measurement. Tighten the flywheel retaining cap screws to a torque of 75 ft.-lbs. when installing flywheel.

OIL PAN

All Models

113. The heavy cast-iron oil pan serves as the tractor frame and attaching point for the tractor front support. To remove the oil pan, support tractor underneath transmission housing and drain the oil pan. On Model MF180, loosen side rail rear cap screws. On all models, remove cap screws securing oil pan to cylinder block, front support and transmission housing and lower the oil pan from block.

Install by reversing the removal procedure.

OIL PUMP

All Models

114. The gear type oil pump is mounted on engine balancer frame and driven by balancer shaft as shown in Fig. 102. Oil pump can be removed after removing oil pan. The thickness of oil pump gears (17 & 18) should be 0.001 greater to 0.004 less than gear pocket depth in pump body (20). Radial clearance of gears in body bores should be 0.0003-0.012. Examine body, gears and plate for wear or scoring and renew any parts which are questionable. Refer to paragraph 115 for overhaul of relief valve.

RELIEF VALVE

All Models

115. The plunger type relief valve is located in pump body (20—Fig. 102). Normal engine oil pressure is 30-60 psi at normal operating speeds and temperature. Relief valve should be set to open at 50-60 psi. Spring (22) should have a free length of 1½-inches and should test approximately 12½ lbs. when compressed to a height of 1 inch.

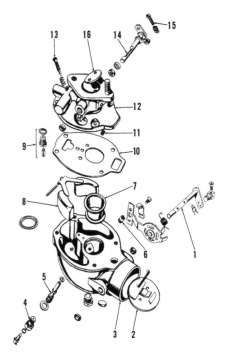

Fig. 108—Exploded view of Marvel Schebler carburetor of the type used on gasoline models.

1. Choke shaft	9. Inlet needle & seat
2. Choke valve	10. Gasket
3. Fuel bowl	11. Idle jet
4. Main needle	12. Throttle body
5. Main nozzle	13. Idle mixture needle
6. Power jet	14. Throttle shaft
7. Venturi	15. Idle stop screw
8. Float	16. Throttle plate

GASOLINE FUEL SYSTEM

CARBURETOR

All Models

116. **ADJUSTMENT.** Marvel-Schebler or Zenith carburetor may be used. Initial adjustments are approximately one turn open for idle mixture adjustment needle and 1½-2 turns open for main adjustment needle. Final adjustments must be made under operating conditions with engine at normal operating temperature. After mixture adjustments have been made, adjust idle speed to 450-500 rpm for models with Continental engine or 725-775 rpm for models with Perkins engine.

117. **OVERHAUL.** (Marvel-Schebler) Refer to Fig. 108 for an exploded view of Marvel-Schebler carburetor of the type used. To disassemble the removed carburetor, first clean outside with a suitable solvent and remove main adjustment needle (4) and idle mixture adjustment needle (13). Remove the

screws which retain throttle body (12) to fuel bowl (3) and lift off throttle body, gasket and venturi (7). Remove float shaft and float, gasket, venturi and inlet needle valve. Withdraw venturi from gasket and discard the gasket. Remove inlet needle valve seat, idle jet (11), power jet (6) and main nozzle (5). Discard the gaskets from nozzle and needle valve seat.

Remove throttle and choke valves, shafts and packing. Bushings for throttle and choke shafts are not provided.

Clean all parts in a suitable carburetor cleaner and rinse in clean mineral solvent. Blow out passages in body and bowl with compressed air and renew any parts which are worn or damaged.

Assemble by reversing the disassembly procedure, using new gaskets and packing. Install throttle valve (16) with marking toward mounting flange on the side away from adjust-

ing needles. Adjust float height to ¼-inch when measured from gasket surface to nearest edge of float, with throttle body inverted. If adjustment is required, carefully bend float arms using a bending tool or needle nose pliers, keeping the two halves of float parallel and equal. Check mixture adjustments after installation, as outlined in paragraph 116.

118. **OVERHAUL (Zenith).** To disassemble the removed carburetor, first clean outside with a suitable solvent. Remove the screws retaining throttle body (4—Fig. 109) to fuel bowl (13) and remove fuel bowl. Remove float shaft (8), float and inlet valve needle. NOTE: Float shaft (8) is a tight fit in slotted side of hinge bracket and should be removed from opposite side.

Remove venturi (10), inlet valve seat and idle jet (6) from throttle body. Remove idle adjusting needle (5). Remove fuel shut-off solenoid (12), discharge jet (15), main jet (11) and well vent jet (14) from fuel bowl. Remove throttle and choke valves, shafts and packing.

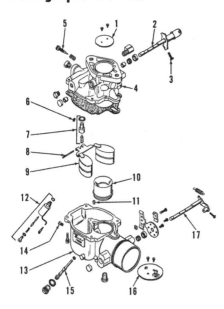

Fig. 109——Exploded view of Zenith carburetor of the type used on some models.

1. Throttle valve	9. Float
2. Throttle shaft	10. Venturi
3. Idle speed screw	11. Main jet
4. Throttle body	12. Shut-off solenoid
5. Idle mixture needle	13. Float chamber
6. Idle jet	14. Well vent jet
7. Inlet needle valve	15. Discharge jet
8. Float shaft	16. Choke valve

Discard all gaskets and packing and clean remainder of parts in a suitable carburetor cleaner. Rinse in clean mineral solvent and blow out passages in body and bowl with compressed air. Renew all gaskets and packing and any other parts which are worn or damaged.

Assemble by reversing the disassembly procedure, using new gaskets and packing. Install throttle valve (1) so beveled edges will fit throttle body bore with throttle closed, with side of throttle plate farthest from mounting flange aligned with idle port. Adjust float height to 13/64 inch measured

USE 3/16" AND 7/32"
DRILL BITS AS GO
AND NO-GO GAUGES

Fig. 110—Use drill bits as Go—No Go gages for float adjustment.

from gasket to nearest edge of float as shown in Fig. 110. Drill bits in $\frac{3}{16}$ and $\frac{7}{32}$ size can be used as GO—No Go gages as shown. If adjustment is required, carefully bend float arms using needle nose pliers or a bending tool, keeping the two halves of float parallel and equal. Check mixture adjustments after installation as outlined in paragraph 116.

LP-GAS SYSTEM

OPERATING ADJUSTMENTS

All Models

119. Initial, dead engine adjustments for the carburetor are 2½ turns open for idle adjustment screw (2—Fig. 111) and 3½ turns open for main adjustment needle (4).

With engine at operating temperature, set throttle stop screw (1) to obtain a slow idle speed of 450 rpm. Final adjustments must be made for maximum power and economy under operating conditions and load.

FUEL TANK AND LINES

All Models

120. The pressure tank is fitted with fuel filter, vapor return, pressure relief, bleeder, and liquid and vapor withdrawal valves which can only be renewed when fuel tank is completely empty. The fuel gage unit consists of a dial face unit which can be renewed at any time, and a float unit which can only be renewed when fuel tank is empty. The safety relief valve is set to open at 312 psi pressure to protect the tank against excessive pressures. Fuel lines, filter, vaporizer, heat exchanger or carburetor can be renewed at any time without emptying fuel tank, if liquid and vapor withdrawal valves are closed and engine allowed to run until fuel is exhausted. U. L. Regulations in most states prohibit any welding or repair on LP-Gas containers, and the tank must be renewed, rather than repaired, in case of damage. The fuel filter (7—Fig. 112) is a "throw-away" type, renewable as a unit.

VAPORIZER

All Models

121. The vaporizer unit (9—Fig. 112) is mounted in water outlet elbow at top of cylinder head. Before removing vaporizer or disconnecting any fuel lines, close both tank withdrawal valves and allow engine to run until fuel is exhausted from vaporizer, fuel lines and carburetor. Drain the radiator, disconnect fuel lines and remove vaporizer. Refer to Fig. 113.

The removed vaporizer assembly can be tested for external or internal leaks without disassembly; proceed as follows:

Connect vaporizer inlet to a controlled source of compressed air, plug the outlet port and completely immerse the unit in water. External leaks will show up as air bubbles. Note especially the areas around vaporizer coil and mounting plate (1), and around diaphragm cover (16). Air bubbles emerging from vent hole (V) in top of diaphragm cover indicates a leaking diaphragm.

To check the vaporizer inlet valve and seat, install a low-pressure test gage in vaporizer outlet port and connect inlet to air pressure. Gage reading should be 9-11 psi and hold

Fig. 111—Zenith LP-Gas carburetor of the type used, showing points of adjustment.

1. Throttle stop screw	4. Main adjustment needle
2. Idle adjustment screw	5. Inlet diaphragm assy.
3. Idle diaphragm assy.	6. Inlet fitting

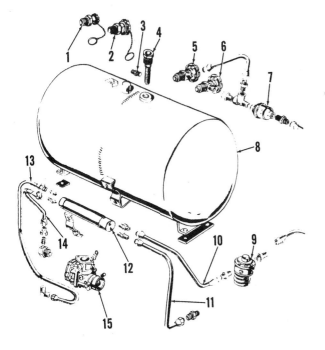

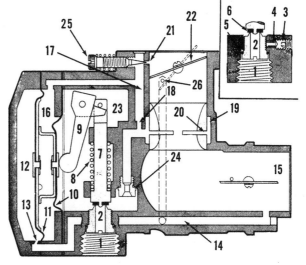

CARBURETOR

All Models

122. The Zenith pressure regulating carburetor serves both as a secondary regulator and as a carburetor.

The fuel valve seat (2—Fig. 114) is adjustable so that position of diaphragm lever (9) can be varied. The seat is locked in position by means of lock screw (3) and nylon plug (4) as shown in inset.

The two diaphragms (10 and 11) control the pressure and flow of incoming fuel to maintain the proper fuel-air mixture.

To remove the carburetor, first close both withdrawal valves and run engine until fuel is exhausted from regulator, lines and carburetor. Disconnect choke and throttle linkage and the fuel inlet line, then unbolt and remove the carburetor.

Remove the diaphragm cover, spacer and diaphragms. Remove idle diaphragm cover and diaphragm; then unbolt and remove idle diaphragm housing from side of carburetor body.

Remove fuel inlet fitting from bottom of carburetor and remove fuel valve seat (2).

NOTE: The valve should be held off its seat by applying light pressure to diaphragm lever (9). Be sure locking screw (3) is first loosened.

Remove the diaphragm lever shaft plug from side of carburetor body, and remove shaft, lever and valve. Remove adjusting needles and main jet. Clean all parts in a suitable solvent and examine for wear or damage.

Fig. 112—Fuel tank and associated parts used on LP-Gas Models.

1. Vapor return valve
2. Fuel filler valve
3. Bleed valve
4. Safety relief valve
5. Vapor withdrawal valve
6. Liquid withdrawal valve
7. Fuel filter
8. Fuel tank
9. Vaporizer
10. Fuel line
11. Coolant line
12. Heat exchanger
13. Fuel line
14. Coolant line
15. Carburetor

steady. If pressure continues to rise, a leaking fuel valve or valve seat "O" ring is indicated, and vaporizer should be overhauled.

To disassemble the vaporizer, remove four alternate screws from diaphragm cover (16) and install aligning studs (Zenith Tool Part No. C161-195). Apply thumb pressure to top of diaphragm cover and remove the remaining diaphragm screws, diaphragm cover (16), spacer (15) and springs (13 and 14). To renew any part of the fuel valve assembly, remove the fuel valve seat (8) using a suitable socket wrench. To renew "O" rings (2 or 3), remove the four screws retaining coil and plate (1) to body, and withdraw the plate.

When reassembling, install fuel valve (6) with long stem toward diaphragm. Make sure all screw holes are aligned in gasket (10), baffle (11), diaphragm (12) and cover (16). Tighten the retaining screws evenly, leaving aligning studs installed until cover is tight. Recheck for leaks after assembly by immersing in water or using a soap solution.

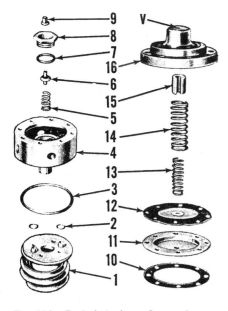

Fig. 113—Exploded view of vaporizer assembly.

1. Coil & plate	10. Gasket
2. "O" ring	11. Baffle
3. "O" ring	12. Diaphragm
4. Body	13. Inner spring
5. Valve spring	14. Outer spring
6. Valve	15. Spacer
7. "O" ring	16. Cover
8. Valve seat	V. Vent hole
9. Follower	

1. Fuel inlet
2. Fuel valve seat
3. Locking screw
4. Locking plug
5. "O" ring
6. Sealing disc
7. Fuel valve
8. Fuel valve spring
9. Diaphragm lever
10. Inner diaphragm
11. Outer diaphragm
12. Outer diaphragm chamber
13. Air passage orifice
14. Air passage
15. Air intake
16. Inner diaphragm chamber
17. Idle fuel passage
18. Idle orifice
19. Annulus
20. Venturi
21. Idle needle seat
22. Throttle fly
23. Pressure chamber
24. Main jet
25. Idle needle
26. Economizer orifice

Fig. 114—Cross-sectional view of pressure regulating carburetor showing main components. Inset shows fuel valve seat locking arrangement.

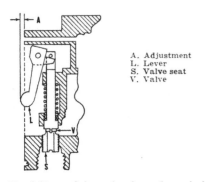

A. Adjustment
L. Lever
S. Valve seat
V. Valve

Fig. 115 — Schematic view of regulating valve assembly used on Zenith Pressure Regulating carburetor.

Reassemble by reversing the disassembly procedure, using new gaskets and seals. When installing fuel valve seat (2), hold valve open by applying pressure to valve lever (9), and tighten seat until lever (9) aligns with Step 2 of Zenith Tool C161-194; or until distance (A—Fig. 115) measures $\frac{1}{16}$-inch. Tighten locking screw (3—Fig. 114) when adjustment has been obtained. Complete the assembly by reversing the disassembly procedure.

Fig. 117—Installed view of fuel lift pump (1) showing hand actuating lever (2).

DIESEL FUEL SYSTEM

The diesel fuel system consists of three basic units; the fuel tank and filters, injection pump and injector nozzles. When servicing any unit associated with the diesel fuel system, the maintenance of absolute cleanliness is of utmost importance. Of equal importance is the avoidance of nicks or burrs on any of the working parts.

Probably the most important precaution that service personnel can impart to owners of diesel powered tractors is to urge them to use an approved fuel that is absolutely clean and free from foreign materials. Extra precaution should be taken to make sure that no water enters the fuel storage tanks. Because of the high pressures and degree of control required of injection equipment, extremely high precision standards are necessary in the manufacture and servicing of diesel components. Extra care in daily maintenance will pay big dividends in long service life and the avoidance of costly repairs.

FUEL FILTERS AND LINES

All Models

123. OPERATION AND MAINTENANCE. Refer to Fig. 116 for a schematic view of fuel flow through filters and injection pump.

NOTE: Actual location of filters may differ somewhat from that shown. The camshaft actuated, diaphragm type fuel lift pump is not shown.

A much greater volume of fuel is circulated within the system than is burned in the engine, the excess serving as a coolant and lubricant for the injection pump. Fuel enters the primary filter (P) through inlet line (I), where it passes through the water trap and first stage filter element. Both lines leading to injection pump (F), and return line (R), are connected to a common passage in secondary filter (S); and separated from primary filter line only by the secondary filter element. The greater volume of filtered fuel is thus recirculated between the secondary filter (S) and injection pump (F). A much smaller quantity of fuel enters the system through inlet line (I) or returns to the tank through line (R), thus contributing to longer filter life.

Inspect the glass bowl at bottom of primary filter (P) daily and drain off any water or dirt accumulation. Drain the primary filter at 100 hour intervals and renew the element each 500 hours. Renew element in secondary filter (S) every 1000 hours. Renew both elements and clean the tank and lines if evidence of substantial water contamination exists.

124. BLEEDING. To bleed the system, make sure tank shut-off valve is open, have an assistant actuate the manual lever (2—Fig. 117) on fuel lift pump (1), and proceed as follows:

Loosen the air vent (1—Fig. 116) on primary filter (P) and continue to

F. Injection pump
I. Inlet line
P. Primary filter
R. Return line
S. Final filter

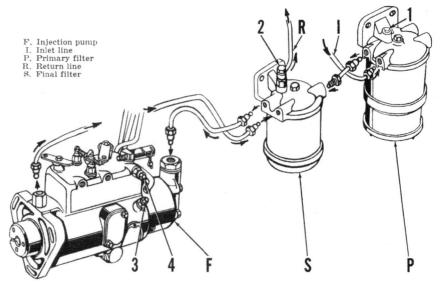

Fig. 116—Schematic view of diesel injection pump, fuel filters and lines. Arrows indicate direction of fuel flow. Bleed screws and proper bleeding order are indicated by the numerical references (1, 2, 3 & 4).

Fig. 118—Exploded view of fuel lift pump.

operate the lift pump until air-free fuel flows from vent plug hole. Tighten plug (1). Loosen vent plugs (2, 3 and 4) on secondary filter and injection pump in the order given, while continuing to operate the lift pump. Tighten each plug as air is expelled and proceed to the next.

NOTE: Air in governor housing relieved by bleed screw (4) will not prevent tractor from starting and running properly; however, condensation in the trapped air can cause rusting of governor components and eventual pump malfunction. Do not fail to bleed governor housing even though the tractor starts and runs properly.

Operate manual lever approximately ten extra strokes after tightening vent plug (4), to expel any air remaining in bleed back lines.

With the fuel supply system bled, push in the stop button, partially open throttle lever and attempt to start the tractor. If tractor fails to fire, loosen

Fig. 119—A suitable injector tester is required to completely test and adjust the injector nozzles.

compression nut at all injector nozzles and turn engine over with starter until fuel escapes from all loosened connections. Tighten compression nuts and start engine.

FUEL LIFT PUMP
All Models

125. The fuel lift pump (Fig. 118) is mounted on right side of engine block as shown in Fig. 117 and driven by the camshaft. All pump parts are available separately. Refer to Fig. 118 as a guide when disassembling and assembling pump. Output delivery pressure should be 2¾-4¼ psi.

INJECTOR NOZZLES
All Models

All models are equipped with C. A. V. multi-hole nozzles which extend through the cylinder head to inject fuel charge into a combustion chamber machined in crown of piston.

WARNING: Fuel leaves the injector nozzle with sufficient force to penetrate the skin. Keep exposed portions of your body clear of nozzle spray when testing.

126. **TESTING AND LOCATING A FAULTY NOZZLE.** If rough or uneven engine operation, or misfiring, indicates a faulty injector, the defective unit can usually be located as follows:

With engine running at the speed where malfunction is most noticeable (usually low idle speed), loosen the compression nut on high pressure line for each injector in turn and listen for a change in engine performance. As in checking spark plugs, the faulty unit is the one which, when its line is loosened, least affects the running of the engine.

If a faulty nozzle is found and considerable time has elapsed since the injectors have been serviced, it is recommended that all nozzles be removed and serviced or that new or reconditioned units be installed. Refer to the following paragraphs for removal and test procedure.

127. **REMOVE AND REINSTALL.** Before loosening any fuel lines, thoroughly clean the lines, connections, injectors and engine area surrounding the injector, with air pressure and solvent spray. Disconnect and remove the leak-off line, disconnect pressure line and cap all connections as they are loosened, to prevent dirt entry into

the system. Remove the two stud nuts and withdraw injector unit from cylinder head.

Thoroughly clean the nozzle recess in cylinder head before reinstalling injector unit. It is important that seating surface be free of even the smallest particle of carbon or dirt which could cause the injector unit to be cocked and result in blow-by. No hard or sharp tools should be used in cleaning. Do not re-use the copper sealing washer located between injector nozzle and cylinder head, always install a new washer. Each injector should slide freely into place in cylinder head without binding. Make sure that dust seal is reinstalled and tighten the retaining stud nuts evenly to a torque of 10-12 ft. lbs. After engine is started, examine injectors for blow-by, making the necessary corrections before releasing tractor for service.

128. **TESTING.** A complete job of testing and adjusting the injector requires the use of special test equipment. Only clean, approved testing oil should be used in the tester tank. The nozzle should be tested for opening pressure, seat leakage, back leakage and spray pattern. When tested, the nozzle should open with a sharp popping or buzzing sound, and cut off quickly at end of injection, with a minimum of seat leakage and controlled amount of back leakage.

Before conducting the test, operate tester lever until fuel flows, then attach the injector. Close the valve to tester gage and pump tester lever a few quick strokes to be sure nozzle valve is not plugged, that four sprays emerge from nozzle tip, and that possibilities are good that injector can be returned to service without overhaul.

NOTE: Spray pattern is not symmetrical with centerline of nozzle tip. The apparently

Fig. 120—Nozzle holes (arrows) are not located an equal distance from nozzle tip.

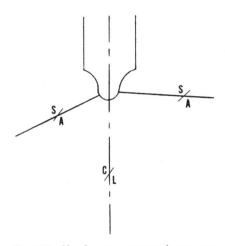

Fig. 121—Nozzle spray pattern is not symmetrical with centerline of nozzle tip.

irregular location of nozzle holes (See Figs. 120 and 121) is designed to provide the correct spray pattern in the combustion chamber.

If adjustment is indicated by the preliminary tests, proceed as follows:

129. OPENING PRESSURE. Open the valve to tester gage and operate tester lever slowly while observing gage reading. Opening pressure should be 2500 psi. If opening pressure is not as specified, remove the injector cap nut (1—Fig. 122), loosen locknut (3) and turn adjusting sleeve (5) as required to obtain the recommended pressure.

NOTE: When adjusting a new injector or an overhauled injector with new pressure spring (7), set the pressure at 2575 psi to allow for initial pressure loss.

130. SEAT LEAKAGE. The nozzle tip should not leak at a pressure less than 2300 psi. To check for leakage, actuate tester lever slowly and as the gage needle approaches 2300 psi, observe the nozzle tip. Hold the pressure at 2300 psi for ten seconds; if drops appear or if nozzle tip is wet, the valve is not seating and injector must be disassembled and overhauled as outlined in paragraph 133.

131. BACK LEAKAGE. If nozzle seat as tested in paragraph 130 was satisfactory, check the injector and connections for wetness which would indicate leakage. If no visible external leaks are noted, bring gage pressure to 2250 psi, release the lever and observe the time for gage pressure to drop from 2250 psi to 1500 psi. For a nozzle in good condition, this time should not be less than six seconds.

A faster pressure drop would indicate a worn or scored nozzle valve piston or body, and the nozzle assembly should be renewed.

NOTE: Leakage of the tester check valve or connections will cause a false reading, showing up in this test as excessively fast leakback. If all injectors tested fail to pass the test, the tester rather than the units should be suspected as faulty.

132. SPRAY PATTERN. If leakage and pressure are as specified when tested as outlined in paragraphs 129 through 131, operate the tester handle several times while observing spray pattern. Four finely atomized, equally spaced, conical sprays should emerge from nozzle tip, with equal penetration into the surrounding atmosphere.

If pattern is uneven, ragged or not finely atomized, overhaul the nozzle as outlined in paragraph 133. NOTE: Spray pattern is not symmetrical with centerline of nozzle tip; refer to Fig. 121.

133. OVERHAUL. Hard or sharp tools, emery cloth, grinding compound, or other than approved solvents or lapping compounds must never be used. An approved nozzle

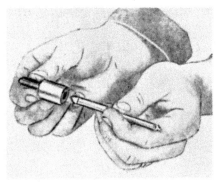

Fig. 124—Clean the spray holes in nozzle tip using a pin vise and 0.009 wire probe.

1. Cap nut
2. Gasket
3. Locknut
4. Tab
5. Adjusting sleeve
6. Adjusting shim
7. Spring
8. Valve spindle
9. Nozzle holder
10. Nozzle valve
11. Nozzle body
12. Nozzle nut
13. Seat washer
14. Dust shield

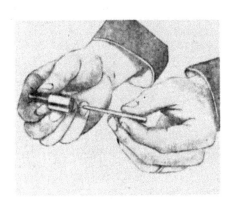

Fig. 122—Exploded view of C. A. V. injector nozzle and holder assembly. Correct opening pressure is indicated on tab (4).

Fig. 123—Clean the pressure chamber in nozzle tip using the special reamer as shown.

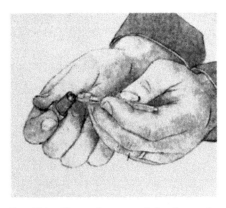

Fig. 125—Clean the valve seats using brass scraper as shown.

cleaning kit is available through any C. A. V. Service Agency and other sources.

Wipe all dirt and loose carbon from exterior of nozzle and holder assembly. Refer to Fig. 122 and proceed as follows:

Secure the nozzle in a soft-jawed vise or holding fixture and remove the cap nut (1). Loosen jam nut (3) and back off the adjusting sleeve (5) to completely unload the pressure spring (7). Remove the nozzle cap nut (12) and nozzle body (11). Nozzle valve (10) and body (11) are matched assemblies and must never be intermixed. Place all parts in clean calibrating oil or diesel fuel as they are removed. Clean the exterior surfaces with a soft wire brush, soaking in an approved carbon solvent if necessary, to loosen hard carbon deposits. Rinse the parts in clean diesel fuel or calibrating oil immediately after cleaning, to neutralize the carbon solvent and prevent etching of the polished surfaces. Clean the pressure chamber of nozzle tip using the special reamer as shown in Fig. 123. Clean the spray

holes in nozzle with an 0.009 (0.24 mm) wire probe held in a pin vise as shown in Fig. 124. Wire probe should protrude from pin vise only far enough to pass through spray holes (approximately $\frac{1}{16}$-inch), to prevent bending and breakage. Rotate pin vise without applying undue pressure.

Clean valve seats by inserting small end of brass valve seat scraper into nozzle and rotating tool. Reverse the tool and clean upper chamfer using large end. Refer to Fig. 125. Use the hooked scraper to clean annular groove in top of nozzle body (Fig. 126.) Use the same hooked tool to clean the internal fuel gallery.

With the above cleaning accomplished, back flush the nozzle by installing the reverse flusher adapter on injector tester and nozzle body in adapter, tip end first. Secure with the knurled adapter nut and insert and rotate the nozzle valve while flushing. After nozzle is back flushed, seat can be polished using a small amount of tallow on end of polishing stick, rotating the stick as shown in Fig. 127.

Light scratches on valve piston and bore can be polished out by careful use of special injector lapping compound only, DO NOT use valve grinding compound or regular commercial polishing agents. DO NOT attempt to reseat a leaking valve using polishing compound. Clean thoroughly and back flush if lapping compound is used.

Reclean all parts by rinsing thoroughly in clean diesel fuel or calibrating oil and assemble valve to body while immersed in the cleaning fluid. Reassemble the injector while still wet. With adjusting sleeve (5—Fig. 122) loose, reinstall nozzle body

(11) to holder (9), making sure valve (10) is installed and locating dowels aligned as shown in Fig. 128. Tighten nozzle cap nut (12—Fig. 122) to a torque of 50 ft.-lbs. Do not overtighten; distortion may cause a valve to stick and overtightening cannot stop a leak caused by scratches or dirt on the lapped mating surfaces of valve body and nozzle holder.

Retest and adjust the assembled injector assembly as outlined in paragraphs 128 through 132.

NOTE: If overhauled injector units are to be stored, it is recommended that a calibrating or preservative oil, rather than diesel fuel, be used for the pre-storage testing. Storage of more than thirty days containing diesel fuel may result in the necessity of recleaning prior to use.

INJECTION PUMP
All Models

The injection pump is a completely sealed unit. No service work of any kind should be attempted on the pump or governor unit without the use of special pump testing equipment and special training. Inexperienced or unequipped service personnel should never attempt to overhaul a diesel injection pump.

134. ADJUSTMENT. The slow idle stop screw (I—Fig. 129) should be adjusted with engine warm and running, to provide the recommended slow idle speed of 725-775 rpm. Also check to make sure that governor arm contacts the slow idle screw (I) and high speed screw (H) when throttle lever is moved to slow and fast positions. Also check to make sure that stop lever arm (L) moves fully

Fig. 126—Use the hooked scraper to clean annular grove in top of nozzle body.

Fig. 127—Polish the nozzle seat using mutton tallow on a wood polishing stick.

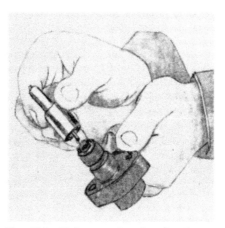

Fig. 128—Make sure locating dowels are carefully aligned when nozzle body is reinstalled.

Fig. 129—Injection pump showing linkage and adjustments.

A. Cable screw	L. Stop lever
B. Clamp screw	N. Nut
H. High speed stop screw	S. Stud
I. Idle speed stop screw	T. Throttle link

Fig. 130—Installed view of injection pump showing timing marks (T) and timing window (W).

Fig. 131—Alignment slot (S) in pump shaft must align with dowel in hub of injection pump drive gear when pump is installed.

Fig. 132—Injection pump with timing cover removed, showing "C" timing mark properly aligned with square end of snap ring. Refer to text.

to operating position when stop button is pushed in, and shuts off the fuel to injectors when stop button is pulled. The high speed stop screw (H) is set at the factory and the adjustment is sealed. Governed speed under load should be 2000 rpm, with a high idle (no load) speed of approximately 2160 rpm. Refer to paragraph 136 for pump timing adjustment.

135. REMOVE AND REINSTALL. Before attempting to remove the injection pump, thoroughly wash the pump and connections with clean diesel fuel or an approved solvent. Disconnect throttle control rod (T—Fig. 129) from governor arm by removing nut (N) and withdrawing stud (S) from arm. Disconnect stop control cable from stop lever (L) and cable housing from bracket by loosening clamp screw (B). Remove the inspection cover from front of timing gear cover, then remove the three screws retaining the injection pump drive gear to pump shaft. Disconnect fuel inlet, outlet and high pressure lines from pump, capping all connections to prevent dirt entry. Check to see that timing marks (T—Fig. 130) align, remove the three flange stud nuts, then withdraw the pump.

Normal installation of injection pump can be accomplished without reference to crankshaft timing marks or internal timing marks on injection pump. Be sure timing scribe lines (T—Fig. 130) are aligned and reverse the removal procedure. Bleed fuel system as outlined in paragraph 124. Check the injection pump timing, if necessary, as outlined in paragraph 136.

136. PUMP TIMING TO ENGINE. The injection pump drive shaft contains a milled slot (S—Fig. 131) which engages a dowel pin (2—Fig. 93) in pump drive gear. Thus, injection pump can be removed and reinstalled without regard to timing position. NOTE: Injection pump gear cannot become unmeshed from idler gear without removal of timing gear cover, therefore timing is not disturbed by removal and installation of pump.

To check the pump timing, shut off the fuel and remove pump timing window (W—Fig. 130) and flywheel timing plug from left side of cylinder block. Turn the crankshaft until the 23° BTDC flywheel timing mark aligns with timing pointer, at which time the "C" timing mark on injection pump rotor should align with straight

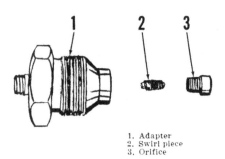

1. Adapter
2. Swirl piece
3. Orifice

Fig. 133—Exploded view of cold starting unit manifold adapter.

edge of snap ring as shown in Fig. 132. The mounting holes in pump mounting flange are elongated to permit minor timing variations. If timing marks cannot be properly aligned by shifting pump on mounting studs, the timing gear must be removed as outlined in paragraph 95 and the gears retimed.

COLD STARTING UNIT
Accessory—All Models

137. Some tractors are optionally equipped with an ether injection type cold weather starting aid consisting of a dash-mounted fuel can adapter, connecting tube, and a manifold adapter (Fig. 133). The only service required is renewal of damaged parts or occasional cleaning if the unit should become inoperative. Proceed as follows:

Disconnect the tubing from manifold adapter housing (1). Blow compressed air through the can adapter and tube to make sure they are open and clean. Remove adapter housing (1) from intake manifold, unscrew orifice plug (3) and withdraw swirl piece (2). Clean all parts in a suitable solvent and drop swirl piece (2) into bore of adapter (1). Apply a small amount of LOCTITE, Grade "A" (Red) to threads of orifice (3), reinstall orifice and tighten to a torque of 25-30 inch-pounds.

NOTE: If orifice (3) cannot be removed, heat adapter (1) in an oven or by other suitable means to a temperature of approximately 500° F. and remove while hot. Heat is necessary to break the previously applied LOCTITE bond. LOCTITE is used to prevent orifice from loosening in adapter and entering engine through manifold.

NON-DIESEL GOVERNOR

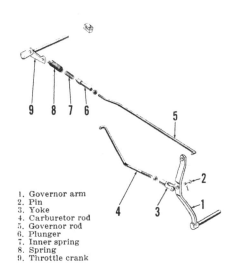

1. Governor arm
2. Pin
3. Yoke
4. Carburetor rod
5. Governor rod
6. Plunger
7. Inner spring
8. Spring
9. Throttle crank

Fig. 134—Exploded view of throttle linkage and associated parts used on gasoline models.

ADJUSTMENT

All Models

138. Recommended governed speeds are as follows:

Low Idle
   Continental Engines ...450-500 rpm
   Perkins Engines ....725-775 rpm
High Idle
   Continental Engines .2200-2250 rpm
   Perkins Engines .....2225-2275 rpm
Loaded Speed
   All Models ....2000 rpm

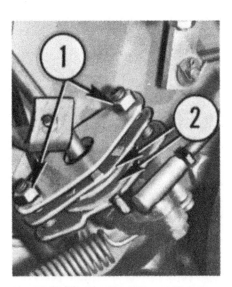

Fig. 135—Throttle hand lever friction is adjusted by tightening nuts (1).

To adjust the governor, refer to Fig. 134 and proceed as follows: Remove pin (2) connecting carburetor rod (4) to governor arm (1). Pull hand throttle down to apply pressure to governor spring, then move governor arm (1) rearward as far as it will go and carburetor rod rearward until throttle is against carburetor stop. Adjust the length of carburetor rod (4) until governor arm (1) must be moved forward $\frac{1}{32}$-inch before pin (2) can be inserted. Start and warm the engine, then adjust slow idle speed by turning stop screw on carburetor.

To adjust the high governed speed, loosen the locknut on governor rod (5) and thread spring plunger (6) on or off governor rod to shorten or lengthen rod as required.

If throttle lever creeps or will not maintain a set position, adjust the pressure on friction disc by tightening the two nuts (1—Fig. 135).

OVERHAUL

Models With Continental Engine

139. To overhaul the governor, first remove timing gear cover as outlined in paragraph 41. The governor lever shaft is supported in needle bearings in timing gear cover. To service the shaft, bearings or seals, remove lock screw from governor fork and withdraw the lever. Bearings and seals can be renewed at this time.

Withdraw governor weight unit from crankshaft. Refer to Fig. 136.

To disassemble the governor weight unit, remove the snap rings (1). Renew governor balls (5), the inner race (4) or outer race (7) if they are worn, scored or discolored.

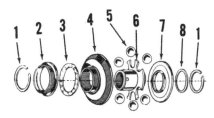

Fig. 136 — Exploded view of crankshaft mounted governor weight unit.

1. Snap rings
2. Fork base
3. Thrust bearing
4. Inner race
5. Governor balls
6. Ball driver
7. Outer race
8. Thrust washer

When installing the timing gear cover, make certain the governor fork falls behind the governor weight unit.

Models With Perkins Engine

140. Governor arm shaft, bearings and associated parts can be inspected and/or overhauled by unbolting and removing governor housing (H—Fig. 137). To remove shaft and weight unit, it is first necessary to remove timing gear cover as outlined in paragraph 62 and governor gear as in para-

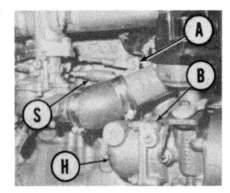

Fig. 137—Installed view of engine governor and associated parts used on tractors equipped with Perkins gasoline engine.

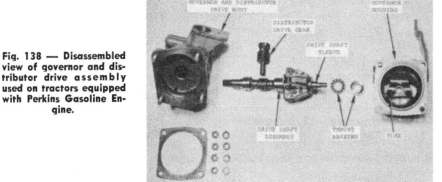

Fig. 138 — Disassembled view of governor and distributor drive assembly used on tractors equipped with Perkins Gasoline Engine.

IGNITION AND

ELECTRICAL

SYSTEM

graph 66. Distributor drive unit which carries governor weights can then be removed as outlined in paragraph 71. Shaft and weight unit is only available as an assembly.

Refer to Fig. 138 for a disassembled view of governor unit and to the following data for overhaul:

Distributor drive shaft
  clearance 0.0008-0.0024

Distributor drive shaft
  end play ....0.004 -0.008

Governor control shaft
  clearance 0.0006-0.0023

COOLING SYSTEM

RADIATOR

All Models

141. All models use a 7 psi pressure type radiator cap. Cooling system capacity is 10½ quarts for non-diesel models and 13½ quarts for diesels.

To remove the radiator, first drain cooling system and remove hood, grille and air cleaner. Disconnect radiator hoses and remove oil cooler radiator from radiator frame; then unbolt radiator, frame and shrouds as a unit.

THERMOSTAT

All Models

142. The by-pass type thermostat is contained in a separate housing located behind the outlet elbow. Thermostat opening temperature is 160° F. on non-diesel models; 168°-176° F. on diesels.

WATER PUMP

All Models

143. Refer to Figs. 139 and 140 for exploded views of water pump and associated parts, and to the accompanying captions for special overhaul notes. Water pump can be removed after draining radiator and removing generator.

DISTRIBUTOR

All Models

144. **TIMING.** Timing marks are located on crankshaft pulley as shown in Fig. 141. Initial (static) timing is 6° BTDC for models with Continental engines or 10° BTDC for models with Perkins engines. Maximum advance timing for Continental engines is 24-28° BTDC at 2000 rpm. For models with Perkins engines, advance timing is 32° @2100 rpm. A stroboscopic timing light is recommended for checking timing.

Firing order is 1-3-4-2 and distributor shaft rotates counter-clockwise when viewed from cap end. Advance mechanism starts to operate at approximately 650 crankshaft rpm on Continental engine or 500 rpm on models with Perkins engines.

145. **OVERHAUL.** Refer to Fig. 142 for an exploded view of the distributor. Centrifugal advance mechanism can be checked for binding or broken springs by turning rotor (21) counter-clockwise and releasing, after removing cap (23). Bushings are not available for housing (5); renew housing and/or shaft (8) if clearance is excessive. Shims (3) are available in thicknesses of 0.005 and 0.010 for adjusting shaft end play which should

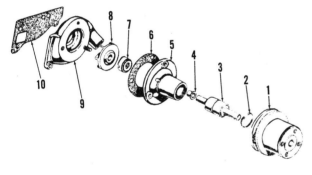

Fig. 139—Exploded view of water pump and associated parts used on models with Continental engine.

1. Fan pulley
2. Snap ring
3. Shaft & bearing
4. Slinger
5. Bearing housing
7. Seal
8. Impeller
9. Body
10. Gasket

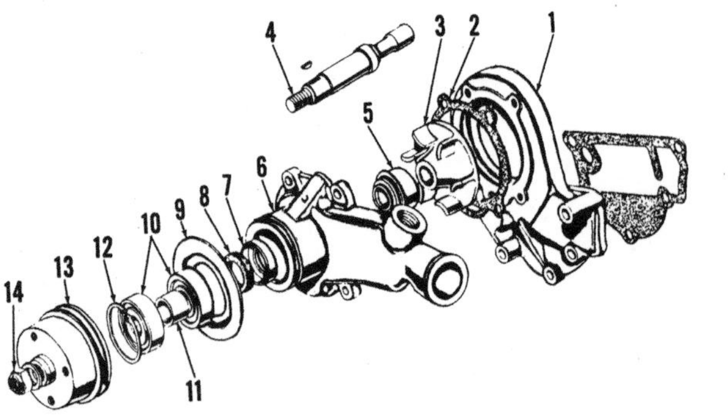

Fig. 140—Exploded view of water pump and associated parts used on Perkins gasoline and diesel models. Pack the area between bearings (10) half full of high melting-point grease when assembling.

1. Impeller housing
2. Gasket
3. Impeller
4. Shaft
5. Seal
6. Shaft housing
7. Retainer
8. Front seal
9. Seal flange
10. Bearings
11. Spacer
12. Snap ring
13. Fan pulley
14. Nut

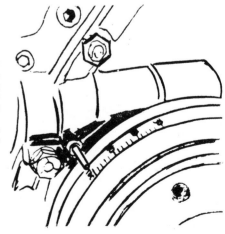

Fig. 141—View of crankshaft pulley and timing gear cover showing timing pointer and timing marks.

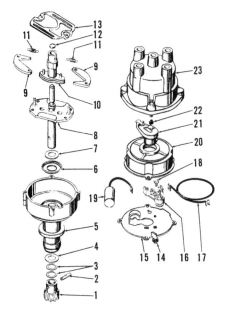

Fig. 142—Exploded view of ignition distributor showing component parts.

1. Drive gear	12. Oiler wick
2. Roll pin	13. Hold down plate
3. Shims	14. Insulator
4. Thrust washer	15. Breaker plate
5. Distributor	16. Point set
housing	17. Primary wire
6. Seal	18. Eccentric screw
7. Thrust washer	19. Condenser
8. Shaft assembly	20. Dust shield
9. Advance weight	21. Rotor
10. Cam assembly	22. Brush
11. Spring	23. Distributor cap

Fig. 144—Exploded view of "DELCOTRON" alternator used on late models.

1. Pulley nut
2. Drive pulley
3. Fan
4. Collar
5. Drive end frame
6. Slinger
7. Bearing
8. Gasket
9. Collar
10. Bearing retainer
11. Stator assembly
12. Rotor assembly
13. Brush holder
14. Capacitor
15. Heat sink
16. Slip ring end frame
17. Felt seal and retainer
18. Needle bearing
19. Negative diode (3 used)
20. Positive diode (3 used)

be 0.002-0.010. Test specifications are as follows:

Delco-Remy 1112647

Breaker contact gap	0.022
Breaker arm spring tension (measured at center of contact)	17-21 oz.
Cam angle (degrees)	31-34

Advance data is in distributor degrees and distributor rpm.

Start advance	0-2 @ 325
Intermediate advance	5-7 @ 700
Maximum advance	9-11 @ 1000

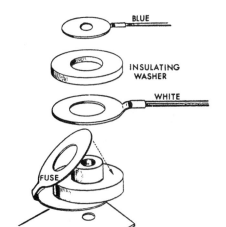

Fig. 143—Exploded view of alternator output terminal components showing output fuse/washer which wraps around terminal insulator. Check the fuse if output circuit is dead.

Delco-Remy 1112693

Breaker contact gap	0.022
Breaker arm spring tension (measure at center of contact)	17-21-oz.
Cam angle (degrees)	31-34

Advance data is in distributor degrees and distributor rpm.

Start advance	0-1 @ 250
Maximum advance	10-12 @ 1050

GENERATOR & REGULATOR
All Models So Equipped

146. Delco-Remy units are used; test specifications are as follows:

Generator

Brush spring tension	28 oz.
Field draw	
Volts	12
Amperes	1.58-1.67
Output (Cold)	
Maximum amperes	25
Volts	14
RPM	3040

Regulator

Cutout Relay	
Air gap	0.020
Point gap	0.020
Closing voltage (range)	11.8-14.0
Adjust to	12.8
Voltage regulator	
Air gap	0.075
Voltage range	13.6-14.5
Adjust to	14.0
Ground polarity	Negative

ALTERNATOR AND REGULATOR
All Models So Equipped

147. **ALTERNATOR.** A "DELCO-TRON" generator (alternator) is used

on late models. Units are negative ground.

The only test which can be made without removal and disassembly of alternator is output test. Output should be approximately 32 amperes at 5000 alternator rpm.

IMPORTANT: Outlet terminal post of alternator is equipped with a fuse/washer as shown in Fig. 143. If fuse burns out, charging output is cut off from white (output) wire. Current still flows to the blue wire leading to voltage regulator terminal. With charging flow to the battery cut off, voltage rises in the blue (control) wire causing the regulator to cut back the charging current. If generator shows no output, check the fuse or connect white wire directly to output terminal. Renew the fuse if damaged, and assemble as shown in Fig. 143.

To disassemble the alternator, first place match marks (M—Fig. 144) on the two frame halves (5 & 16), then remove the four through-bolts. Pry frame apart with a screwdriver between stator frame (11) and drive end frame (5). Stator assembly (11) must remain with slip ring end frame (16) when unit is separated.

NOTE: When frames are separated, brushes will contact rotor shaft at bearing area. Brushes MUST be cleaned of lubricant if they are to be re-used.

Clamp the iron rotor (12) in a protected vise only tight enough to permit loosening pulley nut (1). Rotor and end frame can be separated after pulley is removed. Check bearing surfaces of rotor shaft for wear or scoring. Examine slip ring surfaces for

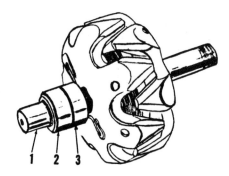

Fig. 145—Removed rotor assembly showing test points to be used when checking for grounds, shorts and opens.

Fig. 146—Exploded view of brush holder assembly. Insert wire in hole (W) to hold brushes up. Refer to text.

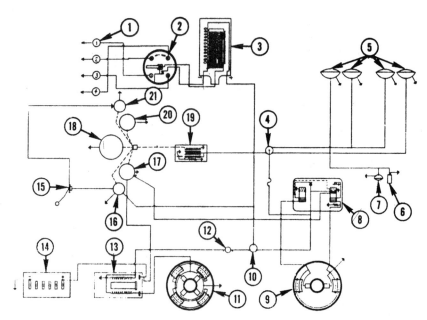

Fig. 147—Wiring diagram typical of that used on early gasoline models equipped with DC generator.

1. Spark plugs	8. Voltage regulator	16. Fuel gage
2. Distributor	9. Generator	17. Ammeter
3. Coil	10. Ignition switch	18. Tachometer
4. Light switch	11. Starter	19. Power supply
5. Headlights	12. Starter safety	20. Temperature
6. Auxiliary light	switch	gage
connector	13. Starter solenoid	21. Oil pressure
7. Work & warning	14. Battery	gage
light	15. Fuel gage	
	sending unit	

scoring or wear and windings for over-heating or other damage. Check rotor for grounded, shorted or open circuits using an ohmmeter as follows:

Refer to Fig. 145 and touch the ohmmeter probes to points (1-2) and (1-3); a reading near zero will indicate a ground. Touch ohmmeter probes to the two slip rings (2-3); reading should be 4.6-5.5 ohms. A higher reading will indicate an open circuit and a lower reading will indicate a short. If windings are satisfactory, mount rotor in a lathe and check runout at slip rings using a dial indicator. Runout should not exceed 0.002. Slip ring surfaces can be trued if runout is excessive or if surfaces are scored. Finish with 400 grit or finer polishing cloth until scratches or machine marks are removed.

Disconnect the three stator leads and separate stator assembly (11—Fig. 144) from slip ring end frame assembly. Check stator windings for grounded or open circuits as follows: Connect ohmmeter leads successively between each pair of stator leads. Readings should be equal and relatively low. A high reading would indicate an open lead. Connect ohmmeter leads to any stator lead and to stator frame. The three stator leads have a common connection in the center of the windings, a reading other than infinity would indicate a grounded circuit. A short circuit within the stator windings cannot be

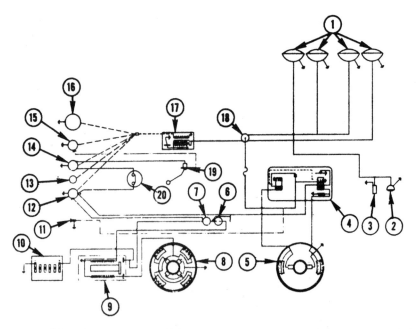

Fig. 148—Wiring diagram typical of that used on early diesel models equipped with DC generator.

1. Headlights	7. Starter switch	14. Fuel gage
2. Work & warning	8. Starter	15. Oil pressure
light	9. Starter solenoid	gage
3. Auxiliary light	10. Battery	16. Tachometer
connector	11. Cigarette lighter	17. Power supply
4. Voltage regulator	12. Ammeter	18. Headlight switch
5. Generator	13. Temperature	19. Sending unit
6. Starter safety	gage	20. Oil pressure
switch		switch

readily determined by test because of the low resistance of the windings.

Three negative diodes (19) are located in slip ring end frame (16) and three positive diodes in heat sink (15). Diode should test at or near infinity in one direction when tested with an ohmmeter, and at or near zero when leads are reversed. Renew any diode with approximately equal meter readings in both directions. Diodes must be removed and installed using an arbor press which contacts only the outer edge of diode. Do not attempt to drive a faulty diode out of end frame or heat sink as shock may cause damage to the other good diodes. If all diodes are being renewed, make certain the positive diodes (marked with red printing) are installed in heat sink and negative diodes (marked with black printing) are installed in end frame.

Brushes are available only in an assembly which includes brush holder (13). Brush springs are available for service and should be renewed if heat damage or corrosion is evident. If brushes are re-used, make sure all grease is removed from surface of brushes before unit is reassembled. When reassembling, install brush springs and brushes in holder, push brushes up against spring pressure and insert a short piece of straight wire through hole (W—Fig. 146) and through end frame (16—Fig. 144) to outside. Withdraw wire only after alternator is assembled.

Capacitor (14) connects to the heat sink and is grounded to end frame. Capacitor protects the diodes from voltage surges.

Remove and inspect ball bearing (7). If bearing is in satisfactory condition, fill bearing ¼-full with Delco-Remy Lubricant No. 1960373 and reinstall. Inspect needle bearing (18) in slip ring end frame. This bearing should be renewed if its lubricant is exhausted; no attempt to re-lubricate the bearing should be made. Press old bearing out toward the inside and new bearing in from outside until bearing is flush with outside of end frame. Saturate felt seal with SAE 20 oil and install seal and retainer assembly.

Reassemble alternator by reversing the disassembly procedure. Tighten pulley nut to a torque of 45 ft.-lbs.

NOTE: A battery powered test light can be used instead of ohmmeter for all electrical checks except shorts in rotor winding; however when checking diodes, test light must not be more than 12 volts.

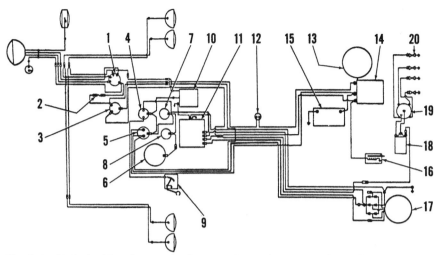

Fig. 149—Typical wiring diagram for late gasoline models. Except for obvious differences, late diesels are similar.

1. Light switch	7. Oil pressure gage	11. Regulator	16. Fuel shut-off
2. Fuse	8. Water temperature	12. Neutral switch	solenoid
3. Starter switch	gage	13. Starter motor	17. Alternator
4. Fuel gage	9. Fuel sender unit	14. Solenoid	18. Coil
5. Ammeter	10. Inverter	15. Battery	19. Distributor
6. Tachometer			20. Spark plugs

148. REGULATOR. A Delco-Remy standard two-unit regulator is used. Except for output (fuse) terminal (Fig. 143), quick disconnect plugs are used at regulator and alternator. Production regulator is riveted to shock mount; service units are shipped less mount and are attached with screws. Test specifications are as follows:

Regulator Model 1119513

Ground polarity ....Negative
Field Relay
   Air Gap 0.015
   Point Opening 0.030
   Closing Voltage Range 3.8-7.2

Voltage Regulator
   Air Gap 0.067*
   Point Opening 0.014
   Voltage Setting:

     @ 65° F 13.9-15.0
     @ 85° F 13.8-14.8
     @ 105° F 13.7-14.6
     @ 125° F 13.5-14.4
     @ 145° F 13.4-14.2
     @ 165° F 13.2-14.0
     @ 185° F 13.1-13.9

*The specified air gap setting is for bench repair only; make final adjustments to obtain specified voltage, with lower contacts opening at not more than 0.4 volt less than upper contacts. Temperature (ambient) is measured ½-inch away from regulator cover and adjustment should be made only when regulator is at normal operating temperature.

STARTING MOTOR

All Models

149. Delco-Remy starting motors are used on all models. Specifications are as follows:

Model 1107329

Brush spring tension
   (minimum) ....35 oz.
No-Load test
   Volts 10.6
   Amperes (w/solenoid) 49-76
   Minimum rpm 6200

Model 1107539

Brush spring tension
   (minimum) ....35 oz.
No-Load test
   Volts 10.6
   Amperes (w/solenoid) 75-100
   Minimum rpm 6450

Model 1108396

Brush spring tension
   (minimum) ....35 oz.
No-Load Test
   Volts 9
   Amperes (w/solenoid) 50-80
   Minimum rpm 5500

Starter drive pinion clearance is not adjustable, however, some clearance must be maintained between end of pinion and starter drive frame; to assure solid contact of the heavy-duty magnetic switch. Normal pinion clearance should be within the limits of 0.010-0.140. Connect a 6-volt battery to solenoid terminals when measuring pinion clearance, to keep armature from turning.

ENGINE CLUTCH

Tractors may be equipped with a flywheel mounted dual stage clutch and continuous power take-off or a split torque clutch and independent power take-off. Refer to the appropriate following paragraphs for adjustment and overhaul procedures.

ADJUSTMENT

All Models

150. **PEDAL FREE PLAY.** Clutch pedal free play is measured between release shaft arm and stop on clutch housing. To check the clearance, depress clutch pedal until resistance is felt, then measure the clearance. Free play should be ⅛-inch for Model MF-175; refer to (C—Fig. 150). On Model MF180, free play (C—Fig. 151) should be $\frac{1}{16}$ inch for models with dual clutch or ⅛ inch for models with split torque clutch.

If adjustment is required, insert a long punch in hole (H—Fig. 151) in throwout shaft and loosen release arm clamp bolt (B). Turn shaft (A) until throwout collar contacts release fingers, reposition release arm with specified clearance (C) and retighten clamp bolt (B). After free play has been adjusted, check linkage as outlined in paragraph 151.

151. **LINKAGE ADJUSTMENT.** Remove the pin securing clutch link (D—Fig. 150 or 151) to release arm (E). Make sure free play is correctly adjusted as outlined in paragraph 150; then shorten or lengthen clutch link (D) until pin can just be inserted with clutch pedal in highest position.

TRACTOR SPLIT

Model MF175

152. To detach engine from transmission assembly, remove rear hood side panels, disconnect battery cables and remove battery. Remove the cap screws securing instrument support and steering housing to transmission case. Disconnect light wires, wires from starter safety switch and "Multipower" shift linkage. Disconnect oil cooler lines at rear end. Support steering column from an overhead hoist and place rolling floor jacks under transmission case and oil pan.

Remove the cap screws securing transmission case to engine and carefully separate the units until clutch shaft is clear of clutch; then lower front of transmission and/or raise rear of engine until steering housing and instrument panel support will clear transmission case and linkage, and roll the units apart.

Join the tractor by reversing the split procedure.

Model MF180

153. To detach engine from transmission assembly, first drain cooling system and remove hood and side panels. Shut off fuel and remove fuel tank. Disconnect wiring harness at front end and oil cooler lines at rear. Disconnect power steering cylinder lines at "Hydramotor" and power steering pressure and return lines at pump. Remove temperature gage sending unit from engine and discon-

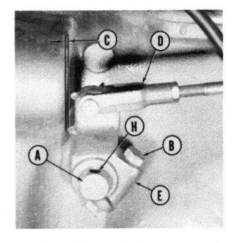

Fig. 151—View of clutch throwout shaft and link used on Model MF180.

A. Throwout shaft
B. Clamp bolt
C. Measure clearance
   (1/16-inch)
D. Pedal link
E. Release arm
H. Hole

nect oil pressure gage tube and "Tractormeter" drive cable.

Support both halves of tractor separately, remove the attaching cap screws and separate the tractor. Join the tractor by reversing the split procedure.

OVERHAUL

Dual Clutch

154. **REMOVE AND REINSTALL.** To remove the clutch, first split tractor as outlined in paragraph 152 or 153. Make up three special "T" bolts by welding a cross bar to ¼-20 X 6 inch threaded rods, then add forcing nuts. Install the "T" bolts in the three holes in outer edge of clutch cover as shown in Fig. 152 and tighten the nuts to compress the springs. Mark

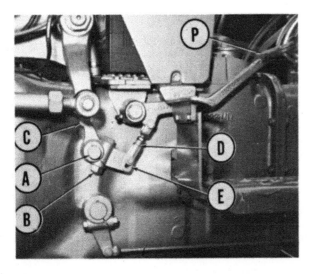

Fig. 150—View of clutch pedal and linkage of the type used on Model MF175.

A. Throwout shaft
B. Clamp bolt
C. Measure clearance
   (⅛-inch)
D. Pedal link
E. Release arm
P. Clutch pedal

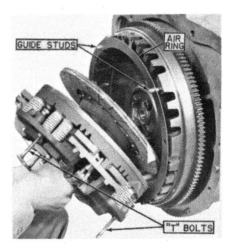

Fig. 152—Special "T" bolts are used for removal and installation of clutch. Guide studs are used when installing.

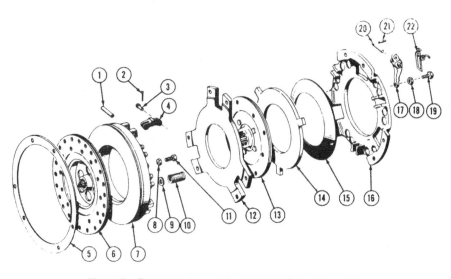

Fig. 153—Exploded view of dual clutch of the type used.

1. Pin
2. Cotter pin
3. Link pin
4. Link
5. Air ring
6. Transmission disc
7. Pressure plate
8. Locknut
9. Insulating washer
10. Clutch spring
11. Adjusting screw
12. Drive plate
13. PTO disc
14. Pressure plate
15. Belleville washer
16. Clutch cover
17. Release lever
18. Lock nut
19. Adjusting screw
20. Pivot pin
21. Retainer pin
22. Torsion spring

the clutch cover (16—Fig. 153), pressure plate (14), drive plate (12) and pressure plate (7), to assure correct assembly and maintain clutch balance; remove the retaining cap screws and lift the clutch assembly from flywheel.

To install the clutch assembly, first install the air ring using guide studs as shown in Fig. 152. Insert the clutch pilot tool through the driven discs and reinstall by reversing the removal procedure. Tighten the retaining cap

screws to a torque of 28-33 ft.-lbs. Adjust the clutch, if necessary, as outlined in paragraph 156.

155. OVERHAUL. To disassemble the removed clutch unit, unhook the three torsion springs (22—Fig. 153) from clutch release levers (17). Back off the locknuts (8) and thread adjustment screws (11) into pressure plate (7) until they bottom. Back off the forcing nuts on the three "T" bolts until Belleville washer is free

and clutch fingers assume the approximate position shown in Fig. 154; then drive the groove pins (21—Fig. 153) into cover until pivot pins (20) can be removed.

NOTE: Groove pins can be removed from the bottom after clutch is disassembled. Do not attempt to drive the pins out of clutch cover or Belleville spring may be damaged.

Remove the pivot pins, back out the forcing nuts on "T" bolts until spring pressure is removed; then remove the "T" bolts and disassemble the clutch.

Thoroughly clean and examine all parts and renew any which are damaged or worn. Linings are available for both driven plates. The Belleville spring (15) is color coded light blue.

Coil springs (10) should test 130-145 lbs. when compressed to a height of 1½-inches. Link pins (1) in pressure plate have one knurled end in early models; late models use a smooth pin which is a press fit in retaining spacer in link (4). If early (knurled) pins are removed, late pins and retaining spacers should be used for reassembly.

Place clutch cover (16) upside down on a bench, then center the Belleville spring (15) in cover groove with convex side up. Place pressure plate (14) on spring, aligning the previously affixed assembly marks. Install the smaller clutch disc (13) hub down, and drive plate (12) with assembly marks aligned. Temporarily hold drive plate and clutch cover together using three $\frac{5}{16}$ X 1½ inch bolts through flywheel mounting holes.

Place the 11-inch pressure plate (7) face down on the bench and reinstall links (4), centering pins (1) in link. Complete the assembly by reversing the disassembly procedure, making sure assembly marks are aligned. When inserting pivot pins (20), make sure holes are aligned, then reinstall groove pins (21). Adjust the clutch as outlined in paragraph 156.

156. ADJUSTMENT. After the clutch assembly has been installed on flywheel, two adjustments are necessary for proper clutch operation. Proceed as follows:

NOTE: A new 11-inch transmission clutch disc is required when adjusting release lever height. If the removed clutch disc will be reused, first install the clutch using a new disc, make the adjustment, then assemble using the partially worn parts, without changing the adjustment.

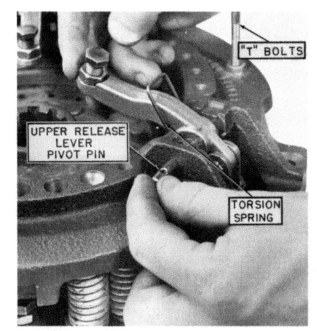

"T" BOLTS

UPPER RELEASE LEVER PIVOT PIN

TORSION SPRING

Fig. 154—Removing upper release lever pivot pin.

Fig. 155—Using the special tool to adjust clutch release lever height.

Fig. 158—Nine springs are used in split torque clutch.

Fig. 159 — Removing finger from clutch cover used on models with split torque clutch.

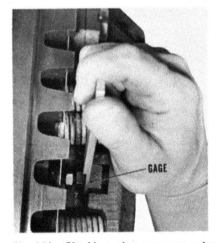

Fig. 156—Checking primary pressure plate free play.

To remove the clutch, first split tractor as outlined in paragraph 152 or 153. Use a suitable aligning tool to install and tighten retaining cap screws to a torque of 30-35 ft.-lbs. Long end of clutch plate hub goes to rear.

To disassemble the removed pressure plate and cover unit, place the assembly in a press as shown in Fig. 160 and apply only enough pressure to relieve tension on pins. Remove the pins and lift off cover as shown in Fig. 157.

Inspect release levers and pins for damage and pressure plate for scoring, heat checks or wear at actuating pin holes. Pressure plate may be refaced if necessary and suitable facilities are available. Inspect pto drive hub in cover for spline wear or looseness. Inspect springs for heat discoloration or other damage. Springs should test 220 lbs at $1\frac{7}{16}$ inch test height.

Assemble by reversing the disassembly procedure, making sure pins are installed with heads leading when flywheel is turning in normal direction of rotation. (See Fig. 160). Clutch fingers should be adjusted to equal height after clutch installation, using adjusting gage MFN202D.

Using Massey-Ferguson Finger Adjusting Tool (MFN-202C) as shown in Fig. 155, adjust all clutch fingers to an equal height within 0.005. When finger height has been adjusted, permanently install clutch cover and 11-inch plate which will be used. Tighten the retaining cap screws to a torque of 28-33 ft.-lbs., then adjust secondary pressure plate clearance to 0.080 as shown in Fig. 156. Adjustment is made by varying the height of adjusting screws (11—Fig. 153). Join the tractor as outlined in paragraph 152 or 153, and adjust the linkage as in paragraphs 150 and 151.

Split-Torque Clutch

157. Models with independent power take-off use a split torque, single disc clutch with power take-off and hydraulic pump driven by a splined hub carried in clutch cover.

Fig. 157—Removing cover from split torque clutch used on models with independent power take-off.

Fig. 160 — Using a press to disassemble split torque clutch. Clutch is assembled with head of pivot pin leading as shown.

DUAL RANGE TRANSMISSION
(Without Multipower)

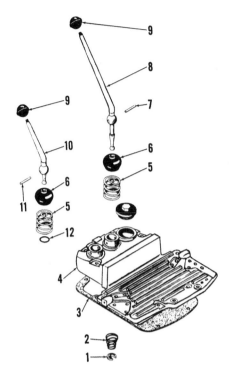

Fig. 161—Exploded view of transmission top cover and shift levers used on Model MF-175.

1. Spring seat
2. Spring
3. Gasket
4. Top cover
5. Spring
6. Cover
7. Pin
8. Gear shift lever
9. Knob
10. Range shift lever
11. Pin
12. "O" ring

ing cap screws and lift off top cover with shift levers attached.

Overhaul shift levers on top cover as outlined in paragraph 160, install by reversing the removal procedure.

160. **OVERHAUL.** Refer to Fig. 161. To remove either shift lever, compress the spring (2) and unseat and remove spring seat (1). Pull up the cover (6), drive out retaining pin (7 or 11); then withdraw shift lever upward out of cover. Range shift lever (10) contains an "O" ring oil seal (12) which fits against a shoulder in lever bore. Assemble by reversing the disassembly procedure.

MAIN DRIVE SHAFT (Clutch Shaft) Model MF175

161. **REMOVE AND REINSTALL.** To remove the main drive shaft (clutch shaft), it is first necessary to remove the complete transmission assembly as outlined in paragraph 158.

Refer to Fig. 162 and remove clutch release bearing and linkage. Remove pto bearing front cover plate (27— Fig. 163) and unseat and remove snap ring (24) and retaining washer (23). Thread forcing screws into bearing

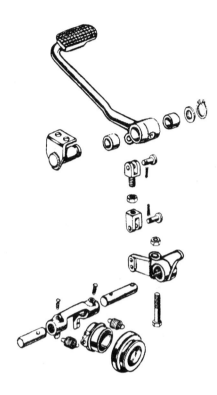

Fig. 162—Exploded view of clutch pedal and release linkage used on Model MF180. Model MF175 is similar.

TRANSMISSION REMOVAL
Model MF175

158. To remove the complete transmission unit, first detach (split) engine from clutch housing as outlined in paragraph 152. Drain transmission and rear axle center housing. Disconnect brake rods and remove step plates. Support transmission and rear axle center housings separately, remove the attaching bolts and lift off the transmission unit. Install by reversing the removal procedure.

TRANSMISSION TOP COVER
Model MF175

159. **REMOVE AND REINSTALL.** To remove the transmission top cover, first split engine from clutch housing as outlined in paragraph 152 or remove the steering gear as outlined in paragraph 18. Remove the attach-

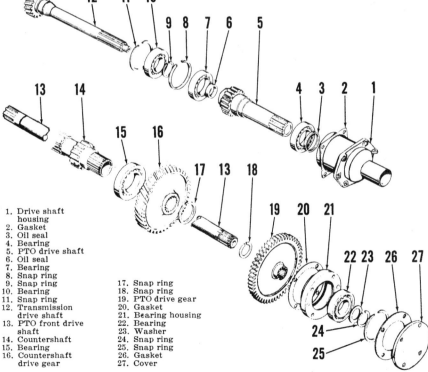

1. Drive shaft housing
2. Gasket
3. Oil seal
4. Bearing
5. PTO drive shaft
6. Oil seal
7. Bearing
8. Snap ring
9. Snap ring
10. Bearing
11. Snap ring
12. Transmission drive shaft
13. PTO front drive shaft
14. Countershaft
15. Bearing
16. Countershaft drive gear
17. Snap ring
18. Snap ring
19. PTO drive gear
20. Gasket
21. Bearing housing
22. Bearing
23. Washer
24. Snap ring
25. Snap ring
26. Gasket
27. Cover

Fig. 163—Exploded view of transmission input shaft and gears used on Model MF175 without "Multipower" transmission.

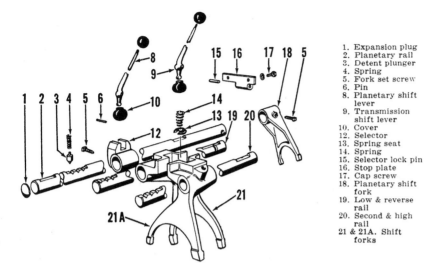

1. Expansion plug
2. Planetary rail
3. Detent plunger
4. Spring
5. Fork set screw
6. Pin
8. Planetary shift lever
9. Transmission shift lever
10. Cover
12. Selector
13. Spring seat
14. Spring
15. Selector lock pin
16. Stop plate
17. Cap screw
18. Planetary shift fork
19. Low & reverse rail
20. Second & high rail
21 & 21A. Shift forks

Fig. 164—Exploded view of shifter rails and forks of the type used on Model MF175. Levers (8 & 9) and associated parts differ on Model MF180; refer to Fig. 169.

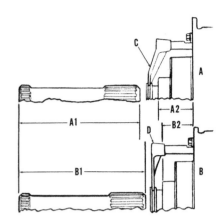

Fig. 166—Schematic method of identifying the late (wide) planetary unit (A) and early (narrow) planetary unit (B). Many parts are not interchangeable.

housing (21) and, tightening screws evenly, remove bearing housing and associated parts. Bump or pull pto front drive shaft (13) rearward until clear of splines in pto drive gear (19), and lower gear to bottom of transmission housing. Remove the retaining cap screws and withdraw main drive shaft and housing assembly (1 through 12) as a unit.

Overhaul the main drive shaft as outlined in paragraph 162 and install by reversing the removal procedure.

162. **OVERHAUL.** To disassemble the removed clutch shaft and associated parts, unseat snap ring (11—Fig.

163) and bump transmission drive shaft (12) and bearing (10) rearward out of housing (1). Unseat and remove snap ring (8) and bump pto clutch shaft (5) and bearings (4 & 7) from housing. Seal (6) in shaft (5) and seal (3) in housing (1) can be renewed at this time. Install seals with lips to rear and use seal protectors when installing shafts. Assemble by reversing the disassembly procedure and install as outlined in paragraph 161.

SHIFTER RAILS AND FORKS
Model MF175

163. To remove the shifter rails and forks, first remove transmission top cover as outlined in paragraph 159

and detach transmission from rear axle center housing.

Unwire and remove the set screws retaining selector and shifter forks to rails, remove detent springs (4—Fig. 164), plungers (3) and stop plate (16); then withdraw shifter rails and forks from transmission case.

Forks (21 and 21A) are interchangeable but rails (19 and 20) are not. Rails should be installed with milled flat to top rear and selector lock grooves to center.

A change has been made in planetary shift rail (2) and fork (18) which corresponds with a change in the width of the dual range planetary gears. On early models which use the narrow planetary gears, rail (2) is $20\frac{11}{16}$ inches in length and fork (18) is straight. On late models using the wide planetary gears, rail (2) is $20\frac{1}{32}$ inches long and fork (18) is offset to rear when installed. Refer also to paragraph 167 and Fig. 166 and be sure the correct parts are installed if any of the parts are renewed. Tractor serial numbers corresponding to the change are not available.

MAIN (Sliding Gear) SHAFT
Model MF175

164. To remove the transmission main shaft (12—Fig. 165), first remove the transmission assembly as outlined in paragraph 158, clutch shaft as in paragraph 161 and shifter rails and forks as in paragraph 163. Remove the four cap screws securing the planetary unit to rear of transmission case and withdraw rear cover plate (27), thrust washer (19) and planet carrier (26). Using two screw drivers, work the planetary ring gear

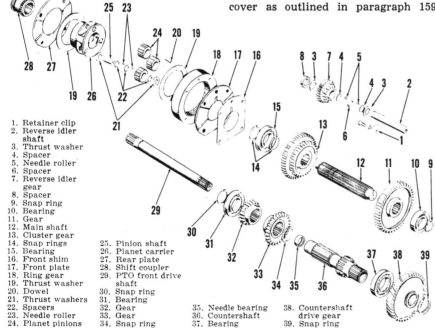

1. Retainer clip
2. Reverse idler shaft
3. Thrust washer
4. Spacer
5. Needle roller
6. Spacer
7. Reverse idler gear
8. Spacer
9. Snap ring
10. Bearing
11. Gear
12. Main shaft
13. Cluster gear
14. Snap rings
15. Bearing
16. Front shim
17. Front plate
18. Ring gear
19. Thrust washer
20. Dowel
21. Thrust washers
22. Spacers
23. Needle roller
24. Planet pinions

25. Pinion shaft
26. Planet carrier
27. Rear plate
28. Shift coupler
29. PTO front drive shaft
30. Snap ring
31. Bearing
32. Gear
33. Gear
34. Snap ring

35. Needle bearing
36. Countershaft
37. Bearing

38. Countershaft drive gear
39. Snap ring

Fig. 165—Exploded view of transmission shafts, gears and associated parts used on Model MF175 without "Multipower" transmission.

(18) and dowels from case. Remove planetary front cover (17) and shim (16).

Remove snap ring (9) from front of mainshaft and bump mainshaft rearward out of front bearing; then withdraw shaft (12) and bearing (15) out from rear while lifting gears (11 & 13) out top opening. To remove front bearing (10) from transmission case, first remove snap ring (39) and slide countershaft drive gear (38) forward off of shaft. Remove main shaft rear bearing (15) toward front of shaft (12) after removing front snap ring (14). Install by reversing the removal procedure.

COUNTERSHAFT
Model MF175

165. To remove the countershaft (36—Fig. 165), first remove mainshaft as outlined in paragraph 164 and proceed as follows:

Remove snap ring (30) from rear of countershaft and snap ring (39) and gear (38) from front end. Use a suitable step plate in the hollow shaft and bump countershaft (36) forward out of transmission case, bearing (31) and gears (32 & 33). Front bearing (37) can be removed from shaft and rear bearing (31) from case at this time. Renew needle bearing (35) in rear of shaft bore if bearing is damaged.

Install by reversing the removal procedure. Use a wooden block or buck up front of shaft by other suitable means while drifting rear bearing (31) on shaft.

REVERSE IDLER ASSEMBLY
Model MF175

166. The reverse idler gear and shaft can be removed after removing the mainshaft as outlined in paragraph 164, however, removal is not required for removal or installation of countershaft.

Reverse idler shaft (2—Fig. 165) is retained by clip (1) which fits a notch in shaft. Reverse idler bearing consists of two rows of loose needle rollers (5) separated by spacers (4 & 6). A total of 56 rollers (5) are used. Assemble by reversing the removal procedure, using Fig. 165 as a guide.

PLANETARY UNIT
Model MF175

167. The planetary unit can be removed as outlined in paragraph 164,

after detaching transmission from rear axle center housing.

Planet pinion shafts (25—Fig. 165) are a tight press fit in planet carrier (26); use a suitable press when removing and installing. Two rows of (27 each) loose needle rollers (23) are used in each planet pinion, separated and spaced by three washers (22). Use viscous grease to stick thrust washers (21) to gears, and completely assemble the bearing before attempting to install the pinion shafts (25).

Two different planetary units have been used. Refer to Fig. 166. The late (wide) unit (A) and early (narrow) unit (B) can be easily identified by the shape of shifter fork (C or D), or by measuring the distance (A2 or B2) from rear of transmission housing to rear edge of planet carrier. The distance (B2) is $3\frac{1}{8}$ inches for early units; or $3\frac{7}{16}$ inches (A2) for late units. Specified length of rear drive shaft is 9.19 inches (B1) for early (narrow) planetary unit or 8.72 inches (A1) for late (wide) unit. Refer also to paragraph 163 for correct length of planetary shift rail to be used with the respective planetary units.

MULTIPOWER TRANSMISSION

The Massey-Ferguson "Multipower" transmission is a modification of the standard dual-range, three speed transmission, providing an additional hydraulically operated high-low range, thus making available a total of twelve forward and four reverse gear speeds. The "Multipower" unit may be shifted while tractor is moving under load, without disengaging the transmission clutch. Power for the "Multipower" clutch is supplied by a hydraulic pump located in rear axle center housing.

OPERATION
All Models

168. Refer to Fig. 167. The "Multipower" unit consists of the low-range drive gears (1 & 2), high range drive gears (3 & 4), the hydraulically actuated "Multipower" clutch (5), the jaw-type over-running clutch (6), the "Multipower" pump (P) and valve (V).

The two pairs of input gears (1 & 2) and (3 & 4) are constantly meshed. When tractor is operating in low range, power flow is through gears (1 & 2), then through the locked jaw-type clutch (6) to the countershaft. The high-range driven gear (4) is splined to the countershaft and drives the high-range output gear (3) faster than the input gear and shaft (1), the slippage occurring in the released "Multipower" clutch (5). NOTE: The tractor will "Free-wheel" or coast in Low-Multipower, as the

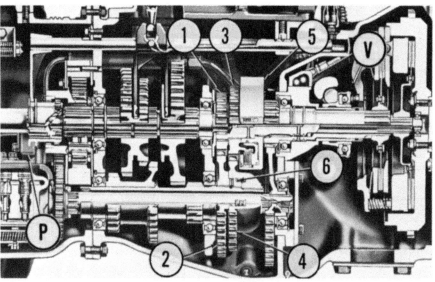

Fig. 167—Cross sectional view of "Multipower" transmission.

P. Multipower pump	1. Low input gear	3. High input gear	5. Multipower clutch
V. Multipower valve	2. Low driven gear	4. High driven gear	6. Jaw clutch

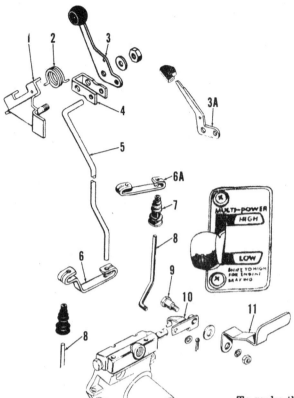

Fig. 168—Exploded view of "Multipower" clutch control linkage and associated parts.

1. Pivot bracket
2. Spring
3. Lever (MF175)
3A. Lever (MF180)
4. Spacer
5. Upper link rod
6. Link (MF175)
6A. Link (MF180)
7. Boot
8. Lower link rod
9. Pivot bolt
10. Shift lever
11. Bracket

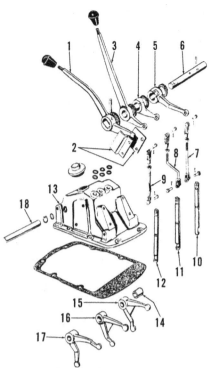

Fig. 169—Transmission top cover and associated parts used on Model MF180.

1. Gear shift lever	10. Plunger
2. Pivot bracket	11. Plunger
3. Range shift lever	12. Plunger
4. Shift arm	13. Top cover
5. Shift arm	14. Spacer
6. Shift shaft	15. Lever
7. Link	16. Lever
8. Link	17. Lever
9. Link	18. Pivot shaft

jaw-type clutch will release if countershaft speed exceeds input speed through gear (2).

When tractor is operating in "High-Multipower" range, clutch (5) is hydraulically engaged, locking gear (3) to input shaft and gear (1), and power flow is through gears (3 & 4) to the countershaft. The low-range driven gear (2) continues to be driven by gear (1) at a speed slower than countershaft speed, the slippage occurring at the jaw-type over-running clutch (6). NOTE: The design of the jaw-type over-running clutch serves as an automatic hill-holder, preventing tractor from rolling backward when clutch is released with "Multipower" control lever in high range.

CAUTION: A tractor equipped with "Multipower" transmission cannot be parked in gear. ALWAYS set the brakes when tractor is stopped.

ADJUSTMENT

All Models

169. Refer to Fig. 168 for an exploded view of "Multipower" linkage and associated parts. The only adjustment required is adjustment of lower control rod (8) to hand shift lever (3 or 3A).

To make the adjustment, move hand lever (3 or 3A) fully upward to "HIGH" position. Loosen clamping bolt in link (6 or 6A) and push lower control rod (8) firmly downward as far as it will go. Retighten clamping bolt to lock the adjustment.

TESTING

All Models

170. To check the "Multipower" pressure on early models, refer to Fig. 170 and proceed as follows:

Tee a suitable pressure gage into line (F) leading from auxiliary pump to filter; with tractor at operating temperature and engine running at 2000 rpm, pressure should be 200-220 psi.

High pressure could indicate a plugged oil filter or cooler; move pressure gage to connection (V) and recheck. If pressure difference between connections (F) and (V) is greater than 40 psi, renew filter cartridge and check oil cooler and lines.

Low pressure in "High-Multipower" range could indicate leakage in clutch passages or clutch, improperly adjusted or malfunctioning regulator valve, or malfunctioning pump. Move selector lever to "Low-Multipower" position and recheck.

If pressure rises to normal when selector lever is moved to "Low-Multipower", check for leakage at "Multipower" clutch or clutch passages.

If equally low pressure is registered in both "Multipower" shift positions,

the trouble is either in the "Multipower" control valve or auxiliary pump. To isolate the control valve as the cause of trouble, disconnect the line (F) leading to oil filter and connect a 2000 psi pressure gage directly to the line leading from transmission

Fig. 170—Left side view of Model MF180 clutch housing showing "Multipower" oil cooler and filter lines.

F. Pump to filter
V. Radiator to valve

Fig. 170A—Pressure gage can be installed in left side cover in IPTO equipped tractors.

case. Start and run tractor ONLY long enough to obtain a pressure reading which should be 500-800 psi. If pressure is as indicated, control valve is at fault; if pressure is low, pump is faulty.

On late models with independent power take-off, pressure gage can be installed in left side cover as shown in Fig. 170A. Measured at the IPTO test point in left side cover, pressure at normal operating speed can vary from approximately 750 psi at 50° F. to 400 psi at 150° F. On all models after introduction of IPTO, pressure measured at transmission return line (F—Fig. 170) should be 250-300 psi.

TRANSMISSION REMOVAL
Model MF175

171. To remove the complete transmission unit, follow the general procedures outlined in paragraph 158 ex-

Fig. 171—Front view of transmission housing showing control valve and associated parts.

1. Control rod	4. Clutch fork
2. Actuating lever	5. Control valve
3. Bracket	6. Input shaft

cept remove transmission top cover and detach "Multipower" pressure line before detaching transmission from rear axle center housing.

Model MF180

172. To remove the complete transmission assembly, first split transmission from engine as outlined in paragraph 153. Remove right step plate and tool box, and left step plate and battery box; then unbolt and remove platform center section. Disconnect brake link rods and clutch link; then unbolt and remove upper foot rests, brake cross shaft and pedals as a unit. Disconnect shift links and remove instrument panel, steering support frame and associated parts. Unbolt and remove transmission top cover, disconnect pressure line from "Multipower" pump, then unbolt and remove transmission from rear axle center housing.

TRANSMISSION TOP COVER
Model MF175

173. Refer to paragraph 159 for removal and installation procedure and to paragraph 160 for overhaul.

Model MF180

174. **REMOVE AND REINSTALL.** To remove the transmission top cover, first drain cooling system and remove hood and side panels. Shut off fuel and remove fuel tank. Disconnect wiring harness at front end, power steering cylinder lines at "Hydramotor", and remove power steering pressure and return lines. Remove right step plate and tool box, left step plate and battery box, and platform center section. Disconnect brake link rods and clutch link, then unbolt and remove upper foot rests, brake cross shaft and pedals as an assembly. Disconnect shift links and remove instrument panel, steering support frame and associated parts as a unit; then unbolt and remove transmission top cover.

Overhaul the cover controls as outlined in paragraph 175. When installing the cover, make sure the levers (15, 16 & 17—Fig. 169) properly engage shift forks and planetary range selector in transmission case and that all plungers shift through the three detent positions, then install and tighten the cover cap screws. When installing steering support and instrument panel, make sure 2nd & 3rd speed link (8) connects the left plunger (10) to center shift arm (4).

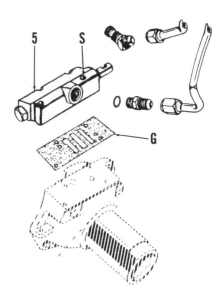

Fig. 172 — Disassembled view of control valve. The bolt which goes in hole (S) retains valve spool and is sealed with a copper washer. Always use a new gasket (G) when reinstalling valve.

Shorten or lengthen the three links (7, 8 and 9) if necessary, until transmission shifts easily and fully to all gear positions. Complete the installation by reversing the removal procedure.

175. **OVERHAUL.** To overhaul the removed transmission cover, refer to Fig. 169. Remove the retaining snap ring and withdraw pivot shaft (18), spacer (14) and levers (15, 16 and 17). Remove links (7, 8 & 9) and withdraw plungers (10, 11 & 12) downward out of housing (13). "O" ring seals are used for plunger bores and bore for pivot shaft (18).

To disassemble shift linkage, drive out the pin securing lever (1) to shaft (6) and withdraw shaft from bore in steering support. Assemble by reversing the disassembly procedure; and install and adjust as outlined in paragraph 174.

CONTROL VALVE
All Models

176. **REMOVE AND REINSTALL.** To remove the "Multipower" control valve, first detach (split) engine from transmission housing as outlined in paragraph 152 or 153. Remove clutch release bearing, release fork and shafts. On Model 175, remove brake cross shaft. Refer to Fig. 171. Disconnect shift linkage (1) and remove shift bracket (3). Disconnect pressure supply tube from valve. Remove the cap screws securing the pto input shaft and retainer to front wall of

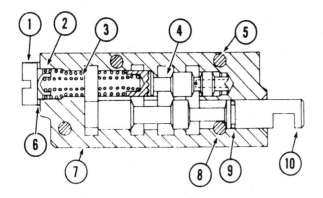

Fig. 173—Cross sectional view of "Multipower" control valve.

1. Regulator valve plug
2. Outer spring
3. Inner spring
4. Regulating valve
5. Short cap screw
6. Washer
7. Valve body
8. Long cap screw
9. "O" ring
10. Valve spool

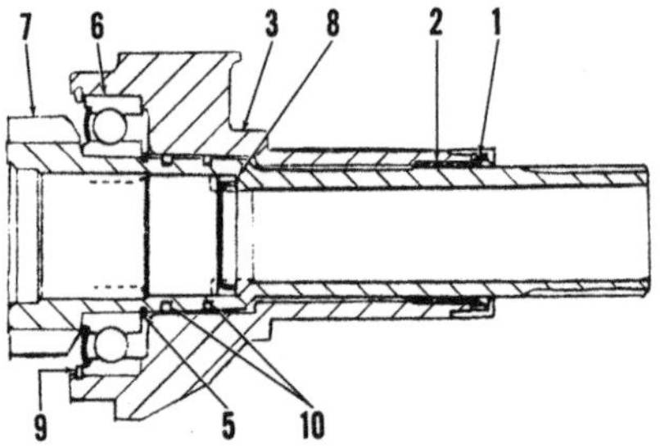

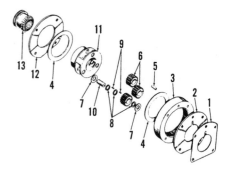

Fig. 175—Cross sectional view of PTO drive shaft and associated parts. Refer to Fig. 174 for parts identification.

transmission housing and withdraw input shaft, retainer and control valve as a unit from transmission housing.

When detaching control valve (5) from input shaft retainer, note that the cap screw which goes in hole (S—Fig. 172) is sealed with a copper washer. This cap screw also retains the control valve spool; do not withdraw screw from hole (S) when removing or installing valve.

Always use a new gasket (G) when installing the valve. Make sure the long mounting cap screw with the sealing washer is installed in hole (S) and tighten all screws to a torque of 36-48 inch pounds.

177. OVERHAUL. To overhaul the removed "Multipower" control valve, refer to Fig. 173 and proceed as follows:

Remove the sealed cap screw (8) and withdraw control valve spool (10). Remove regulator valve plug (1), springs (2 & 3) and valve (4).

Clean all parts in a suitable solvent, discard "O" ring (9) and carefully examine parts for wear, scoring or other damage. All parts are available individually.

When reassembling the valve unit, make sure regulating spool (4) is installed with drilled end and identifying notch to closed end of housing

bore as shown, and complete the assembly by reversing the disassembly procedure using a new "O" ring (9). Install valve in tractor as outlined in paragraph 176, adjust the linkage as in paragraph 169 and test as in paragraph 170.

PTO DRIVE SHAFT AND INPUT SHAFT SEALS

All Models

178. Remove the pto input shaft and housing assembly as outlined in paragraph 176. To disassemble the removed unit, refer to Figs. 174 and 175.

Unseat and remove the large snap ring (9) from groove in rear of housing (3), and bump the pto input shaft (7) and bearing (6) rearward out of housing. Bearing can be removed from the shaft after removing snap ring (5). Seals (1 & 8) are both installed with sealing lips to rear. Inner seal (8) must be carefully positioned in bore of shaft (7), to clear the shoulder of transmission input shaft when unit is installed. (Relative installed position of transmission input shaft is indicated by broken lines, Fig. 175). A special removing tool (MFN 850) and installing tool (MFN 849) are available from Massey-Ferguson, Inc. for service on the seal.

Assemble by reversing the disassembly procedure.

SHIFTER RAILS AND FORKS

All Models

179. The transmission shifter rails and forks used on "Multipower" models are identical to those used on standard model. Shifter rails can be removed as outlined in paragraph 163, however, on Model MF180, transmission must first be removed as in paragraph 172.

NOTE: Before transmission can be detached from rear axle center housing, it will first be necessary to remove hydraulic lift cover or transmission top cover, and disconnect the "Multipower" oil supply line.

REAR PLANETARY UNIT

All Models

180. REMOVE AND REINSTALL. To remove the rear planetary unit, first drain the transmission and hydraulic system reservoir and remove the hydraulic lift cover as outlined in paragraph 221. Reaching through top cover opening, disconnect the "Multipower" pump pressure line leading

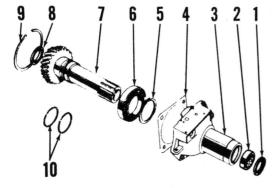

Fig. 174—PTO drive shaft and associated parts used on models with "Multipower" transmission.

1. Oil seal
2. Bushing
3. Housing
4. Gasket
5. Snap ring
6. Bearing
7. PTO drive shaft
8. Oil seal
9. Snap ring
10. Sealing rings

Fig. 176—Exploded view of rear planetary unit and associated parts of the type used on all models.

1. Front shim
2. Front plate
3. Ring gear
4. Thrust washer
5. Dowel
6. Planet pinions
7. Thrust washers
8. Spacers
9. Needle rollers
10. Pinion shaft
11. Planet carrier
12. Rear plate
13. Shift coupler

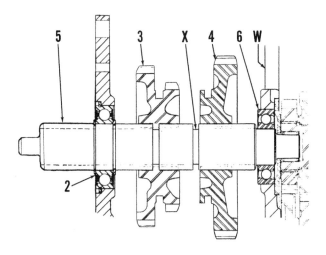

W. Housing wall
X. Interlock groove
2. Bearing
3. Cluster gear
4. Low gear
5. Main shaft
6. Bearing

to oil cooler. On Model MF180, disconnect brake actuating rods. On all models, support transmission and rear axle center housing separately, remove the securing flange bolts and separate the transmission from rear axle center housing.

Remove range shift fork and coupler and the four cap screws securing planetary unit to rear of transmission case and withdraw rear cover plate (12—Fig. 176), thrust washer (4) and planet carrier (11). Using two screw drivers, work the planetary ring gear (3) and dowels (5) from case. Remove planetary front cover (2) and shim (1).

Rear planetary unit used on "Multipower" models is identical to unit used on standard models; refer to paragraph 167 for overhaul procedures. Reinstall the unit by reversing the removal procedure.

TRANSMISSION GEARS AND SHAFTS

All Models

181. For service on any of the transmission gears or shafts except pto input gear or rear planetary unit, it is first necessary to remove the complete transmission assembly as outlined in paragraph 171 or 172, remove the rear planetary unit and shifter rails and forks as previously outlined, then proceed as outlined in the appropriate following paragraphs.

182. **MAIN (OUTPUT) SHAFT.** To remove the main (output) shaft and gears after shift mechanism and rear planetary unit have been removed, refer to Fig. 177 and proceed as follows:

Move the low speed sliding gear (4) forward and insert a large blade

screwdriver or similar flat tool in interlock groove (X) in shaft; then insert a large pry bar between low gear (4) and housing wall (W). Pry shaft, gears and bearings rearward until bearings are free of housing bores. Slide the shaft through the gears to bump front bearing (6) from shaft. Remove shaft and rear bearing rearward while lifting front bearing and gears out top opening.

When reinstalling, place the cluster gear (3) on shaft with smaller gear to front, and low gear (4) on shaft with shift fork groove toward cluster gear. Use a thin spacer plate against housing wall (W) as a support and bump front bearing (6) into position on shaft; then bump the assembled shaft forward into bearing bores of transmission case.

183. **TRANSMISSION INPUT SHAFT AND "MULTIPOWER" CLUTCH.** To remove the transmission main (input) shaft and multipower clutch unit, first remove the transmission main (output) shaft as outlined in paragraph 182 and the "Multipower" control valve and input housing as in paragraph 176. Move input shaft forward slightly and remove thrust spacer (7—Fig. 178), then withdraw input shaft (9B) rearward out of case while lifting clutch (13), gears and associated parts out top opening.

Input shaft (9B) contains a needle roller bearing (8) which pilots on front end of main (output) shaft (5), and cast iron sealing rings (10). Examine sealing rings and needle bearing, and inspect the polished bearing and sealing surfaces of shaft for wear or other damage.

Overhaul the removed "Multipower" clutch as outlined in paragraph 184, and assemble by reversing the disassembly procedure.

184. **"MULTIPOWER" CLUTCH.** To disassemble the removed "Multipower" clutch unit (13—Fig. 178), place unit on a clean bench with overdrive pinion (11) up. Apply slight pressure to clutch retainer plate (2—Fig. 180) and remove snap ring (1) with a narrow blade screwdriver. Completely disassemble the clutch and examine the component parts for wear or scoring. Renew the piston sealing rings (7 & 8) whenever clutch is disassembled.

When installing the piston, carefully compress the outer sealing ring (7), using a narrow blade screwdriver or similar tool, and work the piston

1. Snap ring
2. Bearing
3. Cluster gear
4. Low gear
5. Main shaft
6. Bearing
7. Spacer
8. Needle bearing
9A. Input gear
9B. Input shaft
10. Sealing rings
11. Overdrive pinion
12. Bushing
13. Multipower clutch
14. Thrust washer
15. Thrust washer

Fig. 178—Exploded view of upper shafts, gears and associated parts.

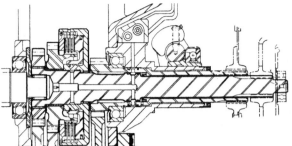

Fig. 179—Cross sectional view of transmission input shaft and "Multipower" clutch assembly.

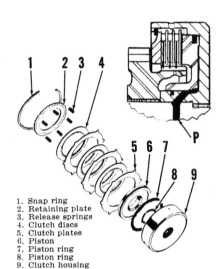

1. Snap ring
2. Retaining plate
3. Release springs
4. Clutch discs
5. Clutch plates
6. Piston
7. Piston ring
8. Piston ring
9. Clutch housing

Fig. 180—Exploded view of "Multipower" clutch and associated parts. Inset shows cross sectional view and pressure passage (P) which enters clutch through the drilled input shafts.

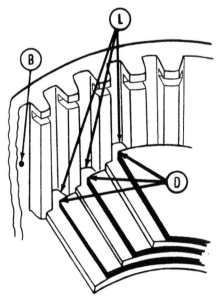

Fig. 181—Cross sectional view of "Multipower" clutch showing recommended method of assembly. Refer to text.

B. Bleed hole
D. Clutch discs
L. Drive lugs on clutch plates

Fig. 182—Cross sectional view of transmission countershaft and jaw-type over-running clutch. Refer to Fig. 183 for parts identification.

into its bore. The inner ring (8) will normally compress because of chamfer in inner bore of piston, if care is used in assembly.

When assembling clutch plates, note that clutch drum (9) contains six bleed holes which are evenly spaced, and that driving plates (5) have six external driving lugs. With piston installed, refer to Fig. 181 and install the first drive plate on top of piston with lugs (L) one spline clockwise from bleed holes (B). Install an internally splined clutch disc (D), then the second plate with drive lug (L) one spline clockwise from lug of first plate. Repeat the procedure for remainder of discs and plates. Place the piston return springs (3—Fig. 180) on the driving lugs of first plate installed, then install retainer plate (2) and snap ring (1).

185. COUNTERSHAFT ASSEMBLY. To remove the countershaft (7—Fig. 182), first remove transmission input shaft as outlined in paragraph 183. Remove the cap screws securing the pto countershaft front bearing plate (16—Fig. 183) and remove the plate. Remove snap ring (17) and spacer (18) on some early models; or snap ring (17—Fig. 182) on late models. On all models, use two ⅜-inch NC cap screws as forcing screws and remove front bearing retainer (21—Fig. 183) and bearing (20) as a unit. Slide pto front drive shaft (25) out rear

of transmission case and countershaft (7), and lift drive gear (23) out top opening.

Use the special Nuday clamping tool (MFN830) or a small "C" clamp to secure the countershaft drive gears (10 & 14) together, then remove snap rings (1 & 15) from each end of countershaft. Insert a step plate of proper size in rear end of countershaft and bump shaft forward slightly until snap ring (5) in front of high speed pinion can be unseated and moved forward on shaft.

With snap ring (5) unseated, insert the step plate in forward end of countershaft and bump shaft rearward until the countershaft driving gears (10 & 14) and over-running clutch (12) can be slipped off forward end of shaft and lifted from transmission.

With drive gears and over-running clutch removed, drive the countershaft forward until free of rear bearing (2); then slide shaft and front bearing forward out of transmission case while lifting the gears (3 and 4) out top opening.

Assemble by reversing the disassembly procedure. The two gears (3 & 4) must be installed with hubs together and the larger gear forward. The countershaft driving gears (10 & 14) and over-running clutch assembly may be installed as a unit by using Nuday Tool MFN-830; or individually if the special tool is not used. If the special tool is used, leave clamping screws slightly loose until splines are engaged, then tighten clamp and leave in place until all snap rings are seated, to minimize bouncing of overdrive gear against snap ring.

186. REVERSE IDLER. The reverse idler assembly is identical to that used in standard transmission. Refer

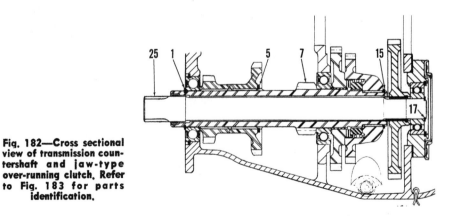

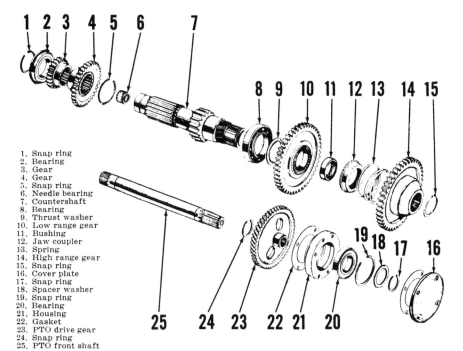

1. Snap ring
2. Bearing
3. Gear
4. Gear
5. Snap ring
6. Needle bearing
7. Countershaft
8. Bearing
9. Thrust washer
10. Low range gear
11. Bushing
12. Jaw coupler
13. Spring
14. High range gear
15. Snap ring
16. Cover plate
17. Snap ring
18. Spacer washer
19. Snap ring
20. Bearing
21. Housing
22. Gasket
23. PTO drive gear
24. Snap ring
25. PTO front shaft

Fig. 183—Exploded view of transmission lower shafts and associated parts. Refer to Fig. 182 for cross sectional assembled view.

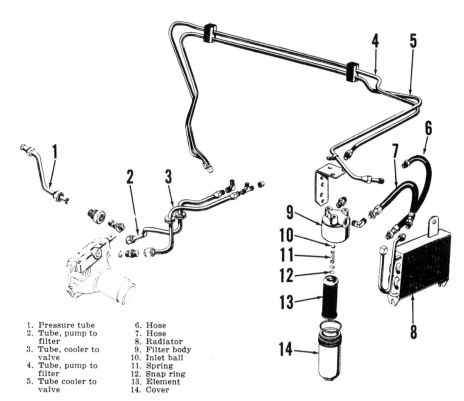

1. Pressure tube
2. Tube, pump to filter
3. Tube, cooler to valve
4. Tube, pump to filter
5. Tube cooler to valve
6. Hose
7. Hose
8. Radiator
9. Filter body
10. Inlet ball
11. Spring
12. Snap ring
13. Element
14. Cover

Fig. 184—Exploded view of "Multipower" oil cooler, filter, lines and associated parts used on all models.

to paragraph 166 for removal and overhaul procedure.

"MULTIPOWER" (Auxiliary) PUMP
All Models

187. Power for the "Multipower" disc clutch is supplied by the gear- type auxiliary pump which mounts on top of the regular hydraulic system pump in rear axle center housing, and also supplies working fluid for external hydraulic cylinders. Refer to paragraph 215 for testing procedure and to paragraph 227 for overhaul.

OIL COOLER AND FILTER
All Models

188. Refer to Fig. 184 for an exploded view of oil cooler, filter and associated parts. Oil filter and oil cooler radiator operate continuously at "Multipower" system pressure and flow. Make sure units are in good condition and connections are tight. Refer to paragraph 170 for test procedure.

DIFFERENTIAL, BEVEL GEARS AND FINAL DRIVE

All models are equipped with a planetary final drive unit located at outer end of rear axle housings. A mechanically actuated, jaw-type differential lock is standard equipment.

DIFFERENTIAL
All Models

189. **REMOVE AND REINSTALL.** The ring gear and differential unit can be removed after removing the complete left final drive unit as outlined in paragraph 198.

Correct carrier bearing pre-load is determined when tractor is assembled, and the correct thickness bearing spacer shield (3—Fig. 185) is selected to provide the proper adjustment. Shield (3) is available in thicknesses of 0.030, 0.035, 0.040, 0.045 and 0.050. An 0.005 thick shim is also available which can be used between shield and bearing cup (4) instead of a thicker shield. Differential bearings should be adjusted to approximately zero end play. The recommended method of checking the adjustment if major parts are renewed, is by use of special tool MFN-245Y as shown in Fig. 186. Checking and/or adjustment is not usually necessary unless major parts are renewed.

Main drive bevel gear backlash should be 0.003-0.019 and is not adjustable. Backlash can be measured through opening in differential housing wall (Fig. 187) after removing hydraulic lift top cover.

1. Oil seal
2. Carrier plate
3. Spacer shield
4. Bearing cup
5. Fork, differential lock
6. Coupling half
7. Bearing cone
8. Coupling half
9. Differential case half
10. Thrust washers
11. Axle gears
12. Differential spider
13. Differential pinions
14. Thrust washers
15. Pilot bearing
16. Main drive bevel gears
17. Differential case half
18. Bearing cup
19. Bearing cone
20. Carrier plate
21. Bearing cone
22. Bearing sleeve
23. Bearing cone
24. PTO gear hub
25. Tab washer
26. Adjusting nut
27. PTO gear
28. Snap ring

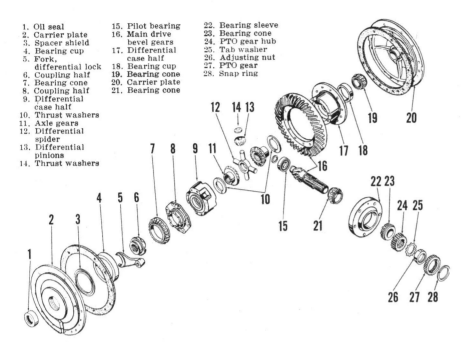

Fig. 185—Exploded view of differential, final drive bevel gears and associated parts.

Fig. 187—Rear axle center housing with hydraulic lift cover removed. Opening permits inspection of main drive bevel gears and checking gear backlash.

190. OVERHAUL. To disassemble the removed differential unit, first place correlation marks on both halves of differential case to insure correct assembly. Remove the eight retaining bolts, lift off the differential lock coupling half (8—Fig. 185) and separate the case halves (9 & 17). Differential pinions (13), spider (12) and side gears (11) can now be removed. Recommended backlash of 0.003-0.008 between the differential pinions and side gears is controlled by the thrust washers (10 & 14).

Grade CV (Blue) LOCTITE is used to secure the bolts and nuts retaining the main drive bevel ring gear to differential case. When reassembling, use LOCTITE of the appropriate grade. Tighten the gear retaining bolts and nuts to a torque of 110-120 ft.-lbs. Tighten the differential case bolts to a torque of 75-85 ft.-lbs.

DIFFERENTIAL LOCK

All Models

191. OPERATION. The mechanically actuated differential lock assembly is standard equipment. When the differential lock foot pedal is depressed, the axle half of coupler is forced inward to contact the differential case half of coupler. If slippage is occuring at one wheel, depressing the pedal will cause the coupler dogs to lock the differential case to the right axle. The differential and both drive wheels then rotate together as a unit. As soon as contact is made by the coupler dogs, the pressure will keep the differential lock engaged and foot pedal may be released. When ground traction on both wheels again becomes equal, coupler dog contact pressure will be relieved and the coupler will automatically disengage Refer to Fig. 188 for an exploded view of the differential lock actuating mechanism and associated parts.

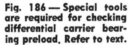

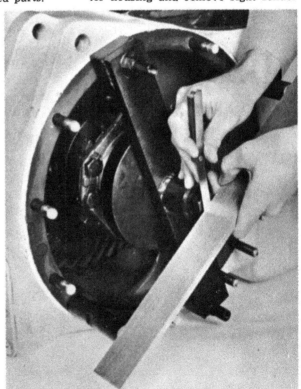

Fig. 186 — Special tools are required for checking differential carrier bearing preload. Refer to text.

192. ADJUSTMENT. The differential lock coupler should be fully engaged when pedal pad (2 or 2A—Fig. 188) clears tractor step plate or platform by ¼-inch. If adjustment is required, loosen the clamp screw (4) and reposition pedal (3 or 3A) on actuating shaft (6). Retighten clamp screw when adjustment is complete.

193. REMOVE AND REINSTALL. To remove the differential lock coupler halves (20 & 22—Fig. 188), first drain transmission and hydraulic reservoir, block up under rear axle center housing and remove right fender

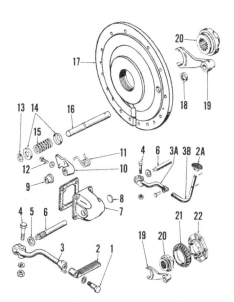

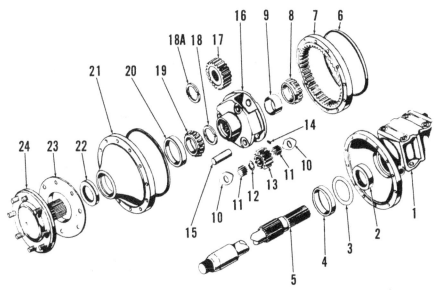

Fig. 188 — Exploded view of differential lock and associated parts.

1. Bolt	10. Actuating cam
2. Pad (MF175)	11. Return spring
2A. Pad (MF180)	12. Set screw
3. Pedal (MF175)	13. Snap ring
3A. Pedal (MF180)	14. Spring guide
3B. Extension	15. Release spring
(MF180)	16. Actuating shaft
4. Clamp bolt	17. Carrier plate
5. Spacer washer	18. Oil seal
6. Pedal shaft	19. Shifter fork
7. Cam housing	20. Sliding coupling
8. Expansion plug	21. Bearing cone
9. Thrust bushing	22. Coupling half

Fig. 190—Exploded view of planetary final drive unit and associated parts of the type used on all models.

1. Axle housing	7. Ring gear	13. Planet pinion	19. Bearing cone
2. Oil seal	8. Bearing cone	14. Lock screw	20. Bearing cup
3. Adjusting shim	9. Bushing	15. Pinion shaft	21. Drive cover
4. Bearing cup	10. Thrust washer	16. Planet carrier	22. Oil seal
5. Axle shaft	11. Needle rollers	17. Sun gear	23. Dust shield
6. Gasket	12. Spacer washer	18. Snap ring	24. Wheel axle
		18A. Segmented ring	

and rear tire and wheel assembly. Remove right lower hitch link and disconnect right brake linkage. Support right rear axle housing assembly in a hoist and remove retaining stud nuts; then slide right rear axle and housing as a unit away from rear axle center housing.

Remove bearing cone (21) from coupling half (22); remove the differential case retaining cap screws and lift off the coupling (22). When installing, tighten differential case cap screws to a torque of 75-85 ft.-lbs.

To remove axle half coupler and actuating mechanism, drive out the

Fig. 189—Installed view of actuating shaft and associated parts. Refer to Fig. 188 for parts identification.

roll pin (P—Fig. 189) and remove the two screws securing carrier plate (17) to axle housing; then slip carrier plate (17), fork (19) and sliding coupler (20—Fig. 188) as a unit from axle. Remove cam housing (7) and brake lever support from axle housing and, working through hole in axle housing, unseat snap ring (13); then withdraw shaft (16), spring (15) and associated parts.

Assemble by reversing the disassembly procedure and adjust as outlined in paragraph 192 after axle housing is reinstalled.

BEVEL GEARS

All Models

194. **BEVEL PINION.** The main drive bevel pinion is available only in a matched set which also includes the bevel ring gear and attaching bolts and nuts. To remove the bevel pinion, first remove the hydraulic lift cover as outlined in paragraph 221 and proceed as follows:

Working through the top opening in rear axle center housing, remove the large cotter pin and collapse and remove the rear driveshaft assembly. Unbolt pinion bearing sleeve (22—Fig. 185) from center housing wall and, using two jack screws in the tapped holes, pull the pinion and sleeve assembly forward out of center housing bore.

To disassemble the unit, remove snap ring (28) and gear (27). Unlock and remove nut (26) and bump pinion out of sleeve (22). If bearing cups in sleeve are damaged, renew the sleeve.

NOTE: This requires sleeve and one bearing cone when substituting late type for early type. Check with parts department when ordering.

When reassembling, tighten the nut (26) to obtain a rolling torque of 18-22 inch-pounds for the pinion shaft bearings. Install pinion assembly by reversing removal procedure.

195. **BEVEL RING GEAR.** The main drive bevel ring gear is only available in a matched set which also includes the pinion and attaching bolts and nuts.

To remove the main drive bevel ring gear, first remove differential assembly as outlined in paragraph 189. Ring gear retaining nuts are installed with LOCTITE, Grade CV, and removal might require a slight amount of heat.

When installing the ring gear, make sure that mating surfaces of ring gear and differential case are thoroughly clean and free from nicks and burrs. Use two drops of LOCTITE, Grade CV (blue) on each attaching bolt and tighten the nuts to a torque of 100-120 ft.-lbs. Runout of ring gear should not exceed 0.002.

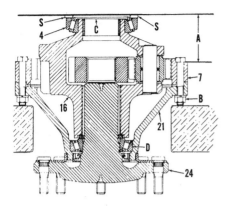

Fig. 191—Using spacers, a straightedge and inside micrometer to determine bearing adjustment as outlined in paragraph 196.

A. Measure distance
B. Bolts
C. Bearing cage
D. Outer bearing
S. Spacer
4. Bearing cup
7. Ring gear
16. Planet carrier
21. Drive cover
24. Wheel axle

REAR AXLE AND FINAL DRIVE
All Models

196. ADJUSTMENT. Planet carrier bearing preload of 0.006-0.015 is adjusted by means of shims (3—Fig. 190) placed in bore of axle housing (1) behind bearing cup (4). Shims are available in thicknesses of 0.005, 0.010 and 0.015. Two methods of adjustment can be used as follows:

Remove wheel axle and planet carrier assembly as outlined in paragraph 199. Remove bearing cup (4) from axle housing (1). Using four bolts of the correct length, bolt planetary ring gear (7) to drive cover (21) as shown in Fig. 191. Support the assembly on drive cover so that wheel axle (24) does not touch, allowing outer carrier bearing (D) to seat.

Position the removed bearing cup (4) over bearing cone as shown. Place two spacers (S) of known and equal thickness on bearing cup to clear the bearing cage (C); then use a straight edge and inside micrometer to measure the distance (A) from straight edge to surface of ring gear flange. Obtain a measurement from each side of bearing, turn straightedge 90° and remeasure; then add the measurements and divide by four to obtain an average. Subtract the thickness of spacers (S) to obtain the distance from bearing cup to ring gear flange. If the distance is 2.948-2.952, no shims (3—Fig. 190) are required. If distance is less than 2.948, add a sufficient quantity or thickness of shims (3) to bring the measurement within the range of 2.948-2.952. Reinstall bearing cup (4) using the proper shims, then reinstall wheel axle and planet car-

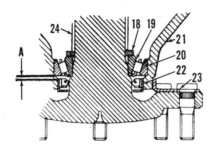

Fig. 192—When installing seal (22), make sure clearance (A) between seal and bearing cup (20) is 0.010. Refer to Fig. 190 for parts identification.

rier assembly as outlined in paragraph 199.

An alternate method of adjustment is to remove the entire final drive unit as outlined in paragraph 198 and stand unit on end with wheel axle up. Remove wheel axle and planet carrier assembly, bearing cup (4) and shim pack (3); then reinstall bearing cup and wheel axle and planetary unit omitting the shim pack. Attach a dial indicator with indicator button contacting wheel axle flange and measure the end play by lifting the wheel axle and planetary unit using two pry bars or screwdrivers. Install a shim pack equal in thickness to the measured end play plus 0.006-0.015, and reassemble by reversing the disassembly procedure.

197. OVERHAUL. Left planetary unit can be overhauled after removing wheel axle and planet carrier assembly as outlined in paragraph 199 and locking the brake to permit withdrawal of main axle shaft (5—Fig. 190). If right main axle shaft is to be removed, the complete final drive unit must be removed as outlined in paragraph 198 to correctly install the differential lock sliding coupling. If main axle shaft is not removed, remainder of right unit may be overhauled after removing wheel axle and planet carrier assembly as outlined in paragraph 199.

To disassemble the removed wheel axle and planet carrier unit, loosen locking set screw (14) and remove planet pinion shaft (15), pinion gear and associated parts from largest opening in planet carrier (16); then slide sun gear (17) out the opening.

Wheel axle (24) is a tight press fit in splines of planet carrier (16), and a three leg puller or press of 20-40 tons capacity is required for removal. If a press is used, remove the remaining planet pinions (13) and support

the planet carrier. DO NOT attempt removal by supporting drive cover (21) in the press.

Remove snap ring (18) or ring segments (18A) from axle shaft, then press axle shaft downward out of bearing cone (19).

The seal bore in drive cover (21) is not shouldered. When installing the seal, refer to Fig. 192 and install seal so that 0.010 clearance exists between inner edge of seal (22) and outer edge of bearing cup (20) as shown at (A).

With seal installed, carefully position drive cover (21) over axle shaft and install and seat outer bearing cone (19). Snap ring (18) and split ring (18A—Fig. 190) are interchangeably used and are variable in thickness. The snap ring (18) is not available for service. Split ring (18A) is available in nine thicknesses from 0.230 to 0.248. With axle and bearing assembled, use the thickest split ring which will seat in snap ring groove.

Press the planet carrier on wheel axle shaft using a suitable press, making sure segments of split ring (18A) properly enter counterbore in carrier hub.

Complete the assembly by reversing the disassembly procedure, adjust carrier bearings as outlined in paragraph 196 and install as in paragraph 198 or 199.

198. R&R COMPLETE FINAL DRIVE. To remove either final drive assembly as a unit, first drain the transmission, suitably support rear of tractor and remove the fender and rear wheel and tire unit. Disconnect the lower lift link and brake actuating lever. On right side of Model MF180, remove upper differential lock pedal link. Support final drive assembly from a hoist, remove the attaching stud nuts and separate the final drive unit from rear axle center housing.

Use only one standard gasket when reinstalling, and tighten the retaining stud nuts to a torque of 50-55 ft.-lbs.

199. R&R WHEEL AXLE & PLANET CARRIER ASSEMBLY. To remove the wheel axle and planet carrier as a unit, suitably support the tractor and drain the final drive planetary housing. Remove the wheel and tire unit and fender assembly. Securely lock the brake on side to be removed. Remove the securing bolt circle and, using suitable hoisting equipment, move wheel axle and drive cover straight out away from drive axle housing.

Install by reversing the removal procedure. Tighten the retaining bolts to a torque of 50-55 ft.-lbs.

BRAKES

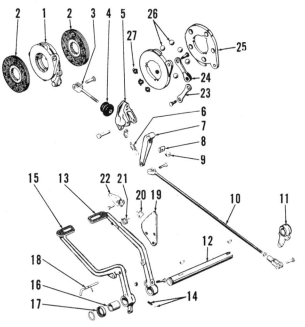

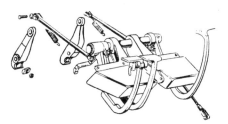

Fig. 193—Exploded view of disc type individual wheel brakes used on Model MF175. Pedal and linkage components for Model MF180 are shown in Fig. 194.

1. Actuating disc
2. Lined discs
3. Brake rod
4. Dust cover
5. Support
6. Spring
7. Brake lever
8. Adjusting block
9. Adjusting nut
10. Brake rod
11. Brake lever
12. Cross shaft
13. Brake pedal
14. Detent
15. Brake pedal
16. Bushing
17. Thrust washer
18. Interlock lever
19. Sector
20. Pawl
21. Spring
22. Latch
23. Link
24. Link
25. Actuating disc
26. Steel balls
27. Return springs

Fig. 194—Partially assembled view of brake cross shaft and linkage used on Model MF-180. Internal brake parts are identical to those shown in Fig. 193.

ADJUSTMENT

All Models

200. To adjust the disc-type brakes, turn each adjusting nut (9—Fig. 193) either way until brake pedal free play is 2½-3 inches when measured at pedal pad. Adjust both pedals equally. On Model MF175, pedal height may be equalized by adjusting link rods (10).

Refer to Fig. 194 for cross shaft and linkage used on Model MF180.

OVERHAUL

All Models

201. To remove the brake assemblies, first remove differential lock

(right side) as outlined in paragraph 193 or complete final drive unit (left side) as in paragraph 198. On left side, remove the retainer screws and lift off the differential carrier plate.

On all models, remove adjusting nut (9—Fig. 193), block (8) and lever support (5); then withdraw lined discs (2) and actuating disc (1) from inner side of axle housing.

Actuating disc assembly (1) can be disassembled after removing brake rod (3) and unhooking the return springs (27). Renew any parts which are questionable and assemble by reversing the disassembly procedure. Adjust the brakes after installation as outlined in paragraph 200.

BELT PULLEY

202. **OVERHAUL.** To overhaul the belt pulley attachment, Refer to Fig. 195 and proceed as follows:

Remove input shaft bearing housing (22) and expansion plug (9). Remove the cotter pin and nut (10) and bump shaft and gear (18) out of housing. Remove pulley (1), nut (2) and hub (5), and bump pinion shaft (17) out of outer bearing and housing.

When reassembling the unit, install pinion shaft (17), bearings and hub (5), and tighten nut (2) until a rolling torque of 2-4 inch pounds is applied to shaft bearings. Secure nut (2) with a cotter pin. Install drive gear (18) and tighten nut (10) until the bevel gears have a backlash of 0.004-0.006; then back off the nut if necessary, until cotter pin can be inserted. Fill to level of filler plug with SAE 90 gear oil.

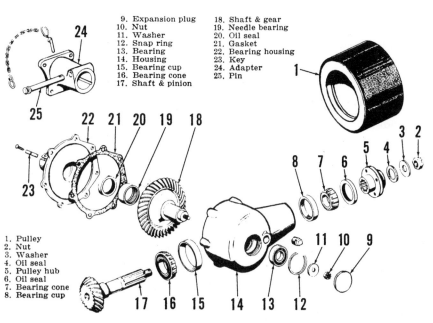

9. Expansion plug
10. Nut
11. Washer
12. Snap ring
13. Bearing
14. Housing
15. Bearing cup
16. Bearing cone
17. Shaft & pinion
18. Shaft & gear
19. Needle bearing
20. Oil seal
21. Gasket
22. Bearing housing
23. Key
24. Adapter
25. Pin

1. Pulley
2. Nut
3. Washer
4. Oil seal
5. Pulley hub
6. Oil seal
7. Bearing cone
8. Bearing cup

Fig. 195—Exploded view of pto mounted belt pulley attachment available for all models.

CONTINUOUS POWER TAKE-OFF

OUTPUT SHAFT

All Models

203. To remove the pto output shaft, drain the transmission and hydraulic system fluid and unbolt rear bearing retainer (17—Fig. 196) from rear axle center housing. Withdraw shaft (11) rear bearing (13) and seal housing (15) as an assembly.

Rear bearing (13), seal (14) or sealing "O" ring (16) can be renewed at this time, as can shaft (11) or seal retainer (15).

Output shaft needle bearing (10) is located in rear axle center housing wall and can be renewed after first removing hydraulic system pump as outlined in paragraph 225.

GROUND SPEED GEARS

All Models

204. To remove the ground speed drive gear, first remove the hydraulic lift cover as outlined in paragraph 221 and proceed as follows:

Working through the top opening in rear axle center housing, collapse and remove the rear drive shaft as-

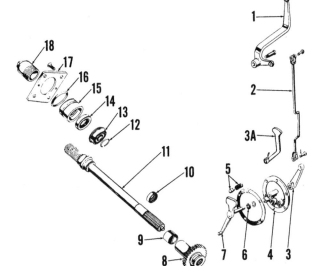

Fig. 196—Exploded view of PTO output shaft and associated parts.

1. Lever (MF180)
2. Link (MF180)
3. Lever (MF180)
3A. Lever (MF175)
4. Shift cover
5. Detent
6. Oil seal
7. Shift fork
8. Gear & coupling
9. Bushing
10. Needle bearing
11. Output shaft
12. Snap ring
13. Bearing
14. Oil seal
15. Seal retainer
16. "O" ring
17. Retainer
18. Cap

sembly. Remove the retaining snap ring and slide the gear forward off of drive pinion.

To remove the driven gear (8—Fig. 196) after drive gear is out, remove the hydraulic pump as outlined in paragraph 225, remove pto shift cover (4) and associated parts, and slide the gear forward out of bushing (9). Assemble by reversing the disassembly procedure.

PTO MAIN DRIVE GEARS

All Models

205. The pto main drive gears and clutch are included in transmission drive train. Refer to the appropriate transmission and clutch section for removal and overhaul procedure.

INDEPENDENT POWER TAKE-OFF

Tractors may be optionally equipped at the factory with an Independent Power Take-Off which is driven by a flywheel mounted "Split-Torque" clutch and controlled by a hydraulically actuated multiple disc clutch contained in rear axle center housing.

OPERATION

All Models So Equipped

206. On models with Independent Power Take-Off, the pto drive shaft is splined into a hub contained in the Split Torque Clutch Cover (paragraph 157) and turns continuously when engine is running.

The IPTO control lever is mounted on left side cover of rear axle center housing as shown in Fig. 197. Moving lever forward disengages the hydraulically actuated multiple disc clutch and engages the brake. Moving lever rearward releases the hydraulic brake and engages the multiple disc IPTO clutch.

Standby hydraulic pressure to actuate the hydraulic clutch and brake is provided by the auxiliary gear type pump which also provides power for the Multipower clutch and/or auxiliary hydraulic system if tractor is so equipped.

REMOVE AND REINSTALL

All Models So Equipped

207. **OUTPUT SHAFT.** To remove the IPTO output shaft, first drain transmission & hydraulic system fluid then unbolt retainer plate (2—Fig. 198) from rear axle center housing. Remove protective cap (1). Insert a punch or similar tool in hole in shaft (8) and withdraw shaft, bearing, retainer and associated parts.

Rear bearing (6), seal (5) or sealing O-ring (3) can be renewed at this time. Front needle bearing (9) is contained in center housing wall and renewal requires splitting tractor and

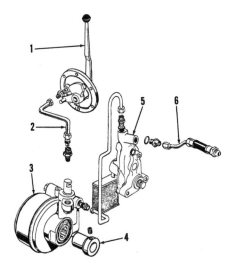

Fig. 197—Schematic view of IPTO clutch and associated parts used on some models.

1. Control lever
2. Pressure line
3. Multiple disc clutch
4. Retainer sleeve
5. Auxiliary pump
6. Multipower pressure hose

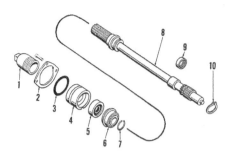

Fig. 198—IPTO output shaft showing component parts.

1. Cap
2. Retainer
3. O-ring
4. Seal retainer
5. Oil seal
6. Output bearing
7. Snap ring
8. Output shaft
9. Needle bearing
10. Snap ring

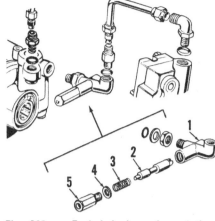

Fig. 201 — Exploded view of regulating valve used on IPTO equipped models without Multipower.

1. Body
2. Spool
3. Spring
4. Lockwasher
5. Plug

Fig. 199—Installed view of IPTO clutch with pump package removed.

Fig. 200—Adjust clutch clearance after installation by moving retainer sleeve (4—Fig. 197).

sleeve (4—Fig. 197) and adjust clearance between sleeve and clutch hub to 0.005-0.015 by moving the sleeve forward or rearward as required. See Fig. 200. Tighten the set screw after adjustment is obtained.

209. IPTO REGULATING VALVE. On "Multipower" transmission models, the IPTO regulating valve is combined with the "Multipower" Control Valve and removal procedure is contained in paragraph 176. On models without "Multipower" transmission, the regulating valve is installed in pressure line leading from pump to clutch valve as shown in Fig. 201. To remove or service the valve, remove hydraulic lift cover as outlined in paragraph 221.

OVERHAUL

All Models So Equipped

210. OUTPUT SHAFT. With output shaft removed from tractor as outlined in paragraph 207, seal retainer (4—Fig. 198) can be withdrawn toward rear. Seal (5) bottoms against shoulder in retainer bore and is installed with lip to front.

Rear bearing (6) is a press fit on shaft and is secured by retaining ring (7).

211. REGULATING VALVE. Fig. 201 shows an exploded view of IPTO regulating valve used on models without "Multipower" transmission. Plug (5) is strongly spring loaded and should

removing clutch pack as outlined in paragraph 208.

Install by reversing the removal procedure. Apply pressure to seal retainer (4) as it enters housing, to prevent shaft being pushed forward out of retainer and damaging seal lip.

208. IPTO CLUTCH UNIT. To remove the IPTO clutch and valve unit, first split tractor and remove the hydraulic pump package as outlined in paragraph 225. Remove pto shift cover (housing side cover), then rock top of clutch unit to left and remove through center housing front opening.

Reinstall by reversing the removal procedure. After hydraulic pumps are installed, loosen set screw in retainer

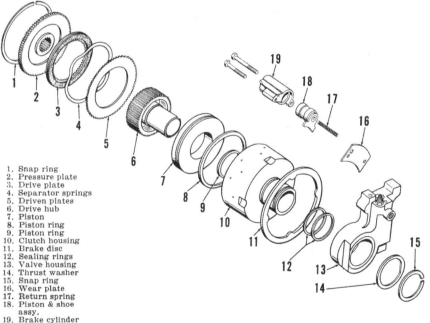

1. Snap ring
2. Pressure plate
3. Drive plate
4. Separator springs
5. Driven plates
6. Drive hub
7. Piston
8. Piston ring
9. Piston ring
10. Clutch housing
11. Brake disc
12. Sealing rings
13. Valve housing
14. Thrust washer
15. Snap ring
16. Wear plate
17. Return spring
18. Piston & shoe assy.
19. Brake cylinder

Fig. 202—Exploded view of IPTO multiple disc clutch and hydraulic brake unit.

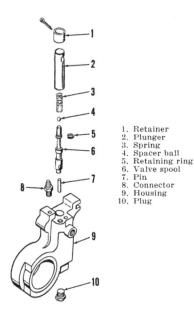

1. Retainer
2. Plunger
3. Spring
4. Spacer ball
5. Retaining ring
6. Valve spool
7. Pin
8. Connector
9. Housing
10. Plug

Fig. 203 — Exploded view of modulating valve, valve housing and associated parts.

be removed with care if disassembly is attempted. All parts are available individually except valve body (1). If body is damaged, renew valve assembly.

On models with "Multipower" transmission, IPTO regulating valve is contained in the "Multipower" control valve and overhaul procedure is contained in paragraph 177.

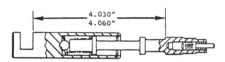

Fig. 204—Collapsed length of modulating valve should be 4.030-4.060 as shown. Length is adjusted by installing a different size spacer ball (4—Fig. 203).

212. IPTO CLUTCH & VALVE UNIT. To disassemble the removed IPTO clutch and valve unit, place the assembly on a bench, valve (front) side down. Remove snap ring (1—Fig. 202), drive plates (3), separator springs (4) and driven plates (5). Lift out drive hub (6). Using two pairs of pliers, grasp strengthening ribs of piston (7) and lift out piston.

Unseat and remove snap ring (15) and lift off valve housing (13) and associated parts. Brake disc (11) will be removed with valve housing. Brake cylinder and associated parts can be disassembled after removing the retaining cap screws.

To disassemble the modulating valve (parts 1 through 6—Fig. 203) apply slight pressure while unseating the internal expanding snap ring (5). Pin (7) is used only in early models.

Spacer ball (4) is available in alternate diameters of $\frac{1}{4}$, $\frac{9}{32}$, $\frac{5}{16}$ and $\frac{11}{32}$ inch. Ball is used to establish collapsed length of modulating valve within the recommended 4.030-4.060 as shown in Fig. 204.

When assembling the clutch unit, use tool kit (MFN—768) for installing the piston as shown at (T—Fig. 205). If assembly tool kit is not available, cut 5 or 6 two-inch pieces of $\frac{3}{16}$-inch rod to serve as a guide for piston installation. With piston in place, install center hub (6—Fig. 202) and, beginning with externally splined separator plate (5), alternately install seven separator plates (5), six wave springs (4) and six friction discs (3). Push down on the last separator plate to compress the wave springs and insert two Allen wrenches (or rod ends) in bleed holes (B—Fig. 205) as shown in Fig. 206, to hold the plate in place. Install the remaining friction disc and wave spring, then install pressure plate (2—Fig. 202) and snap ring (1). The restraining Allen wrenches can be removed at this time.

NOTE: If new friction discs are installed, they should be soaked for 30 minutes in transmission and hydraulic fluid before installation.

Install the assembled clutch unit as outlined in paragraph 208.

Fig. 207—Front view of assembled IPTO clutch unit.

1. Modulating valve
2. Snap ring
3. Thrust washer
4. Clutch housing
5. Brake disc

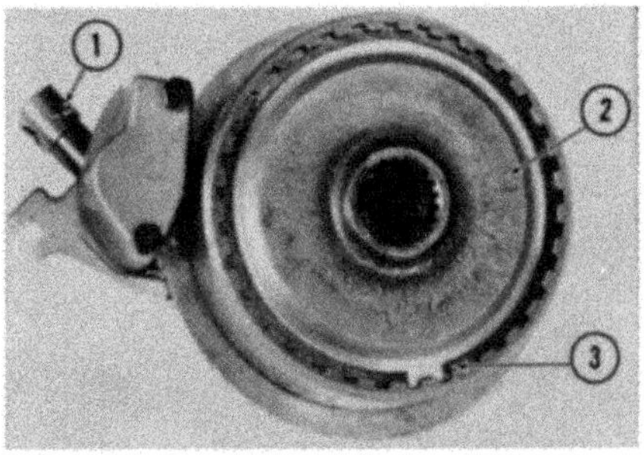

Fig. 208 — Rear view of assembled IPTO clutch unit.

1. Cotter pin
2. Pressure plate
3. Snap ring

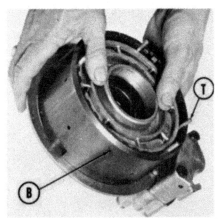

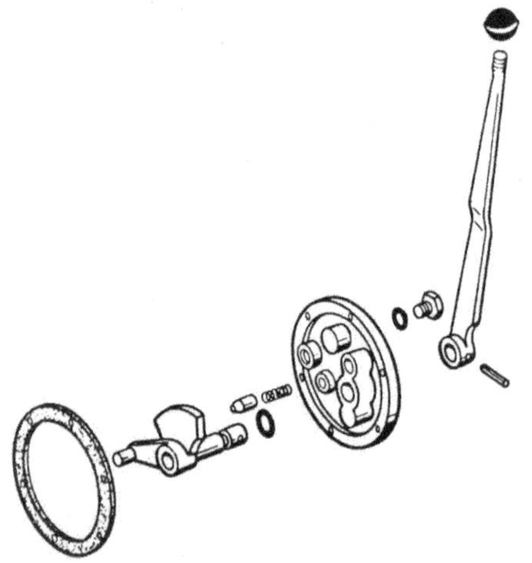

Fig. 205—Use assembly tool kit (T) to install piston as shown. Oil bleed holes (B) are used in assembly as outlined in text and shown in Fig. 206.

Fig. 206—Insert Allen wrenches or rod ends in bleed holes (B—Fig. 205) to hold separator springs compressed during clutch assembly. Refer to text.

Fig. 209 — Exploded view of IPTO cover showing lever components and detent plunger.

Fig. 210—Move IPTO control lever forward to disengage IPTO clutch and rearward to engage clutch.

213. HYDRAULIC PUMP. Power for the IPTO clutch and brake is supplied by the gear type auxiliary pump which also supplies the "Multipower" transmission unit and/or auxiliary hydraulic system if tractor is so equipped. Refer to paragraph 227 for overhaul procedures.

HYDRAULIC

SYSTEM

The hydraulic system consists of a pto driven piston type pump which is submerged in the operating fluid; and a single acting ram cylinder enclosed in the same housing. A control valve is located in the pump unit which meters the operating fluid at pump inlet. The rockshaft position can be automatically controlled by compression or tension on the upper implement attaching link, by a cam on rockshaft ram arm, or by pressure in the ram cylinder; to control rockshaft height and/or to transfer implement weight to rear tires for additional traction.

A gear-type auxiliary pump supplies pressure for the "Multipower" clutch and pressure and flow for remote hydraulic cylinder applications.

The transmission lubricant is the operating fluid for the hydraulic system. Massey-Ferguson M-1127 Fluid is recommended. System capacity is 8 U. S. Gallons.

TROUBLE SHOOTING
All Models

214. SYSTEM CHECKS. Before attaching an implement to tractor, start the engine, move response control lever to "Fast" position and position control lever to "Transport" position. With engine running at slow idle speed, check to make sure that rockshaft moves through full range of travel as draft control lever is moved to "Down" and "Up" positions. Using the draft control lever, stop and hold the movement with lower links in an approximately horizontal position. Lever should be centered between the sector marks on quadrant.

Attach an overhanging implement such as a plow to the links. With draft control lever in "UP" position and engine running at slow idle speed, raise and lower the implement a little at a time, using the position control lever. Implement should move in response to the lever and hold steady after completion of movement, through full range of rockshaft travel. Move lever to "Transport" position and scribe a line across lift arm hub and lift cover. Move position control lever past the transport stop to "Constant Pumping" position. The scribed lines should be separated $\frac{1}{16}$-$\frac{1}{8}$ inch.

Hold the implement clear of the ground using "Draft Control" lever. If draft control spring is properly adjusted, system should respond to the application of pressure or lifting force to rear of implement, lowering when pressure is applied and raising when rear of implement is lifted.

Check pressure and flow as outlined in paragraph 215. If system fails to perform as indicated, adjust as outlined in the appropriate following paragraphs and/or overhaul the system.

215. PRESSURE AND FLOW. Two different hydraulic pumps have been used. On early tractors, relief valve setting was 2350 psi and rated delivery was 4.5 gpm @ 2000 engine rpm & 1500 psi. On late models, relief pressure has been increased to 2700-3200 psi and pump delivery to 4.8 gpm. On models with pressure control, the system relief valve is not used and the pressure control mechanism serves in the added capacity of maximum pressure relief.

The gear-type auxiliary pump may be any of several types, depending on date of manufacture and type of auxiliary uses. Only those with auxiliary hydraulic system affect the hydraulic lift system adjustments and tests. Refer also to paragraph 170 for pressure tests on "Multipower" and IPTO hydraulic system circuits. On early models with auxiliary hydraulic system and "Multipower", the gear-type pump supplied a priority flow of 2 gpm to the "Multipower" control valve and 7 gpm to the auxiliary hydraulic system at 2000 engine rpm. Late models with auxiliary hydraulic system plus "Multipower" and/or In-

Fig. 211—Right side view of operator's platform and associated parts on Model MF180.

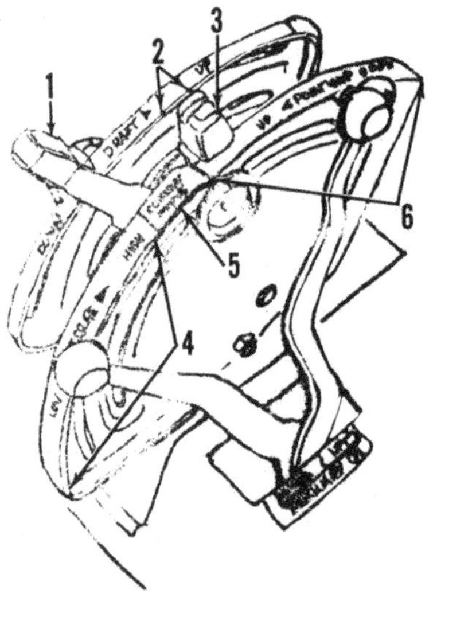

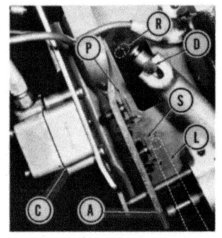

Fig. 212—Hydraulic lift control quadrant showing operating areas for the various functions.

1. Draft control lever
2. Sector marks
3. Position control lever (Hydralever)
4. Pressure control sector
5. Constant pumping sector
6. Position control sector

dependent Power Take-Off use a dual section, gear type pump, with the small set of gears supplying 4 gpm to the "Multipower"/IPTO control circuits and the large set of gears supplying 8 gpm to the auxiliary hydraulic system. A combining valve is optionally available to combine the flow from both hydraulic system pumps for auxiliary or remote hydraulic use.

On models with Pressure Control, main pump pressure can be tested by installing a suitable pressure gage in pipe plug opening (P—Fig. 216) in hydraulic lift cover. With gage installed and engine running at slow idle speed, move draft control lever (Fig. 212) to transport position and Hydralever (3) to extreme "Low" end of pressure range (4). Gage pressure should read approximately 2 psi.

Move Hydralever (3) up the quadrant a little at a time while watching

gage. Pressure reading should increase at an even rate as lever is moved until lever reaches "Constant Pumping" range. Maximum reading should be 2350 psi on some early models or 2950-3100 psi on most models. If pressure is not as indicated, adjust as outlined in paragraph 218 or overhaul as in paragraph 224.

Auxiliary pump pressure and flow can be tested by installing suitable test equipment in breakaway couplings. On models equipped with combining valve, pressure and flow of both systems can be tested by connecting a flow meter to breakaway couplings. Tests should be conducted at rated engine speed (2000 rpm) with transmission and hydraulic fluid at operating temperature, proceed as follows:

Move inner quadrant control lever to "Down" in Position Control Range or close the combining valve. Move auxiliary valve lever to "Raise" position and check auxiliary pump flow which should be approximately 7 gpm for early models or 8 gpm for models with late pump. Move inner quadrant control lever to "Constant Pumping" position and/or open the combining valve and check the combined flow.

Close the restrictor valve on flow meter while observing pressure and flow. Relief pressure of the main hydraulic pump will be indicated when flow drops. To check the relief pressure of auxiliary pump, isolate the main pump by closing the combining valve.

ADJUSTMENTS

All Models

216. **CONTROL SPRING ADJUSTMENT.** To check the master control spring, disconnect spring link as shown in Fig. 213 and check for end play by pushing and pulling on the clevis. If end play is present, loosen the Allen head set screw (S) in side of housing and unscrew the retainer nut (N), using the special spanner

wrench (FT-358). Withdraw the master control spring assembly as shown in Fig. 214. Remove the groove pin (P) and turn clevis on control spring plunger until spring is snug but can be rotated by hand pressure. Align one of the slots and insert groove pin.

Reinstall control spring assembly and turn retainer nut into housing until all end play is eliminated.

Fig. 215—Internal hydraulic control linkage showing points of final adjustment. Dotted lines indicate approximate positions of draft response cam (R) and valve control lever (L).

A. Control arm
C. Control diaphragm
D. Dashpot adjusting screw
L. Valve control lever
P. Pressure control screw
R. Response cam
S. Adjusting screw

Fig. 213 — Master control spring can be checked by disconnecting the link from lift cover, then pushing and pulling on clevis. To make an adjustment or remove the spring, loosen set screw (S) and turn nut (N).

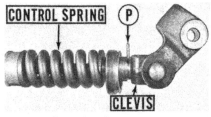

Fig. 214—Control spring assembly removed from cover. Remove pin (P) to turn clevis for spring adjustment.

Fig. 216—Lever support bracket and internal linkage showing points of adjustment.

A. Adjusting screw
L. Locknut
N. Locknut
P. Port plug
S. Adjusting screw
T. Adjusting tube

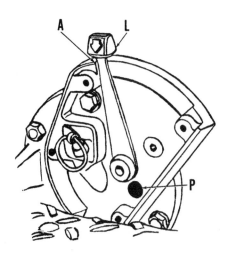

Fig. 217—To adjust the draft response, remove control cover as shown. Move lever (L) until distance (A) from rear stop measures 3/16-inch and remove plug from port (P) to make the adjustment.

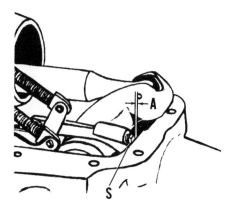

Fig. 218—Turn stop screw (S) until clearance (A) measures 0.160-0.180. Draft control lever must be in transport position.

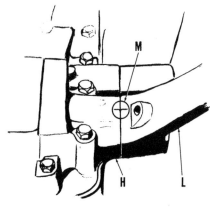

Fig. 219—Place chisel marks (M) on lift arm (L) and cover (H) with rockshaft in transport position. Refer to text.

NOTE: End play will exist if nut (N—Fig. 213) is either too tight or too loose. Tighten Allen head set screw (S) to a torque of 40-65 inch-pounds and complete the assembly by reversing the disassembly procedure.

217. **VALVE SYNCHRONIZATION.** The control valve, located in main pump body, must be synchronized with control linkage located on top cover whenever either unit is removed or as a check when trouble exists. To synchronize the linkage, drain the system down until the side cover containing the response control can be removed, then remove the cover.

Start and idle the engine, move position control lever (3—Fig. 212) to "Transport" end of position control quadrant (6). Center the draft control lever (1) between the two sector marks (2) on edge of draft control quadrant. Reaching through side opening, move valve control lever (L—Fig. 215) forward and allow lift arms to lower, then release the lever and allow it to rest against adjusting screw (S). Thread adjusting screw (S) into control arm (A) until lift arms start to raise, then back out until lift arms are stationary.

Check for proper response to draft lever setting by moving the lever to "Up", "Down" and neutral position. Pressure control adjustment must be check as outlined in paragraph 218 and the necessary readjustments made AFTER valve synchronization is changed.

218. **PRESSURE ADJUSTMENT.** Main pump relief pressure is controlled by the same valve which regulates the weight transfer (Pressure Control) hitch. Adjustment must be made after valves are synchronized as outlined in paragraph 217 or if pressures were incorrect when tested as in paragraph 215.

To make the adjustment, remove plug (P—Fig. 216) from either side of lift cover and install a suitable pressure gage. Move draft control lever to transport position and hydra-lever (position control lever) to "Constant Pumping" position. Start and idle the engine, then turn pressure control screw (P—Fig. 215) in or out until lever (L) starts to hunt or waver between intake and exhaust. At this point, back screw (P) out until gage pressure rises and holds steady. The setting is correct when diaphragm plunger of control (C) is in contact with screw (P) and the highest even pressure is obtained.

Gage pressure should be 2350 psi for some early models or 3100 psi for most models. If it is not, turn pressure adjusting tube (T—Fig. 216) either way as required until the correct reading is obtained.

219. **RESPONSE ADJUSTMENT.** To adjust the response control, remove response control lever and plug (Model MF175) or tool box assembly and plug (Model MF180). Refer to Fig. 217 Move lever (L) toward the slow position until distance (A) from rear stop measures $\frac{3}{16}$-inch. Reaching through plug port (P), loosen the dashpot plunger adjusting screw (D—Fig. 215) slightly to release the

plunger rod, then retighten screw to a torque of 2-3 ft.-lbs.

NOTE: Dashpot plunger adjustment screw (D—Fig. 155) may be either an Allen head or hex head screw.

220. **INTERNAL LINKAGE ADJUSTMENT.** To adjust the internal lift linkage, first check to make sure the transport position is marked as shown in Fig. 219, then remove hydraulic lift cover as outlined in paragraph 221. Check and/or adjust the control spring as outlined in paragraph 216.

Invert the lift cover in a fixture or on a bench, blocking up the cover so that rockshaft can be moved to the normal raised "Transport" position. Move draft control lever to full "Up" position and, using Massey-Ferguson special gage MFN-124 or an 11/64-inch drill bit as a gage, adjust the clearance (A—Fig. 218) to 0.160-0.180 by turning stop screw (S) as required.

Position the rockshaft in transport position by using the gage (G—Fig. 220) or aligning the marks (M—Fig. 219) NOTE: Two different lift covers have been used; before positioning the gage (G—220), measure front flange thickness of cover. If front flange is ⅞-inch thick, side of gage marked ".470" must face ram arm. If front flange is 1⅛-inch thick, side marked ".636" must face ram arm.

On Model MF180, use a ¼-inch bolt with two nuts and locate the draft and position control arms as shown in Fig. 221, by inserting the bolt through link holes in both arms then tightening the nuts on each side of outer bracket; thus holding control arms in a fixed position.

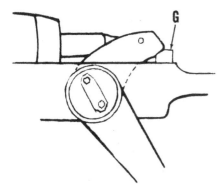

Fig. 220—Use gage block (G) or chisel marks (M—Fig. 159) to position rockshaft for linkage adjustment. Refer to text.

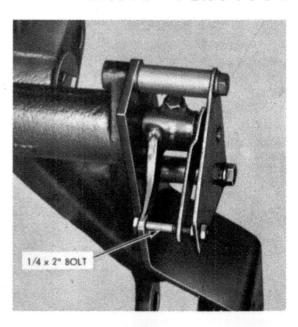

Fig. 221—On Model MF-180, use a bolt and two nuts as shown to position control levers.

1/4 x 2" BOLT

On Model MF175, move hydralever to "Constant Pumping" position and draft control lever between the sector marks (2—Fig. 212) on edge of quadrant. On all models, block the dashpot piston in fully extended position, apply approximately 3 lbs. pressure to end of vertical lever and measure the clearance between lever and end of dashpot piston as shown at (A—Fig. 222). Clearance should be 0.000-0.002 on Model MF175 or 0.002-0.004 on Model MF180. To adjust the clearance, loosen locknut (L—Fig. 216) and turn adjusting screw (A).

With draft control linkage adjusted, proceed as follows: On MF175, move both the hydralever and draft control lever to transport position. On all models, apply approximately 3 lbs. pressure to end of vertical lever and measure clearance between lever and dashpot piston as shown at (A—Fig. 222) Clearance should be 0.000-0.002 on all models. Adjust by loosening locknut (N—Fig. 216) and turning adjusting screw (S).

HYDRAULIC LIFT COVER
All Models

221. **REMOVE AND REINSTALL.** On Model MF180 it will first be necessary to remove operator's platform, seat, seat console and associated parts as follows: Remove tool box and frame from right side of transmission housing, battery and battery box from right side; then unbolt and remove platform center section. Remove the seat. Disconnect wires from light and ignition switch, quadrant linkage, and auxiliary hydraulic system pressure, return and bleed lines. Remove knob from response control lever and push lever through slot in console. Attach a suitable hoist to seat frame, remove the bolts securing seat frame

to tractor and lift off seat frame and console assembly as a unit.

On Model MF175, disconnect auxiliary hydraulic system pressure, return and bleed lines and unbolt breakaway couplings from fender bracket. Remove the seat, then lift off auxiliary hydraulic valve and console as an assembly.

On all models, drain hydraulic system fluid down until response side cover can be removed and remove the cover; then, reaching through side cover opening, spread control valve lever arms (L—Fig. 223) and remove roller (R). Remove transfer plate (3—Fig. 224) or accessory valve, and stand pipe (2). Disconnect upper lift links from lift arms and control beam (6) from clevis (16). Remove the attaching cap screws, then using suitable hoisting equipment, carefully remove the cover.

Keep cover level when installing, and make sure that linkage support bracket is positioned between arms (L—Fig. 223) of valve lever. Tighten the cover retaining cap screws to a torque of 50-55 ft.-lbs., reinstall valve lever roller (R) and complete the installation by reversing the removal procedure. Check valve synchronization after cover is installed, as outlined in paragraph 217.

222. **OVERHAUL.** Refer to Figs. 224, 225 and 226 for exploded views of lift cover and linkage. Lift cover also contains the draft response dashpot and pressure control valve which may be removed and overhauled at this time as outlined in paragraphs 223 and 224.

Lift arms, ram arm and rockshaft all have master splines for correct

Fig. 222—To make either adjustment, block draft response plunger in fully extended position and apply light pressure as shown by arrows; then measure clearance (A). Refer to text.

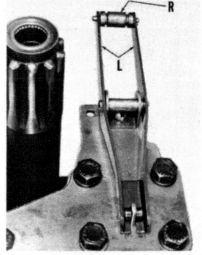

Fig. 223—Control lever (L) and roller (R) assembled to pump unit.

assembly. Rockshaft bushings are a slip fit in top cover bores.

To remove the control quadrant and associated parts, first remove the countersunk set screw from gasket surface of cover, then withdraw the unit as an assembly. With quadrant out, control cams can be withdrawn after loosening the retaining set screw and withdrawing the pivot shaft. NOTE: Guide rods for control cam return springs are drilled near loose end to assist in disassembly and re-assembly; compress the springs and insert cotter pins through drilled holes to retain the springs.

Remove master control spring and associated parts as outlined in paragraph 216. Refer to paragraph 223 for overhaul of response control dashpot and to paragraph 224 for service on pressure control valve and associated parts. Assemble the cover by reversing the disassembly procedure and adjust as outlined in paragraphs 216 through 220.

223. DRAFT RESPONSE DASH-POT. The draft response dashpot assembly (Fig. 228) attaches to lever bracket (30—Fig. 226) and can be removed after removing lift cover as outlined in paragraph 221.

To disassemble the dashpot, loosen adjusting screw (18 or 18A—Fig. 228) and remove adjustment rod (20) and spring (19). Invert the housing and remove plunger (17), ball (16), needle (15) and spring (14). Depress piston rod and guide (8) and remove snap ring (7), then remove spring (9), piston (10) and spring (11). Expansion plugs or guide (12) can be removed from body (13) if renewal is indicated.

Assemble by reversing the disassembly procedure, using Fig. 228 as a guide. Adjust as outlined in paragraph 219 after tractor is assembled.

224. PRESSURE CONTROL VALVE. The pressure control valve shown exploded in Fig. 229 serves the dual purpose of providing pressure relief for the main hydraulic pump during normal operation; and providing weight transfer for added traction when using some types of pulled or mounted implements.

Pressure line (2) is connected to the ram cylinder passage, and cylinder pressure acts against servo piston (7) which is held in the closed position by the variable rate spring (3). The spring is compressed to the maximum when hydralever is in "Position Control" or "Constant Pumping" sectors of quadrant and is released at a

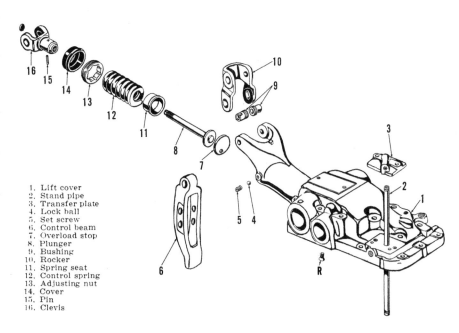

1. Lift cover
2. Stand pipe
3. Transfer plate
4. Lock ball
5. Set screw
6. Control beam
7. Overload stop
8. Plunger
9. Bushing
10. Rocker
11. Spring seat
12. Control spring
13. Adjusting nut
14. Cover
15. Pin
16. Clevis

Fig. 224—Hydraulic lift cover showing master control spring and associated parts. Countersunk screw (R) retains control quadrant.

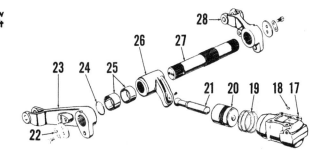

Fig. 225—Exploded view of ram cylinder, rockshaft and associated parts.

17. Cylinder
18. "O" ring
19. Piston rings
20. Piston
21. Connecting rod
22. Lock clip
23. Lift arm
24. "O" ring
25. Bushings
26. Ram arm
27. Rockshaft
28. Lift arm

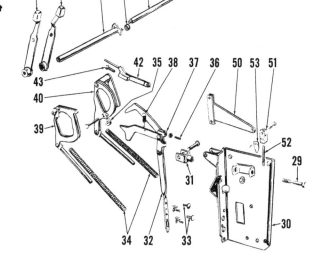

Fig. 226—Hydraulic lift internal control linkage.

29. Adjusting bolt
30. Support bracket
31. Pivot yoke
32. Control arm
33. Adjusting screws
34. Springs
35. Cam roller
36. Adjusting screw
37. Position control arm
38. Return spring
39. Position control cam
40. Draft control cam
42. Draft control rod
43. Stop screw
44. Cam roller
45. Draft control shaft
46. Cam roller
47. Position control shaft
48. Lever
49. Lever

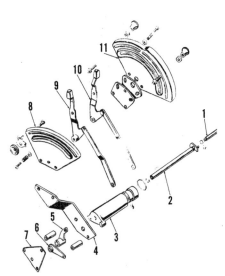

Fig. 227—Control quadrant and associated parts used on Model MF180.

1. Draft control shaft
2. Position control shaft
3. Support
4. Bracket
5. Lever
6. Lever
7. Bracket
8. Quadrant
9. Lever
10. Lever
11. Quadrant

uniform rate as hydralever is moved toward "Low" end of "Pressure" sector on quadrant. When pressure valve is correctly adjusted and operating properly, pressure in the ram cylinder circuit should be approximately 20 psi with hydralever at extreme "Low" end of "Pressure" quadrant, and should increase at a steady rate as lever is moved, maximum pressure is obtained as hydralever moves into "Constant Pumping" sector of quadrant.

The servo piston (7) and matched sleeve (8), connect the control diaphragm assembly (Fig. 230) to control passage of pressure valve. When

servo piston is against stop pin in valve body (9—Fig. 229), the diaphragm passage is open to reservoir. As pressure builds up in ram cylinder circuit and piston (7) moves upward against spring pressure, the diaphragm exhaust passage is closed. When the pressure limit is reached, fluid passes servo piston to diaphragm passage, extending the diaphragm plunger to move the main hydraulic pump control arm to neutral.

Relief valve (10) is pre-set at a slightly higher pressure than diaphragm return spring (2—Fig. 230). The valve provides a safety relief passage if diaphragm plunger malfunctions or system is improperly adjusted.

To disassemble the pressure control valve, refer to Fig. 229 and proceed as follows: Withdraw adjusting tube (1), spring (3) and guide (4), then unscrew and remove pilot (5). Using needle nose pliers, carefully withdraw servo piston (7) and remove piston sleeve (8), using a hooked wire. Relief valve (10) and spring (11) can be removed after removing snap ring (12). Servo piston (7) and sleeve (8) are available only as a matched set. All other parts are available individually. Examine all parts carefully and

renew any which are worn, scored or otherwise damaged. Assemble the pressure control valve by reversing the disassembly procedure. Adjust as outlined in paragraph 218 after lift cover in reinstalled.

MAIN HYDRAULIC PUMP

All Models

225. **REMOVE AND REINSTALL.** To remove the main hydraulic pump, first drain the transmission and hydraulic system reservoir and remove the hydraulic lift cover as outlined in paragraph 221. Reaching through top cover opening, disconnect the "Multipower" pressure line from fitting (1—Fig. 231). Disconnect brake actuating rods. Support transmission and rear axle center housing separately, remove the securing flange bolts and separate the transmission from rear axle center housing. Remove auxiliary pump pressure line (2) and the two pump mounting dowels (P—Fig. 232); and slide the main and auxiliary pumps as a unit, forward out of rear axle center housing.

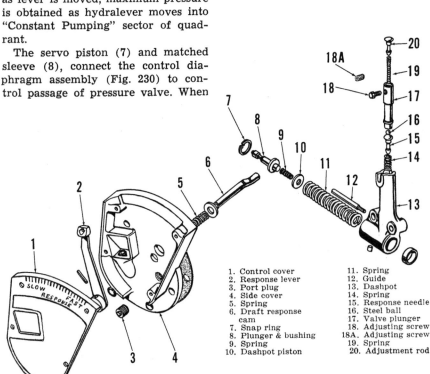

1. Control cover
2. Response lever
3. Port plug
4. Side cover
5. Spring
6. Draft response cam
7. Snap ring
8. Plunger & bushing
9. Spring
10. Dashpot piston
11. Spring
12. Guide
13. Dashpot
14. Spring
15. Response needle
16. Steel ball
17. Valve plunger
18. Adjusting screw
18A. Adjusting screw
19. Spring
20. Adjustment rod

Fig. 228—Draft response dashpot, control cover and associated parts.

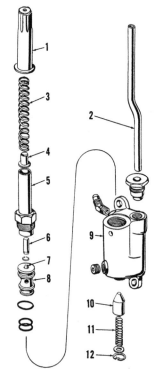

Fig. 229—Exploded view of pressure control valve and associated parts.

1. Adjusting tube
2. Pressure tube
3. Pressure spring
4. Spring guide
5. Pilot
6. Plunger
7. Piston
8. Piston sleeve
9. Body
10. Valve
11. Spring
12. Snap ring

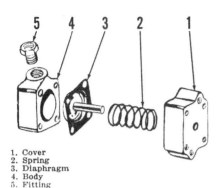

Fig. 230—Exploded view of pressure control diaphragm.

1. Cover
2. Spring
3. Diaphragm
4. Body
5. Fitting

Fig. 231—View of front side of rear axle center housing with early type hydraulic pumps installed.

1. "Multipower" pressure fitting
2. Auxiliary pressure tube
3. Auxiliary pump
4. Idler gear
5. Drive gear
6. Main hydraulic pump
7. Shim pack

Fig. 232—Side of rear axle center housing showing hydraulic pump positioning dowel pin (P).

Fig. 233—On some models it may be necessary to reposition pump package and remove elbow (1) when removing the pump unit. Late pump is shown.

NOTE: On some models with auxiliary hydraulic system, it is necessary to reposition pump and remove elbow (1—Fig. 233) before pump package can be withdrawn.

A number of changes have been made in main hydraulic pump, auxiliary pump and auxiliary pump mounting brackets. On models using bracket (8—Fig. 243), drive gear backlash is controlled by shims (7). Shims are available in thicknesses of 0.002, 0.005 and 0.010 and recommended backlash is 0.002-0.005.

On late models, the gear train mounting plates serve as pump support. Refer to Fig. 234. Gear train must be disassembled to remove pump. Idler gear (9) contains 22 loose needle bearings (10). Gear backlash is not adjustable; renew parts concerned if backlash exceeds 0.015 between any two gears. Tighten retaining bolts to a torque of 30-35 ft.-lbs.

Fig. 235—Hydraulic pump with valve lever removed. Refer to Figs. 236 and 237 for parts identification.

Install by reversing the removal procedure. Check all adjustments as outlined in paragraphs 217 through 220, after tractor is reassembled.

226. **OVERHAUL.** To disassemble the removed hydraulic pump, pull straight out on actuating lever (41—Fig. 235) to free lever from retaining stud (40), then lift the lever off of actuating rod (12) and rollers (13). Remove coupling (22—Fig. 237) or drive gear (22A). Remove nut (1) and unbolt and remove pump cover (24). Remove actuating pin (14—Fig. 236) and rollers (13) and disconnect oscillator (6) from cam followers (27—Fig. 236). Control valve and oscillator assembly (2 through 17—Fig. 236) can then be withdrawn as an assembly from the pump.

Fig. 234—Exploded view of auxiliary pump drive gear train and associated parts used on late models.

1. Main hydraulic pump
2. Auxiliary pump
3. Snap ring
4. Idler shaft
5. Mounting bracket
6. Pump gear
7. Snap ring
8. Thrust washer
9. Idler gear
10. Needle roller
11. Spacer
12. Bushing
13. Drive gear
14. Bracket

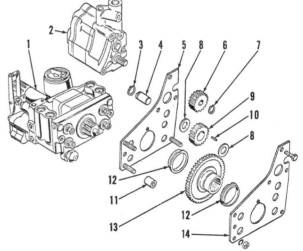

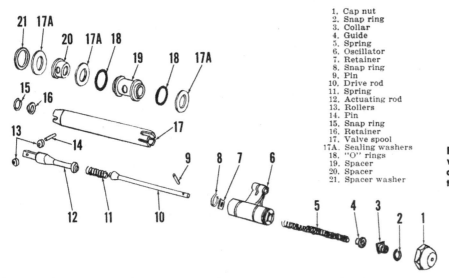

1. Cap nut
2. Snap ring
3. Collar
4. Guide
5. Spring
6. Oscillator
7. Retainer
8. Snap ring
9. Pin
10. Drive rod
11. Spring
12. Actuating rod
13. Rollers
14. Pin
15. Snap ring
16. Retainer
17. Valve spool
17A. Sealing washers
18. "O" rings
19. Spacer
20. Spacer
21. Spacer washer

Fig. 236—Exploded view of hydraulic pump control valve and camshaft driven oscillator which prevents valve from sticking. Nut (1) is shown in proper relation to pump housings in Fig. 237.

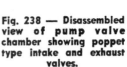

Fig. 238 — Disassembled view of pump valve chamber showing poppet type intake and exhaust valves.

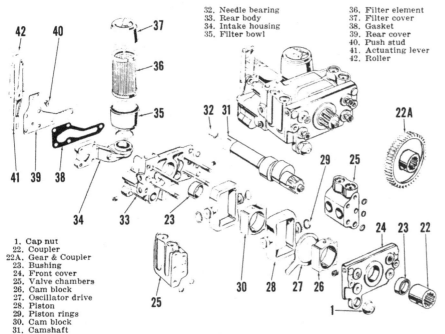

32. Needle bearing
33. Rear body
34. Intake housing
35. Filter bowl
36. Filter element
37. Filter cover
38. Gasket
39. Rear cover
40. Push stud
41. Actuating lever
42. Roller

1. Cap nut
22. Coupler
22A. Gear & Coupler
23. Bushing
24. Front cover
25. Valve chambers
26. Cam block
27. Oscillator drive
28. Piston
29. Piston rings
30. Cam block
31. Camshaft

Fig. 237—Exploded view of main hydraulic system pump and associated parts. Coupler (22) is used only on Model MF175 without "Multipower."

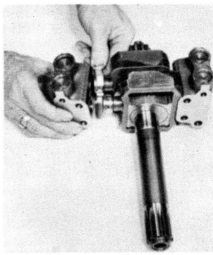

Fig. 239—Use a suitable ring compressor as shown when installing side chamber.

A. Auxiliary port
P. Priority port
1. Priority valve plug
2. Priority relief valve
3. Passage plug
4. Auxiliary relief valve
5. Inlet line
6. Inlet screen

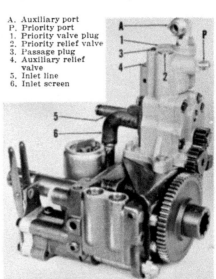

Fig. 240—Main and auxiliary pumps removed from tractor.

The control valve is serviced as a matched assembly which includes the spool (17) and three sealing washers (17A). To disassemble the control valve spool, first remove retaining ring (2) and pin (9) and remove oscillator (6) and associated parts; then remove snap ring (15), retainer (16), actuating rod (12), spring (11) and oscillator drive rod (10) from spool. To remove the sealing washers (17A), unbolt and remove rear cover (39—Fig. 237) and intake housing (34) from rear body (33) and push

sealing washers (17A—Fig. 236), spacers (19 & 20), "O" rings (18) and spacer washer (21) from rear body. Assemble by reversing the disassembly procedure, using Figs. 236 and 237 as a guide.

Withdraw side chambers (25—Fig. 237) from pistons (28). Side chambers contain poppet type inlet and exhaust valves as shown in Fig. 238. Valves and associated parts can be removed for cleaning or parts renewal after removing the retaining snap rings. Use a suitable ring compressor as

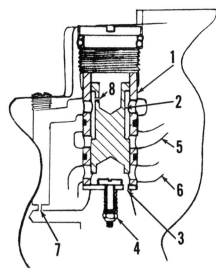

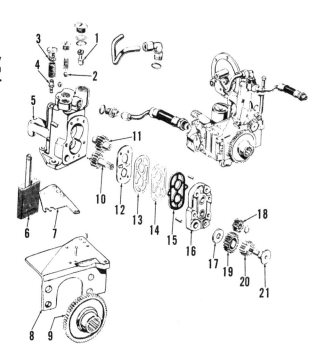

Fig. 243—Exploded view of the gear-type Cessna auxiliary pump and associated parts.

1. Flow divider spool
2. Priority relief valve
3. Adjusting shims
4. Auxiliary relief valve
5. Pump body
6. Intake screen
7. Mounting shim pack
8. Mounting bracket
9. Gear & coupler
10. Driven gear
11. Follow gear
12. Diaphragm
13. Gasket
14. Protector
15. Seal
16. Cover
17. Washer
18. Drive gear
19. Idler gear
20. Needle rollers
21. Idler stud

Fig. 241—Cross sectional view of priority (divider) valve and associated parts.

1. Valve spool
2. Valve spool
3. Auxiliary orifice
4. Adjustable orifice
5. Priority passage
6. Auxiliary passage
7. Priority orifice
8. Balance orifice

shown in Fig. 239 when installing pistons in valve chambers. Tighten pump retaining stud nuts to a torque of 30-35 ft.-lbs.

AUXILIARY PUMP

All Models

227. A number of different pumps have been used. Refer to Figs. 243 through 245 for exploded view and to the following for service information.

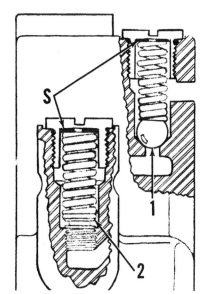

Fig. 242 — Cross sectional view of relief valves located in auxiliary pump body.

1. Priority relief valve
2. Auxiliary relief valve
S. Adjusting shims

228. CESSNA DUAL DELIVERY. Refer to Fig. 240 for view of pumps removed from tractor. Auxiliary pump intake is through suction screen (6) from the sump, and through mainfold tube (5) which carries the auxiliary valve return flow. Auxiliary pump flow is divided within the pump body, the smaller priority flow emerging through port (P) and flowing through the system filter and oil cooler, then to the multipower control valve and clutch. The larger auxiliary flow leaves the pump at port (A) and returns to pump intake through tube (5) where it is recirculated.

Refer to Fig. 241 for a cross sectional view of the balanced type priority valve. All fluid must pass through the two fixed orifices (3 & 7) and the adjustable, spring loaded valve (4) before leaving the pump through priority passage (5) or auxiliary passage (6). The flow divider valve spool (2) moves as required to regulate the priority flow which may vary from approximately 1 gpm at engine idle speed to 2 gpm at rated speed. Auxiliary flow may vary from approximately ¾ gpm at engine idle speed to 6.8 gpm at rated speed.

Fig. 242 shows a cross sectional view of the two relief valves located in pump body. Both valves are shim-adjusted using the interchangeable shims (S). Valve (1) is located in the priority passage and operates in a safety capacity only, the 500-800 psi setting being much higher than the

200 psi regulated pressure of the "Multipower" unit. Auxiliary relief valve (2) is adjusted to 2500-2600 psi and serves as the main relief valve for the auxiliary hydraulic system.

Refer to Fig. 243 for an exploded view of the pump and associated parts. The sleeve for priority valve (1) and seat for poppet type relief valve (4) are factory installed, and removal is not recommended during pump service. With the exception of priority valve (1) and pump body assembly (5), all parts shown are available individually. Tighten the cap screws retaining cover plate (16) to a torque of 7-10 ft.-lbs. when assembling the pump.

229. WARNER MOTIVE DUAL SECTION. Refer to Fig. 244. To disassemble the pump, remove cap screws and stud nut securing end cover (23); cover will be forced off by pressure of spring (22). Do not remove high pressure adjusting screw (24); screw is staked in place and setting should not be changed. Pry rear body (16) from front body (6) using a suitable tool in the notches provided. If pump housings show evidence of wear or scoring in gear pocket area, renew the pump. If housings are serviceable, gears, relief valves and seals can be removed. When assembling the pump, omit high pressure relief valve spring (22) and its washer, and use the body cap screws to draw bodies together over dowel pins. Remove screws and cover

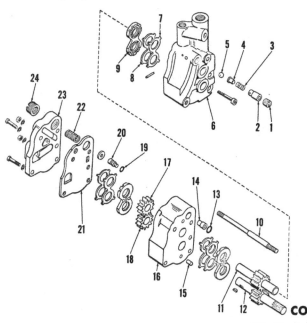

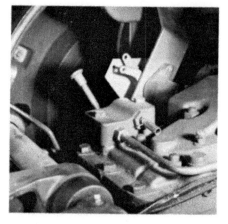

Fig. 244—Exploded view of WARNER MOTIVE Dual Section pump used on late models with auxiliary hydraulics.

1. Adjusting plug
2. Seat
3. Spring
4. Poppet
5. Ball
6. Front body
7. Seal
8. Seal
9. Pressure plate
10. Dowel stud
11. Follow gear
12. Driven gear
13. O-ring
14. Seat
15. Dowel
16. Rear body
17. Follow gear
18. Driven gear
19. O-ring
20. Piston
21. Plate
22. Spring
23. End cover
24. Adjusting screw

Fig. 247—Installed view of the combining valve available as an accessory.

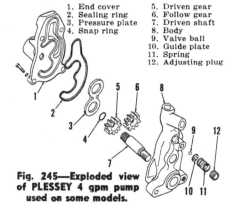

1. End cover
2. Sealing ring
3. Pressure plate
4. Snap ring
5. Driven gear
6. Follow gear
7. Driven shaft
8. Body
9. Valve ball
10. Guide plate
11. Spring
12. Adjusting plug

Fig. 245—Exploded view of PLESSEY 4 gpm pump used on some models.

COMBINING VALVE

All Models

232. Refer to Fig. 247 for an installed view and to Fig. 249 for an exploded view. The valve combines the flow of the main and auxiliary hydraulic pumps providing a fluid flow of approximately 12 gpm for use of hydraulic equipment.

HYDRAULIC ACCESSORIES

All Models

233. Refer to Fig. 250 for an exploded view of the double acting remote cylinder and to Fig. 251 for quick hitch coupler. Other types of hydraulic accessories may be used.

and install relief valve spring. Tighten to a torque of 18-20 ft.-lbs. When installing low pressure relief valve (parts 2 through 5), tighten adjusting plug (1) until it bottoms, back out four full turns and stake in place. This procedure should establish relief pressure within the recommended range of 650-800 psi.

230. PLESSEY 4 GPM. Refer to Fig. 245. Pressure plate (3) is hydraulically loaded to control gear and clearance. Pressure plate and the two pumping gears are available as a matched kit. End cover (1) is available separately but pump body (8) is not. Bearings, shaft, seal and pressure relief valve components are available individually.

AUXILIARY VALVE

All Models

231. Refer to Fig. 246 and Fig. 248 for exploded views. The single remote valve is similar to one half of the dual valve shown.

Fig. 246—Dual auxiliary valve, manifold plates and associated parts.

1. Valve
2. Valve lines
3. Bleed line
4. Manifold
5. Side cover
6. Pump inlet line
7. Manifold
8. Pump auxiliary line

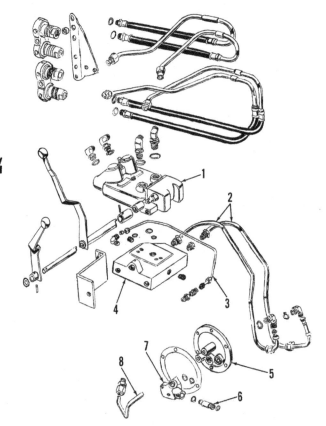

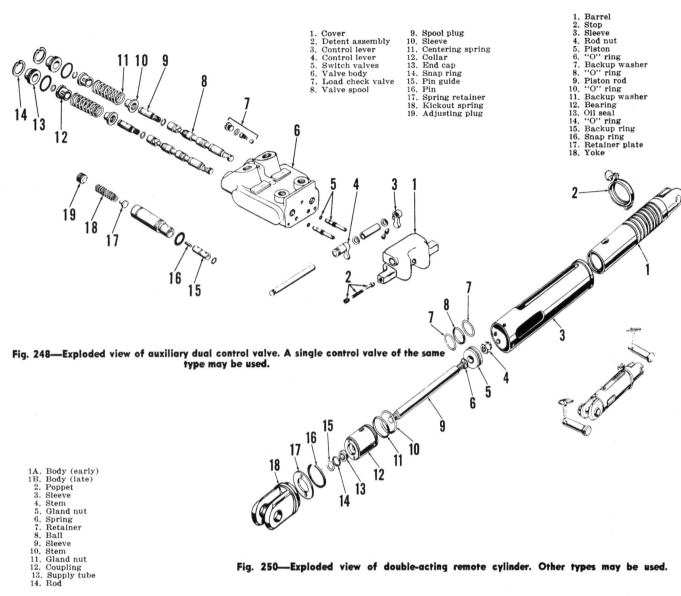

1. Cover
2. Detent assembly
3. Control lever
4. Control lever
5. Switch valves
6. Valve body
7. Load check valve
8. Valve spool
9. Spool plug
10. Sleeve
11. Centering spring
12. Collar
13. End cap
14. Snap ring
15. Pin guide
16. Pin
17. Spring retainer
18. Kickout spring
19. Adjusting plug

1. Barrel
2. Stop
3. Sleeve
4. Rod nut
5. Piston
6. "O" ring
7. Backup washer
8. "O" ring
9. Piston rod
10. "O" ring
11. Backup washer
12. Bearing
13. Oil seal
14. "O" ring
15. Backup ring
16. Snap ring
17. Retainer plate
18. Yoke

Fig. 248—Exploded view of auxiliary dual control valve. A single control valve of the same type may be used.

1A. Body (early)
1B. Body (late)
2. Poppet
3. Sleeve
4. Stem
5. Gland nut
6. Spring
7. Retainer
8. Ball
9. Sleeve
10. Stem
11. Gland nut
12. Coupling
13. Supply tube
14. Rod

Fig. 250—Exploded view of double-acting remote cylinder. Other types may be used.

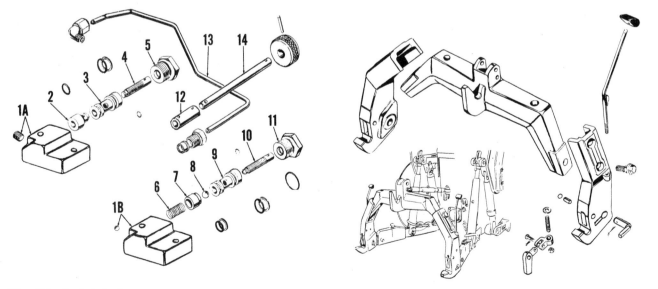

Fig. 249—Exploded view of combining valve. Early type unit is shown at top-left.

Fig. 251—Exploded view of quick-hitch coupler available as an accessory.

MASSEY-FERGUSON

Models ■ MF 205 ■ MF 210 ■ MF 220

Previously contained in I & T Shop Manual No. MF-37

MASSEY-FERGUSON

SHOP MANUAL

MODELS MF205, MF210 and MF220

Tractor serial number is located on plate attached to steering gear cover.

Engine serial number is stamped in machined pad on left side of engine near rear.

Transmission serial number is stamped in machined pad on left side.

INDEX (By Starting Paragraph)

	MF205		MF210		MF220	
	2W Drive	4W Drive	2W Drive	4W Drive	2W Drive	4W Drive
BRAKES	200	200	200	200	200	200
CLUTCH	75	75	75	75	75	75
COOLING SYSTEM	60	60	60	60	60	60
DIESEL FUEL SYSTEM						
Bleeding	41	41	41	41	41	41
Filters	40	40	40	40	40	40
Nozzles	46	46	46	46	46	46
Pump	43	43	43	43	43	43
Speed Adjustment	44	44	44	44	44	44
Timing	42	42	42	42	42	42
DIFFERENTIAL						
Front	....	8	....	8	....	8
Rear	132	135	138	141	138	141
ELECTRICAL SYSTEM						
Alternator	65	65	66	66	66	66
Starter	67	67	68	68	68	68
ENGINE						
Assembly R&R	10	10	10	10	10	10
Camshaft	22	22	22	22	22	22
Connecting Rods	26	26	26	26	26	26
Crankshaft	29	29	29	29	29	29
Cylinder Head	11	11	11	11	11	11
Flywheel	32	32	32	32	32	32
Oil Pan	34	34	34	34	34	34
Pistons, Rings & Cylinders	24	24	24	24	24	24
Tappet Adjustment	12	12	12	12	12	12
Timing Gears	21	21	21	21	21	21
Valves	14	14	14	14	14	14
FINAL DRIVE						
Front	....	5	....	5	....	5
Rear	150	150	151	151	151	151
FRONT SYSTEM						
Axle	1	2	1	2	1	2
Split	4	4	4	4	4	4
Steering Gear	3	3	3	3	3	3
FRONT WHEEL DRIVE	....	5	....	5	....	5
HYDRAULIC SYSTEM	300	300	300	300	300	300
POWER TAKE OFF	250	250	250	255	250	255
TRANSMISSION						
Inspection	85	96	110	120	110	120
Remove and Reinstall	88	99	112	122	112	122

CONDENSED SERVICE DATA

	MF205 2W Drive	MF205 4W Drive	MF210 2W Drive	MF210 4W Drive	MF220 2W Drive	MF220 4W Drive
GENERAL						
Engine Make	——Toyosha——		——Toyosha——		——Toyosha——	
Engine Model	——S107——		——P126——		——S148——	
Number of Cylinders	——2——		2		2	
Bore-mm	——88——		92		97	
Stroke-mm	——88——		95		100	
Displacement-CC	——1070——		1263		1477	
Compression Ratio	——23:1——		23:1		23:1	
TUNE-UP						
Valve Tappet Gap-Cold						
Exhaust-mm	——0.35——		——0.35——		——0.35——	
Inlet-mm	——0.30——		——0.30——		——0.30——	
Governed Speeds-Engine RPM						
Low Idle	——700-750——		——750-800——		——750-800——	
High Idle	——2450-2500——		——2650-2750——		——2650-2750——	
Loaded	——2400——		——2500——		——2500——	
Horsepower at Pto						
Shaft	——16——		——21——		——26——	
Battery—						
Volts	——12——		——12——		——12——	
Ground Polarity	——Neg.——		——Neg.——		——Neg.——	
CAPACITIES						
Cooling System (Liters)	——4——		——5——		——5——	
Crankcase (Liters) **	——3.7——		——4.7——		——4.7——	
Front Differential (Liters)	....	1.4	....	1.7	....	2.7
Front Gear Case (Liters) each side	....	0.4	....	0.4	....	1.8
Fuel Tank (Liters)	——27——		——27——		——27——	
Transmission & Hydraulic (Liters)	16	17	21	25.7	21	27

**With filter.

FRONT AXLE AND STEERING SYSTEM

FRONT AXLE

Two Wheel Drive Models

1. Refer to Fig. 1 for exploded view of fixed tread axle and to Fig. 2 for adjustable front axle. Toe-in should be 0-4 mm. Front wheel bearings should be removed, cleaned, inspected and renewed if damaged or packed with new grease after each 500 hours of operation.

Four Wheel Drive Models

2. The front axle of four wheel drive

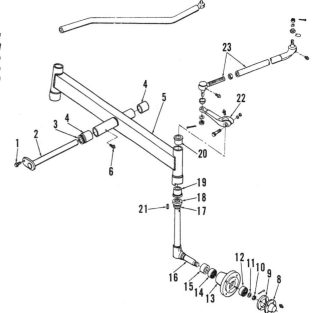

Fig. 1—Exploded view of fixed tread front axle used on MF205 models. Refer to Fig. 2 for adjustable axle used on MF210 and MF220 models.

1. Pin retaining screw
2. Pivot pin
3. Shim
4. Bushing (2 used)
5. Axle
6. Grease fitting
8. Bearing cap
9. Gasket
10. Nut
11. Washer
12. Bearing
13. Hub
14. Bearing
15. Seal
16. Spindle
17. Seal
18. Thrust bearing
19. Bushing (same as 20)
20. Bushing
21. Key
22. Steering arm
23. Tie rod

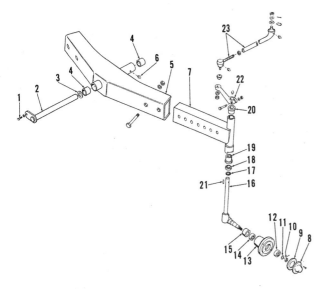

sion output splined shaft, then install drive shaft (4), gasket (2) and cover (3). Start all three drive shaft cover retaining screws enough to hold cover, gasket and shaft in position while installing axle. Install new "O" rings (5 and 9). Loop the two "O" rings (9) around carriers (6 and 21) as shown in insets, then position carriers (6 and 21) on pivot pins of housing (18). Wire can be used to hold carriers in position while installing axle assembly. Move the axle assembly into position, while sliding pinion splines into drive shaft splines and aligning bolt holes in carriers with mounting holes in front support. Tighten the four screws attaching carrier brackets (6 and 21) to 77-89 N·m. Shim (20) can be installed if necessary to limit end play of axle between carrier brackets to 2-4 mm. Carefully pull "O" rings (9) down into "V" grooves after axle is installed. Reattach tie rods to steering arms and reinstall wheels. Screws attaching wheels to axles should be tightened to 77-90 N·m torque.

Fig. 2 — Exploded view of adjustable axle used on MF210 and MF220 models. Refer to Fig. 1 for legend except:
7. Axle extension.

models includes the differential and axle pivot, axle housings, drive shafts, universal joints and final drive. For service to individual units, refer to the appropriate paragraphs 5 through 8. Tie rods should be adjusted to set wheels within limits of 5 mm toe-in to 5 mm toe-out.

To remove the complete front drive axle assembly, first support tractor behind axle and remove both front wheels. Disconnect both tie rods from steering arms. Support axle level with floor to prevent tipping and move front drive lever to "disengage" position. Loosen the three screws which attach drive shaft cover at rear, then remove the four screws which attach axle pivot (carrier) brackets (6 and 21 – Fig. 8) to front support. Carefully lower the axle

until it can be moved forward out of drive shaft splines. Lift off both axle pivot (carrier) brackets, then complete lowering and removing axle assembly. The drive shaft and housing should be removed so that new gasket (2) and "O" ring (5) can be installed when assembling.

When assembling, observe the following: Position coupling (1) on transmis-

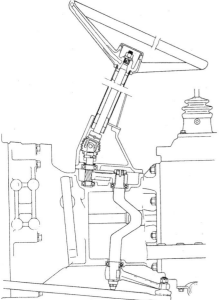

Fig. 3 — Cross-section drawing of steering gear typical of all models. Refer to Fig. 4 for exploded view.

Fig. 4 — Exploded view of clutch housing and steering gear.

1. Steering wheel
2. Bushings
3. Steering shaft
4. Snap rings
5. Universal joint
6. Pins
7. Snap rings
8. Steering seal
9. Steering gear cover
10. Oil seal
11. Snap ring
12. Steering pinion
13. Bearing
14. Retaining nut
15. Quadrant
16. Bearing
17. Steering arm shaft
18. Bearings
19. Snap ring
20. Key
21. Steering arm
22. Nut
23. Tie rod
24. Steering wheel shaft housing

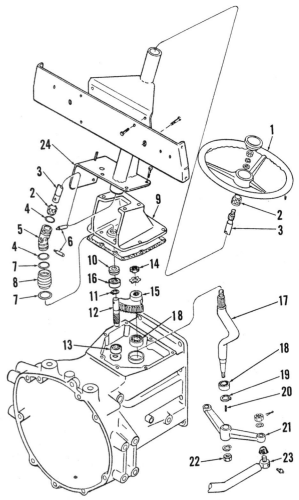

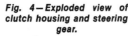

STEERING GEAR

All Models

3. The steering ratio is 10:1 for two wheel drive models and 13:1 for tractors with four wheel drive. Except for numbers of teeth on gears (12 and 15 – Fig. 4), steering gears are alike for all models.

To service steering gear, it is necessary to first remove fuel tank, instrument panel and steering wheel as follows: Shut fuel off, then disconnect fuel line from tank. Disconnect battery, then detach wires and controls from instrument panel. Remove screws and nuts attaching fuel tank and instrument panel to steering gear cover. Remove the four screws that attach steering wheel shaft housing (24) to the steering gear cover (9). Remove retaining ring from lower end of steering universal joint (5), use a punch to drive stepped pin (6) out, then separate universal joint from steering shaft. Lift the complete steering wheel and instrument panel from tractor. Remove the six attaching screws, then lift steering gear cover (9) from tractor. The pinion and quadrant gears (12 and 15) can be removed after removing quadrant gear retaining nut (14). The steering arm shaft (17) can be removed after splitting tractor between engine and clutch housing.

When reassembling, front wheels must be straight ahead and steering pinion (12) must be in center of quadrant gear teeth (15). Flat side of quadrant gear retaining nut (14) must be up and tab of washer should be bent up into notch of retaining nut. Pack steering gear housing with grease before assembling.

FRONT SPLIT

All Models

4. To separate the front axle and support from front of engine, remove hood, drain cooling system, disconnect battery, then remove battery. Disconnect coolant hoses from engine then unbolt and remove radiator support bracket and radiator. Remove cotter pin from lower end of steering shaft, then remove steering arm from steering shaft. Attach chain hoist to engine lift supports, then raise tractor enough to block up under transmission housing. On four-wheel-drive models, loosen the three screws which attach drive shaft cover at rear. On all models, remove the six nuts which attach front frame to engine, then roll front axle away from tractor.

When installing reverse removal procedure. Tighten front frame to engine

nuts to 83-103 N·m and steering arm nut to 29-39 N·m.

FOUR WHEEL DRIVE

FRONT FINAL DRIVE

5. Each front final drive can be disassembled without disturbing other parts of the drive system. Support weight of tractor front end and remove the front wheel. Remove drain plug (13 – Fig. 5), then unbolt cover (3) from housing (15). Carefully bump cover from studs and bearing (6) from inner bore, then withdraw cover, stub axle and reduction gear from housing. Remove cotter pin then remove nut (8) and washer (7) from inner end of stub axle. Bump inner end of axle with a soft hammer while supporting cover (3) to remove bearing (6), gear (5), bearing (4) and cover (3) from axle (1). Oil seal (2) and bearings (4 and 9) can be removed from cover if renewal is required. Drive gear (11) can be removed after removing

snap ring (10).

Clean all gasket surfaces and coat gasket (3G) with sealer before assembling. Assemble cover (3), bearings (4 and 9), gear (5), bearing (6), washer (7) and nut to stub axle (1). Tighten nut (8) then install cotter pin. **DO NOT** loosen nut to install cotter pin. Tighten nuts and screws which attach cover (3) to housing (15) to 23-27 N·m torque. Tighten wheel retaining screws to 77-90 N·m torque.

AXLE HOUSING, KING PIN, AXLE SHAFTS AND UNIVERSAL JOINTS

6. **REMOVE AND REINSTALL.** The differential can remain installed on tractor while removing and installing axle housing. To remove one axle housing and associated parts from one side, raise tractor and remove wheel. Remove drain plug (13 – Fig. 5) and allow oil to drain. Block axle, which is to remain installed, up so that axle to be removed is lower. Refer to paragraph 5 to remove Front Final Drive if so desired. The king pin housing (15) can be removed if desired, leaving housing (22) attached to

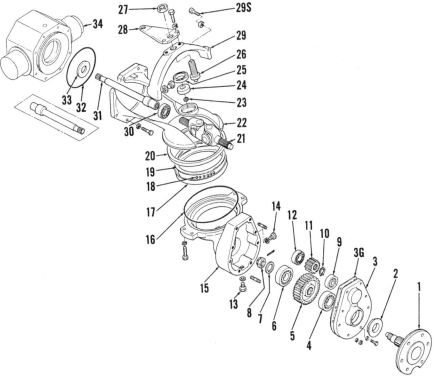

Fig. 5 — Exploded view of four wheel drive axle housing and related parts. Refer to Fig. 8 for exploded view of differential housing (34).

1. Axle	10. Snap ring	19. Bearing race	28. Steering arm
2. Oil seal	11. Gear	20. Seal	29. Support
3. Cover	12. Bearing	21. Universal joint	29S. Stop screw
3G. Gasket	13. Drain plug	22. Housing	30. Bearing
4. Bearing	14. Fill plug	23. Bearing plug	31. Drive shaft
5. Gear	15. King pin housing	24. Bearing	32. "O" ring
6. Bearing	16. "O" ring	25. Seal	33. Thrust washer
7. Washer	17. Bearing race	26. Adjusting bolt	34. Differential housing
8. Nut	18. Bearing balls	27. Locknut	
9. Bearing	(4.70 mm)		

differential housing (34).

To remove complete assembly, detach tie rod, support housing, then unbolt axle housing (22) from differential housing. Carefully slide axle (31) and housing (22) away from differential housing (34).

NOTE: The screws attaching right axle housing to differential housing also attach differential housing cover. Be careful to separate between axle housing and cover. Be careful to slide axle shaft (31) out together with housing (22) unless king pin is to be disassembled. If axle (31) pulls out of bearing (30), it may be necessary to remove the king pin housing (15) so that shaft (31) can be installed in bearing (30) and universal joint (21) can be installed in end of shaft (31).

To reinstall, be sure that opposite axle is raised to permit more room on side which is being installed. Install new "O" ring (32) on differential housing, be sure that thrust washer (33) is positioned on axle shaft (31), then carefully slide axle assembly onto differential housing. Tighten attaching screws to 47-55 N·m. Attach tie rod end to steering arm. Stop screws (29S) should be adjusted to 50.8mm from center of support (29) and locked with nut.

7. OVERHAUL. Service to king pin bearings and universal joints can be accomplished without separating axle housing from differential housing; however, removal is required for service to bearing (30–Fig. 5) and axle shaft (31).

Refer to paragraph 5 and remove the front final drive (1 through 9). Remove snap ring (10) and withdraw gear (11). Unbolt support (29) from housing (15), then carefully remove housing (15). There are 105 loose bearings (18) in each race that can easily be dropped when removing housing. Carefully remove seal (20), bearing races (19 and 17), balls (18) and "O" ring (16).

Inspect all parts carefully and renew all that are questionable. Install bearing inner ring (19) using a soft hammer to tap lightly into position. Install seal (20) using race (17) as an assembly tool. Install outer race (17) and the 105 balls (18), being careful not to fold seal under race (17). Install bearing (30) with sealed side towards outside (universal joint side). Position "O" ring (16) in groove of housing (15) and bearing (12) in bore of housing. Install universal joint (21) through bearing (12), then install gear (11) and snap ring (10). Position shaft (31) in bearing (30), then locate housing (15) on housing (22) with universal joint spline meshing with shaft (31) coupling. Install support (29) and tighten screws to 77-89 N·m. Adjust bearing (25) preload by turning bolt (26) to 3.38 N·m

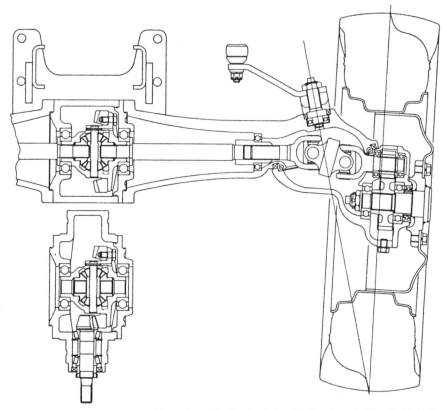

Fig. 6 — Cross-section drawings of four wheel drive front axle. Inset is top view of differential housing showing drive pinion.

torque and lock with nut (27) torqued to 10 N·m. Rolling torque of housing (15) to housing (22) should be 3.38 N·m with bolt (26) correctly adjusted.

Refer to paragraph 5 for notes about assembling final drive and to paragraph 6 for installing axle housing assembly.

DIFFERENTIAL

8. R&R AND OVERHAUL. Refer to paragraph 2 and remove the axle assembly from tractor, then remove axle assemblies as outlined in paragraph 6. Refer to Fig. 8 and bump bearings (23), differential carrier (31), ring gear (24) and right side cover (33) out of housing (18). Be careful not to mix, damage or lose shims (22L and 22R) behind both bearings (23).

The pinion nut (10) is staked in position and is secured with "Loctite". Hold pinion shaft and remove nut (10). Be careful to notice position of shims (15 and 15M) between bearings (13) and

behind teeth of pinion gear (17). Shims (15) adjust preload of bearings (13); shims (15M) control mesh position of pinion (17). Bump pinion (17) inward to remove. Drive pin (25) out, then remove shaft (26) and parts (27, 28, 29 and 30). Bearings (13), spacer (16), seal collar (12), seal (11) and associated parts can be removed for inspection or renewal.

NOTE: On some models, the pinion shaft (17), bearings (13), spacer (16), seal (11) and related parts may be contained in a housing that is attached to differential housing (18) with four cap screws. On these models, mesh position is controlled by varying the thickness of shims (15M) between pinion housing and differential housing. Adjustment of bearings (13) is still controlled by shims (15) between bearings.

Observe the following when assembling differential on all models: Ring gear (24) and pinion (17) are available only as a matched set. Threads of screws which

Fig. 7 — View of four wheel drive front axle showing location of fill plugs and drain plugs.

attach ring gear (24) to carrier (31) should be coated with "Loctite" before tightening to 49-58 N·m torque.

Assemble pinion (17) and associated parts (11 through 16) in housing (18) using shims (15) that were removed. Tighten nut (10) to 58-78 N·m torque, then check torque required to turn pinion shaft in bearings. If rolling torque is not within range of 0.58-0.98 N·m, add or decrease thickness of shims (15) between bearings (13) as required. Recheck rolling torque after changing shims and be sure that nut (10) is tightened to correct torque when checking. Shims (15M) between pinion gear and front bearing are same as shims (15); however, shims at location (15M) adjust mesh position.

Assemble differential carrier (31) and bearings (23) in housing (18) using shims (22L and 22R) that are removed. Install cover (33) using at least three evenly spaced screws. Flat washers will be necessary if original long screws are used to hold cover on. With cover retaining screws tightened to 47-55 N·m torque, ring gear and differential carrier should be rotated by 4.5-22.2 Newtons (light hand) pressure. Add or remove shims (22L or 22R) as necessary to provide correct bearing adjustment. Be sure sufficient thickness of shims (22L) are installed at left side if pinion (17) is installed during this check.

To check backlash, install pinion (17) with shims (15M) that were removed, then tighten nut (10) to 58-78 N·m tor-

que. Install differential assembly (23 through 31) with shims (22L and 22R) necessary for correct bearing adjustment as previously checked. Install cover (33) using at least three evenly spaced screws torqued to 47-55 N·m. Mount a rocker type dial indicator as shown in Fig. 9 with probe through hole in housing cover and against head of ring gear retaining screw. Hold pinion shaft from turning, then turn differential (at inner bearing race) and measure backlash. Correct backlash is 0.13-0.18 mm. To increase backlash, remove shims (22R – Fig. 8) and install at (22L). Transfer shims from left side (22L) to right side to decrease backlash. Moving shims from one side to the other will not change bearing adjustment, but will change backlash.

To check mesh position of pinion (17) with ring gear (24), proceed as follows: Install pinion (17), with shims (15M) that were removed, then tighten nut (10) to 58-78 N·m torque. Coat teeth of ring gear (24) with mechanics bluing, then install differential assembly (23 through 31) with shims (22L and 22R) necessary for correct bearing adjustment. Install cover (33) using at least three evenly spaced screws torqued to 47-55 N·m. Rotate pinion through bluing on ring gear, then remove differential from housing and check mesh pattern of gear teeth as indicated by marks in bluing. Refer to Fig. 10. Correct mesh position is shown at (A). Mesh pattern (B) indicates that ring gear is too far away

from pinion and can usually be corrected by moving shims from left side (22L – Fig. 8) to right side (22R). Mesh pattern (C – Fig. 10) is caused by too little backlash (ring gear too close to pinion). Condition resulting in pattern (C) should be changed by moving shims from right side (22R – Fig. 8) to left side (22L). Pattern (D – Fig. 10) is caused by pinion (17 – Fig. 8) too far from center of ring gear and is corrected by adding shims at (15M). Mesh pattern (E – Fig. 10) is caused by pinion too far into ring gear and is corrected by removing some shims (15M – Fig. 8). Always recheck mesh position using fresh coating of mechanics blue to be sure that shims (15, 15M, 22L and 22R) are all correctly positioned before final reassembly. Use new "O" ring (32) and coat threads for nut (10) with "Loctite" upon final assembly. Stake nut (10) in three places after tightening to 58-78 N·m torque.

Fig. 9 – A rocker type dial indicator can be mounted as shown to measure backlash. Refer to text.

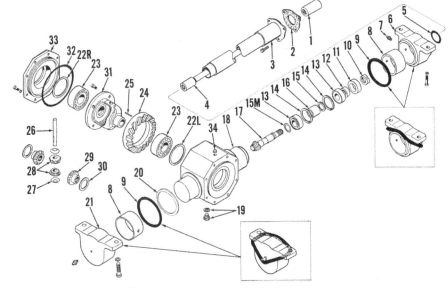

Fig. 8 – Exploded view of differential housing and related parts. Refer to text for installation of seal rings (9).

1. Coupling	10. Retaining nut	18. Housing	27. Thrust washer
2. Gasket	11. Oil seal	19. Drain plug	28. Pinions
3. Housing	12. Seal collar	20. Shim	29. Side gears
4. Drive shaft	13. Bearing	21. Carrier bracket	30. Thrust washer
5. "O" ring	14. Snap ring	22L. & 22R. Shims	31. Differential carrier
6. Carrier bracket	15. Shim	23. Bearings	32. "O" ring
7. Grease fitting	15M. Shim	24. Ring gear	33. Cover
8. Bushing	16. Spacer	25. Pin	34. Breather
9. "O" rings	17. Pinion gear	26. Shaft	

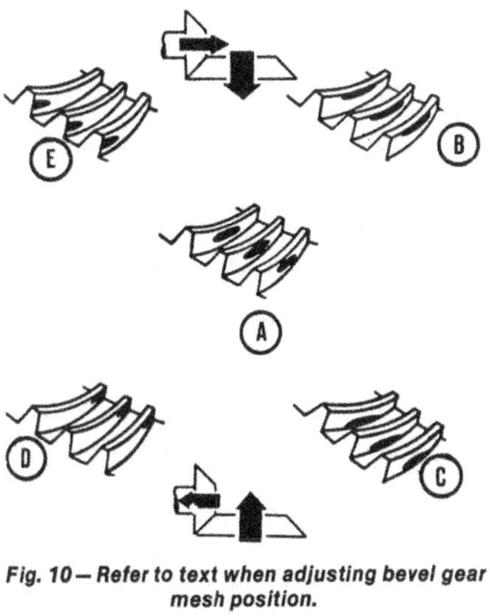

Fig. 10 – Refer to text when adjusting bevel gear mesh position.

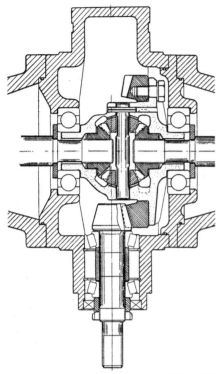

Fig. 11—Cross-section of front differential assembly.

ENGINE AND CLUTCH

R&R ENGINE WITH CLUTCH

All Models

10. To remove the engine and clutch as a unit, first drain cooling system and, if engine is to be disassembled, drain engine oil. Remove hood, disconnect battery, then remove battery. Disconnect hydraulic lines at pump, then plug these lines and openings to prevent entrance of dirt and oil spillage. Disconnect wires from alternator, starter and oil pressure sensor. Disconnect coolant hoses from engine, remove radiator support brackets, then unbolt and remove radiator. Remove the two screws attaching starter motor then withdraw starter. Remove the water temperature sensor from water outlet (or thermostat) housing. Disconnect tachometer drive cable from behind oil filter and fuel shut-off an throttle linkage from injection pump. Disconnect glow plug wires and injector bleed-off line. Be sure fuel is shut off, then disconnect fuel line from injection pump. Remove cotter pin from lower end of steering shaft, then remove steering arm from steering shaft. Attach chain hoist to engine lift supports, then raise tractor enough to block up

under transmission housing. Remove the six nuts which attach front frame to engine, then roll front axle and frame away from tractor. Remove nuts and cap screws attaching engine to clutch housing, then separate engine from clutch housing. Be careful when separating engine and clutch housing, because clutch housing is aluminum and easily damaged.

When reassembling, reverse removal procedure and observe the following tightening torques: Clutch housing to engine cap screws and nuts should be tightened to 63-73 N·m torque. Starter attaching screws should be tightened to 39-45 N·m torque. Tighten front frame to engine nuts to 83-103 N·m. Steering arm nut should be torqued to 29-39 N·m and cotter pin installed.

CYLINDER HEAD

All Models

11. **REMOVE AND REINSTALL.** To remove cylinder head, remove hood, drain coolant and disconnect upper radiator hose from thermostat housing or outlet housing. Remove high pressure injection lines, disconnect leak-off line and remove injectors. Be sure to cap all openings in fuel lines and in attaching parts to prevent entrance of dirt. Detach engine lubrication line from left rear of

engine block and loosen connection on rear of cylinder head. Be careful not to damage oil line while removing and installing cylinder head. Disconnect wire for glow plug, remove exhaust pipe and muffler. Detach air intake hose from inlet manifold and remove air cleaner assembly. Remove rocker arm cover and on models so equipped disconnect temperature sending unit. Remove the two nuts and two long head bolts retaining rocker arm assembly, then lift rocker arm assembly from engine. Remove push rods, remove the ten remaining head bolts, then carefully lift cylinder head from engine.

Refer to appropriate paragraphs 14, 15 and 16 for servicing valves, guides and springs. Check cylinder head for flatness using a straight edge and feeler gage. Discretion is recommended when checking for warpage, but generally 0.05 mm limit can be observed when checking from front to rear and from side to side. If cylinder head is resurfaced, to eliminate warpage, be sure that head of valve is recessed 0.4-0.5 mm from gasket surface of cylinder head (D – Fig. 16). Swirl chamber inserts should be tight in head and should be retained by pin. Swirl chambers can be pressed from head via hole for glow plug. After cleaning bore in head, swirl chambers should be chilled by freezing before installation in head. Be sure that retaining pin is installed, then allow

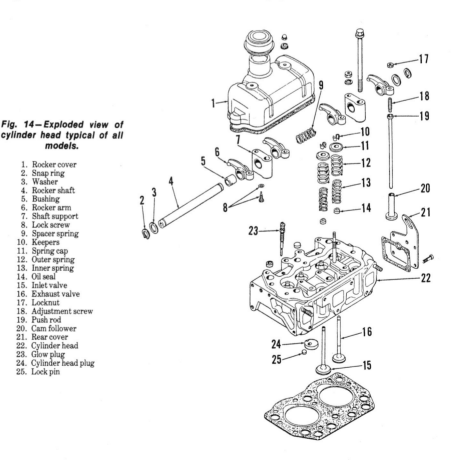

Fig. 14—Exploded view of cylinder head typical of all models.

1. Rocker cover
2. Snap ring
3. Washer
4. Rocker shaft
5. Bushing
6. Rocker arm
7. Shaft support
8. Lock screw
9. Spacer spring
10. Keepers
11. Spring cap
12. Outer spring
13. Inner spring
14. Oil seal
15. Inlet valve
16. Exhaust valve
17. Locknut
18. Adjustment screw
19. Push rod
20. Cam follower
21. Rear cover
22. Cylinder head
23. Glow plug
24. Cylinder head plug
25. Lock pin

swirl chamber to warm up to room temperature. Check for tightness before reinstalling head on engine.

Cylinder head gasket should be installed with tabs of cylinder sealing rings toward top (head). Gasket is coated with sealer on both sides and additional sealer is not recommended. The oil line should be attached loosely to back of cylinder head before assembling, because of limited clearance between cylinder head and fuel tank. Be careful not to damage oil line when assembling and be sure that oil line is correctly routed before attaching cylinder head.

Install the ten short head bolts and tighten in sequence shown in Fig. 15 to initial torque of 39-49 N·m. Install the push rods, then locate rocker arm assembly over studs, install the two long head bolts to 39-49 N·m and stud nuts to 29-39 N·m. Gradually tighten cylinder head retaining screws in approximately 10 N·m increments to final torque of 88-108 N·m using sequence shown in Fig. 15. Refer to paragraph 12 and adjust valve clearance, then install rocker arm cover.

TAPPET GAP ADJUSTMENT

All Models

12. Recommended valve clearance (tappet gap) is 0.3 mm for inlet valves; 0.35 mm for exhaust valves. Check and adjust valve clearance with engine cold

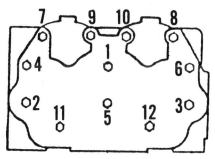

Fig. 15 — Tighten cylinder head retaining screws in sequence shown.

and not running. Tighten locknuts to 39-44 N·m after adjusting valve clearance.

VALVES AND SEATS

All Models

14. Intake and exhaust valves are not interchangeable. Valve seats can be reconditioned using 30, 45 and 60 degree stones. Preferred order of grinding is 60 degree, 30 degree then 45 degree. Seat width (A – Fig. 16) should be 1.4 mm. Valve face angle should also be 45 degree. Renew valve if valve margin (E) is less than 1.0 mm. Valve should be recessed 0.4-0.5 mm from gasket surface of head as shown at (D). Valve stem diameter is 7.945-7.96 mm for inlet valves; 7.920-7.935 mm for exhaust valves. Overall length of new inlet and exhaust valves is 125 mm. End of stem should not be ground away more than 0.5 mm. Install new valve stem seals whenever valves are serviced.

VALVE GUIDES

All Models

15. Inlet and exhaust guides are semi-finished and must be reamed after installation in cylinder head. Valve guide should be pressed into bore until top of guide is flush with gasket surface for rocker arm cover. Finished diameter of guide should be 8.000-8.018 mm to provide stem to guide clearance of 0.065-0.095 mm for exhaust valve; 0.040-0.065 mm for inlet valves. Refer to paragraph 14 for refinishing valve seat after renewing guide. All valves are equipped with cup type valve stem seals which should be renewed whenever valves are serviced.

VALVE SPRINGS

All Models

16. Valve springs are interchangeable

for inlet and exhaust valves. Some models may not be equipped with inner valve springs. Renew springs which are distorted, heat discolored or fail to meet test specifications which follow:
Outer Spring –
   Free length . . . . .56.6 mm
   Installed length . . . . .50.0 mm
   Test pressure N. at mm. 118 N. at 50.0
Inner Spring –
   Free length 55.0mm
   Installed length 48.0mm
   Test pressure N. at mm. . .49 N. at 48.0

ROCKER ARMS

All Models

17. Inlet and exhaust valve rocker arms are interchangeable. Rocker arms are lubricated by oil delivered to the rocker arm shaft via drilled passage in cylinder head and the oil tube attached to rear of cylinder head. Refer to paragraph 11 for tightening rocker shaft support retaining stud nuts and cap screws. Refer to paragraph 12 for adjusting valve clearance.

CAM FOLLOWERS

All Models

18. All models are equipped with mushroom type cam followers which can only be removed from below after removing camshaft.

Check surface of cam follower that contacts camshaft and renew followers if worn. Also check camshaft carefully if follower shows wear and renew camshaft if chipped, broken, scored or otherwise worn. Refer to paragraph 22 for cam follower to cylinder block bore specifications.

VALVE TIMING

All Models

19. Valves are properly timed when timing marks are aligned as shown in

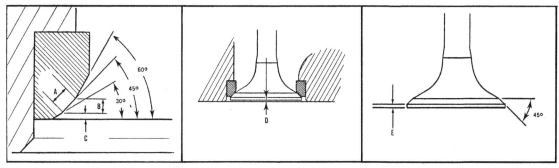

Fig. 16 — Drawings of valves and seats showing desired dimensions.

A. 1.4 mm
B. 1.0 mm
                   C. 0.5 mm
                                                   D. 0.4-0.5 mm
                                                   E. 1.5 mm

Fig. 18. Because of the uneven number of teeth on idler gear (2), the timing marks will only align every 14th revolution of the crankshaft. Alignment can be checked by turning crankshaft (1) until keyway is vertical and checking mark on camshaft gear (4). If mark on gear (4) is not toward idler gear (2), turn crankshaft one revolution until keyway is again vertical. Remove idler gear (2), then reinstall with marks aligned on all gears (1,2,3 and 4). Idler gear is retained to idler shaft by snap ring on some models, while on other models, idler gear is retained by cap screw which should be tightened to 42-49 N·m torque.

TIMING GEAR COVER

All Models

20. To remove the timing gear cover, first drain cooling system, remove hood and disconnect battery. Remove air cleaner, radiator, alternator, alternator bracket and crankshaft pulley. Detach hydraulic tubes, cover all openings, then remove hydraulic pump. The cap screws and nuts retaining timing gear cover can now be removed, then cover can be separated from engine front plate. Notice that some of the timing gear cover screws are provided with nuts and that gasket will stick cover to engine front plate. Be careful not to damage the aluminum timing gear cover.

Crankshaft front oil seal should be pressed into bore with closed side out and flush with front surface of cover or until it bottoms in bore. Pack seal with light, clean grease to assure that seal is immediately lubricated. Apply oil to crankshaft pulley seal surface before installing. Tighten crankshaft pulley retianing nut to 98-117 N·m.

TIMING GEARS

All Models

21. Refer to Fig. 19 for an exploded view of timing gears and associated parts. Backlash between idler gear (2) and either camshaft gear (4) or crankshaft gear (1) should be 0.05-0.20 mm. Backlash between idler gear (2) and injection pump (3) should be 0.1-0.3 mm. Install new gears if backlash between any two meshing gears exceeds 0.5 mm.

Camshaft end play should be checked before removing camshaft gear. Recommended end play is 0.05-0.2 mm. Camshaft gear can be removed using a puller without removing camshaft. Crankshaft gear is 0.028-0.061 mm interference (press) fit on crankshaft. Refer to Fig. 18 for view of timing marks. Camshaft retaining screw should be tightened to 49-54 N·m torque. Injection pump gear

retaining nut should be tightened to 59-69 N·m torque. Tighten idler gear retaining screw to 42-49 N·m torque. Crankshaft pulley nut (5–Fig. 19) should be tightened to 98-117 N·m.

CAMSHAFT

All Models

22. To remove the camshaft, first drain engine oil and coolant then remove

Fig. 18 — View of timing gear marks correctly aligned. Crankshaft gear (1), idler gear (2), injection pump gear (3) and camshaft gear (4).

engine as outlined in paragraph 10. Refer to paragraphs 76, 32 and 30 for removing the clutch, flywheel and rear seal cover assembly. Remove rocker arm cover, the two stud nuts and two cylinder head screws that retain rocker shaft supports, then lift rocker shaft assembly and push rods from engine. Refer to paragraphs 20 and 21 for removing the timing gear cover and timing gears. Refer to paragraph 43 and remove injection pump. Unbolt and

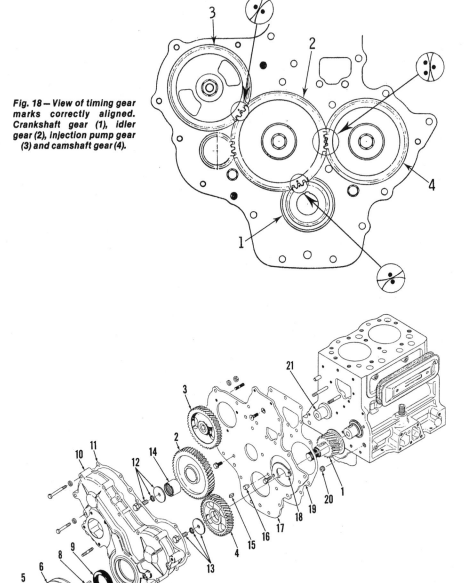

Fig. 19 — Exploded view of timing gear cover, gears and related parts for MF210 and MF220 models. Other models are similar.

1. Crankshaft gear	7. Timing pointer	12. Screw and washers	17. Front plate
2. Idler gear	8. Key	13. Screw and washers	18. Thrust plate
3. Injection pump gear	9. Seal	14. Needle bearing	19. Gasket
4. Camshaft gear	10. Timing gear cover	15. Key	20. Plug
5. Nut	11. Gasket	16. Oil injector	21. Idler shaft
6. Pulley			

remove water pump, front plate, rear plate, oil pan and oil pump. Unbolt thrust plate (18–Fig. 19) hold cam followers up (or invert engine) then withdraw camshaft.

Camshaft journal diameter should be 49.950-49.975 mm and should have 0.021-0.102 mm clearance in bushings. If clearance exceeds 0.12 mm, install new camshaft and/or bearings. Cam lobe height new is 37.276 mm. Renew camshaft lobe if worn or nicked. Carefully inspect face of cam followers and install new follower if scuffed or worn. Clearance of cam follower stem in block bores should be 0.04-0.078 mm. Renew cam follower if clearance in block exceeds 0.12 mm. Camshaft end play should be 0.05-0.20 mm and is limited by thrust plate (18–Fig. 19). End play can only be measured after camshaft, thrust plate, drive gear (4) and retaining screw and washers (13) are assembled and correctly torqued to 49-54 N·m.

Assemble in reverse of removal procedure. Oil pump retaining cap screws should be tightened to 12-15 N·m and oil screen retaining screws should be tightened to same (12-15 N·m) torque. Clean mating surface of cylinder block and oil pan, then coat surface evenly with an RTV sealant. Install oil pan and tighten retaining screws to 29-34 N·m torque. Install rear plate and refer to paragraph 30 for installing rear seal and cover. Install front cover plate, refer to paragraph 21 for installing timing gears and to paragraph 20 for installing timing gear cover. Install push rods and rocker arm assembly tightening rocker shaft retaining stud nuts for 29-39 N·m and the two long head bolts to 88-108 N·m torque. Check to be sure that all other head bolts are also torqued to 88-108 N·m. Correct tightening sequence is shown in Fig. 15. Refer to paragraph 12 and adjust valve clearance. Refer to paragraph 10 for reinstalling engine.

ROD AND PISTON UNITS

All Models

23. Connecting rod and piston units are removed from above after removing the engine as outlined in paragraph 10, the cylinder head as in paragraph 11 and oil pan as in paragraph 34. Unbolt and remove oil pump. Remove connecting rod cap and push piston and rod assembly up out of cylinder block. Be sure that cylinder ridge is removed before pushing piston out.

Connecting rod cap parting line is slanted to offset cap bolts toward crankcase side opening. Parting line is serrated for precise alignment and to remove shear stress from bolt shanks. Tighten connecting rod cap screws to a

torque of 49-58 N·m for MF205 models, 59-67.8 N·m for all other models. On all models, the special connecting rod bolts do not use spring washers but contact special machined seat in connecting rod cap. Side clearance on journal should be 0.1-0.3 mm.

PISTON, RINGS AND CYLINDER

All Models

24. The cam ground aluminum alloy pistons are fitted with two compression rings and one oil ring on MF205 and MF220 models, while MF210 models have three compression rings and one oil ring. Pistons and rings are available in standard size and 0.5 mm oversize. Examine cylinder bore carefully and rebore to next available oversize if taper exceeds 0.1 mm. Cylinder head surface of block must be flat within 0.1 mm. If cylinder block is machined, be sure that pistons do not project more than 0.5 mm at TDC. Refer to the following specification data:

MF205
Piston diameter*,
   std. ....87.925-87.945 mm
Cylinder bore standard
   diameter ....88.000-88.022 mm
Piston to cylinder
   clearance ....0.10-0.14 mm
Ring side clearance in groove–
   Top ring ....0.04-0.08 mm
   Second & bottom rings .0.03-0.07 mm
Ring end gap (all rings) .....0.2-0.4 mm

MF210
Piston diameter*,
   std. ....91.880-91.900 mm
Cylinder bore standard
   diameter ....92.000-92.022 mm
Piston to cylinder
   clearance ....0.10-0.14 mm
Ring side clearance in groove–
   Top ring ....0.04-0.08 mm
   Second, third & bottom
    rings ....0.03-0.07 mm
Ring end gap (all rings) .....0.2-0.4 mm

MF220
Piston diameter*,
   std. ....96.905-96.925 mm
Cylinder bore standard
   diameter ....97.000-97.022 mm
Piston to cylinder
   clearance ....0.10-0.14 mm
Ring side clearance in groove–
   Top ring ....0.05-0.09 mm
   Second & bottom
    rings ....0.03-0.07 mm
Ring end gap (all rings) .....0.2-0.4 mm

*On all models, measure piston diameter at bottom of skirt and at right angles to piston pin.

Install chrome plated ring in top

groove. Second compression ring should be installed with groove on inside circumference toward top. Third compression ring, if used should have groove on outside circumference toward bottom. Marked sides of piston rings should be toward top. Always stagger ring end gaps before installing piston and rod assembly in cylinder bore.

PISTON PIN

All Models

25. The full floating piston pin is a tight fit in piston bosses at room temperature. Piston should be heated to 70-80 degrees C. before pressing pin out. Be sure to remove retaining snap rings. Maximum allowable clearance of pin in rod bushing is 0.04 mm. New pin bushing may be pressed into rod if clearance is excessive. It will be necessary to ream bushing after installation to provide recommended clearance of 0.011-0.038 mm. Refer to the following specification data:

MF205
Piston pin
   diameter ....24.990-24.996 mm
Rod bushing
   diameter ....25.007-25.028 mm
Pin bore in piston....24.989-24.996 mm

MF210 & MF220
Piston pin
   diameter ....27.990-27.996 mm
Rod bushing
   diameter ....28.007-28.028 mm
Pin bore in piston....27.989-27.999 mm

CONNECTING RODS AND BEARINGS

All Models

26. Connecting rod crankpin bearing inserts are available in standard size, 0.25 mm undersize and 0.50 mm undersize. Refer to the following specification data:

MF205
Piston pin bushing
   I.D. ....25.007-25.028 mm
    Pin clearance.....0.011-0.038 mm
Crankpin journal std.
   diameter ....52.960-52.980 mm
    Journal fillet ....2.00-2.50 mm
    Maximum out-of-round or
     taper ....0.03 mm
Connecting rod bearing to crankpin
clearance –
    Desired ....0.020-0.085 mm
    Wear limit ....0.085 mm
Connecting rod side clearance on
   crankpin ....0.1-0.3 mm
Rod bolt torque ....49-58 N·m

MF210 & MF220

Piston pin bushing
I.D. ....28.007-28.028 mm
  Pin clearance......0.011-0.038 mm
Crankpin journal std.
  diameter ....57.960-57.980 mm
    Journal fillet ....2.00-2.50 mm
    Maximum out-of-round or
      taper ....0.03 mm
Connecting rod bearing to crankpin
  clearance –
    Desired ....0.020-0.085 mm
    Wear limit ....0.085 mm
Connecting rod side clearance on
  crankpin ....0.1-0.3 mm
Rod bolt torque ....59-67.8 N·m

When installing connecting rod cap retaining screws, notice that a special screw is used which enters a special machined seat in rod cap. It is suggested that the special screws be checked by magnetic partical inspection then visually examined. Renew any connecting rod screw, which is found defective during magnetic check, if indented not parallel to axis of screws, if indentations under head of screw extend into screw shank (regardless of depth), or if longitudinal indentations (including seams) exceed 0.2 mm in depth.

CRANKSHAFT AND MAIN BEARINGS

All Models

29. The crankshaft is supported in three slip-in, precision type main bearings. To remove the main bearing inserts, it is necessary to remove the engine from tractor as outlined in paragraph 10, then remove the flywheel, rear seal housing, and rear engine plate. Remove the timing gear cover, water pump, injection pump, timing gears and front engine plate. Remove oil pan and oil pump, mark main bearing caps and block to indicate position (1, 2 and 3) of caps, then unbolt and remove main bearing caps. If connecting rod caps are removed, the crankshaft can be removed.

Main journal standard diameter is 69.96-69.98 mm and standard size inserts as well as 0.25 and 0.50 mm undersizes are available. Main journals should be reground to undersize or renewed if out-of-round or taper exceeds 0.03 mm. If reground to undersize, fillet radii of 2.0-2.5 mm should be maintained. Main bearing journal clearance should be 0.02-0.085 mm with wear limit of 0.15

mm. Crankshaft end play should be 0.1-0.3 mm and is controlled by thrust bearing inserts at center main bearing. Thrust bearing inserts are on cap and in block for MF210 and MF220 models, but only on center main cap of MF205 models. Thrust bearing inserts are located on dowel pins in cap to prevent rotation. Thicker (0.25 mm) thrust bearings are available to limit crankshaft end play.

When assembling, be sure to assemble main bearing caps in original location. Notice that arrow cast into caps points toward rear. Main bearing cap retaining special screws seat on main bearing cap to provide lock without using spring washer. Tighten main bearing cap screws to 59-78 N·m torque.

CRANKSHAFT REAR OIL SEAL

All Models

30. The lip type crankshaft rear oil seal (4 – Fig. 25) is contained in rear oil seal cover (5). Seal can be renewed after splitting tractor between engine and clutch housing, then removing clutch, flywheel and seal cover. Closed side of seal (4) should be toward rear and lip should be toward front. Cover (5) is piloted only in cylinder block (top). Tighten flywheel retaining screws to 44-49 N·m.

FLYWHEEL

All Models

32. Flywheel is positively located on crankshaft by a dowel pin and six screws. Make sure mating surfaces of flywheel and shaft are clean and free of dirt, rust and burrs. Tighten retaining cap screws evenly to 44-49 N·m.

Pilot bearing is a press fit in center of flywheel. Starter ring gear is a tight shrink fit at outside of flywheel. Clutch surface of flywheel must be smooth, flat and not show signs of overheating. Surface may be machined, but no more than 1mm should be removed before renewing flywheel.

OIL PAN

All Models

34. The oil pan is located below the engine block, between the front and rear plates which are attached to the engine block. Removal of the oil pan requires removal of the front and rear plates as follows:

Drain engine oil and coolant, then remove the engine as outlined in paragraph 10. Refer to paragraph 20 to

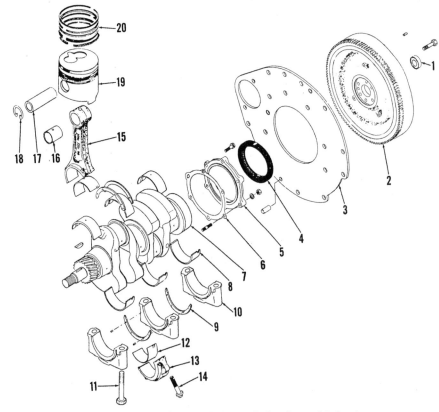

Fig. 25 – Exploded view of typical crankshaft and associated parts.

1. Pilot bearing	6. Gasket	11. Screw	16. Bushing
2. Flywheel	7. Crankshaft	12. Rod insert	17. Piston pin
3. Rear plate	8. Main bearing insert	13. Rod cap	18. Retaining ring
4. Rear seal	9. Thrust bearing	14. Screw	19. Piston
5. Seal cover	10. Main cap	15. Connecting rod	20. Rings

remove timing gear cover and paragraph 21 to remove timing gears. Remove water pump, then unbolt and remove the front plate. Refer to paragraph 76 and remove clutch assembly. Unbolt and carefully remove flywheel and refer to paragraph 30 and remove crankshaft rear oil seal cover. Unbolt and remove engine rear plate, then unbolt and remove oil pan.

Clean mating surfaces of oil pan completely, then coat surface evenly with an RTV sealant. Install oil pan and torque pan to block retaining screws to 29-34 N·m. Reinstall rear plate and refer to paragraph 30 for installation of rear seal and cover. Install front plate, refer to paragraph 21 for installing timing gears and to paragraph 20 for installation of timing gear cover. Refer to paragraph 10 for reinstalling engine.

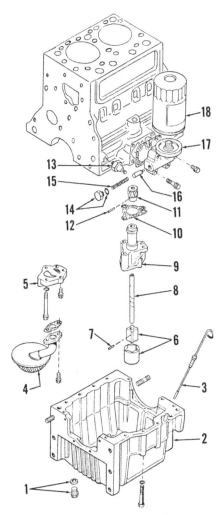

Fig. 28 — Exploded view of typical oil pump and related parts.

1. Drain plug	10. Gasket
2. Oil pan	11. Drive gear
3. Dip stick	12. Pin
4. Inlet	13. Pressure switch
5. Cover	14. Plug
6. Pump rotor set	15. Relief valve spring
7. Pin	16. Relief valve
8. Pump shaft	17. Filter housing
9. Pump body	18. Oil filter

OIL PUMP

All Models

35. The rotor type oil pump is driven by the gear in center of valve camshaft. The oil pump can be removed after removing oil pan as outlined in paragraph 34. Oil pump can be unbolted and withdrawn.

Tip clearance of rotors should be 0.07-0.12 mm. If tip clearance exceeds 0.2 mm renew rotors. Side clearance of rotors should be 0.03-0.06 mm with wear limit of 0.1 mm. Clearance between outer rotor and pump body bore should be 0.1-0.18 mm with wear limit of 0.2 mm.

When assembling pump, use a small quantity of sealer on flange of housing, then tighten pump cover retaining screws evenly to 4.4-5.9 N·m torque. Install pump to bottom of engine block and tighten retaining screws to 12-15 N·m. Oil screen retaining screws should be tightened to same (12-15 N·m) torque. Refer to paragraph 34 for installation of oil pan.

OIL CONTROL HOUSING AND RELIEF VALVE

All Models

36. The oil control housing (17 – Fig. 28) is located on left side of engine block and includes the mounting for oil filter (18) and relief valve (14, 15 and 16). The oil filter cartridge contains a built in by-pass check valve and the complete filter cartridge element should be renewed at least every 100 hours of operation. Relief valve opening pressure should be 490-539 kPa. Low pressure indicator light should be turned on by switch (13) when oil pressure is less than 48.26 kPa.

DIESEL FUEL SYSTEM

All models are equipped with a two plunger Diesel KIKI type PES.K injection pump, pintle nozzles and a swirl chamber combustion chamber.

Because of extremely close tolerances and precise requirements of all diesel components, it is of utmost importance to use clean fuel and that careful maintenance be practiced at all times. Unless necessary special tools are available, service on injectors and injection pumps should be limited to removal, installation and exchange of complete assemblies. It is impossible to re-calibrate an injection pump or reset an injector without proper specifications, equipment and training.

FUEL FILTERS AND LINES

All Models

40. **OPERATION AND MAINTENANCE.** The fuel system includes a fuel filter that should be renewed at regular intervals and lubrication oil in injection pump crankcase should be checked and changed at regular intervals.

Filter life depends more upon careful maintenance than it does on hours or conditions of operation. Necessity for careful filling with CLEAN fuel cannot be over-stressed. To minimize contamination of diesel fuel system, the following precautions are recommended.

Fill fuel tank after use and before storage, to eliminate presence of humid air in tank and reduce contamination due to condensation.

Check fuel strainer at bottom of fuel tank and drain off any water **at least every 50 hours or once each week** (more often if trouble is suspected). Allow fuel to drain until clean fuel flows. Make sure that sediment bowl fills before attempting to start engine to avoid entrance of air in injection pump. If bowl doesn't fill, open shut-off valve and loosen bowl slightly by loosening finger nut on retaining bail. Tighten finger nut when fuel has filled bowl.

Renew filter element **at least every 400 hours** or immediately if water contamination is discovered. Bleed filter and lines as outlined in paragraph 41 after new filter is installed. **Do not attempt to start engine until air is bled from system or injection pump may be damaged.**

The injection pump is lubricated by 110 ml of the same type of oil as used in engine. Oil in the injection pump should be checked at **100 hour or 2 week intervals** and should be drained and filled with new oil every **200 hours, every 2 months** or when seasonal changes in engine oil are required. Oil drain plug is shown at (3 – Fig. 40), level plug at (2) and fill plug at (1).

Fig. 40 — View of typical fuel injection pump showing fill plug (1), level plug (2) and drain plug at (3).

41. BLEEDING. The fuel system should be bled if fuel tank is allowed to run dry, if fuel lines, filter or other components within the system have been disconnected or removed or if engine has not operated for long period of time. If the engine fails to start or if it starts, then stops, the cause could be air in the system, which should be removed by bleeding.

To bleed, make sure tank has sufficient amount of fuel, then open shut-off valve at bottom of tank. If sediment bowl doesn't fill, loosen bowl slightly by loosening finger nut until fuel fills bowl. Use small wrench (not screwdriver) to loosen bleed screw (4–Fig. 41) 1½-2 turns counter-clockwise. Allow air bubbles and fuel to drain from bleed screw until air bubbles completely disappear, then close bleed screw. **Do not completely remove bleed screw and do not crank engine or attempt to start with bleed screw open.**

Partially open throttle and attempt to start engine. If engine does not start, loosen high pressure fuel lines at injectors and continue to crank engine until fuel escapes from all loosened connections. Tighten compression nuts and start engine.

INJECTION PUMP

All Models

The injection pump is a completely sealed unit and no service work or disassembly other than that specified should be attempted without necessary special equipment and training. Specified jobs must be performed under conditions of strict cleanliness. Minute dust particles can damage the diesel fuel system.

42. TIMING TO ENGINE. Start of injection should occur at 20 degrees BTDC. Firing order is front (No. 1) cylinder, followed by 540 degrees of crankshaft rotation, rear (No. 2) cylinder injection, 180 degrees of crankshaft rotation, then front cylinder injection again.

MF205 models have two marks on crankshaft pulley (180 degrees apart) and degree marks on timing gear cover as shown in Fig. 42.

MF210 and MF220 models have a pointer on timing gear cover and notches on crankshaft pulley. Two notches in crankshaft pulley (1–Fig. 43) indicates TDC for number 1 (front) cylinder when aligned with pointer. A single mark on pulley (2) located 180 degree from the double notches is TDC mark for rear cylinder. A single notch located 20 degrees before the double notches is used for setting injection pump spill timing.

To set injection spill timing, turn crankshaft in normal direction until front piston is beginning compression stroke, then align 20 degree BTDC mark as indicated in Fig. 42 or Fig. 43. Be sure that front cylinder is on compression stroke. Shut fuel off at tank shut-off valve and check that tank is filled with fuel. Remove high pressure injector line to front cylinder injector, then remove retainer (2–Fig. 44) and delivery valve spring (1), then reinstall retainer without spring. Loosen the three injection pump retaining nuts, move throttle

Fig. 42 – View of MF205 injection timing marks on cover and notch in crankshaft pulley. Refer to Fig. 43 for other models.

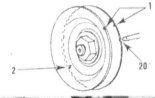

Fig. 43 – View of timing pin and pulley used on MF210 and MF220 models. Refer to text for setting injection timing.

lever to full open position and open fuel tank shut-off valve. Move top of injection pump toward engine until fuel flows from delivery valve holder. Rotate pump back until fuel flow just stops, then tighten the three pump retaining nuts to maintain pump position. Turn fuel off, remove retainer (2) reinstall spring (1), then tighten retainer to 34 N·m torque. Install high pressure line from pump to injector nozzle and leave nozzle end loose. Refer to paragraph 41 and bleed system including high pressure line. Tighten injection line fittings to 20-25 N·m torque.

43. REMOVE AND REINSTALL. To remove the injection pump, first remove the timing gear cover as outlined in paragraph 20. Remove injection pump drive gear retaining nut, and use suitable puller to remove injection pump gear. Shut fuel off at tank, remove high pressure injector lines and disconnect supply line to injection pump. Cover all fuel openings to prevent entrance of dirt, then unbolt and remove injection pump.

When assembling, install injection pump and lines but do not tighten. Refer to paragraph 19 and align all timing marks while installing injection pump gear. Torque idler gear retaining screw to 42-49 N·m and injection pump gear retaining nut to 59-69 N·m. Refer to paragraph 20 for installation of timing gear cover. Refer to paragraph 42 and time injection pump to engine. Bleed the fuel system including the high pressure fuel lines to injectors as described in paragraph 41.

44. ENGINE SPEED. Be sure that injection timing is correctly set as described in paragraph 42 and that engine is otherwise operating correctly. Low idle speed should be 700-750 rpm for MF205 models, 750-800 rpm for other models. Rated full load speed should be 2400 rpm for MF205 models, 2500 rpm for other models. High idle no load speed should be 2450-2500 rpm for

Fig. 41 – View of typical injection pump showing bleed screw (4). Refer to text for bleeding procedure.

Fig. 44 – View of pump with retainer removed showing spring (1). Refer to text for setting spill timing.

MF205 models, 2650-2750 rpm for other models. Seals on speed adjusting stop screws should not be broken by unauthorized personnel.

INJECTOR NOZZLE

All Models

All models use a straight pintle nozzle as shown in Fig. 45.

46. **TESTING AND LOCATING A FAULTY NOZZLE.** If engine is missing and fuel system is suspected as being the cause of trouble, system can be checked by loosening each injector line connection in turn, while engine is running at slow idle speed. If engine operation is not materially affected when injector line is loosened, that cylinder is missing. Remove and test (or install a new or reconditioned unit) as outlined in appropriate following paragraphs.

47. **REMOVE AND REINSTALL.** Before removing an injector or loosening injector lines, thoroughly clean injector, lines and surrounding area using compressed air and a suitable solvent. Remove high pressure line leading from pump to injector unit and disconnect the bleed line from injector. Unscrew nozzle nut (3 – Fig. 45) and withdraw nozzle from cylinder head.

Clean exterior surfaces of nozzle assembly and bore in cylinder head, being careful to keep all surfaces free of dust, lint and other small particles until nozzle is installed. Tighten nozzle holder into cylinder to 59-78 N·m torque. Refer to paragraph 41 for bleeding information.

48. **TESTING.** A complete job of testing and adjusting the injector requires use of special test equipment. Only clean, approved testing oil should be

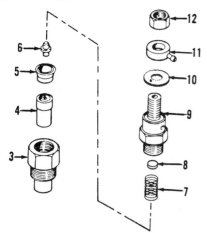

Fig. 45 — Exploded view of injection nozzle.

3. Nozzle nut	8. Spring shim
4. Nozzle	9. Nozzle holder
5. Spacer	10. Gasket
6. Push rod	11. Banjo fitting
7. Spring	12. Nut

used in tester tank. Nozzle should be tested for opening pressure, seat leakage, back leakage and spray pattern. When tested, nozzle should open with a high-pitched buzzing sound, and cut off quickly at end of injection with a minimum of seat leakage and a controlled amount of back leakage.

Before conducting test, operate tester lever until fuel flows, then attach injector. Close valve to tester gage and pump tester lever a few quick strokes to be sure nozzle valve is not stuck, and that possibilities are good that injector can be returned to service without disassembly.

WARNING: Fuel leaves injector nozzle with sufficient force to penetrate the skin. Keep exposed portions of your body clear of nozzle spray when testing.

49. OPENING PRESSURE. Open valve to tester gage and operate tester lever slowly while observing gage reading. Opening pressure should be 11800-12700 kPa (1710-1840 psi).

Opening pressure is adjusted by adding or removing shims in shim pack (8 – Fig. 45).

50. SPRAY PATTERN. Spray pattern should be well atomized and slightly conical, emerging in a straight axis from nozzle tip. If pattern is wet, ragged or intermittent, nozzle must be overhauled or renewed.

55. **OVERHAUL.** Hard or sharp tools, emery cloth, grinding compound or other than approved solvents or lapping compounds must never be used. An approved nozzle cleaning kit is available through a number of specialized sources.

Wipe all dirt and loose carbon from exterior of nozzle and holder assembly. Refer to Fig. 45 for exploded view and proceed as follows:

Secure nozzle in a soft jawed vise or holding fixture and remove nut (3). Place all parts in clean calibrating oil or diesel fuel as they are removed, using a compartmented pan and using extra care to keep parts from each injector together and separate from other units which are disassembled at the time.

Clean exterior surfaces with a brass wire brush, soaking in an approved carbon solvent if necessary, to loosen hard carbon deposits. Rinse parts in clean diesel fuel or calibrating oil immediately after cleaning to neutralize the solvent and prevent etching of polished surfaces.

Clean nozzle spray hole from inside using a pointed hardwood stick or wood splinter, then scrape carbon from pressure chamber using hooked scraper. Clean valve seat using brass scraper, then polish seat using wood polishing stick and mutton tallow.

Reclean all parts by rinsing thoroughly in clean diesel fuel or calibrating oil and assemble while parts are immersed in cleaning fluid. Make sure adjusting shim pack is intact. Tighten nozzle retaining nut (3) to a torque of 61-75 N·m. Do not overtighten, distortion may cause valve to stick and no amount of overtightening can stop a leak caused by scratches or dirt. Retest assembled injector as previously outlined.

GLOW PLUGS

All Models

56. Glow plugs are parallel connected with each glow plug grounding through mounting threads. Start switch is provided with a "GLOW" position which can be used to energize the glow plugs for faster warm up. Normal starting position also energizes the glow plugs, but does not cause "GLOW" lamp on instrument panel to light. If indicator light fails to glow when start switch is held in "GLOW" position an appropriate length of time, check for loose connections at switch, indicator lamp, resistor connections, glow plug connections and ground. A test lamp can be used at glow plug connection to check for current to glow plug.

COOLING SYSTEM

All models use a pressurized cooling system which raises coolant boiling point. An impeller type centrifugal pump is used to provide forced circulation, and a thermostat is used (on all except MF205 models) to stabilize operating temperature.

RADIATOR

All Models

60. Radiator cap pressure valve is set to open at 88.2 kPa (12.8 psi) on all models. Cooling system capacity is 4.0 liters for MF205 models; 5.0 liters for other models.

To remove radiator, first drain coolant and remove hood. Remove air cleaner hose, disconnect coolant hoses and remove radiator and fan shroud as an assembly. Install by reversing removal procedure.

THERMOSTAT

MF210·MF220 Models

61. On models so equipped, the by-

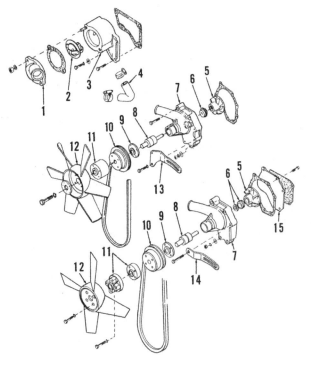

Fig. 62 — Exploded view of thermostat housing and water pump. Lower water pump is used on MF205 models.

1. Water outlet
2. Thermostat
3. Thermostat housing
4. Thermostat by-pass hose
5. Impeller
6. Seal
7. Pump housing
8. Shaft and bearing
9. Hub
10. Pulley
11. Spacer
12. Fan
13. Alternator bracket
14. Alternator bracket
15. Rear cover

nected and engine running should be approximately 27 volts AC. If AC test is within limits but DC test was not, rectifier/regulator unit is probably faulty. Alternator coil winging can be checked with ohmmeter with engine stopped. Continuity should exist across alternator leads but not between leads and alternator frame.

Rectifier/regulator (1 – Fig. 65A) can be checked after detaching connector (2 and 3). Black wire on rectifier/regulator is used to ground unit to tractor frame and a good ground is necessary for proper control. Check continuity with an ohmmeter between red wire terminal and each blue wire terminal. It will be necessary to reverse ohmmeter leads but continuity should exist with one connection, but not with leads reversed. Install new rectifier/regulator if continuity does not exist in one direction and the DC and AC charging tests also indicate that rectifier/regulator is probably at fault.

pass type thermostat is located in outlet elbow. Thermostat should begin to open at 71 degrees C. (160°F.) and be completely open at 79 degrees C. (174°F.).

WATER PUMP

All Models

62. Refer to Fig. 62 for exploded view of water pump. To remove pump, first drain coolant. Loosen, then remove water pump/alternator belt. Unbolt the fan and allow it to lower against radiator, then unbolt and remove water pump.

The shaft and bearing assembly (8) is pressed into pump body as is seal (6). Part of seal assembly is bonded to impeller of some models and is renewable only with impeller.

Installation is reverse of removal procedure. Tighten water pump/alternator belt by moving alternator in slotted bracket (13 or 14). Correct belt tension will permit belt deflection of 8-10 mm when pressed with thumb (approximately 147 N.) between water pump pulley and alternator pulley.

ELECTRICAL SYSTEM

ALTERNATOR AND REGULATOR

MF205 Models

65. The Kokusan Denki alternator uses a separate rectifier/regulator located behind the instrument panel. The rectifier/regulator is a sealed unit and service is limited to renewal of the unit.

The alternator rotor (1 – Fig. 65) contains permanent magnets and the drive pulley. The stator coils (4) are mounted on the plate (bearing housing). No brushes are used and most service to alternator will be limited to mechanical damage to coils or to bearing failure.

With engine running and charging system operating properly, voltage across battery terminals should be 15 volts DC. Voltage across alternator leads, with alternator leads discon-

MF210-MF220 Models

66. The Hitachi 15 amp alternator incorporates regulator and rectifier (diode) units. Individual components can be renewed if necessary. Length of brushes new is 14.5 mm and wear limit is 7 mm. Rotor coil resistance should be 10 ohms at 20 degrees C. Resistance of each of the three phases of stator coil (7 – Fig. 66) should be 0.11 ohm at 20 degrees C. Slip ring outside diameter is 31 mm when new. Run out limit of slip rings is 0.3 mm and should be corrected to within 0.05 mm but slip ring diameter should not be less than 30 mm after correcting. Front bearing (12) is 6202B and rear bearing (8) is 6201SD.

When assembling, brush holder (4) should be tightened to 1.6-2.0 N·m torque; diode (5), regulator (3), "B" and "E" terminals should all be tightened to 3.1-3.4 N·m; front bearing retainer screws (10) should be torqued to 1.6-2.0 N·m and pulley nut (21) should be tightened to 34-39 N·m torque. Threads

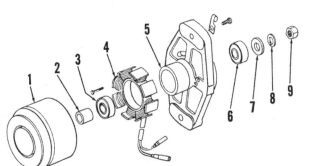

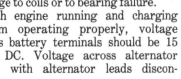

Fig. 65 — Exploded view of alternator used on MF205 models.

1. Rotor and pulley
2. Spacer
3. Bearing
4. Stator
5. Housing
6. Bearing
7. Washer
8. Lock washer
9. Nut

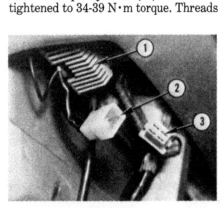

Fig. 65A — View under MF205 instrument panel showing regulator/rectifier unit (1) and connector (2 and 3). Refer to text for checking.

of through bolts (6) should be coated with "Loctite" before tightening to 3.1-3.4 N·m torque.

STARTER

MF205 Models

67. The Hitachi starter used on these models has a gear reduction built into the nose of the starter. Refer to Fig. 67.

Removal of the starter is accomplished in usual manner. To disassemble, remove the two screws which attach solenoid (1), then withdraw solenoid, lever, spring and plunger as shown in Fig. 67A. Remove through bolts (6 – Fig. 67), then separate rear cap (7), armature (9), field coil assembly (8) and brush assembly (12). Front cover (22) is attached to center housing (13) with three screws. One of the three screws is located inside center housing (13).

MF210-MF220 Models

68. Hitachi S12-31 or S12-61 starter is used on these models. Refer to the following test specifications.

S12-31
No Load Test
    Volts 8.2
    Amperes . . . . Less than 90
    Rpm . . . . More than 4000
Loaded Test
    Amperes 450
    Torque . . . . .13 N·m
    Rpm . . . . More than 1200

S12-61
No Load Test
    Volts 8.5
    Amperes . . . . Less than 90
    Rpm . . . . More than 4000
Loaded Test
    Amperes 420
    Torque . . . . .15 N·m
    Rpm . . . . More than 1200

Insulation between commutator segments should be undercut 0.5-0.8 mm for both models. Brush length new is 19.5 mm and minimum limit is 11.5 mm. With solenoid (15 – Fig. 68) engaged, clearance between pinion (9) and stopper (10) should be 0.3-1.5 mm. Spacer plates of 0.4 and 0.8 mm thickness can be installed between seal (14) and solenoid (15) to decrease clearance between pinion and stopper. Tighten solenoid mounting screws (18) and nut for pivot pin (13) should be tightened to 3.43-4.90 N·m. Screws attaching brush holder (2) to rear cover (1) should be tightened to 1.67-2.35 N·m and through bolts (3) should be tightened to 6.37-8.83 N·m torque. Screws attaching field coil cores to housing (4) should be tightened to 29.42-34.32 N·m torque.

CIRCUIT DESCRIPTION

69. Refer to Fig. 69 for wiring schematic of MF205 models and to Fig.

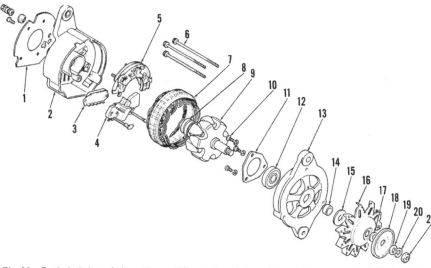

Fig. 66 – Exploded view of alternator, rectifier and regulator unit used on MF210 and MF220 models.

1. Dust cover	7. Stator assembly	12. Ball bearing	17. Washer
2. Cover	8. Ball bearing	13. Front cover	18. Pulley flange
3. Regulator	9. Rotor	14. Spacer	19. Washer
4. Brush assembly	10. Screws	15. Washer	20. Lock washer
5. Diode assembly	11. Bearing retainer	16. Fan	21. Nut
6. Screws			

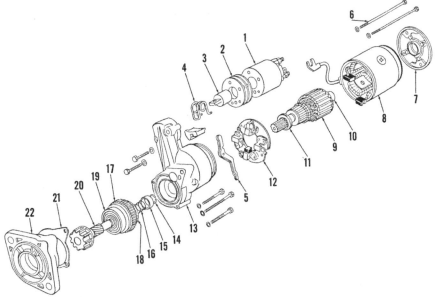

Fig. 67 – Exploded view of starter assembly used on MF205 models.

1. Solenoid		17. Clutch pinion
2. Gaskets	7. End cap	12. Brush holder	18. Ball bearing
3. Plunger	8. Stator	13. Starter housing	19. Ball bearing
4. Spring	9. Armature	14. Retaining ring	20. Pinion
5. Lever	10. Ball bearing	15. Collar	21. Dust cover
6. Screws	11. Ball bearing	16. Spring	22. Gear cover

Fig. 67A – View showing proper relation of spring (4), solenoid (1) and lever (5).

70 for diagram typical of MF210 and MF220 models. Negative terminal of battery is grounded on all models.

CLUTCH

ADJUSTMENT

All Models

75. Clutch pedal free play should be 20-30 mm on all models. To adjust pedal free play, change length of clutch control rod at clutch arm end. Refer to Fig. 75.

REMOVE AND REINSTALL

All Models

76. **CLUTCH SPLIT.** To detach (split) tractor between engine and clutch housing, first remove hood, disconnect battery and wedge blocks between frame and front axle to prevent engine from tilting. On MF210 and MF220 models, drain engine coolant and detach temperature sending unit from engine. On four-wheel-drive models, remove drive shaft to front wheels. On all models, support rear of engine and front of clutch housing. Shut fuel off under fuel tank and detach fuel shut-off rod from pump. Disconnect wire to glow plugs, injector bleed off line, tachometer cable, wiring harness and steering drag links (at rear). Remove right side step plate, then remove the two hydraulic lines between hydraulic pump (at front) and hydraulic system (at rear). Cover all tubes, hoses and ports to prevent entry of dirt. Remove all bolts and cap screws, which attach engine to clutch housing, then carefully separate engine from clutch housing. Be careful to assure that each part of tractor is adequately supported. Clutch housing is aluminum and is easily damaged.

When assembling, check alignment carefully, reverse disassembly procedure and observe the following torque values. Tighten cap screws and nuts, which attach engine to clutch housing to 63-73 N·m torque. Starter attaching screws should be tightened to 39-45 N·m torque.

77. Clutch pressure plate assembly can be removed from flywheel by removing screws which attach plate to flywheel.

When installing, long side of clutch friction disc hub should be toward rear. Use removed clutch shaft or suitable pilot shaft to align friction disc while attaching pressure plate to flywheel. Tighten all six retaining screws evenly

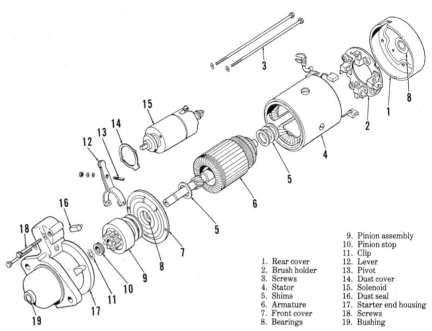

9. Pinion assembly
10. Pinion stop
11. Clip
12. Lever
13. Pivot
14. Dust cover
15. Solenoid
16. Dust seal
17. Starter end housing
18. Screws
19. Bushing

1. Rear cover
2. Brush holder
3. Screws
4. Stator
5. Shims
6. Armature
7. Front cover
8. Bearings

Fig. 68 — Exploded view of starter used on MF210 and MF220 models.

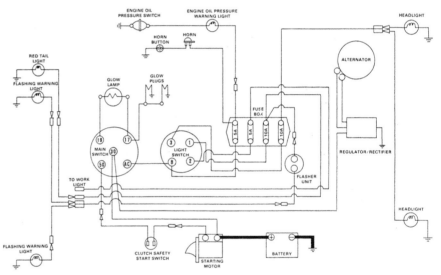

Fig. 69 — Wiring diagram for MF205 models.

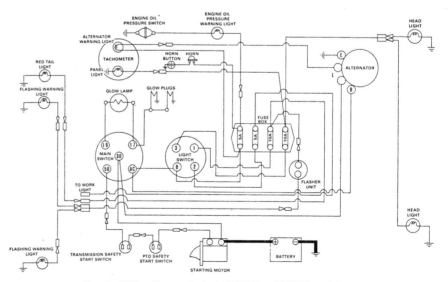

Fig. 70 — Wiring diagram typical of MF210 and MF220 models.

to 6.8-9.5 N·m.

Refer to paragraph 80 for clutch shaft and release bearing service and to paragraph 75 for clutch pedal free play adjustment.

OVERHAUL

All Models

78. Inspect the removed clutch disc and pressure plate assembly as follows. Position a straight edge across friction surface of pressure plate and measure warpage by measuring gap with a feeler gage. Gap (warpage) must be less than 0.02 mm. Measure thickness of clutch fingers which contact throw-out bearing. Install new clutch pressure plate if clutch fingers are less than 1.1 mm thick or if they show any damage. Friction surface of disc should be within 1.0 mm of perfectly flat. Refer to paragraph 32 for inspection of flywheel.

CLUTCH SHAFT AND RELEASE BEARING

All Models

80. Separate clutch housing from

Fig. 75 — View of clutch linkage showing adjustment points. Clevis (2) can be turned to change length of clutch rod (1). Measured free play should be within limits of 20-30 mm.

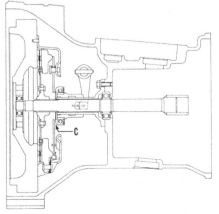

Fig. 79 — Chamfered side (C) of release bearing should be toward clutch fingers.

engine as outlined in paragraph 76. The clutch release (throw-out) bearing (2–Fig. 80) and housing (3) can be removed after removing the two retaining spring clips (1). Bearing is packed at time of manufacture with heat resistant grease. **DO NOT** wash bearing in solvent or water. Old bearing can be pressed from housing and new bearing can be pressed on if renewal is required. Chamfered side of release bearing must be toward clutch fingers (engine) as shown in Fig. 79.

Remove the three screws (7–Fig. 80) which attach pilot sleeve (8) to front of clutch housing, then withdraw clutch shaft (12), pilot sleeve (8) and bearing (11). Install new bearing if questionable. Examine end of shaft (12) and renew shaft and pilot bearing in flywheel if end of shaft is damaged.

TRANSMISSION

(MF205 TWO WHEEL DRIVE)

Two wheel drive MF205 models are equipped with a six speed transmission shown in Fig. 85.

INSPECTION

85. The shift cover together with the shift rails and forks can be unbolted from top of transmission housing. Check

Fig. 80 — Exploded view of clutch shaft, release bearing and associated parts.

1. Clips
2. Release bearing
3. Housing
4. Release fork
5. Pins
6. Release shaft
7. Screws
8. Pilot sleeve
9. Snap rings
10. Washers
11. Bearing
12. Clutch shaft

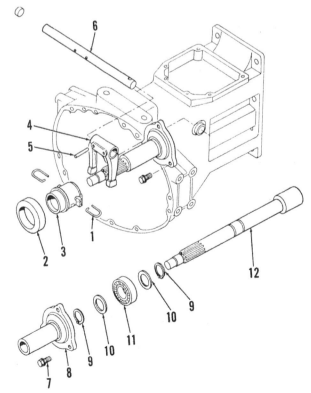

to be sure that forks are firmly attached to shift rails and check condition and operation of detents in rails. Gears and shafts can be visually inspected through opening at top. Transmission, center housing (differential and final drive), rear axles and hydraulic system share a common sump, which should be drained as described in paragraph 86. Check filter, drained oil and transmission compartment for evidence of metal particles.

LUBRICATION

86. Transmission, center housing (differential and final drive), rear axles and hydraulic system share a common sump. The drain plug is located on bottom of center housing. Be sure to allow sufficient time for oil to drain from interconnected compartments. Remove strainer from under cover (S–Fig. 86). Strainer can be cleaned with diesel fuel and reinstalled if not damaged. Be sure to use new gasket between flange of strainer and center housing.

Fill system through fill plug opening (F) in top cover with Massey-Ferguson Permatran (M-1129A) or equivalent. Capacity is 16 liters; however, correct amount will be to level of plug (L) on side of housing. Be sure to allow sufficient time for oil to pass between compartments and be sure that tractor is level before checking at level plug (L).

Check for leaks after oil has warmed

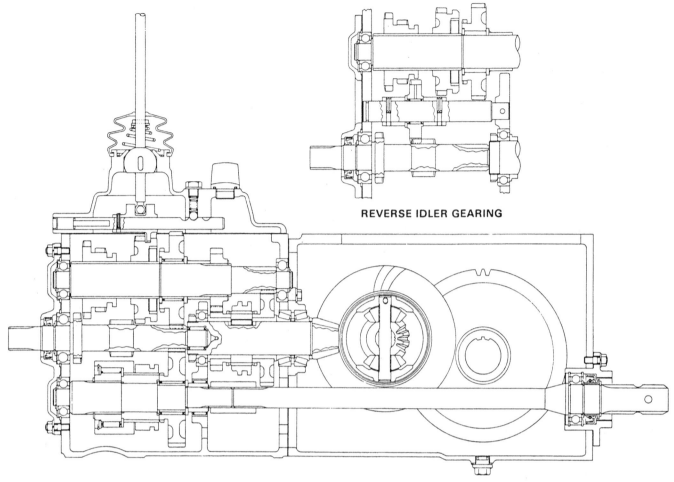

Fig. 85 — Cross-section of MF205 Two Wheel drive transmission.

REVERSE IDLER GEARING

up especially around strainer cover and drain plug.

REMOVE AND REINSTALL

88. To detach (split) tractor between clutch housing and front of transmission, proceed as follows: Drain oil from transmission and center housing as described in paragraph 86. Disconnect battery cables and rear light wiring harness. Detach pedal return springs for both brake pedals and clutch pedal, then disconnect clutch operating rod. Remove right step plate, and disconnect hydraulic lines at rear. Wedge blocks between frame and front axle to prevent tipping. Block up under rear of engine and under any front end weights. Support transmission with movable floor jack or other suitable method that will permit moving rear of tractor away from engine and clutch housing. Remove screws which attach front of transmission to clutch housing, then carefully separate by moving rear of tractor straight back away from clutch housing.

To remove transmission, first detach clutch and brake pedal support bracket from bottom of transmission. Support

the rear of tractor (center housing) sufficiently to prevent tipping. Attach a suitable support to transmission, then unbolt and remove transmission from center housing.

When reinstalling, tighten 12 mm retaining stud nuts to 46-54 N·m torque. Tighten pedal support retaining screws to 17.6-20.6 N·m torque. Refer to paragraph 86 for lubrication requirements.

OVERHAUL

90. **SHIFTER COVER.** Remove shifter cover and inspect shift forks and rails for wear or damage. Detent balls and springs can be removed after removing plugs (11 – Fig. 90) Both roll pins (20 and 19) must be removed from each shift fork (16 and 18) before rails can be removed from forks. When assembling, be sure to install double roll pins (19 and 20) in each shift fork.

91. **TOP (COUNTER) SHAFT.** To remove the top shaft, first remove transmission as outlined in paragraph 88. Unbolt and remove shifter cover from top of transmission and aluminum

cover (1 – Fig. 91) from front. Remove snap rings (3 and 18) and washers (4 and 17) from ends of shaft, then remove both bearings (5 and 16). Release snap ring (15) from groove and move toward rear of shaft. Move shaft and gears forward in housing until rear end of shaft can be lifted out top opening.

92. **REVERSE IDLER.** Remove set screw (26 – Fig. 92) then drive roll pins (22) from each of the spacer collars (21). Slide the idler gear shaft (20) toward front and remove parts (21, 23, 24 and 25) through top opening.

93. **INPUT SHAFT.** The input shaft can be withdrawn from front of housing after removing the aluminum cover from front of transmission housing. Be careful that bearing or parts of damaged bearing (45 – Fig. 93) are removed if allowed to fall to bottom of housing.

Be especially careful when installing input shaft, because rollers of bearing (45) may be loose enough in retainer (cage) to catch and prevent entrance of input shaft. A suggested method of installation is to position bearing (45) on end of shaft (30) using a light grease to hold bearing rollers in against inner race and on shaft. Care must still be exercis-ed to assure correct assembly.

94. PINION GEAR AND SHAFT. To remove the pinion gear and shaft (47 – Fig. 93) and related parts, it is necessary to remove nut (31). Nut is left-hand thread and secured using "Loctite". To remove, secure pinion in a suitable holding fixture or by carefully blocking gears. The nut (31) can not be completely removed until bearing retainer (42) is loosened allowing shaft to move toward rear. Be careful not to damage or lose pinion positioning shims (41). It may be necessary to bump shaft towards rear if front bushing (33) for gear (34), or front bearing (32) is tight on shaft. Be extremely careful not to damage bearing (45) or bearing bore in end of shaft (47).

Refer to paragraph 95 for removal and reinstallation of pto drive shaft.

Original thickness of shims (40) can be reinstalled if pinion shaft bearings (38 and 43) and bearing housing (42) are reassembled. If any of these parts are renewed or if adjustment of bearings (38 and 43) is questioned, proceed as follows: Assemble parts (38, 39, 40, 42 and 43) on special dummy pinion with nut (31) or assemble all parts (31 through 40, 42, 43 and 44) on shaft (47), outside of housing. Tighten nut (31) temporarily to 88 N·m, clamp bearing housing (42) in vise, then check rolling torque of shaft (47) in bearings. If rolling torque is more than 1.17 N·m, then increase thickness of shims (40) and recheck. If

rolling torque of pinion shaft is less than 0.58 N·m; decrease thickness of shims (40) and recheck. Shims (40) are available in 0.10, 0.15, 0.20 and 1.00 mm thicknesses. After correct thickness of shims (40) has been determined, pinion shaft (47) and associated parts (31 through 46), can be assembled in transmission housing. Apply "Loctite" to pinion shaft threads, then tighten nut (31) to final torque of 78.45-98.07 N·m.

Thickness of shims (41) determines the mesh position of pinion with differential ring gear. Original thickness of shims (41) can be considered correct if original pinion shaft (47) and bearing housing (42) are reinstalled. If mesh position was questioned, refer to paragraph 133 for

Fig. 86 — View showing fill plug (F), level (L) and strainer cover (S).

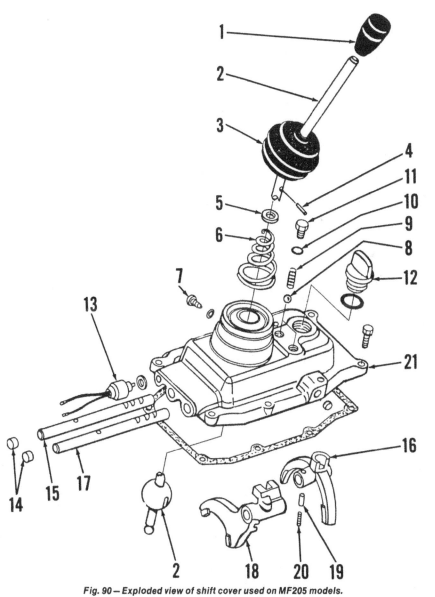

Fig. 90 — Exploded view of shift cover used on MF205 models.

1. Knob	6. Spring	11. Plug	16. Low & 2nd shift fork
2. Lever	7. Lock screw	12. Fill plug	17. Third & reverse shift rail
3. Boot	8. Detent ball	13. Neutral safety switch	18. Third & reverse shift fork
4. Pin	9. Detent spring	14. Plugs	19. Pin (outer)
5. Washer	10. Gasket	15. Low & 2nd shift rail	20. Pin (inner)

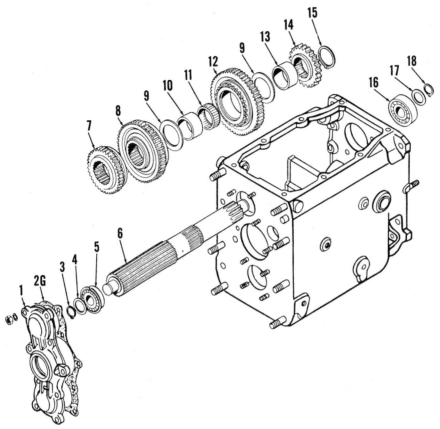

Fig. 91 — Exploded view of MF205 transmission top shaft and related parts used with two wheel drive.

1. Cover	6. Top shaft	10. Bearing inner ring	14. Gear (23 teeth)
2G. Gasket	7. Third & reverse gear	11. Needle bearing	15. Snap ring
3. Snap ring	8. Second gear	12. Low gear	16. Bearing
4. Washer	9. Thrust washers	13. Spacer	17. Thrust washer
5. Bearing			18. Snap ring

checking and adjusting mesh position and backlash.

Complete assembly by reversing disassembly procedure.

95. **PTO DRIVE SHAFT.** The pto drive shaft (51 – Fig. 95) and related parts can be removed after the pinion shaft has been removed. Remove snap ring (63) from rear of shaft. Slide shaft toward front while moving overrunning clutch toward rear, then dislodge snap ring (53) from groove. Move snap ring toward rear of shaft and withdraw shaft from front.

Overrunning clutch (54) can be checked by turning the inner splined hub while holding outer surface. Unit should rotate one direction smoothly, but should lock up when attempting to turn in opposite direction.

When assembling, observe the following: Recess in hub (55) for thrust washer (57) should be toward rear. Chamfer on thrust washer (57) should be toward front. Shift fork slot in collar (56) should

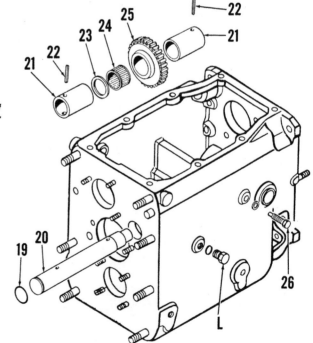

Fig. 92 — Exploded view of MF205 reverse idler assembly.

L. Level screw (Fig. 86)
19. "O" ring
20. Reverse idler shaft
21. Spacers
22. Roll pins
23. Thrust washer
24. Needle bearing
25. Reverse idler
26. Retaining screw

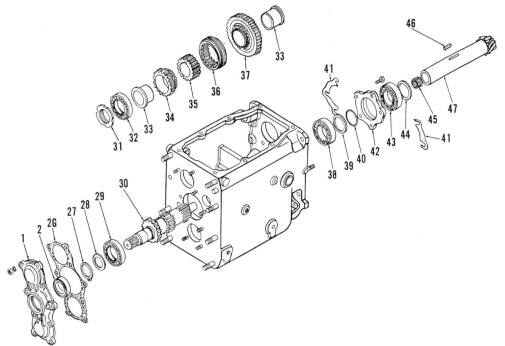

Fig. 93 — Exploded view of input and pinion shafts used on MF205 two wheel drive tractors.

1. Cover	30. Input shaft	36. Range coupling	42. Bearing retainer
2. Oil seal	31. Retaining nut	37. Low gear assembly	43. Bearing
2G. Gasket	32. Ball bearing	38. Bearing	44. Thrust washer
27. Snap ring	33. Gear bushings	39. Spacer	45. Bearing
28. Thrust washer	34. High gear	40. Shims	46. Key
29. Ball bearing	35. Hi-Lo range hub	41. Shims	47. Pinion shaft

be toward rear. Select thickness of thrust washer (61) necessary to provide end play of 0.1-0.3mm after assembly of parts (53 through 63) on shaft (51). Selection of thrust washer (61) can be accomplished by trial assembly on shaft before assembling parts (53 through 63) in housing. Refer to paragraph 94 for assembly of pinion shaft and related parts.

TRANSMISSION (MF205 FOUR WHEEL DRIVE)

Four wheel drive MF205 models are equipped with a six speed transmission shown in Fig. 96.

INSPECTION

96. The shift cover together with the shift rails and forks can be unbolted from top of transmission housing. Check to be sure that forks are firmly attached to shift rails and check condition and operation of detents in rails. Gears and shafts can be visually inspected through opening at top. Transmission, center housing (differential and final drive), rear axles and hydraulic system share a common sump, which should be drained

as described in paragraph 97. Check filter, drained oil and transmission compartment for evidence of metal particles.

LUBRICATION

97. Transmission, center housing (differential and final drive), rear axles and hydraulic system share a common sump. The drain plug is located on bottom of center housing. Be sure to allow sufficient time for oil to drain from interconnected compartments. Remove strainer from under cover (S – Fig. 97). Strainer can be cleaned with diesel fuel and reinstalled if not damaged. Be sure to use new gasket between flange of strainer and center housing.

Fill system through fill plug opening (F) in top cover with Massey-Ferguson Permatran (M-1129A) or equivalent. Capacity is 17 liters, however, correct amount will be to level of plug (L) on side of housing. Be sure to allow sufficient time for oil to pass between compart-

Fig. 95 — Exploded view of pto drive shaft used on MF205 two wheel drive tractors.

48. Snap ring
49. Thrust washer
50. Bearing
51. Front shaft
52. Bearing
53. Snap ring
54. Pto clutch
55. Hub
56. Sleeve
57. Thrust washer
58. Pto (first) gear
59. Bearing
60. Inner race
61. Thrust washer
62. Collar
63. Snap ring
64. Coupling

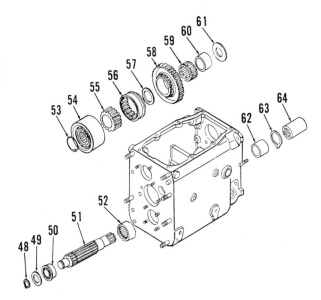

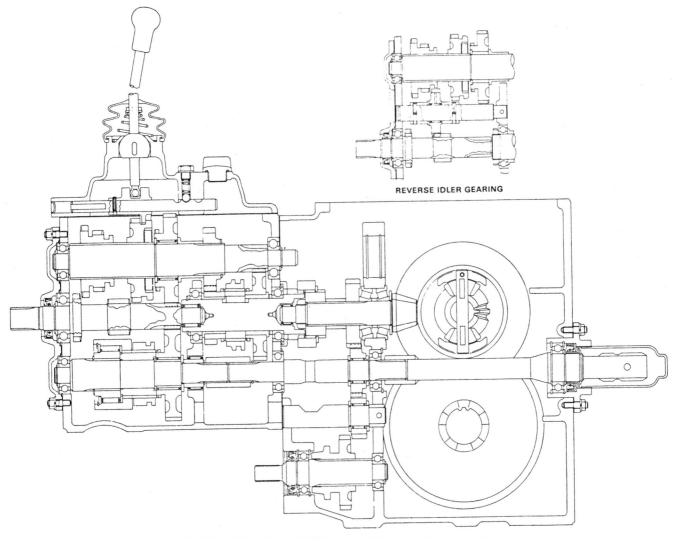

Fig. 96 — Cross-section of MF205 four wheel drive transmission assembly.

REVERSE IDLER GEARING

ments and be sure that tractor is level before checking at level plug (L).

Check for leaks after oil has warmed up especially around strainer cover and drain plug.

REMOVE AND REINSTALL

99. To detach (split) tractor between clutch housing and front of transmission, proceed as follows: Drain oil from transmission and center housing as described in paragraph 97. Disconnect battery cables and rear light wiring harness. Detach pedal return springs for both brake pedals and clutch pedal, then disconnect clutch operating rod. Remove right step plate, and disconnect hydraulic lines at rear. Wedge blocks between frame and front axle to prevent tipping. Move front drive lever to "disengage" position, then loosen the three screws which attach drive shaft cover at rear. Block up under rear of engine and under any front end weights. Support transmission with movable

floor jack or other suitable method that will permit moving rear of tractor away from engine and clutch housing. Remove screws which attach front of transmission to clutch housing, then carefully separate by moving rear of tractor straight back away from clutch housing. The drive shaft and housing should be removed so that new gasket and "O" ring can be installed when assembling.

To remove transmission, first detach clutch and brake pedal support bracket from bottom of transmission. Support the rear of tractor (center housing) sufficiently to prevent tipping. Attach a suitable support to transmission, then unbolt and remove transmission from center housing.

When assembling, observe the following: Position coupling on transmission output splined shaft, then install front wheel drive shaft, gasket and cover. Start all three drive shaft cover retaining screws enough to hold cover, gasket and shaft in position. Carefully move rear of tractor with transmission for-

ward engaging engine, clutch and front drive shaft. Tighten 12 mm retaining stud nuts to 46-54 N·m torque. Tighten pedal support retaining screws to 17.6-20.6 N·m torque. Refer to paragraph 97 for lubrication requirements.

OVERHAUL

100. **SHIFTER COVER.** Remove shifter cover and inspect shift forks and rails for obvious wear or damage. Detent balls and springs can be removed after removing plugs (11–Fig. 100). Both roll pins (20 and 19) must be removed from each shift fork (16 and 18) before rails can be removed from forks. When assembling, be sure to install double roll pins (19 and 20) in each shift fork.

101. **TOP (COUNTER) SHAFT.** To remove the top shaft, first remove transmission as outlined in paragraph 99. Unbolt and remove shifter cover

from top of transmission and aluminum cover (1 – Fig. 101) from front. Remove snap rings (3 and 18) and washers (4 and 17) from ends of shaft, then remove both bearings (5 and 16). Release snap ring (15) from groove and move toward rear of shaft. Move shaft and gears forward in housing until rear end of shaft can be lifted out top opening.

102. **REVERSE IDLER.** Remove set screw (26 – Fig. 102) then drive roll pins (22) from each of the spacer collars (21). Slide the idler gear shaft (20) toward front and remove parts (21, 23, 24 and 25) through top opening.

103. **INPUT SHAFT.** The input shaft can be withdrawn from front of housing after removing the aluminum cover from front of transmission housing. Be careful that bearing or parts of damaged bearing (45 – Fig. 103) are removed if

allowed to fall to bottom of housing.

Be especially careful when installing input shaft, because rollers of bearing (45) may be loose enough in retainer (cage) to catch and prevent entrance of input shaft. A suggested method of installation is to position bearing (45) on end of shaft (47) using a light grease to hold bearing rollers in against inner race and on shaft. Care must still be exercised to assure correct assembly.

104. **SUB GEAR SHAFT AND GEARS.** To remove the sub gear shaft (47 – Fig. 103) and related parts, first remove top shaft as outlined in paragraph 101, reverse idler as outlined in paragraph 102 and input shaft as

described in paragraph 103. Remove snap ring (31), then unbolt retainer (42). Withdraw shaft (47) while withdrawing parts (31 through 42). It may be necessary to bump shaft toward rear if front bushing (33), for gear (34), or front bearing (32) is tight on shaft. Be extremely careful not to damage bearing (45) or bearing bore in end of shaft (47).

Reassemble by reversing disassembly procedure. Thrust washer (40) is available in thicknesses of 3.80, 4.10, 4.40 and 4.70 mm. Select thrust washer (40) of correct thickness to limit shaft end play to 0.1-0.3 mm. Thrust washer (40) of correct thickness can be selected by assembling parts (31 through 41) on shaft (47) before assembling in housing.

Fig. 97 – View showing fill plug (F), level (L) and strainer cover (S).

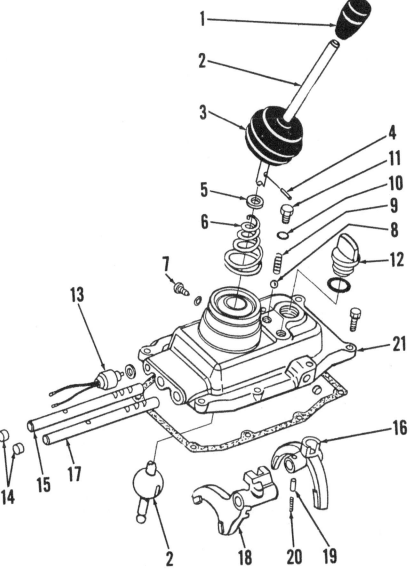

Fig. 100 – Exploded view of shift cover used on MF205 models.

1. Knob	6. Spring	11. Plug	16. Low & 2nd shift fork
2. Lever	7. Lock screw	12. Fill plug	17. Third & reverse shift rail
3. Boot	8. Detent ball	13. Neutral safety switch	18. Third & reverse shift fork
4. Pin	9. Detent spring	14. Plugs	19. Pin (outer)
5. Washer	10. Gasket	15. Low & 2nd shift rail	20. Pin (inner)

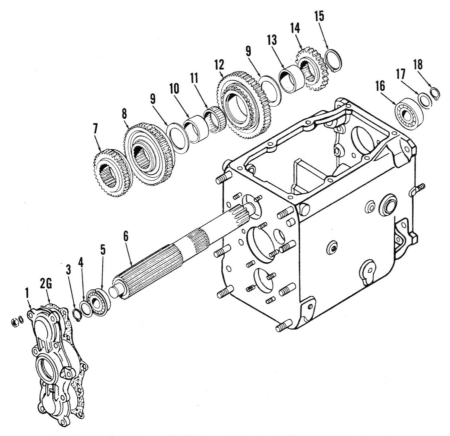

Fig. 101 — Exploded view of MF205 transmission top shaft and related parts.

1. Cover		14. Gear (23 teeth)
2G. Gasket	6. Top shaft	10. Bearing inner ring	15. Snap ring
3. Snap ring	7. Third & reverse gear	11. Needle bearing	16. Bearing
4. Washer	8. Second gear	12. Low gear	17. Thrust washer
5. Bearing	9. Thrust washers	13. Spacer	18. Snap ring

Upon final assembly, be sure to use correct thickness thrust washer. Complete assembly by reversing disassembly procedure.

105. **PTO DRIVE SHAFT.** The pto drive shaft (51–Fig. 105) and related parts can be removed after the sub gear shaft has been removed. Remove snap ring (63) on rear of shaft. Slide shaft toward front while moving overrunning clutch toward rear, then dislodge snap ring (53) from groove. Move snap ring toward rear of shaft and withdraw shaft from front.

Overrunning clutch (54) can be checked by turning the inner splined hub while holding outer surface. Unit should rotate one direction smoothly, but should lock up when attempting to turn in opposite direction.

When assembling, observe the following: Recess in hub (55) for thrust washer (57) should be toward rear. Chamfer on thrust washer (57) should be toward front. Shift fork slot in collar (56) should be toward rear. Select thickness of

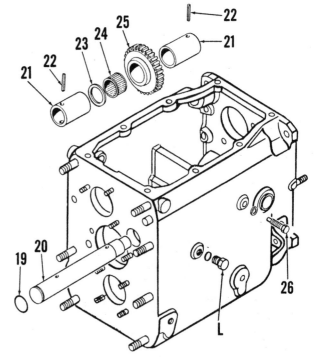

Fig. 102 — Exploded view of MF205 reverse idler assembly.

L. Level screw (Fig. 97)
19. "O" ring
20. Reverse idler shaft
21. Spacers
22. Roll pins
23. Thrust washer
24. Needle bearing
25. Reverse idler
26. Retaining screw

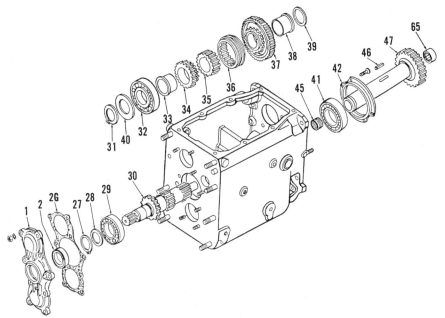

Fig. 103 — Exploded view of input and sub gear shafts used on MF205 four wheel drive models.

1. Cover	30. Input shaft	36. Range coupling	42. Bearing retainer
2. Oil seal	31. Snap ring	37. Low gear assembly	45. Bearing
2G. Gasket	32. Bearing	38. Bearing	46. Key
27. Snap ring	33. Gear bushing	39. Spacer	47. Sub gear shaft
28. Thrust washer	34. High gear	40. Shims	65. Bearing
29. Ball bearing	35. Hi-Lo range hub	41. Bearing	

thrust washer (61) necessary to provide end play of 0.1-0.3 mm after assembly of parts (53 through 63) on shaft (51). Selection of thrust washer (61) can be accomplished by trial assembly on shaft before assembling parts (53 through 63) in housing. Refer to paragraph 104 for assembly of sub gear shaft and related parts.

TRANSMISSION (MF210 and MF220 TWO WHEEL DRIVE)

Two wheel drive MF210 and MF220 models are equipped with a transmission which provides twelve forward speeds. The transmission consists of a main section providing four forward speeds and one reverse speed and a sub transmission section which provides three output ratios. Refer to Fig. 110.

INSPECTION

110. The shift cover with shift rails and forks can be unbolted and removed from top of transmission housing. Check to be sure that forks are firmly attached to shift rails and check condition and operation of detents for shift rails. Gears and shafts can be visually in-

spected through opening in top. Transmission, center housing (differential and final drive), rear axles and hydraulic system share a common sump which should be drained as described in paragraph 111. Check filter, drained oil and transmission compartment for evidence of metal particles.

LUBRICATION

111. Transmission, center housing (differential and final drive), rear axles and hydraulic system share a common sump. The drain plug is located on bot-

tom of center housing. Be sure to allow sufficient time for oil to drain from interconnected compartments. Remove strainer (S – Fig. 111) from under cover shown in left view for MF210 models or from bottom of separate filter housing as shown in right view for MF220 models. The strainer can be cleaned with diesel fuel and reinstalled if not damaged. Be sure to use new gasket between flange of strainer and housing when reassembling.

Fill system through fill plug opening (F) in top cover of all models with Massey-Ferguson Permatran (M-1129A)

Fig. 105 — Exploded view of pto drive shaft used on MF205 four wheel drive tractors.

48. Snap ring
49. Thrust washer
50. Bearing
51. Front shaft
52. Bearing
53. Snap ring
54. Pto clutch
55. Hub
56. Sleeve
57. Thrust washer
58. Pto (first) gear
59. Bearing
60. Inner race
61. Thrust washer
62. Collar
63. Snap ring
64. Coupling

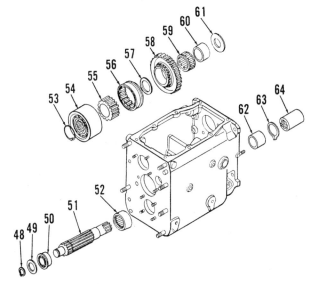

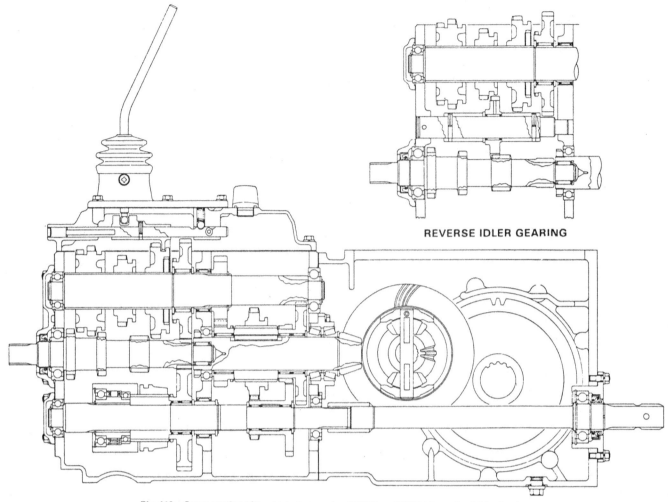

REVERSE IDLER GEARING

Fig. 110 — Cross-section of transmission used on MF210 and MF220 two wheel drive tractors.

or equivalent. Capacity is 21 liters; however, correct amount will be to level of plug (L) on right side of housing. Be sure to allow sufficient time for oil to pass between compartments and be sure that tractor is level before checking at level plug (L).

Check for leaks after oil has warmed up especially around strainer cover and drain plug.

REMOVE AND REINSTALL

112. To detach (split) tractor between clutch housing and front of transmission, proceed as follows: Drain oil from transmission and center housing as described in paragraph 111. Disconnect battery cables and rear light wiring harness. Detach pedal return springs for both brake pedals and clutch pedal, then disconnect clutch operating rod. Remove step plates from both sides and disconnect hydraulic lines at rear. Wedge blocks between frame and front axle to prevent tipping. Block up under rear of engine and under any front end weights. Support transmission with

movable floor jack or other suitable method that will permit moving rear of tractor away from engine and clutch housing. Remove screws which attach front of transmission to clutch housing, then carefully separate by moving rear of tractor straight back away from

Fig. 111 — Views showing location of fill plug (F), level plug (L) and strainers (S) for MF210 and MF220 models.

clutch housing.

To remove transmission, first detach clutch and brake pedal support bracket from bottom of tranmission. Support the rear of tractor (center housing) sufficiently to prevent tipping. Attach a suitable support to transmission, then

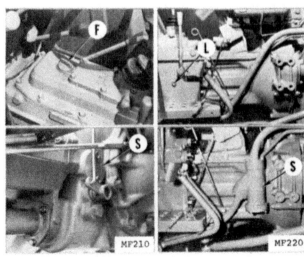

unbolt and remove transmission from center housing.

When reinstalling, tighten the retaining stud nuts to 63-73 N·m torque. Tighten pedal support retaining screws to 17.6-20.6 N·m torque. Refer to paragraph 111 for lubrication requirements.

OVERHAUL

113. **SHIFTER COVER.** Remove complete shift cover from transmission, then unbolt and remove control cover (7–Fig. 113). Be careful not to lose or damage detent springs or balls (8) as they are removed. Turn shift cover upside down and drive out fork lock pins (16). Make sure that all three shift rails are in neutral detent position, then withdraw one of the shift rails. Remove one of the two remaining shift rails, then withdraw last shift rail. Do not lose or damage balls (13) or pin (14) which interlock the three shift rails. Refer to Fig. 113A.

Begin assembly of shifter cover by installing reverse shift rail (20–Fig. 113) and fork (17) in cover. Drive fork lock pins (16) in from top. Position reverse shift rail in neutral then install two of the interlocking balls (13–Fig. 113A). Install first and second shift rod (21–Fig. 113) and fork (18), being careful not to disturb the two interlocking balls located between reverse rail and first/second rail. Interlocking pin (14) can be installed through hole for plug (15) if plug is removed. Drive fork lock pin in from top. Be sure that interlocking pin (14–Fig. 113A) is correctly located, then install two remaining interlocking balls between center shift rail and rail for third and fourth gear. Be sure that the two previously installed shift rails are not twisted and are in neutral position, then install third/fourth shift fork (19–Fig. 113) and shift rail (22). Install shift fork lock pin from top opening, then check interlocking action of the three shift rails. It should be possible to move only one of the shift rails at a time from neutral position.

Be sure that transmission and shift cover are both in neutral when installing cover. Check engagement of forks with gears before finishing assembly.

114. **TOP (COUNTER) SHAFT.** To remove the top shaft, first remove transmission as outlined in paragraph 112. Unbolt and remove shifter cover from top of transmission and aluminum cover (1–Fig. 114) from front. Remove snap ring (20) and washer (19) from rear of shaft, then remove bearing (18). Release snap ring (17) from groove and move toward rear of shaft. Slide (or

bump) shaft (6) forward out of housing while withdrawing gears and associated parts out top opening. Bearing (14) is pressed into bore of housing.

Assemble by reversing disassembly procedure. Insert shaft (6) through front while guiding gears and associated parts onto shaft. Refer also to Fig. 110.

115. **REVERSE IDLER.** Remove set screw (21–Fig. 115) then drive roll pins (23) from each of the spacer collars (26). Slide the idler gear shaft (25) toward

front and remove parts (26, 27, 28 and 29) through top opening.

When assembling, select washer (27) of correct thickness to limit end play of gear (29) to 0.1-0.3 mm. Thrust washer (27) is available in thicknesses of 2.90, 3.20 and 3.50 mm. Selection of thrust washer is easier if parts (23, 26, 27, 28 and 29) are assembled on shaft (25) before installing in housing.

116. **INPUT SHAFT.** The input shaft (33–Fig. 116) can be withdrawn from

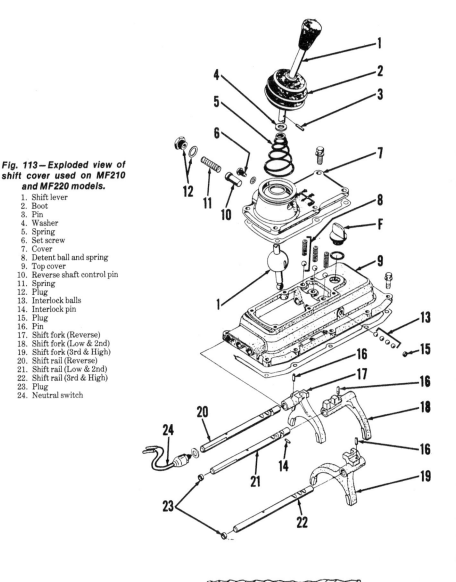

Fig. 113 – Exploded view of shift cover used on MF210 and MF220 models.

1. Shift lever
2. Boot
3. Pin
4. Washer
5. Spring
6. Set screw
7. Cover
8. Detent ball and spring
9. Top cover
10. Reverse shaft control pin
11. Spring
12. Plug
13. Interlock balls
14. Interlock pin
15. Plug
16. Pin
17. Shift fork (Reverse)
18. Shift fork (Low & 2nd)
19. Shift fork (3rd & High)
20. Shift rail (Reverse)
21. Shift rail (Low & 2nd)
22. Shift rail (3rd & High)
23. Plug
24. Neutral switch

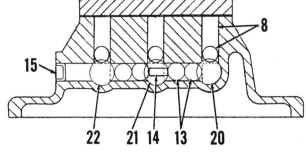

Fig. 113A – Cross-section of shift rails and interlocking balls (13) and pin (14). Refer to Fig. 113 for legend.

front of housing after removing the aluminum cover (1) from front of transmission housing. Be careful that bearing or parts of damaged bearing (52) are removed if allowed to fall to bottom of housing.

Be especially careful when installing input shaft, because rollers of bearing (52) may be loose enough in retainer (cage) to catch and prevent entrance of input shaft. A suggested method of installation is to position bearing (52) on end of shaft (33) using a light grease to hold bearing rollers in against inner race and on shaft. Care must still be exercised to assure correct assembly.

117. PINION GEAR AND SHAFT. To remove the pinion gear and shaft (53–Fig. 116) and related parts, it is necessary to remove nut (34). Nut is left-hand thread and secured using "Loctite". Further disassembly is easier if pto and range shift forks (5 and 15–Fig. 117) are removed. To remove shift forks, first remove detent plugs, springs and balls (6 and 16). Remove shaft lock screw (19) and lock plate (20), then withdraw shafts (7 and 17) and lift out forks (5 and 15).

To remove nut (34–Fig. 116), secure pinion in a suitable holding fixture or by carefully blocking gears. The nut (34) can not be completely removed until bearing retainer (49) is loosened allowing shaft to move toward rear. Be careful not to damage or lose pinion positioning shims (51). It may be necessary to bump shaft toward rear if bearing collar (39 or 44) or front bearing (35) is tight on shaft. Be extremely careful not to damage bearing (52) or bearing bore in end of shaft (53).

Refer to paragraph 118 for removal and reinstallation of pto drive shaft.

Original thickness of shims (48) can be reinstalled if pinion shaft bearings (46) and bearing housing (49) are reassembled. If any of these parts are renewed or if adjustment of bearings (46) is questioned, proceed as follows: Assemble parts (46, 47, 48 and 49) on special dummy pinion with nut (34) or assemble all parts (34 through 50) on shaft (53), outside of housing. Tighten nut (34) temporarily to 88 N·m, clamp bearing housing (49) in vise, then check rolling torque of shaft in bearings using a torque wrench. If rolling torque is more than 1.17 N·m, then increase thickness of shims (48) and recheck. If rolling torque of pinion shaft is less than 0.6 N·m, decrease thickness of shims (48) and recheck. Shims (48) are available in 0.10, 0.15, 0.20 and 1.00 mm. thicknesses. After correct thickness of shims (48) has been determined, pinion shaft and associated parts (34 through 54), can be assembled in transmission housing. Ap-

ply "Loctite" to pinion shaft threads, then tighten nut (34) to final torque of 78.45 – 98.07 N·m.

Thickness of shims (51) determines the mesh position of pinion with differential ring gear. Original thickness of shims

(51) can be considered correct if original pinion shaft (53) and bearing housing (49) are reinstalled. If mesh position was questioned, refer to paragraph 139 for checking and adjusting mesh position and backlash.

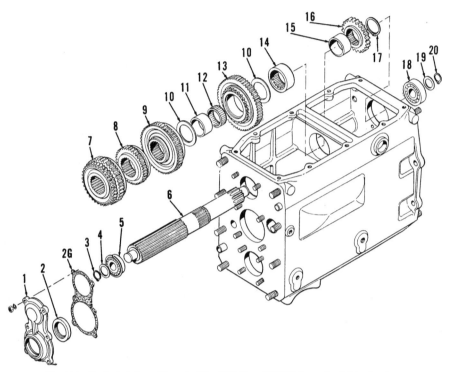

Fig. 114 — Exploded view of top shaft for MF210 and MF220 two wheel drive tractors.

1. Cover	6. Top shaft	11. Inner bearing race	16. Gear (25 teeth)
2. Seal	7. Gear (3rd & High)	12. Bearing	17. Snap ring
2G. Gasket	8. Gear (Reverse)	13. Gear (Low)	18. Bearing
3. Snap ring	9. Gear (2nd)	14. Bearing	19. Thrust washer
4. Thrust washer	10. Thrust washer	15. Inner bearing race	20. Snap ring
5. Bearing		

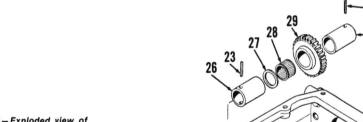

Fig. 115 — Exploded view of reverse idler gear and related parts used on MF210 and MF220 two wheel drive models.

21. Lock screw
22. Washer
23. Pin
24. "O" ring
25. Idler shaft
26. Spacer
27. Shim
28. Bearing
29. Reverse idler gear

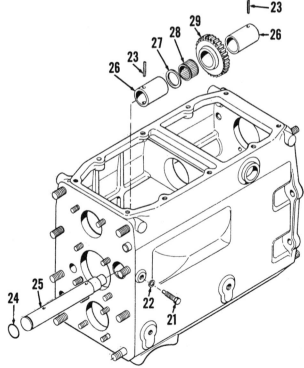

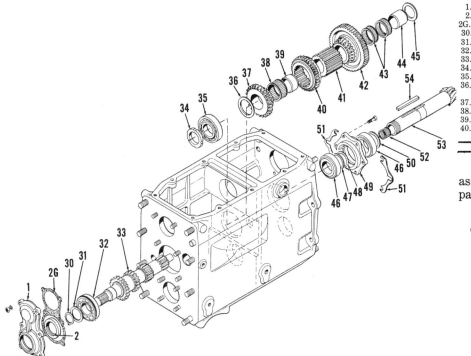

1. Cover	41. Hub
2. Seal	42. Main gear
2G. Gasket	(24 and 55 teeth)
30. Snap ring	43. Bearing
31. Thrust washer	44. Inner race
32. Bearing	45. Thrust washer
33. Input shaft	46. Bearing
34. Nut	47. Drive pinion spacer
35. Bearing	48. Shim
36. Thrust washer (same	49. Retainer
as 45)	50. Thrust washer
37. Gear (23 teeth)	51. Shims
38. Bearing	52. Bearing
39. Inner race	53. Pinion shaft
40. Sliding gear	54. Key

assembly of pinion shaft and related parts.

TRANSMISSION (MF210 and MF220 FOUR WHEEL DRIVE)

Four wheel drive MF210 and MF220 models are equipped with a transmission which provides twelve forward speeds and three reverse speeds. The transmission consists of a main section providing four forward speeds and one reverse

Fig. 116 – Exploded view of input and pinion shafts used on MF210 and MF220 two wheel drive tractors.

Complete assembly by reversing disassembly procedure.

118. **PTO DRIVE SHAFT.** The pto drive shaft (59–Fig. 118) and related parts can be removed after the pinion shaft has been removed. Dislodge snap rings (70 and 76) from grooves in shaft. Move snap rings toward rear and withdraw shaft from floor.

Overrunning clutch (60) can be checked by turning the inner splined hub while holding outer surface. Unit should rotate one direction smoothly, but should lock up when attempting to turn in opposite direction.

When assembling, observe the following: Flat face of hub (61) should be toward front. Chamfer on thrust washer (66) inside diameter should be toward front. Shift fork slot in collar (62) should be toward front. Select thickness of thrust washer (66) necessary to provide end play of 0.1-0.3 mm after assembly of parts (60 through 70) on shaft (59). Selection of thrust washer (66) can be accomplished by trial assembly on shaft before assembling parts (60 through 70) in housing. Refer to paragraph 117 for

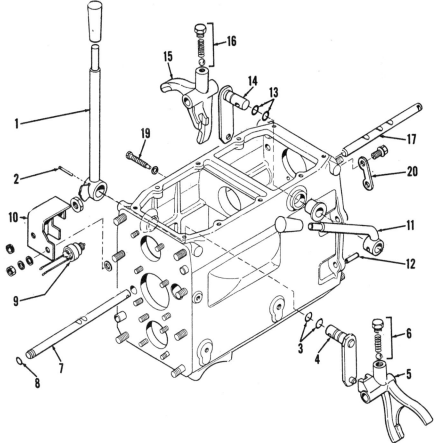

1. Pto shift lever	11. Hi-lo shift lever
2. Pin	12. Pin
3. "O" rings	13. "O" rings
4. Shift lever	14. Shift lever
5. Shift fork	15. Range shift fork
6. Detent assembly	16. Detent assembly
7. Shift rail	17. Shift rail
8. "O" ring	18.
9. Pto safety switch	19. Retainer screw for
10. Cover	rail (7)
	20. Lock plate

Fig. 117 – Exploded view of pto and hi-lo range shift controls used on MF210 and MF220 two wheel drive models.

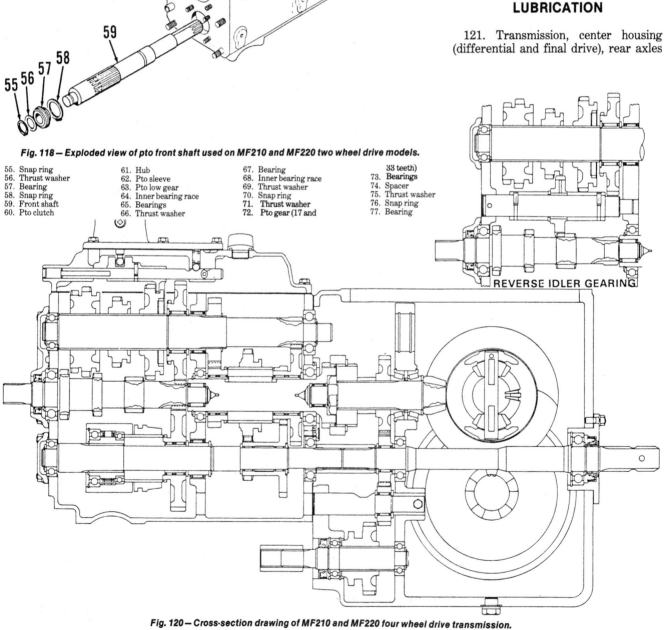

speed and a sub transmission section which provides three output ratios. Refer to Fig. 120.

INSPECTION

120. The shift cover with shift rails and forks can be unbolted and removed from top of transmission housing. Check to be sure that forks are firmly attached to shift rails and check condition and operation of detents for shift rails. Gears and shafts can be visually inspected through opening in top. Transmission, center housing (differential and final drive), rear axles and hydraulic system share a common sump which should be drained as described in paragraph 121. Check filter, drained oil and transmission compartment for evidence of metal particles.

LUBRICATION

121. Transmission, center housing (differential and final drive), rear axles

Fig. 118 — Exploded view of pto front shaft used on MF210 and MF220 two wheel drive models.

55. Snap ring	61. Hub	67. Bearing	33 teeth)
56. Thrust washer	62. Pto sleeve	68. Inner bearing race	73. **Bearings**
57. Bearing	63. Pto low gear	69. Thrust washer	74. Spacer
58. Snap ring	64. Inner bearing race	70. Snap ring	75. Thrust washer
59. Front shaft	65. Bearings	71. **Thrust washer**	76. Snap ring
60. Pto clutch	66. Thrust washer	72. Pto gear (17 and	77. Bearing

REVERSE IDLER GEARING

Fig. 120 — Cross-section drawing of MF210 and MF220 four wheel drive transmission.

and hydraulic system share a common sump. The drain plug is located on bottom of center housing. Be sure to allow sufficient time for oil to drain from interconnected compartments. Remove strainer (S–Fig. 121) from under cover shown in left view for MF210 models or from bottom of separate filter housing as shown in right view for MF220 models. The strainer can be cleaned with diesel fuel and reinstalled if not damaged. Be sure to use new gasket between flange of strainer and housing when reassembling.

Fill system through fill plug opening (F) in top cover of all models with Massey-Ferguson Permatran (M-1129A) or equivalent. Capacity for MF210 models is 25.7 liters; capacity for MF220 models is 27 liters. Correct amount of fluid will be to level of plug (L) on right side of housing. Be sure to allow sufficient time for oil to pass between compartments and be sure that tractor is level before checking at level plug (L).

Check for leaks after oil has warmed up especially around strainer cover and drain plug.

REMOVE AND REINSTALL

122. To detach (split) tractor between clutch housing and front of transmission, proceed as follows: Drain oil from transmission and center housing as described in paragraph 121. Disconnect battery cables and rear light wiring harness. Detach pedal return springs for both brake pedals and clutch pedal, then disconnect clutch operating rod. Remove step plates from both sides and disconnect hydraulic lines at rear. Wedge blocks between frame and front axle to prevent tipping. Move front drive lever to "disengage" position, then loosen the three screws which attach drive shaft cover at rear. Block up under rear of engine and under any front end weights. Support transmission with movable floor jack or other suitable method that will permit moving rear of tractor away from engine and clutch housing. Remove screws which attach front of transmission to clutch housing, then carefully separate by moving rear of tractor straight back away from clutch housing. The drive shaft and housing should be removed so that new gasket and "O" ring can be installed when reassembling.

To remove transmission, first detach clutch and brake pedal support bracket from bottom of transmission. Support the rear of tractor (center housing) sufficiently to prevent tipping. Attach a suitable support to transmission, then unbolt and remove transmission from center housing.

When assembling, observe the following: Position coupling on transmission output splined shaft, then install front wheel drive shaft, gasket and cover. Start all three drive shaft cover retaining screws enough to hold cover, gasket and shaft in position. Carefully move rear of tractor with transmission forward engaging engine, clutch and front drive shaft. Tighten retaining stud nuts to 63-73 N·m torque. Tighten pedal support retaining screws to 17.6-20.6 N·m torque. Refer to paragraph 121 for lubrication requirements.

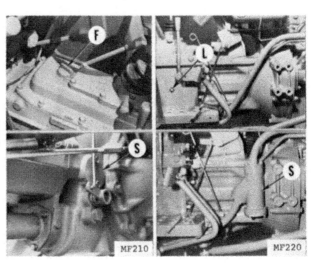

Fig. 121—Views showing location of fill plug (F), level plug (L) and strainers (S) for MF210 and MF220 models.

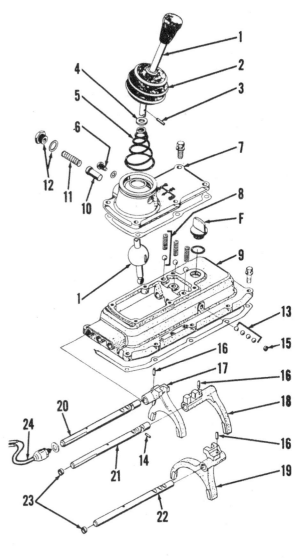

Fig. 123—Exploded view of shift cover used on MF210 and MF220 models.

1. Shift lever
2. Boot
3. Pin
4. Washer
5. Spring
6. Set screw
7. Cover
8. Detent ball and spring
9. Top cover
10. Reverse shaft control pin
11. Spring
12. Plug
13. Interlock balls
14. Interlock pin
15. Plug
16. Pin
17. Shift fork (Reverse)
18. Shift fork (Low & 2nd)
19. Shift fork (3rd & High)
20. Shift rail (Reverse)
21. Shift rail (Low & 2nd)
22. Shift rail (3rd & High)
23. Plug
24. Neutral switch

OVERHAUL

123. SHIFTER COVER. Remove complete shift cover from transmission, then unbolt and remove control cover (7–Fig. 123). Be careful not to lose or damage detent springs or balls (8) as they are removed. Turn shift cover upside down and drive out fork lock pins (16). Make sure that all three shift rails are in neutral detent position, then withdraw one of the shift rails. Remove one of the two remaining shift rails, then withdraw last shift rail. Do not lose or damage balls (13) or pin (14) which interlock the three shift rails. Refer to Fig. 123A.

Begin assembly of shifter cover by installing reverse shift rail (20–Fig. 123) and fork (17) in cover. Drive fork lock pins (16) in from top. Position reverse shift rail in neutral then install two of the interlocking balls (13–Fig. 123A). Install first and second shift rod (21–Fig. 123) and fork (18), being careful not to disturb the two interlocking balls located between reverse rail and first/second rail. Interlocking pin (14) can be installed through hole for plug (15) if plug is removed. Drive fork lock pin in from top. Be sure that interlocking pin (14–Fig. 123A) is correctly located, then install two remaining interlocking balls between center shift rail and rail for third and fourth gear. Be sure that the two previously installed shift rails are not twisted and are in neutral position, then install third/fourth shift fork (19–Fig. 123) and shift rail (22). Install shift fork lock pin from top opening, then check interlocking action of the three shift rails. It should be possible to move only one of the shift rails at a time from neutral position.

Be sure that transmission and shift cover are both in neutral when installing cover. Check engagement of forks with gears before finishing assembly.

124. TOP (COUNTER) SHAFT. To remove the top shaft, first remove transmission as outlined in paragraph 122. Unbolt and remove shifter cover from top of transmission and aluminum cover (1–Fig. 124) from front. Remove snap ring (20) and washer (19) from rear of shaft, then remove bearing (18). Release snap ring (17) from groove and move toward rear of shaft. Slide (or bump) shaft (6) forward out of housing while withdrawing gears and associated parts out top opening. Bearing (14) is pressed into bore in housing.

Assemble by reversing disassembly procedure. Insert shaft (6) through front while guiding gears and associated parts onto shaft. Refer also to Fig. 120.

125. REVERSE IDLER. Remove set screw (21–Fig. 125) then drive roll pins (23) from each of the spacer collars (26). Slide the idler gear shaft (25) toward front and remove parts (26, 27, 28 and 29) through top opening.

When assembling, select washer (27) of correct thickness to limit end play of gear (29) to 0.1-0.3 mm. Thrust washer (27) is available in thicknesses of 2.90, 3.20 and 3.50 mm. Selection of thrust washer is easier if parts (23, 26, 27, 28 and 29) are assembled on shaft (25) before assembling in housing.

126. INPUT SHAFT. The input shaft (33–Fig. 126) can be withdrawn from front of housing after removing the aluminum cover (1) from front of transmission housing. Be careful that bearing or parts of damaged bearing (52) are removed if allowed to fall to bottom of housing.

Be especially careful when installing input shaft, because rollers of bearing (52) may be loose enough in retainer (cage) to catch and prevent entrance of input shaft. A suggested method of installation is to position bearing (52) on end of shaft (33) using a light grease to hold bearing rollers in against inner race and on shaft. Care must still be exercised to assure correct assembly.

127. SUB GEAR SHAFT AND GEARS. To remove the sub gear shaft (53–Fig. 126) and related parts, first remove top shaft as outlined in paragraph 124, reverse idler as outlined in paragraph 125 and input shaft as described in paragraph 126. Disassembly is easier if pto and range

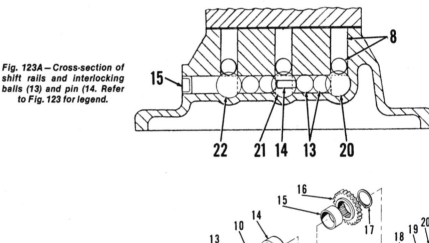

Fig. 123A – Cross-section of shift rails and interlocking balls (13) and pin (14. Refer to Fig. 123 for legend.

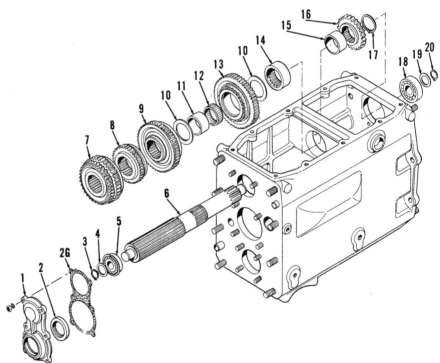

Fig. 124 – Exploded view of top shaft for MF210 and MF220 four wheel drive tractors.

1. Cover	6. Top shaft	11. Inner bearing race	16. Gear (25 teeth)
2. Seal	7. Gear (3rd & High)	12. Bearing	17. Snap ring
2G. Gasket	8. Gear (Reverse)	13. Gear (Low)	18. Bearing
3. Snap ring	9. Gear (2nd)	14. Bearing	19. Thrust washer
4. Thrust washer	10. Thrust washer	15. Inner bearing race	20. Snap ring
5. Bearing			

shift forks (5 and 15 – Fig. 127) are removed. To remove shift forks, first remove detent plugs, springs and balls (6 and 16). Remove shaft lock screw (19) and lock plate (20), then withdraw shafts (7 and 17) and lift out forks (5 and 15).

Remove snap ring (34 – Fig. 126), then unbolt retainer (49). It may be necessary to bump shaft toward rear if bearing collar (39 or 44) or front bearing (35) is tight on shaft. Be extremely careful not to damage bearing (52) or bearing bore

in end of shaft (53).

Refer to paragraph 128 for removal and reinstallation of pto drive shaft.

Reassemble by reversing disassembly procedure. Thrust washer (50) is available in several different thicknesses. Select thrust washer (50) to limit shaft end play to 0.1-0.3 mm. Correct thickness of thrust washer (50) can be determined by assembling parts (35 through 46) on shaft (53) before assembling in housing. Upon final assembly, be sure to use correct thickness thrust washer. Complete assembly by reversing disassembly procedure.

128. PTO DRIVE SHAFT. The pto drive shaft (59 – Fig. 128) and related parts can be removed after the pinion shaft has been removed. Dislodge snap rings (70 and 76) from grooves in shaft. Move snap rings toward rear and withdraw shaft from front.

Overrunning clutch (60) can be checked by turning the inner splined hub while holding outer surface. Unit should rotate one direction smoothly, but should lock up when attempting to turn in opposite direction.

When assembling, observe the following: Flat face of hub (61) should be toward front. Chamfer on thrust washer (66) inside diameter should be toward front. Shift fork slot in collar (62) should be toward front. Select thickness of

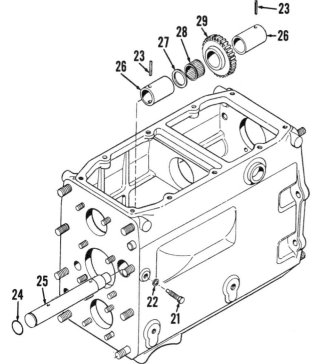

Fig. 125 – Exploded view of reverse idler gear and related parts used on MF210 and MF220 four wheel drive models.

21. Lock screw
22. Washer
23. Pin
24. "O" ring
25. Idler shaft
26. Spacer
27. Shim
28. Bearing
29. Reverse idler gear

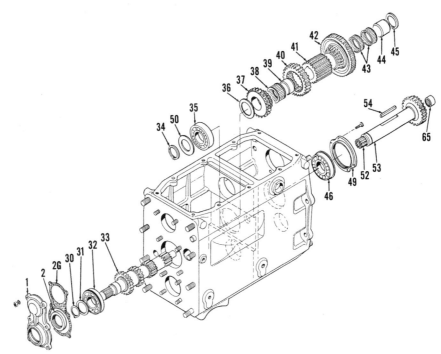

Fig. 126 – Exploded view of input and sub gear shafts used on MF210 and MF220 four wheel drive tractors.

1. Cover	34. Snap ring	41. Hub
2. Seal	35. Bearing	42. Main gear (24 and
2G. Gasket	36. Thrust washer	54 teeth)
30. Snap ring	37. Gear (23 teeth)	43. Bearings
31. Thrust washer	38. Bearing	44. Inner bearing race
32. Bearing	39. Inner bearing race	45. Thrust washer
33. Input shaft	40. Sliding gear	46. Bearing

49. Retainer
50. Shim
52. Bearing
53. Sub gear shaft
54. Key
65. Bearing

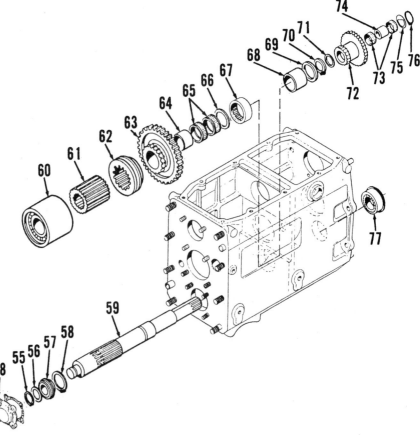

1. Pto shift lever	11. Hi-lo shift lever
2. Pin	12. Pin
3. "O" rings	13. "O" rings
4. Shift lever	14. Shift lever
5. Shift fork	15. Range shift fork
6. Detent assembly	16. Detent assembly
7. Shift rail	17. Shift rail
8. "O" ring	19. Retainer screw for
9. Pto safety switch	rail 7
10. Cover	20. Lock plate

and brake assemblies attached. On four wheel drive models, the axle center housing contains the differential drive pinion as well as the differential. The front wheel drive idler and driven gears are also included in the center housing.

AXLE CENTER HOUSING

All Models

130. **SPLIT TRANSMISSION FROM CENTER HOUSING.** To separate tractor between rear of transmission and front of axle center housing, proceed as follows: Drain oil from transmission and center housing as outlined in paragraph 86 or 111 for two-wheel drive models; paragraph 97 or 121 for four wheel drive tractors. On four wheel drive models, move front drive lever to "disengage" position, then loosen the three screws which attach drive shaft cover at rear. On two wheel drive models, unbolt rear bearing re-

Fig. 127 — Exploded view of pto and hi-lo range shift controls used on MF210 and MF220 four wheel drive tractors.

thrust washer (66) necessary to provide end play of 0.1-0.3 mm after assembly of parts (60 through 70) on shaft (59). Selection of thrust washer (66) can be accomplished by trial assembly on shaft before assembling parts (60 through 70) in housing. Refer to paragraph 127 for assembly of sub gear shaft and related parts.

AXLE CENTER HOUSING AND DIFFERENTIAL

The axle center housing on two wheel drive tractors contains the differential assembly with the final drive assemblies

55. Snap ring	
56. Thrust washer	68. Inner bearing race
57. Bearing	69. Thrust washer
58. Snap ring	70. Snap ring
59. Front shaft	71. Thrust washer
60. Pto clutch	72. Pto gear (17 and 33
61. Hub	teeth)
62. Pto sleeve	73. Bearings
63. Pto low gear	74. Spacer
64. Inner bearing race	75. Thrust washer
65. Bearings	76. Snap ring
66. Thrust washer	77. Bearing
67. Bearing	78. Cover

Fig. 128 — Exploded view pto front shaft used on MF210 and MF220 four wheel drive models.

tainer, then withdraw pto shaft. On all models, disconnect battery cables and rear wiring harness. Detach pedal return springs for both brake pedals and clutch pedal, then disconnect clutch operating rod. Remove right step plate and disconnect hydraulic lines at rear. Wedge blocks between frame and front axle to prevent tipping. Block up under transmission and under any front end weights. Support center housing, unbolt rear of transmission from center housing; then, carefully separate by moving center housing back away from transmission housing. The drive shaft and housing should be completely removed from four wheel drive models so that new gasket and "O" ring can be installed when assembling.

When assembling, observe the following: On four wheel drive models, position coupling on transmission output splined shaft, then install front wheel drive shaft, gasket and cover. Start all three drive shaft cover retaining screws

enough to hold cover, gasket and shaft in position. Carefully move axle center housing forward, engaging sub gear, pto shaft and front drive shaft.

On MF205 models tighten the 12 mm retaining stud nuts to 46-54 N·m torque. On MF210 and MF220 models, tighten the retaining stud nuts to 63-73 N·m torque. On all models, tighten pedal support retaining screws to 17.6-20.6 N·m torque. Refer to appropriate paragraph 86 (MF205 two wheel drive), 97 (MF205 four wheel drive), 111 (MF210 and MF220 two wheel drive) or 121 (MF210 and MF220 four wheel drive) for lubrication requirements.

131. **REMOVE AND REINSTALL.** To remove the axle center housing, refer to paragraph 150 or 151 and remove the final drive housing from each side of center housing. Refer to paragraph 130 and complete removal of axle center housing.

DIFFERENTIAL AND BEVEL GEARS

MF205 Models With Two Wheel Drive

132. **R&R AND OVERHAUL.** The differential carrier (37–Fig. 132) and related parts can be removed from axle center housing after removing both final drive (axle) housings as outlined in paragraph 150 and separating axle center housing from transmission as outlined in paragraph 130.

Shims (43) are used to adjust carrier bearings and backlash between bevel pinion and ring gear (35). Refer to paragraph 94 for removal of the bevel pinion gear and shaft. Differential pinion gear shaft (42) can be removed after driving pin (36) out toward ring gear side. Hole is provided in opposite side of carrier (37) to facilitate removal. Ring gear is attached to carrier with cap screws. Tighten all of the ring gear re-

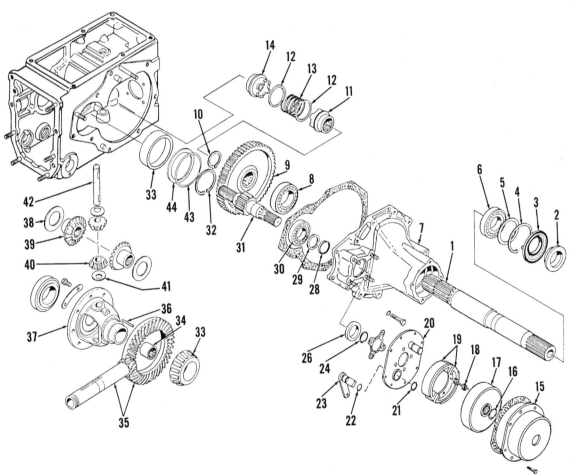

Fig. 132 — Exploded view of axle center housing and final drive used on MF205 two wheel drive models.

1. Axle	8. Bearing	15. Cover	22. "O" ring	31. Final drive pinion	38. Thrust washers
2. Seal ring	9. Gear	16. Snap ring	23. Actuating cam	32. Snap ring	39. Side gears
3. Seal	10. Snap ring	17. Brake drum	24. Snap ring	33. Bearing cup & cone	40. Pinion gears
4. Snap ring	11. Lock coupling	18. Return springs	26. Seal	34. Pinion bearing	41. Thrust washers
5. Shims	12. Thrust washers	19. Brake shoes	28. Snap ring	35. Bevel pinion and ring gear	42. Shaft
6. Bearing	13. Spring	20. Backing plate	29. Thrust washer	36. Pin	43. Shims
7. Axle housing	14. Lock coupling	21. Snap ring	30. Bearing	37. Carrier	44. Thrust ring

taining screws to 48.8-58.3 N·m torque, then lock in place by bending lock tab around head of screw. Ring gear and pinion shaft (35) are available only as a matched set.

The differential can be reinstalled by reversing removal procedure. Refer to paragraph 133 for adjustment.

133. **ADJUSTMENT.** Refer to paragraph 94 for selection of shims (40 – Fig. 93). Shims (41) used to position bevel pinion must be selected to provide correct pinion position relative to bevel ring gear.

Adjustment of differential carrier

bearings and bevel gear backlash is accomplished by varying thickness of shims (43 – Fig. 132). Shims for left side are not the same as for right side but both are available in thicknesses of 0.10, 0.15, 0.20 and 0.50 mm. Select shims that provide 0 to 0.10 mm play in carrier bearings. If side play is 0, be sure that rolling torque required to rotate differential does not exceed 0.79 N·m. Retainer plates may be fabricated to attach across bore for bearings (33), thrust plates (44) and shims (43) to facilitate selection of shims without installing axle housings (7).

After correctly setting carrier bearing

adjustment, attach axle center housing to transmission housing and check backlash between bevel pinion and ring gear. A dial indicator may be used to check backlash at outer diameter of ring gear teeth which should be 0.1-0.2 mm. If backlash exceeds 0.2 mm, remove some shims (43) from left side, install the same thickness on right side and recheck. To increase backlash, move ring gear away from bevel pinion by removing shims from right side and adding same thickness to left side.

The differential carrier assembly (37) with ring gear must be assembled in axle center housing, the bevel pinion must be

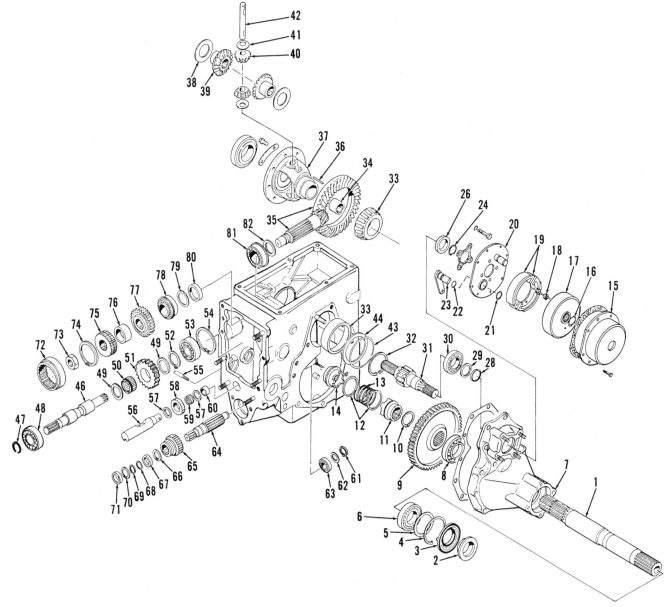

Fig. 135 — Exploded view of axle center housing and final drive used on MF205 four wheel drive tractors. Refer to Fig. 132 for legend except the following.

46. Pto countershaft			70. Snap ring	76. Spacer
47. Snap ring	53. Bearing	59. Bearing	71. Seal	77. Gear (32 teeth)
48. Bearing	54. Snap ring	60. Spacer	72. Sleeve	78. Bearing
49. Thrust washers	55. Retaining screw	61. Snap ring	65. Gear	73. Nut	79. Shim
50. Bearing	56. Idler shaft	62. Thrust washer	66. Snap ring	74. Snap ring	80. Spacer
51. Gear (24 teeth)	57. Thrust washers	63. Bearing	67. Bearing	75. Hub	81. Bearing
52. Snap ring	58. Gear (20 teeth)	64. Front wheel drive shaft	68. Thrust washer		82. Shim
			69. Snap ring		

installed in transmission housing and the transmission housing must be attached to the axle center housing in order to check mesh position. Backlash should be adjusted within limits of 0.10-0.20 mm (measured at ring gear teeth) before checking mesh position. Mesh position is determined by coating ring gear teeth with mechanics' bluing, then turning coated teeth through pinion. Refer to Fig. 136. Correct mesh position is shown at (A). Mesh position (B) indicates the ring gear is too far away from pinion (too much backlash) and can usually be corrected by removing shims from left side carrier bearing and installing same thickness of shims on right side. Notice that shims (43–Fig. 132) for left and right sides are not the same. Mesh pattern (C–Fig. 136) is caused by too little backlash and can usually be corrected by removing some shims from right side and installing similar thickness on left. Pattern (D) is caused by pinion too far from center of ring gear and is corrected by adding thickness to shims at (41–Fig. 93). Mesh pattern (E–Fig. 136) is caused by pinion too far into ring gear and is corrected by removing some thickness of shims (41–Fig. 93). Shims (41) are available in thicknesses of 0.10, 0.20 and 0.30 mm. Always recheck mesh position using fresh coating of mechanics' bluing to be sure that shims are correct before final assembly.

MF205 Models With Four Wheel Drive

135. **REMOVE AND REINSTALL.** The differential carrier (37–Fig. 135) and related parts can be removed from axle center housing after removing both final drive (axle) housings as outlined in paragraph 150. To remove the bevel pinion, refer to paragraph 130 and separate axle center housing from transmission.

Be careful not to lose or damage shims (43) which are used to adjust carrier bearings and backlash between bevel pinion and ring gear. Also, be careful not to damage bearing (65–Fig. 103) when separating or joining transmission and axle center housing.

Differential pinion gear shaft (42–Fig. 135) can be removed after driving pin (36) out toward ring gear side. Hole is provided in opposite side of carrier (37) to facilitate removal. Ring gear is attached to carrier with cap screws. Tighten all of the ring gear retaining screws to 48.8-58.3 N·m torque, then lock in place by bending lock tab around heads of screws. Ring gear and pinion shaft (35) are available only as a matched set.

Remove nut (73–Fig. 135), hub (75), spacer (76) and gear (77). Pinion can be withdrawn from rear. Be careful not to lose or damage bearing cup/cone (78), bearing adjustment shims (79), spacer (80), bearing cup/cone (81) or shims (82).

Original thickness of shims (79) can be reinstalled if original bearings (78 and 81), spacer (80) and housing are reinstalled. If any of these parts are renewed or if adjustment is questioned, refer to paragraph 136.

Original thickness of shims (82) can be reinstalled if original bearing (81), housing, bevel pinion and ring gear (35) are reinstalled. If any of these parts are renewed, determine correct thickness of shims (79), then adjust mesh position as outlined in paragraph 136.

136. **ADJUSTMENT.** To determine correct thickness of shims (79–Fig. 135), assemble parts (73 through 82) in housing and on pinion shaft. Tighten nut (73) to 88 N·m torque then check rolling torque of pinion shaft in bearings (78 and 81). If rolling torque is more than 1.17 N·m, then increase thickness of shims (79). If rolling torque of pinion shaft is less than 0.58 N·m; decrease thickness of shims (79) and recheck. Shims (79) are available in thicknesses of 0.10, 0.15, 0.20 and 1.00 mm. After correct thickness of shims (79) has been determined, mesh position of bevel gears must be checked and adjusted if necessary by changing thickness of shims (82).

Adjustment of differential carrier bearings and bevel gear backlash is accomplished by varying thickness of shims (43). Shims for left side are not the same as for right side but both are available in thicknesses of 0.10, 0.15, 0.20 and 0.50 mm. Select shims that provide 0 to 0.10 mm play in carrier bearings. If side play is 0, be sure that some

backlash exists between bevel pinion and ring gear and that rolling torque required to rotate differential does not exceed 0.79 N·m. Retainer plates may be fabricated to attach across bore for bearings (33), thrust plates (44) and shims (43) to facilitate selection of shims without installing axle housings (7).

After correctly setting carrier bearing adjustment, check backlash between bevel pinion and ring gear. A dial indicator may be used to check backlash at outer diameter of ring gear teeth which should be within limits of 0.1-0.2 mm. If backlash exceeds 0.2 mm, remove some shims (43) from left side, install the same thickness on right side and recheck. To increase backlash, move ring gear away from bevel pinion by removing shims from right side and adding same thickness to left side.

The differential carrier assembly (37) with ring gear and the bevel pinion must be installed in housing in order to check mesh position. Backlash should be within limits of 0.10-0.20 mm (measured at ring gear teeth) when checking mesh position. Mesh position is determined by coating ring gear teeth with mechanics' bluing, then turning through pinion. Refer to Fig. 136. Correct mesh position is shown at (A). Mesh position (B) indicates that ring gear is too far away from pinion (too much backlash) and can usually be corrected by removing shims from left side carrier bearing and installing shims on right side. Notice that shims (43–Fig. 135) for left and right sides are not the same. Mesh pattern (C–Fig. 136) is caused by too little backlash and can usually be corrected by removing some shims from right side and installing similar thickness on left. Pattern (D) is caused by pinion too far from center of ring gear and is corrected by adding thickness to shims at (82–Fig. 135). Mesh pattern (E–Fig. 136) is caused by pinion too far into ring gear and is corrected by removing some thickness from shims at (82–Fig. 135). Shims (82) are available in thicknesses of 3.00, 3.10, 3.20, 3.30, 3.40, 3.50 and 3.60 mm. Always recheck mesh position using fresh coating of mechanics' bluing to be sure that shims are correct before final assembly.

After correct thickness of shims (79 and 82–Fig. 135) is determined, coat threads for nut (73) with "Loctite" and tighten to 88 N·m torque.

MF210 and MF220 Models With Two Wheel Drive

138. **R&R AND OVERHAUL.** The differential carrier (37–Fig. 138) and related parts can be removed from axle center housing after removing both final drive axle housings as outlined in

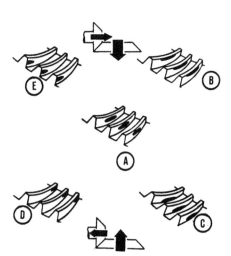

Fig. 136 — Refer to text when setting bevel gear mesh position.

paragraph 151 and separating axle center housing from transmission as outlined in paragraph 130.

Shims (43) are used to adjust carrier bearings and backlash between bevel pinion and ring gear (35). Refer to paragraph 117 for removal of the bevel pinion gear and shaft. Differential pinion gear shaft (42) can be removed after driving pin (36) out toward ring gear side. Hole is provided in opposite side of carrier (37) to facilitate removal. Ring gear is attached to carrier with cap screws. Tighten all of the ring gear retaining screws to 48.8-58.3 N·m torque, then lock in place by bending lock tab around head of screw. Ring gear and pinion shaft (35) are available only as a matched set.

The differential can be reinstalled by reversing removal procedure. Refer to paragraph 139 for adjustment.

139. ADJUSTMENT. Refer to

paragraph 117 for selection of shims (48–Fig. 116). Shims (51) used to position bevel pinion must be selected to provide correct pinion position relative to bevel ring gear.

Adjustment of differential carrier bearings and bevel gear backlash is accomplished by varying thickness of shims (43–Fig. 138). Housings (27) can not be removed with axle housings (7) installed, but can be moved back far enough to add or remove shims for adjustment. Shims (43) are available in thicknesses of 0.10, 0.15, 0.20, 0.30 and 0.50 mm. Select shims that provide 0 to 0.10 mm play in carrier bearings. If side play is 0, be sure that rolling torque required to rotate differential does not exceed 0.79 N·m.
If carefully done, shims can be selected by assembling gently without shims and measuring gap between flange of housing (27–Fig. 138) and axle housing.

After correctly setting carrier bearing

adjustment, attach axle center housing to transmission housing and check backlash between bevel pinion and ring gear. A dial indicator may be used to check backlash at outer diameter of ring gear teeth which should be 0.1-0.2 mm. If backlash exceeds 0.2 mm, remove some shims (43) from left side, install the same thickness on right side and recheck. To increase backlash, move ring gear away from bevel pinion by removing shims from right side and adding same thickness to left side.

The differential carrier assembly (37) with ring gear must be assembled in axle center housing, the bevel pinion must be installed in transmission housing and the transmission housing must be attached to the axle center housing in order to check mesh position. Backlash should be adjusted within limits 0.10-0.20 mm (measured at ring gear teeth) before checking mesh position. Mesh position is determined by coating

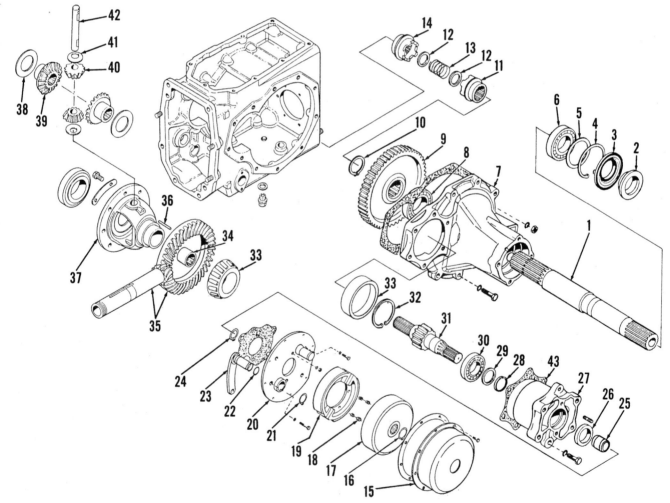

Fig. 138 – Exploded view of axle center housing and final drive used on MF210 and MF220 two wheel drive tractors.

1. Axle	9. Gear	16. Snap ring	23. Actuator cam	30. Bearing	37. Carrier
2. Seal ring	10. Snap ring	17. Brake drum	24. Snap ring	31. Final drive pinion	38. Thrust washers
3. Seal	11. Lock coupling	18. Return springs	25. Spacer	32. Snap ring	39. Side gears
4. Snap ring	12. Thrust washer	19. Brake shoes	26. Seal	33. Bearing cup & cone	40. Pinion gears
5. Shims	13. Spring	20. Backing plate	27. Pinion housing	34. Pinion bearing	41. Thrust washers
6. Bearing	14. Lock coupling	21. Snap ring	28. Snap ring	35. Bevel pinion and ring gear	42. Shaft
7. Axle housing	15. Cover	22. "O" ring	29. Thrust washer	36. Pin	43. Shims
8. Bearing					

ring gear teeth with mechanics' bluing, then turning coated teeth through pinion. Refer to Fig. 136. Correct mesh position is shown at (A). Mesh position (B) indicates that ring gear is too far away from pinion (too much backlash) and can usually be corrected by removing shims from left side carrier bearing and installing same shims on right side.

Mesh pattern (C – Fig. 136) is caused by too little backlash and can usually be corrected by removing some shims from right side and installing on left. Pattern (D) is caused by pinion too far from center of ring gear and is corrected by adding thickness to shims (51 – Fig. 116). Mesh pattern (E – Fig. 136) is caus-

ed by pinion too far into ring gear and is corrected by removing shims (51 – Fig. 116). Shims (51) are available in thicknesses of 0.10, 0.20 and 0.40 mm. Always recheck mesh position using fresh coating of mechanics' bluing to be sure that shims are correct before final setting.

MF210 and MF220 Models With Four Wheel Drive

141. **R&R AND OVERHAUL.** The differential carrier (37 – Fig. 141) and related parts can be removed from axle

center housing after removing both final drive axle housings as outlined in paragraph 151 and separating axle center housing from transmission as outlined in paragraph 130.

Be careful not to lose or damage shims (43) which are to adjust carrier bearings and backlash between bevel pinion and ring gear. Also, be careful not to damage bearing (65 – Fig. 126) when separating or joining transmission and axle center housing.

Differential pinion gear shaft (42 – Fig. 141) can be removed after driving pin (36) out toward ring gear side. Hole is provided in opposite side of carrier (37) to facilitate removal. Ring

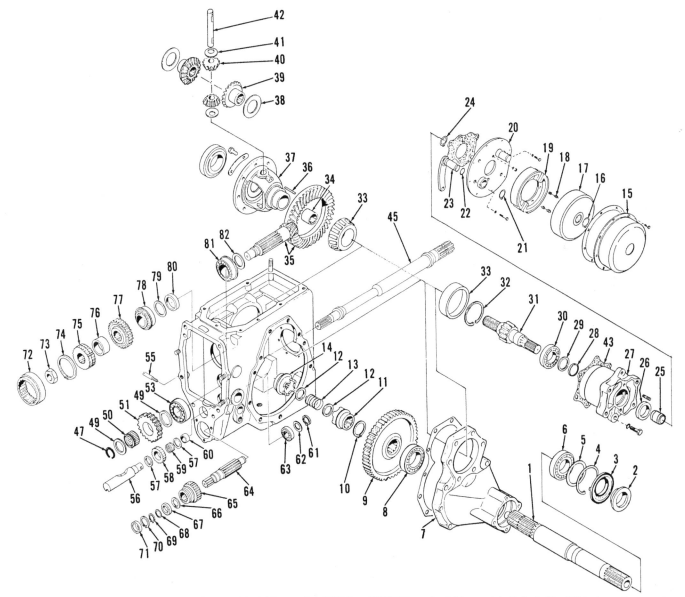

Fig. 141 — Exploded view of axle center housing and final drive used on MF210 and MF220 four wheel drive models. Refer to Fig. 138 for legend except the following.

45. Pto output shaft	55. Retaining screw	61. Snap ring	66. Snap ring	72. Sleeve
47. Snap ring	56. Idler shaft	62. Thrust washer	67. Bearing	73. Nut
49. Thrust washer	57. Thrust washer	63. Bearing	68. Thrust washer	74. Snap ring
50. Bearing	58. Gear (21 teeth)	64. Front wheel	69. Snap ring	75. Hub
51. Pto gear (28 teeth)	59. Bearing	drive shaft	70. Snap ring	76. Spacer
53. Bearing	60. Spacer	65. Gear	71. Seal	77. Front drive gear

78. Bearing
79. Shim
80. Spacer
81. Bearing
82. Shim

gear is attached to carrier with cap screws. Tighten all of the ring gear retaining screws to 48.3-58.3 N·m torque, then lock in place by bending lock tab around heads of screws. Ring gear and pinion shaft (35) are available only as a matched set.

Remove nut (73), hub (75), spacer (76) and gear (77). Pinion can be withdrawn from rear. Be careful not to lose or damage bearing cup/cone (78), bearing adjustment shims (79), spacer (80), bearing cup/cone (81) or shims (82).

Original thickness of shims (79) can be reinstalled if original bearings (78 and 81), spacer (80) and housing are reinstalled. If any of these parts are renewed or if adjustment is questioned, refer to paragraph 142.

Original thickness of shims (82) can be reinstalled if original bearing (81), housing, bevel pinion and ring gear (35) are reinstalled. If any of these parts are renewed, determine correct thickness of shims (79), then adjust mesh position as outlined in paragraph 142.

142. ADJUSTMENT. To determine correct thickness of shims (79 – Fig. 141), assemble parts (73 through 82) in housing and on pinion shaft. Tighten nut (73) to 88 N·m torque then check rolling torque of pinion shaft in bearings (78 and 81). If rolling torque is more than 1.17 N·m, then increase thickness of shims (79). If rolling torque of pinion shaft is less than 0.58 N·m; decrease thickness of shims (79) and recheck. Shims (79) are available in thicknesses of 0.10, 0.15, 0.20 and 1.00 mm. After correct thickness of shims (79) has been determined, mesh position of bevel gears must be checked and adjusted by changing thickness of shims (82).

Adjustment of differential carrier bearings and bevel gear backlash is accomplished by varying thickness of shims (43 – Fig. 141). Housings (27) can not be removed with axle housings (7) installed, but can be moved back far enough to add or remove shims for adjustment. Shims (43) are available in thicknesses of 0.10, 0.15, 0.20, 0.30 and 0.50 mm. Select shims that provide 0 to 0.10 mm play in carrier bearings. If side play is 0, be sure that rolling torque required to rotate differential does not exceed 0.79 N·m. If carefully done, shims can be selected by assembling gently without shims and measuring gap between flange of housing (27 – Fig. 141) and axle housing.

After correctly setting carrier bearing adjustment, check backlash between bevel pinion and ring gear. A dial indicator may be used to check backlash at outer diameter of ring gear teeth which should be 0.10-0.20 mm. If backlash exceeds 0.2 mm, remove some shims (43)

from left side, install on right side and recheck backlash. To increase, move ring gear away from bevel pinion by removing shims from right side and adding same thickness to left side.

The differential carrier assembly (37) with ring gear and the bevel pinion must be installed in housing in order to check mesh position. Backlash should be within limits of 0.10-0.20 mm (measured at ring gear teeth) when checking mesh position. Mesh position is determined by coating ring gear teeth with mechanics' bluing, then turning through pinion. Refer to Fig. 136. Correct mesh position is shown at (A). Mesh position (B) indicates that ring gear is too far away from pinion (too much backlash) and can usually be corrected by removing shims from left side carrier bearing and installing same shims on right side.

Mesh pattern (C – Fig. 136) is caused by too little backlash and can usually be corrected by removing some shims from right side and installing on left. Pattern (D) is caused by pinion too far from center of ring gear and is corrected by adding thickness to shims at (82 – Fig. 141). Mesh pattern (E – Fig. 136) is caused by pinion too far into ring gear and is corrected by removing some thickness from shims at (82 – Fig. 141). Shims (82) are available in thicknesses of 2.70, 2.80, 2.90, 3.00, 3.10, 3.20 and 3.30 mm. Always recheck mesh position using fresh coating of mechanics' bluing to be sure that shims are correct before final assembly.

After correct thickness of shims (79 and 82 – Fig. 141) is determined, coat threads for nut (73) with "Loctite" and tighten to 88 N·m torque.

FINAL DRIVE

AXLE, HOUSING AND FINAL REDUCTION GEARS

All MF205 Models

150. REMOVE AND REINSTALL. To remove final drive assembly, first remove brake cover, drum, brake shoes and backing plate as described in paragraph 202. Remove seat pan and lower lift linkage. Remove all of the housing retaining screws, then carefully separate final drive housing (7 – Fig. 132 or Fig. 135) from center housing. Final drive housing should pull straight away from center housing; however, sealer is used on gasket and separation may be difficult. Other final drive can be similarly removed. Be careful not to lose or damage shims (43) and thrust washer (44). Do not mix cups for bearings (33), thrust washers (44) or shims (43) from

opposite sides. Shims (43) are used to adjust differential carrier bearings and to adjust backlash between pinion and ring gear. Refer to paragraph 133 if backlash or bearing adjustment is required on two wheel drive models, paragraph 136 if backlash or bearing adjustment is required on four wheel drive models. On all models, a strap retainer can be fabricated and attached across opening in center housing to keep shims (43), thrust washers (44), bearings and differential from falling out while final drive housing is removed.

On four wheel drive models, further disassembly will necessitate separation of center housing from transmission housing. Refer to paragraph 130.

On all models, be sure to correctly position differential lock couplings (11 and 14 – Fig. 132 or Fig. 135), washers (12) and spring (13) before installing axle housing (7). End play of axles (1) should be within limits of 0.1 to 0.3 mm. End play is adjustable by adding or removing shims (5) which are available in thicknesses of 0.40 and 0.60 mm. Shims should be divided evenly between left and right axles.

All MF210 and MF220 Models

151. REMOVE AND REINSTALL. To remove final drive assembly, first remove brake cover, drum brake shoes and backing plate as described in paragraph 202. Remove seat pan and lower lift linkage. Remove all of the housing retaining screws, then carefully separate final drive housing (7 – Fig. 138 or Fig. 141) from center housing. Notice that even though all of the screws are removed, differential pinion housing (27) can not be removed before housing (7) and final drive gear (9). Final drive should pull straight away from center housing; however, sealer is used on gasket and separation may be difficult. Other final drive can be removed similarly. Be careful not to let differential assembly fall as second drive assembly is removed. Differential assembly can be withdrawn through side opposite differential lock fork. Further disassembly of four wheel drive models will necessitate separation of center housing from transmission housing as outlined in paragraph 130.

Shims (43) located under pinion housings are used to adjust differential carrier bearings and to adjust backlash between pinion and ring gear. Refer to paragraph 139 or 142 if backlash or bearing adjustment is required.

Be sure to correctly position differential lock couplings (11 and 14), washers (12) and spring (13) before installing axle housings (7). End play of axles (1) should be within limits of 0.1 to 0.3 mm. End play is adjustable by adding or removing shims (5) which are available in thicknesses of 0.40 and 0.60 mm. Shims should be divided evenly between left and right axles.

BRAKES

ADJUST

All Models

200. Brake pedals should have approximately 20-25 mm travel and pedals should align when equal pressure is applied to the two pedals. Adjustment can be accomplished by shortening or lengthening actuating rods. Loosen locknut and turn adjustment turnbuckle located in middle of each rod.

R&R AND OVERHAUL

All Models

202. Service to the brakes is more easily accomplished if wheel and fender is removed. Refer to Fig. 202 or 203. Remove cover (15) and snap ring (16), then withdraw brake drum (17) from pinion shaft splines. Remainder of disassembly will depend upon repair necessary but will be evident after examination of unit and reference to Fig. 202 or 203. Refer to the following specification data.

MF205

Lining Thickness-New	4 mm
Minimum Thickness	2 mm
Drum ID-New	130 mm
Maximum Limit	132 mm

MF210 and MF220

Lining Thickness-New	5 mm
Minimum Thickness	3 mm
Drum ID-New	145.2 mm
Maximum Limit	147.2 mm

Backing plate (20) must be removed before renewing seal (26). When reassembling, be sure that snap ring (24) is installed.

POWER TAKE OFF

OUTPUT SHAFT

MF210 and MF220 Two Wheel Drive Models and All MF205 Models

250. **REMOVE AND REINSTALL.**

The output shaft can be withdrawn after unbolting bearing and seal housing from rear of axle center housing. Bearing and seal can be renewed after removing re-

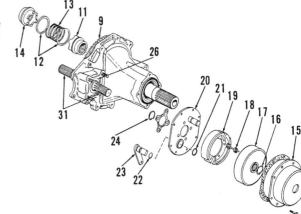

Fig. 202—Exploded view of brake assembly used on MF205 models.

9. Final drive gear
11. Coupling
12. Thrust washers
13. Spring
14. Coupling
15. Cover
16. Snap ring
17. Brake drum
18. Return spring
19. Brake shoes
20. Backing plate
21. Snap ring
22. "O" ring
23. Actuator cam
24. Snap ring
31. Final drive pinion

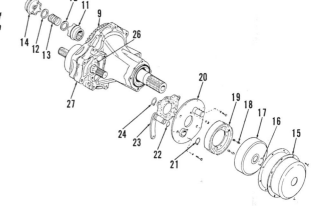

Fig. 203—Exploded view of brake assembly used on MF210 and MF220 Models.

9. Final drive gear
11. Coupling
12. Thrust washers
13. Spring
14. Coupling
15. Cover
16. Snap ring
17. Brake drum
18. Return springs
19. Brake shoes
20. Backing plate
21. Snap ring
22. "O" ring
23. Actuator cam
24. Snap ring
26. Oil seal
27. Pinion housing

taining snap rings. Refer to Fig. 250 for MF205 models with two wheel drive; Fig. 251 for MF205 models with four wheel drive or Fig. 252 for MF210 and

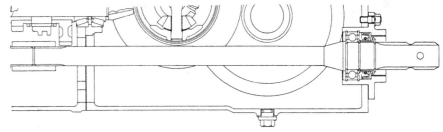

Fig. 250—View of pto rear shaft for MF205 two wheel drive models.

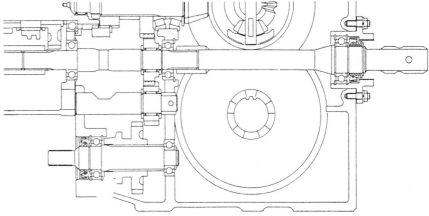

Fiç   1—View of pto rear shaft for MF205 four wheel drive models.

MF220 models with two wheel drive. Apply oil or light grease to seal upon assembly for initial lubrication.

MF210 and MF220 Four Wheel Drive Models

255. **REMOVE AND REINSTALL.** The output shaft can be removed after separating axle center housing from transmission housing as outlined in paragraph 130. Refer to Fig. 141. Remove snap ring (47), thrust washer (49), bearing (50), gear (51) and second thrust washer (49). Unbolt rear bearing and seal retainer from rear of axle center housing, then withdraw output shaft. Apply oil or light grease to seal upon assembly for initial lubrication.

PTO INTERMEDIATE SHAFT

MF205 Four Wheel Drive

260. **REMOVE AND REINSTALL.** The pto intermediate shaft (46 – Fig. 135) can be removed after separating axle center housing from transmission housing as outlined in paragraph 130.

PTO FRONT SHAFT

All Models

265. The pto front shaft of all models is the lower shaft of transmission. Refer to paragraph 95 for MF205 two wheel drive models; paragraph 105 for MF205 four wheel drive models; paragraph 118 for MF210 and MF220 two wheel drive models; paragraph 128 for MF210 and MF220 four wheel drive models.

HYDRAULIC LIFT SYSTEM

FLUID AND FILTER

300. The transmission, center housing (differential and final drive), rear axles and hydraulic system share a common sump. Lubricant should be maintained at level of plug (L – Fig. 300 or Fig. 301). Oil should be drained, strainer should be cleaned and system should be filled with new oil after every 500 hours of operation or yearly which ever occurs first. Approximate capacity is 16 liters for MF205 two wheel drive models; 17 liters for MF205 four wheel drive models; 21 liters for MF210 and MF220 two wheel drive models; 25.7 liters for MF210 four wheel drive models; 27 liters for MF220 four wheel drive models. When filling, allow sufficient time for oil to pass between compartments and be sure that

tractor is level before checking at level plug (L). Use only Massey-Ferguson Permatran (M-1129A) or equivalent oil.

HYDRAULIC PUMP

305. **REMOVE AND REINSTALL.** To remove hydraulic pump, first remove side shield, drain radiator and disconnect lower hose from radiator. Detach the two oil tubes at the pump and insert plugs. Mark relative location of pump on timing gear cover. Unbolt and remove pump. The pump drive gear thrust washer (9 – Fig. 305) is a very close fit in the timing gear cover and it may be necessary to tap pump lightly with a soft faced hammer to dislodge pump.

After installing new or rebuilt pump, operate without load and engine at idle rpm for 30 minutes then gradually increase speed and load. During this break-in, check pump periodically for any abnormally high temperature.

Fig. 252 – View of pto rear shaft for MF210 and MF220 two wheel drive models.

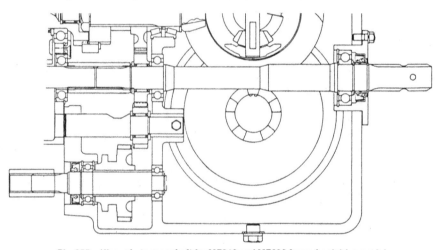

Fig. 255 – View of pto rear shaft for MF210 and MF220 four wheel drive models.

Fig. 300 – View of hydraulic and transmission fluid fill plug (F), level plug (L) and strainer location (S) for MF205 models.

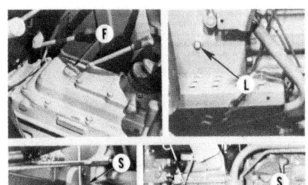

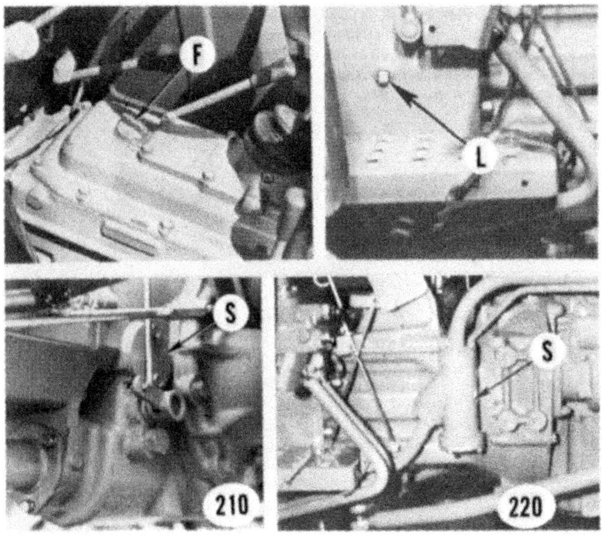

Fig. 301 – View of MF210 and MF220 fill plug (F), level plug (L) and strainer locations.

slide through bushing and seal (1). Install pump driven gear (7) with marked end of shaft out. Complete assembly and apply oil liberally. Before installing pump, rotate pump shaft by hand and check for any stiffness or roughness. If not smooth, disassemble and check condition of all parts, especially "O" rings (3).

RELIEF VALVE

310. **ADJUST.** The hydraulic pressure relief valve is located on right side of lift cover. Relief pressure can be checked as follows: Remove plug (10 – Fig. 311) from relief valve body, install adapter fitting (1 – Fig. 310) and attach a 2000 psi pressure gage. To force relief valve to operate, loosen locknut, then turn off-center pin (2) until hydraulic control lever will move back far enough to overload pressure relief valve. On certain models it may be necessary to remove the rear bolt for the fender mounted quadrant so that lever can move back far enough. Relief pressure should be 11721 kPa (1700 psi) with

306. **OVERHAUL.** Scribe mark across pump end cap and body so that parts can be aligned in original position when reassembling. Remove nut retaining pump drive gear, then use suitable puller to remove gear. Remove six retaining screws, then separate end cap (11 – Fig. 306) and pump body (2). If original parts are to be reassembled, lay parts in order to assure assembly in exact location as removed.

When assembling, be sure to always install new "O" rings and seal. Lip of seal (1) should be toward inside and should be pressed in until flush with outer surface of pump body. Position "O" rings (3) on inner bushings (4) and use light grease (petroleum jelly) to hold "O" rings in place while bushings are being installed in body (2). Be sure that "O" rings (3) do not fall out of position while assembling

or pressure balancing system will be lost. Lubricate inner bushings (4) with oil and slide into position in pump body (2). No force should be required to install bushings. Be sure that "O" rings (3) do not fall out of position while assembling. Lubricate shaft of gear (8) and carefully

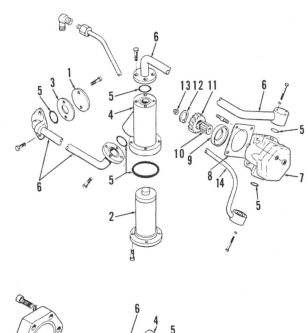

Fig. 305A – View of pump and filter used on MF220 models. Refer to Fig. 305 for legend except strainer housing (4). "O" rings (5) are not the same size.

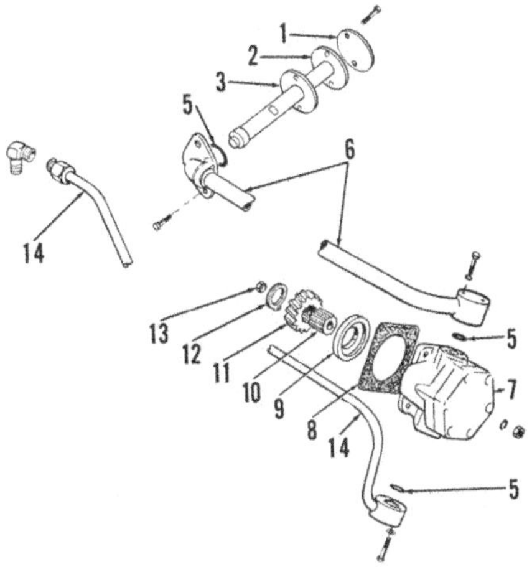

Fig. 305 – View of hydraulic pump typical of MF205 and MF210 models.

1. Cover
2. Strainer
3. Gasket
5. "O" ring
6. Suction tube
7. Pump
8. Gasket
9. Thrust washer
10. Hub
11. Gear
12. Snap ring
13. Nut
14. Pressure tube

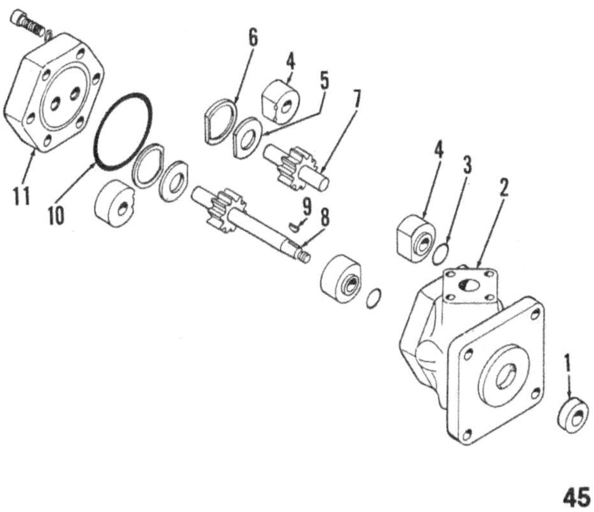

Fig. 306 – Exploded view of hydraulic pump shown at (7 – Fig. 305 and Fig. 305A).

1. Seal
2. Body
3. "O" ring
4. Bushings
5. Support ring
6. Sealing ring
7. Driven gear
8. Drive shaft and gear
9. Woodruff key
10. "O" ring
11. Pump cover

engine at 1285 rpm. Relief valve setting is changed by removing plug (3–Fig. 311) and adding or removing shims (5). Each shim will change relief pressure approximately 1379 kPa (200 psi), but pressure should not be set higher than 11800 kPa (1710 psi).

Tighten plug (3) to 34-44 N·m torque; tighten plug (10) to 7-9 N·m torque.

CONTROL VALVE

320. R&R AND OVERHAUL. The control valve can be unbolted and removed after removing the lift cover from axle center housing. Lift valve spool (4–Fig. 321) is not available separately. Clean all parts and check for nicks, burrs, scores or eroded valve seats. Be sure that all parts are completely clean and lubricated when assembling.

Plugs (9 and 10) should be tightened to 67-88 N·m; seat (19) should be tightened to 49-59 N·m torque. Nut (1) should be tightened to 18-22 N·m torque and adjusting locknuts (20) should be tightened together to 18-22 N·m torque. Refer to paragraph 335 for adjustment procedure.

LIFT ARMS AND RAM

330. R&R AND OVERHAUL. Unbolt and remove lift housing from top of axle center housing. Unbolt cover (8–Fig. 330) and remove cover (8), piston (18)

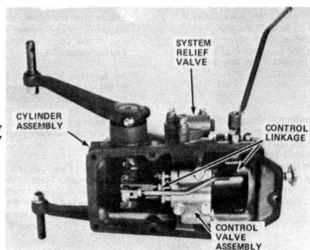

Fig. 320 – View of lift cover removed from axle center housing.

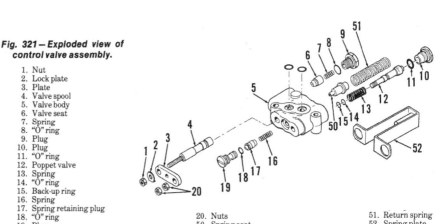

Fig. 321 – Exploded view of control valve assembly.

1. Nut	
2. Lock plate	
3. Plate	
4. Valve spool	
5. Valve body	
6. Valve seat	
7. Spring	
8. "O" ring	
9. Plug	
10. Plug	
11. "O" ring	
12. Poppet valve	
13. Spring	
14. "O" ring	
15. Back-up ring	
16. Spring	
17. Spring retaining plug	
18. "O" ring	
19. Plug	

20. Nuts   51. Return spring
50. Spring seat   52. Spring plate

Fig. 310 – View showing fitting (1) installed in relief valve port for attaching pressure gage. Refer to Fig. 311 and text.

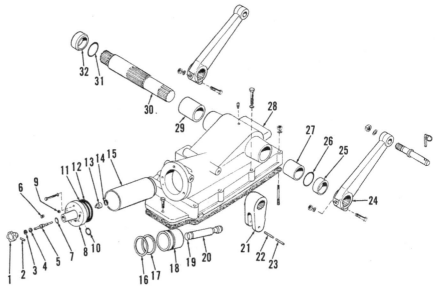

Fig. 330 – Exploded view of lift cover cylinder and lift arms.

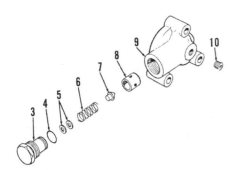

Fig. 311 – Exploded view of relief valve. Refer to text for adjustment procedure.

3. Plug	7. Relief valve
4. "O" ring	8. Valve seat
5. Shims	9. Housing
6. Spring	10. Plug

1. Lowering rate adjusting knob	8. Cover	16. Lip packing	24. Lift arm
2. Pin	9. Plug	17. Back-up ring	25. Collar
3. Snap ring	10. "O" ring	18. Piston	26. "O" ring
4. Washer	11. "O" ring	19. "O" ring	27. Bushing
5. Lowering adjustment screw	12. "O" ring	20. Piston rod	28. Cover
6. Plug	13. Valve	21. Lever arm	29. Bushing
7. "O" ring	14. Snap ring	22. Roll pin	30. Lift arm shaft
	15. Cylinder	23. Roll pin	31. "O" ring
			32. Collar

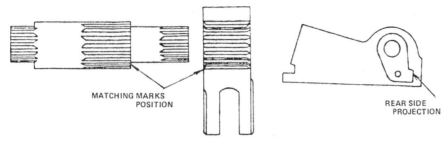

Fig. 331 — View of lift arm shaft and internal lift arm showing correct installation.

and liner (15). Remove clamp screws then remove lift arms (24). Lift arm shaft (30) can be removed from right side.

Inspect all sliding surfaces for burrs, nicks and wear. Polish surface or renew worn part. All "O" rings and lip seals

should be renewed to insure proper sealing.

Observe the following when reassembling. Coat all "O" rings and seals with oil prior to installation and

assembly. Install lift arm shaft (30) through housing bores and hydraulic arm (21), with marks aligned and projection on arm toward rear as shown in Fig. 331. Install lip seal and back-up ring on piston as shown in Fig. 332.

The lowering speed control valve (Fig. 333) controls the volume of oil leaving lift cylinder.

ADJUSTMENT

335. **CONTROL VALVE LINKAGE.** Pull plate (3 – Fig. 335) against nuts (20), but do not compress spring in poppet valve. Measure distance (A) which should be 10 mm. If distance is incorrect, be sure that inner spring is not compressed when measuring, then correct measurement by turning nuts (20). Nut (1) should not need to be changed.

After distance (A) is set at 10 mm, check length of turnbuckle (C – Fig. 336). Two different length of turnbuckles have been used on MF220 models; 58 mm and 68 mm. Measure distance (B – Fig. 335) with plate held against nuts (20) but not preloading spring in poppet valve. Correct distance (B) is 89mm for 58mm turnbuckle, 99mm for 68mm turnbuckle. If distance (B) is incorrect, loosen locknut (22) and turn turnbuckle (21) in or out of plate to set correct adjustment. Tighten locknut (22) after adjustment is complete. Check to be sure that working link will slide freely into turnbuckle slot. Assemble spring guide and spring compression bracket. Position a 10 mm block between plate and valve housing (A – Fig. 337), push working link (25) toward control valve then vary the length of position control rod by turning turnbuckle (26) until the holes align. Install special screw and castle nut finger tight then install cotter key.

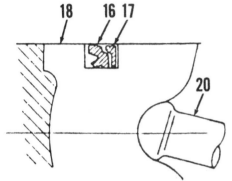

Fig. 332 — View of piston (18) and rod (20) with seal ring (16) and back-up ring (17) installed.

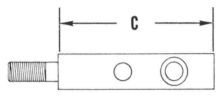

Fig. 336 — Two different lengths of turnbuckles (C) necessitate different adjustment dimensions. Refer to text.

Fig. 337 — View showing correct settings. Special tool is shown at (A) for setting distance of 10 mm. Distance (B) from housing to end of turnbuckle should be 89 mm for short turnbuckle 99 mm for longer turnbuckle. Clearance (C) between nut and plate should be 0.6 mm.

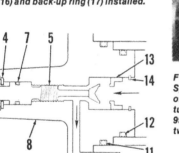

Fig. 333 — Cross-section showing lowering rate adjustment valve. Refer to Fig. 330 for legend.

Fig. 338 — Exploded view of lift cover, valves and linkage used on MF220 models. Other models are similar.

21. Turnbuckle
25. Working link
26. Turnbuckle
27. Rod
28. Lever shaft
29. Washer
30. Thrust washer (2.0 mm)
31. Thrust washer (1.0 mm)
32. Adjuster
33. Pin
34. Stop screw
35. Draft linkage
36. Lift control lever
37. Relief valve assembly
38. "O" rings (same as 44)
39. "E" ring
40. Link guide
41. Pin
42. Oil deflector
43. Control valve
44. "O" rings (same as 38)

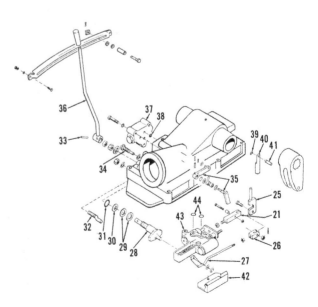

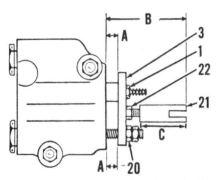

Fig. 335 — View of control valve showing measurement points and points of adjustment. Refer to text.

340. **DRAFT CONTROL LINKAGE.**
To check, attach weight or implement to
lift arms and turn lowering speed knob
to fast setting. Raise and lower lift links
fully using the position control lever,
first with draft lever fully forward, then
with draft lever fully to rear. If weight
moves more slowly with draft lever forward, shorten external draft link
(1 – Fig. 340 or Fig. 341). Change length
of draft link ½ turn at a time, then
recheck until no lowering speed difference is noted, then shorten linkage an
additional ½-turn, then tighten locknut
and install washers and cotter pins.
Nuts (2 – Fig. 340) can be adjusted to
vary lever friction.

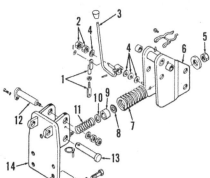

Fig. 341 – Refer to text for adjusting draft control
link (1).

Fig. 340 – Exploded view of draft control linkage.
Link (1) is also shown in Fig. 341. Lower end of
link (1) attaches to external lever of draft linkage
(35 – Fig. 338).

1. Draft link	8. Cushion washer
2. Nuts	9. Cushion spring
3. Draft lever	10. Cushion washer
4. Friction washers	11. Inner spring
5. Nut	12. Pivot shaft
6. Sensor bracket	13. Clevis pin
7. Outer spring	14. Draft control bracket

MASSEY-FERGUSON

Models ■ MF 2675 ■ MF 2705

Previously contained in I & T Shop Manual No. MF-38

SHOP MANUAL

MASSEY-FERGUSON

MODELS

MF2675   **MF2705**

Tractor serial number is stamped on left side of instrument console.
Engine serial number is stamped on left side at injection pump base.

INDEX (By Starting Paragraph)

BRAKES
Adjust parking brake 88A
Adjustment 88
Brake pistons & discs 91
Fluid & bleeding 89
Overhaul master cylinder 90

CLUTCH
Adjustment 62
Overhaul 64A
Release linkage 15A
Tractor split 63

COOLING SYSTEM
Radiator 55
Thermostat 56
Water pump 57

ELECTRICAL SYSTEM
Alternator 58
Circuit description 61
Starting motor 60
Wiring diagram . . . . Figs. 74 & 75

ENGINE
Assembly R&R 15
Cam followers 20
Camshaft 26
Connecting rods & bearings 30
Crankshaft & bearings 31
Crankshaft rear oil seal 32
Cylinder head 16
Flywheel 33

ENGINE CONT.
Oil pan 34
Oil pump 35
Pistons, sleeves & rings 28
Piston pins 29
Rocker arms 21
Timing cover & gears 24
Valves & seats 17
Valve clearance 22
Valve guides 18
Valve springs 19
Valve timing 23

FINAL DRIVE
Differential & bevel gear 82
Differential lock valve 127
Rear axle & final drive 85

FRONT SYSTEM
Axle 1
Flow control valve 8
Steering cylinder 7
Steering valve 12

FUEL SYSTEM
Bleeding 39
Fuel filters & lines 38
Fuel lift pump 40
Injection pump 49
Injector nozzles 41

HYDRAULIC SYSTEM
Auxiliary control valves 125
Auxiliary hydraulic pump 123
Brake & steering tests 104
IPTO/Differential lock 127
Lift system 128
Lubrication 97
Main pump 121
Sensing linkage 138
Tests and Adjustments 98

POWER TAKE-OFF
Clutch 93
Output housing & gears 96
Output shaft 95
Valve 127

TRANSMISSION
Front cover R&R 68
Front cover overhaul 69, 70
Main transmission overhaul 74
Transmission brake 66

24-SPEED INPUT ASSY.
Diagnosis 65
R&R 68
Overhaul 70
Shift valve 65A

TURBOCHARGER
Troubleshooting 52
Remove & reinstall 53
Overhaul 54

CONDENSED SERVICE DATA

GENERAL

	2675	2705
Engine Make......................	Perkins	
Engine Model	A6.354.4	AT6.354.4
Number of Cylinders	6	
Bore-Inches......................	3.875 inches	
Stroke-Inches	5 inches	
Displacement-Cu.-In.	354	
Compression Ratio	16:1	15.5:1
Main Bearings, No. of	7	
Cylinder Sleeves..................	Dry	
Forward Speeds	8 or 24	
Reverse Speeds	6	

TUNE-UP

	2675	2705
Firing Order	1-5-3-6-2-4	
Valve tappet gap – cold		
Intake (inch)	0.008	
Exhaust (inch)	0.018	
Valve Face Angle	45°	
Valve Seat Angle	45°	
Injection timing –		
Crankshaft Degrees	28	
Injectors –		
Opening Pressure, psi..............	2939	3086
Spray Hole Dia., inch	0.0124-0.0128	0.0116-0.0120
Governed Speeds –		
Engine rpm		
Low Idle	825-875	
High Idle	2675-2825	
Loaded	2500	
540 pto		
Low Idle	223-237	
High Idle	725-766	
Loaded	678	
1000 rpm pto		
Low Idle	395-419	
High Idle	1279-1351	
Loaded	1196	
Horsepower at pto Shaft*	103.29	122.20
Battery**		
Volts...........................	12	
Ground Polarity	Negative	
Capacity Amp/hr.................	95	

*According to Nebraska Test.
**Each battery – two batteries are used, connected in parallel.

CONDENSED SERVICE DATA CONT.

SIZES – CAPACITIES – CLEARANCES

	2675	2705
Crankshaft Main Journal–		
Diameter, inch	2.9984-2.9992	
Bearing Clearance, inch	0.0033-0.0046	
Crankshaft Crankpin–		
Diameter, inch	2.4988-2.4996	
Bearing Clearance, inch	0.0019-0.0032	
Crankshaft End Play, Inch	0.002-0.015	
Camshaft Journal Diameters–		
Front, inch	1.9965-1.9975	
Second, inch	1.9865-1.9875	
Third, inch	1.9765-1.9775	
Fourth, inch	1.9665-1.9675	
Camshaft Bearing Clearance–		
Front, inch	0.0025-0.0045	
Second, Third And Fourth, inch	0.0025-0.0055	
Camshaft End Play, inch	0.004-0.016	
Cooling System-Quarts	31	
Crankcase Oil-Quarts Including filter(s)	19	20
Transmission, Differential and Hydraulic Lift–		
Gallons	30	

FRONT SYSTEM

Refer to Fig. 1 for view of both Western Style (non-adjustable) and Row Crop (wide-adjustable) axle units. Both tractors are available with either style axle.

AXLE ASSEMBLY

All Models

1. **REMOVE AND REINSTALL.** To remove axle assembly on either model, support tractor behind front axle, remove wheels and disconnect tie rod ends from steering arms (10–Fig. 1). Remove weights from front end of tractor while supporting with a suitable jack. Remove steering cylinder (C–Fig. 3) as outlined in paragraph 7. Remove cap screw (13–Fig. 1) from end of pivot pin (14), then remove lockwire and pivot pin set screw (16). Apply light pressure to axle with floor jack, withdraw pivot pin and lower axle assembly to floor. Do not lose washers (18) or shims (22).

To reinstall axle assembly, reverse removal procedure. Use one washer (18) at each end, but shims (22) should be installed at rear of axle only. Install proper shims between axle and support so

maximum clearance does not exceed 0.010-inch. Shims are available in sizes 0.010, 0.020, 0.040 and 0.080-inch. Pivot pin set screw (16) should be tightened to a torque of 85-95 ft.-lbs.

Fig. 1 – Exploded view of non-adjustable (Western-W), wide-adjustable (Row Crop—R) front axles and associated parts. Thrust washer (2) must be installed with grooves up.

1. Spindle
2. Grooved thrust washer
3. Smooth thrust washer
4. Thrust bearing
5. Bushing
6. Axle extension
8. Shim
9. Felt seal
10. Steering arm
11. Lockwasher
12. Nut
13. Cap screw
14. Pivot pin
15. Axle support
16. Set screw
17. Bushing
18. Washer
19. Cap screws
20. Nuts
22. Shims
23. Engine support

AXLE EXTENSIONS

All Models

2. **REMOVE AND REINSTALL.** To

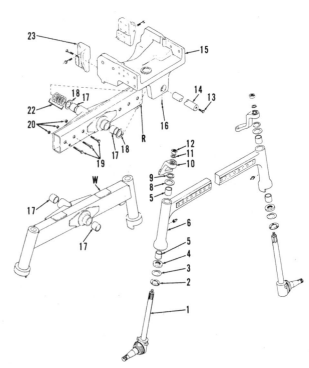

remove either axle extension (6 – Fig. 1), first support tractor properly and remove wheel on side requiring service. Remove steering arm nut (12), remove spindle (1) from axle housing and move steering arm (10) out of the way. Remove cap screws (19) retaining extension, then slide axle extension out of axle center housing. To reinstall axle extension, reverse removal procedure. Tighten retaining cap screws (19) and nuts (20) to a torque of 400-450 ft.-lbs. Tighten steering arm retaining nut (12) to a torque of 225-250 ft.-lbs.

SPINDLES AND BUSHINGS

All Models

3. **REMOVE AND REINSTALL.** To renew spindles (1 – Fig. 1), support tractor properly, remove wheel and hub assembly on side requiring service, then remove steering arm retaining nut (12) and remove spindle (1) from axle housing. Heat bushing area of housing to loosen Loctite seal, then remove bushings (5) using suitable tools. Press new bushings into axle housing using a suitable locking sealer to help retain bushings. Install thrust washer (2) onto spindle (1) with grooved face up, then install smooth face thrust washer (3). Pack thrust bearing (4) with a suitable grease and install bearing onto spindle with closed side up. Apply grease to splines of spindle, then install spindle into axle housing. Install steering arm (10) aligning "X" marks on spindle shaft (1) and arm (10). Install washer (11) and nut (12), tightening to a torque of 225-250 ft.-lbs. Raise spindle shaft (1) with a floor jack and measure gap between axle and spindle arm (10) as shown in Fig. 2. Install shims (8 – Fig. 1) so gap will be 0.005-0.030 inch. Remove spindle arm, install proper shims (8), then reinstall nut (12) and lockwasher (11). Tighten nut to a torque of 225-250 ft.-lbs. Reinstall wheel hub and tighten hub re-

taining nut to a torque of 60 ft.-lbs., then back nut off a maximum of one flat. Lock by bending tab on lockwasher and reinstall wheel(s).

TIE RODS AND TOE-IN

All Models

4. **ADJUST.** Automotive type tie rod ends are used. Recommended toe-in is 0-9/16-inch. To adjust, loosen jam nut on outer end of left tie rod. Remove clamps and bolts, then rotate left tie rod tube until correct toe-in is obtained. Install and tighten clamp bolts on outer ends of each tie rod. Tighten jam nut.

POWER STEERING SYSTEM

All models are equipped with power steering. Refer to Fig. 3 for view of steering hydraulic system. Hydraulic oil for the power steering is supplied from the high pressure side of the main hydraulic pump. This section covers power steering control valve (6), flow control valve assembly (FV) and steering cylinder (C). Refer also to hydraulic system section for information on testing main and auxiliary hydraulic pump, hydraulic valves and operating fluid and steering system tests.

FLUID AND BLEEDING

All Models

5. The power steering system is lubricated by the operating fluid. Refer to paragraph 97 for fluid type and checking procedure.

To bleed system first make sure reservoir has been initially filled, then start engine and set throttle at 800-1000 rpm. Rotate steering wheel fully in both directions until no air can be felt passing through the steering control assembly. Stop engine, and check for leaks, then restart engine, set throttle at 1200 rpm, operate steering in both directions and hold against stops until relief valve becomes noisy. Repeat this procedure at least three times in both directions.

TROUBLESHOOTING

All Models

6. If malfunction of steering system exists, first make an operational check of another hydraulic unit on tractor to eliminate hydraulic system as cause of trouble. The procedure for checking the hydraulic system is given beginning with paragraph 98.

STEERING CYLINDER

All Models

7. **R&R AND OVERHAUL.** To remove steering cylinder (C – Fig. 3), first

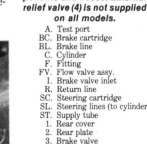

Fig. 3 — View showing external hydraulic system components and hoses. External relief valve (4) is not supplied on all models.

- A. Test port
- BC. Brake cartridge
- BL. Brake line
- C. Cylinder
- F. Fitting
- FV. Flow valve assy.
- I. Brake valve inlet
- R. Return line
- SC. Steering cartridge
- SL. Steering lines (to cylinder)
- ST. Supply tube
- 1. Rear cover
- 2. Rear plate
- 3. Brake valve
- 4. External relief valve
- 5. Check valve (port sv)
- 6. Steering control unit
- 7. Manifold block

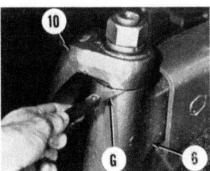

Fig. 2 — Measuring clearance between steering arm (10) and axle arm (6). Proper clearance is 0.005-0.030 inch.

G. Feeler gage
6. Axle arm   10. Steering arm

remove hose type clamps that secure shield to cylinder. Remove shield and disconnect the steering hoses; plug hoses to keep system clean. Remove retaining pins, then remove cylinder from axle mounting. Reinstall cylinder by reversing removal procedure. New safety wire should be attached to retaining pins.

After removing steering cylinder, drain trapped oil and thoroughly clean exterior using a suitable solvent. Remove snap ring (12 – Fig. 4), then pull rod (18) with bearing and piston assembly from barrel of cylinder (6).

NOTE: If assembly sticks, light hydraulic pressure may be applied at head end port of cylinder (6). Do not use air pressure to release assembly.

Secure rod end (15) in a vise, apply heat to nut (21) to break Loctite seal, then unscrew nut from rod (18). Remove piston (8) with seals (19 and 20). Remove bearing (9) with seals (10, 11 and 17). Clean all metal components in a suitable solvent and blow dry. Inspect metal components for excessive wear, scoring or other damage. Check all cylinder orifices for obstructions and clean as necessary. A hone may be used to remove scratches and glaze on inner cylinder wall. Renew all "O" rings, backup washers, piston and bearing seals. Renew bushings (3) if worn excessively.

Install rod seal (17) into inside diameter of bearing (9) with lip of seal facing away from wiper seal (10). Install wiper seal into bearing (9) with lip side of seal facing away from rod seal (17). Install "O" ring (11) into outer groove of

bearing (9) and "O" ring (19) into inner groove of piston (8). Install pressure seals (20) onto piston outside diameter making sure that lips of seals face away from each other. Refer to Fig. 5 for cross-sectional view of seal installation.

Apply clean hydraulic oil to components, then slide bearing (9) with seals onto rod assembly (18) making sure counterbore side of bearing (9) faces away from piston (8). Install piston (8) with seals onto rod (18) with counterbore side of piston facing retaining nut (21). Apply a locking type sealer to threads of rod (18), install piston retaining nut (21), tighten to a torque of 270-305 ft.-lbs., then stake nut securely in place and allow sealer to cure. If rod end (15) had been removed, reinstall by following procedure given in last sentence. Apply clean hydraulic oil to barrel (6) then carefully insert assembly into barrel until snap ring (12) can be installed. When snap ring is in place (sharp edge toward cylinder closed end), pull rod end (15) out until bearing (9) contacts snap ring. Fill hydraulic system as described in paragraph 97 and refer to paragraph 5 for proper bleeding procedure.

FLOW CONTROL VALVE

All Models

8. **R&R AND OVERHAUL.** The flow control valve (FV – Fig. 3) is mounted on the left side of engine just above starter. Refer to paragraph 104 for test procedure. Before removing valve, clean exterior thoroughly to prevent entrance of

dirt. Mark hydraulic lines at valve to aid in reassembly, remove lines from valve then remove valve from tractor. Install valve by reversing removal procedure.

Scribe indentification marks on cartridges (BC & SC – Fig. 8) and hose fittings to aid in reassembly.

NOTE: Cartridges (BC & SC), seals (S – Fig. 7) and back-up rings (R) are not interchangeable. Care must be taken when removing them from valve.

After removing steering cartridge, brake cartridge and their respective

Fig. 6 – View showing types of cartridges used in flow control valve assembly. Valves and internal components are not interchangeable.

F. Flow/control cartridge   FR. Flow/relief cartridge

Fig. 7 – View showing correct installation of seals, back-up rings and filter spring (certain models) onto flow cartridge. Installation is same on both types of cartridges.

R. Back-up rings
S. Seals   5. Filter spring

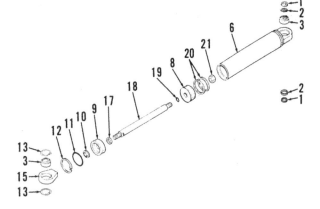

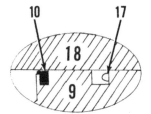 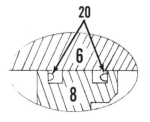

Fig. 4 – Exploded view of steering cylinder used.

1. Spacer
2. Seal
3. Bushing
6. Cylinder
8. Piston
9. Bearing
10. Wiper seal
11. "O" ring
12. Snap ring
13. Snap ring
15. Rod end
17. Rod seal
18. Rod
19. "O" ring
20. Piston seals
21. Nut

Fig. 5 – Cross-sectional view showing proper installation of seals (10, 17 and 20) in piston (8) and bearing (9).

10. Wiper seal
17. Rod seal
20. Piston seals

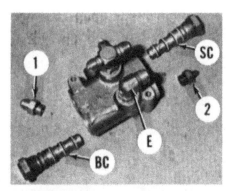

Fig. 8 – View of flow control valve showing location of check fitting, brake fitting and their respective cartridges.

BC. Brake cartridge
E. Elbow
SC. Steering cartridge
1. Check fitting
2. Brake fitting

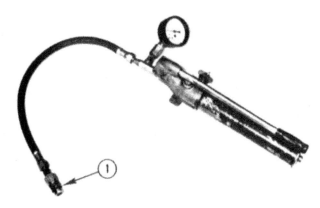

Fig. 9 – Pressure testing check fitting (1). Fitting must be pressurized to 2000 psi.

seals and back-up rings from bores of valve, remove check fitting (1–Fig. 8) and "O" ring, brake fitting (2) and "O" ring, then remove remaining hose fittings.

Valve body may be equipped with either screw-in (using Loctite) or pressed-in and staked type orifices, which are visible after removing check and brake fittings. Orifices should be inspected for correct inside diameter and condition.

NOTE: Steering port "SV" orifices are 0.178-inch in diameter. Brake port orifice is usually 0.081-inch in diameter, however some valves will have a 0.106-inch diameter orifice.

If press-in type orifices are damaged or inside diameter is incorrect, valve body will require renewal. Before attempting removal of screw-in type orifices, check for parts availability. Pressure test check fitting (1–Fig. 8) for leakage as shown in Fig. 9. After making sure orifices in valve body are securely in place, reinstall flow cartridges in their respective bores.

FLOW CONTROL CARTRIDGES

All Models

9. **REMOVE AND REINSTALL.** Remove flow control cartridges after first removing flow control valve as described in paragraph 8. Reverse removal procedure to reinstall cartridges.

NOTE: Flow control cartridges cannot be switched. Seals (S – Fig. 7) and back-up rings (R) are different sizes and care must be taken not to mix them during disassembly or reassembly. Refer to Fig. 6 for view of types of cartridges used.

10. **OVERHAUL FLOW TYPE.** To overhaul flow type cartridge (F–Fig. 6), first press down on spool (2–Fig. 10) and remove snap ring (1); carefully slide spool (2) and spring (3) from cartridge housing, then remove seals (S–Fig. 7) and back-up rings (R). Remove spring type filter (5–Fig. 10) if so equipped and

make sure orifice (4), located under spring filter, is not clogged. Clean components in a suitable solvent and blow dry, then renew any that are worn or damaged excessively. Renew seals (S–Fig. 7) and back-up rings (R) using Fig. 7 as an assembly guide. Reinstall components by reversing removal procedure.

11. **OVERHAUL FLOW/RELIEF TYPE.** To overhaul flow/relief type cartridge (FR–Fig. 6), first press down on spool (2–Fig. 11) and remove snap ring (1); carefully slide spool (2) and spring (3) from cartridge housing, then remove seals (S–Fig. 7), back-up rings (R) and filter spring (5). Before disassembling relief portion of valve, mark plug (P–Fig. 12) in relation to spool (2) to aid in setting correct pressure during reassembly. Record the number of turns required to remove plug from spool. Remove internal relief valve components from spool. Inspect parts for damage and renew as necessary. Inspect plug (P) orifice and cartridge housing orifice (4–Fig. 10) to make sure they are not plugged. Install new seal (S–Fig. 12) and back-up ring (R) onto plug (P), then place steel ball (B) over plug orifice. Insert spring (6) and poppet (7) into spool (2) then carefully thread plug and ball into spool. Steel ball must go into recessed area of poppet. Tighten plug same number of turns as recorded during disassembly. Place spring (3–Fig. 11) over spool (2), insert into cartridge housing and secure with snap ring (1). Install filter spring (5–Fig. 7), then install new seals (S) and back-up rings (R) in their proper grooves. Reinstall cartridge into valve body.

POWER STEERING CONTROL VALVE

All Models

12. **R&R AND OVERHAUL STEERING CONTROL VALVE.** To remove steering control valve (6–Fig. 3), first identify hoses (to simplify reinstallation), then detach from valve. Re-

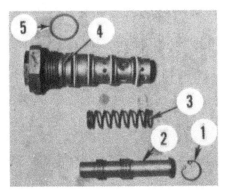

Fig. 10 – Disassembled view of flow control cartridge.

1. Snap ring
2. Spool
3. Spool spring
4. Orifice
5. Filter spring

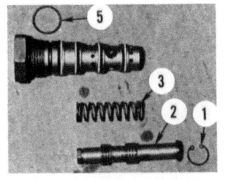

Fig. 11 – Disassembled view of flow/relief cartridge. Refer to Fig. 12 for disassembled view of spool assembly.

1. Snap ring
2. Spool
3. Spring
5. Filter spring

Fig. 12 – Disassembled view of flow/relief cartridge and spool assembly.

B. Ball
P. Plug
R. Back-up ring
S. Seal
2. Spool
3. Spring
5. Filter spring
6. Spring
7. Poppet

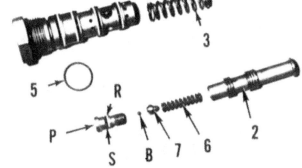

move panel from top of instrument cluster. Remove bolts securing steering control mounting bracket then remove bracket and valve assembly. To reinstall unit, reverse removal procedure. Make sure roll pin through input shaft (30–Fig. 14) of control unit fits correctly into steering shaft. Refer to paragraph 5 for proper fluid and bleeding procedure.

13. After removing control valve, clean exterior with a suitable solvent and blow dry. Remove roll pin from input shaft (30–Fig. 14). Install a fitting in one of the four ports in housing (11) and clamp fitting in a vise so that input shaft (30) is pointing downward. Remove cap screws and end cover (1).

NOTE: Lapped surfaces of end cover (1), commutator set (3 and 4), manifold (5), stator-rotor set (6), spacer plate (9) and

housing (11) must be protected from scratching, burring or any other damage as sealing of these parts depends on their finish and flatness.

Remove seal retainer (8) and seal (7), then carefully remove washer (2), commutator set (3 and 4) and manifold (5). Lift off the spacer plate (9), drive link (10) and stator-rotor set as an assembly. Separate spacer plate and drive link from stator-rotor set. If end cover (1) or pin is damaged, cover must be renewed, since pin is not serviced separately.

Remove unit from vise, then clamp fitting in vise so that input shaft is pointing upward. Place a light mark on flange of upper cover (35) and housing (11) to aid in reassembly. Unbolt upper cover from valve body, remove input shaft, upper cover and valve spool assembly. Remove and discard seal ring (20). Slide upper

cover assembly from input shaft and remove spacer (21). Remove shims (26) from cavity in upper cover or from face of thrust washer (25) and note number and size of shims for ease of reassembly. Remove retaining ring (40), brass washer (37) and seal (36). Retain seal ring (39) and retaining ring (40).

Remove retaining ring (27), thrust washer (25) and thrust bearing (24) from input shaft. Drive out pin from input shaft (30) then withdraw torsion bar (18) and spacer (28). Place end of valve spool on top of bench and rotate input shaft until drive link (29) falls free, then rotate input shaft clockwise until actuator ball (31) is disengaged from helical groove in input shaft. Withdraw input shaft and remove actuator ball. Do not remove actuator ball retaining spring (from around valve) unless renewal is required.

Remove plug (17) and recirculating ball from valve body.

Thoroughly clean all parts in a suitable solvent, visually inspect parts and renew any showing excessive wear, scoring or other damage. Refer to Fig. 15 and use a suitable "L" shaped tool to remove plug(s) (13) and "O" ring(s) (12) from four bolt hole side of housing (11). Tip housing downward and 7/32-inch diameter steel ball will fall out. Use a micrometer to measure thickness of the commutator ring (3–Fig. 14) and commutator (4). If commutator ring is 0.0015-inch or more thicker than commutator, renew the matched set.

Place stator-rotor set (6) on lapped surface of end cover (1). Make certain

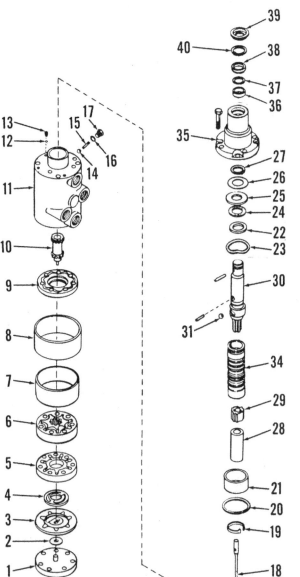

Fig. 14 – Exploded view of power steering control valve assembly.

1. End cover with pin
2. Washer
3. Commutator ring
4. Commutator
5. Manifold
6. Stator-rotor set
7. Seal
8. Seal retainer
9. Spacer plate
10. Drive link
11. Housing assy.
12. "O" ring
13. Plug
14. Recirculating ball
15. Roll pin
16. "O" ring
17. Plug
18. Torsion bar
19. Ball retaining spring
20. Seal ring
21. Spacer
22. Thrust washer
23. Wave spring
24. Bearing
25. Thrust washer
26. Shim(s)
27. Retaining ring
28. Spacer
29. Drive link
30. Input shaft
31. Actuator ball
34. Valve spool
35. Upper cover
36. Shaft seal
37. Brass washer
38. Stepped spacer
39. Dust seal
40. Retaining ring

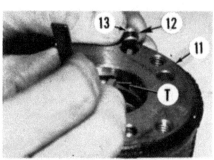

Fig. 15 – Removing plug (13) and "O" ring (12) using a "L" shaped tool (T) from the four bolt hole end of steering housing (11).

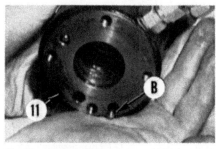

Fig. 16 – View showing removal of 7/32 inch ball (B) after removing plug and "O" ring. Refer to Fig. 15.

that vanes and vane springs are installed correctly in slots of the rotor.

NOTE: Arched back of springs must contact vanes.

Position lobe of rotor in valley of stator as shown in Fig. 17. Center opposite lobe on crown of stator, then using feeler gage, measure clearance between rotor lobe and stator. If clearance exceeds 0.006-inch, renew stator-rotor assembly. Use a micrometer to measure thickness of stator and rotor. If stator is 0.002-inch or more thicker than rotor, renew assembly. Stator, rotor and vanes are available only as an assembly.

Before reassembling, wash all parts in clean solvent and air dry. All parts, unless otherwise indicated, are assembled dry. Install plug (13 – Fig. 14) and new "O" ring (12) in housing. Install recirculating ball (14) and plug (17) with new "O" ring (16) in valve body and tighten plug to a torque of 10-14 ft.-lbs. Clamp fitting (installed in valve body port) in a vise so that top end of valve body is facing upward. Slide wave spring (23) over input shaft (30), followed by thrust washer (22), thrust bearing (24) and thrust washer (25). Secure with snap ring (27). Reinstall 7/32-inch diameter steel ball in four bolt hole end of housing, then install plug(s) (13) and "O" ring (12). If actuator ball retaining spring (19) was removed, install new retaining spring in spool (34). Place actuator ball (31) in its seat from inside of valve spool (34). Insert input shaft into valve spool, engaging helix and actuator ball with a counter-clockwise motion. Use mid-section of torsion bar (18) as a gage between end of valve spool and thrust washer as shown in Fig. 18, to insure assembly is in neutral position. Place assembly in a vertical position with end of input shaft resting on a bench. Insert drive link (29 – Fig. 14) into valve spool until drive link is fully engaged on input shaft spline. Remove torsion bar gage. Install spacer (28) in valve spool, over drive link (29). Install torsion bar into valve spool. Align cross-holes in torsion bar and input shaft with a pin punch, then install pin into input shaft (30). Pin must be pressed into shaft until end of pin is about 1/32-inch below flush. Place spacer (21) over spool and carefully install spool assembly into valve body. Position original shim(s) (26) on thrust washer (25) if the original input shaft and cover are to be used. Lubricate new seal ring (20), place seal ring in upper cover (35) and install upper cover assembly. Align match marks on cover flange and valve body, then install cap screws finger tight. Tighten a worm drive type hose clamp around cover flange and valve body to align outer diameters, (as shown in Fig. 19), then tighten cap screws to a torque of 18-22 ft.-lbs.

NOTE: If either input shaft (30 – Fig. 14) or upper cover (35) or both have been renewed, the following procedure for shimming must be used.

With upper cover installed (with original shims) as previously outlined, invert unit in vise so that input shaft is pointing downward. Grasp input shaft, pull downward and prevent it from rotating. Engage drive link (29) splines in valve spool and rotate drive link until end of spool is flush with end of valve body. Withdraw drive link and check alignment of drive link slot to torsion bar pin. Install drive link, engaging its slot with torsion bar pin. Check relationship of spool end to body end. If end of spool is protruding from body and is within 0.0025-inch of flush, no additional shimming is required. If not within 0.0025-inch of flush, remove cover and add or remove shims (26) as necessary. Shims are available in 0.0028, 0.005, 0.010 and 0.030-inch sizes. Reinstall cover and recheck spool to valve body position.

With drive link installed, place spacer plate (9) on valve body with plain side up. Install stator-rotor set over drive link splines and align cap screw holes. Make certain vanes and vane springs are properly installed. Install manifold (5) with circular slotted side up and align cap screw holes with stator, spacer and valve body. Install commutator ring (3) with slotted side up, then install commutator (4) over drive link end with counter-bore for washer (2) facing out. Make certain that link end is engaged in the smallest elongated hole in commutator. Install seal (7) and retainer (8). Apply a few drops of hydraulic fluid on commutator (4) and manifold (5). Use a small amount of grease to stick washer (2) in position over pin in end cover (1). Install end cover making sure that pin engages center hole in commutator. Align holes and install cap screws. Alternately and progressively tighten cap screws while rotating input shaft. Final tightening torque should be 18-22 ft.-lbs.

Relocate unit in vise so input shaft is up. Lubricate seal (36) and carefully work seal over shaft and into bore with lip toward inside. Install brass washer (37), then install stepped spacer (38) with flat side up. Install retaining ring (40) with rounded edge inward.

Remove unit from vise and remove fitting from port. Turn unit on its side with hose ports upward. Pour clean hydraulic fluid into inlet port, rotate input shaft until fluid appears at outlet port, then plug all ports until unit is installed on tractor.

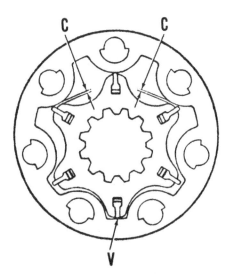

Fig. 17 – With rotor positioned in stator as shown, clearance "C" must not exceed 0.006 inch. Refer to text.

V. Rotor vanes

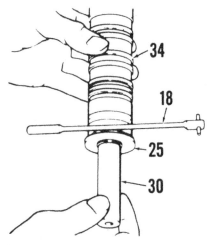

Fig. 18 – Use torsion bar as shown between thrust washer and end of spool, to establish neutral position. Refer to Fig. 14 for parts identification.

Fig. 19 – A large hose clamp "C" may be used as shown to align cover (35) to housing (11) before tightening cap screws.

DIESEL ENGINE AND COMPONENTS

All models are equipped with a Perkins six cylinder diesel engine having a bore of 3⅞-inches, a stroke of 5-inches and a displacement of 354 cubic inches. Engine used in Model 2675 is naturally aspirated while engine for Model 2705 is turbocharged. Compression ratio of naturally aspirated engine is 16.0:1; turbocharged engine compression ratio is 15.5:1.

ENGINE AND CLUTCH

All Models

15. **R&R ASSEMBLY.** To remove engine and clutch as a unit, first drain cooling system and if engine is to be disassembled, drain oil pan.

NOTE: Be careful to prevent entry of dirt or other foreign particles into oil, hydraulic, air conditioning or fuel lines that must be disconnected for removal of engine.

Disconnect batteries and remove hood and all hood side panels. Disconnect air conditioning hoses (AC – Fig. 78) and pull them through support panel grommets. Disconnect air conditioner compressor wire, then unbolt compressor and lay it on top of the radiator. Remove radiator braces to muffler support. Disconnect air cleaner aspirator hose from bottom of muffler. Disconnect windshield washer hose between reservoir and firewall (at firewall), then pull hose free so it can be removed with front end assembly.

Identify (to aid in reassembly), then disconnect power steering valve (6 – Fig. 3) oil lines. Pull hoses through grommets in support panel, then disconnect lower ends of lines. Disconnect oil cooler lines from support panel fittings. Disconnect wiring at oil pressure switch, starter motor, alternator, water temperature sending unit and main electrical wiring connector at firewall. Route wiring to avoid interference when separating tractor.

Disconnect radiator hoses from radiator and air intake tube from intake manifold or turbocharger. Disconnect turbocharger oil cooler lines on 2705 models. On all models, unbolt cooling fan and lay it in radiator shroud.

After properly supporting tractor front end assembly and front of transmission, insert suitable wedges between front axle and front axle support to keep front end from tipping. Remove eight cap screws attaching front axle support to engine adapter plates and oil pan. Separate front end assembly from engine (Fig. 20).

CAUTION: Observe joint while loosening bolts to make sure that tractor and engine are supported evenly. Adjust supports if mating surfaces are not parallel while separating.

Refer to Fig. 20A and remove heat shield (1), muffler, muffler support elbow, exhaust pipe, elbow (2), sleeve (3), air conditioner mounting bracket (4), support angle (5) and muffler mounting bracket (6). Remove support panel, then disconnect heater hoses and hose clamp from engine.

Disconnect tachometer drive cable at engine, pull throttle cable out of lower support bracket, then remove bracket. Close fuel tank shut-off valves and disconnect fuel lines (1 and 2 – Fig. 21). Remove cap screws that secure flow control valve (FV – Fig. 3) to left side of engine and remove battery cable from starter.

15A. Remove clutch release fork by disconnecting clutch release cable from release arm side of transmission. If equipped with 24-speed transmission, disconnect shift cable (C – Fig. 22), complete with mounting bracket and links, from shift valve (V). On all models, disconnect linkage rod (L) from clutch shaft arm. Remove inspection cover on bottom of transmission. Cut safety wires (W – Fig. 23), and remove release fork lock bolts (1). Slide release fork shafts out each side of transmission housing. Shift valve (V – Fig. 22) may have to be removed so right release fork shaft can be removed. Remove release fork through inspection hole.

Remove top link support (Fig. 106A) of three point linkage at rear of tractor to release spring tension on pto shaft. Support engine with suitable overhead hoist, then remove cap screws and bolts

Fig. 20 – Removing front end as a unit for engine work.

Fig. 20A – Some of the components that are removed before separating engine from transmission.
1. Heat shield
2. Exhaust elbow
3. Sleeve
4. Air conditioner compressor bracket
5. Support angle
6. Muffler mounting bracket

Fig. 21 – Disconnect fuel lines (1 and 2) after closing fuel tank shut-off valves.

securing engine to transmission. Insert two guide studs to aid in engine-transmission alignment during disassembly and reassembly.

NOTE: It may be necessary to remove cab floor panel so upper cap screws can be reached.

Separate engine from transmission (Fig. 24), making sure mating surfaces remain parallel while separating and that all wires and tubes are disconnected and positioned so they will not be damaged.

15B. When reinstalling engine, observe the following: Lubricate transmission input shaft splines and clutch release bearing bushings with a light coat of a suitable lithium base grease. Carefully move engine back against transmission while aligning splines of clutch with transmission input shaft. If splines do not align, **DO NOT** force together, but turn flywheel slowly while sliding together. Install cap screws and bolts securing engine to transmission. Tighten screws and bolts to the following torques:

½-inch cap screws . . . . 63-95 ft.-lbs.
⅝-inch cap screws . . . . . 130-200 ft.-lbs.
⅝-bolts & nuts . . . . 173-260 ft.-lbs.
¾-inch cap screws . . . . . 240-350 ft.-lbs.
⅞-inch cap screws . . . . . 300-400 ft.-lbs.
1-inch cap screws . . . . . 440-580 ft.-lbs.

Fig. 24 — Attach hoist as shown when removing engine.

Install cab floor panel. Remainder of procedure is reverse of disassembly.

CYLINDER HEAD

All Models

16. To remove cylinder head, first drain cooling system, disconnect batteries, remove hood and side panels, turbocharger heat shield and oil lines, windshield washer hose (at firewall), power steering valve (6 – Fig. 3) oil lines and oil cooler pipes at support. Pull steering valve hoses and windshield washer tubes through support.

Disconnect wiring for headlights, ether starting solenoid, air restriction indicator, water level indicator switches, water temperature sending unit, alternator, air conditioner compressor, engine oil pressure switch and starter. Carefully work main harness through rear support panel.

Disconnect air conditioner lines at rear support, then unbolt compressor and lay it out of the way; remove ether start tube and air inlet tube. Remove air cleaner aspirator hose (at bottom of muffler), then pull hose out of front support panel. Remove radiator hoses and top radiator brace assembly. Remove turbocharger oil cooler line on models so equipped.

Unbolt air conditioner condenser and hydraulic oil cooler from radiator.

Remove drive belts, fan shroud and radiator. Remove muffler and air conditioner supports, cooling fan-idler pulley and mounting bracket assembly. Disconnect heater hoses at water pump; disconnect engine oil cooler water line at water pump on 2675 models. On all models, remove water pump and backplate. Disconnect tach drive cable and pull it through rear support panel, then remove rear support panel. Disconnect fuel shut-off cable and spring, pull throttle cable out of lower support and remove lower support bracket. Disconnect upper end of oil line (I – Fig. 3). Remove throttle cable bellcrank/mounting bracket assembly, fuel leak off line and manifold, fuel lines, filter and mounting bracket.

Remove fuel injectors and injector

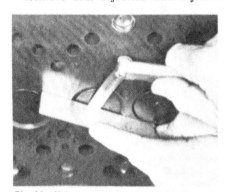

Fig. 26 — Using a straight edge and feeler gage to check valve head height. Refer to text.

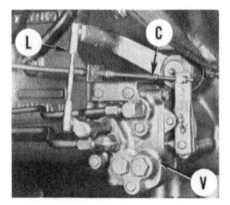

Fig. 22 — View of shift valve and related linkage. Shift valve may require removal before right shift fork shaft can be removed.

C. Shift cable
L. Linkage rod
V. Shift valve

Fig. 23 — View of clutch release fork bolts after removal of inspection cover.

W. Safety wire
1. Fork bolts

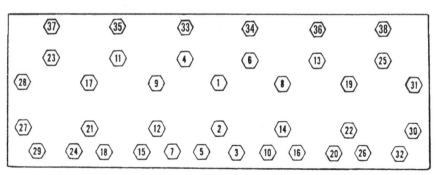

Fig. 25 — Tighten the cylinder heat nuts in sequence shown.

sealing rings. Remove turbocharger, turbocharger oil supply and drain line, intake and exhaust manifolds, valve cover, rocker arm shaft and pushrods. Loosen head nuts in reverse of sequence in Fig. 25, then remove cylinder head from engine block.

The cylinder head can be resurfaced providing the injector nozzle protrusion does not exceed 0.184-inch for turbocharged models or 0.136-inch for non-turbocharged models. Original thickness of cylinder head is 3.735-3.765-inches and normally surfacing of less than 0.012-inch will be possible. After resurfacing cylinder head, it will be necessary to reseat valves to correct depth as described in paragraph 17 and shown in Fig. 27.

Cylinder head gasket is marked "TOP FRONT" for proper installation. Head gasket should be installed dry. Tighten cylinder head fasteners in sequence (1 to 32–Fig. 25), to a torque of 100 ft.-lbs. Tighten remaining screws (33 to 38) to a torque of 29.5 ft.-lbs. Adjust valve clearance after head is installed as outlined in paragraph 22.

VALVES AND SEATS

All Models

17. On 2675 models, both intake and

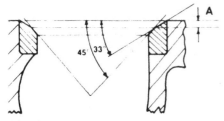

Fig. 27—Valve seat width and location is changed using 33 and 45 degree stones. Depth (A) should be 0.106-0.110 inch for new valves and seats.

exhaust valves seat directly in the cylinder head. Renewable inserts are used for all valve seats on 2705 models. On all models, valve heads should be recessed a specified amount into the cylinder head and inserts should be installed or renewed when specifications are exceeded. Clearance can be checked using a straightedge and feeler gage as shown in Fig. 26. Production clearance is 0.050-inch for both intake and exhaust valves. Renew valve seat and/or valve if recess exceeds 0.050-inch.

New valve seat inserts should be machined as shown in Fig. 27. Depth (A) of 0.106-0.110-inch should provide correct amount of valve recess (refer to Fig. 26) with new valve. All valve seats should be flared using a 33 degree stone and reseated using a 45 degree stone. Correct valve depth and the nozzle protrusion should be checked after resurfacing cylinder head. Refer to paragraph 16. Valve face and seat angles should be 45 degrees for both intake and exhaust valves.

VALVE GUIDES

All Models

18. Intake valve stem to guide wear limit is 0.005-inch and exhaust valve stem to guide wear limit is 0.006-inch. The valve guides are not interchangeable for intake and exhaust. Exhaust valve guides are slightly longer than intake valve guides. Press new guide into head until distance from top of guide to head surface is 19/32-inch.

Inside diameter of new guides should be 0.375–0.376-inch and provide desired diametral clearance of 0.0015–0.0045-inch for intake valve stem and 0.0027–0.0047-inch for exhaust valve stem. Be sure to resurface valve seats after new guides are installed.

VALVE SPRINGS

All Models

19. Springs, retainers, locks and valve stem seals are interchangeable between intake and exhaust valves. Springs should be installed with close (damper) coils toward cylinder head. Renew springs or other components if they are distorted, discolored, or fail to meet the test specifications which follow:

INNER SPRING:
Lbs. Test @
1.340-inches . . . . 20-1/8 – 23-5/16

OUTER SPRING:
Lbs. Test @
1.410-inches . . . . 39½ – 43-11/16

CAM FOLLOWERS

All Models

20. The mushroom type tappets (cam followers) operate in machined bores in the cylinder block and can be renewed after removing camshaft as outlined in paragraph 26. The 0.7475–0.7485-inch diameter tappets should have 0.0015–0.00375-inch diametral clearance in crankcase bores.

ROCKER ARMS

All Models

21. Fig. 29 shows a partially assembled view of the rocker arm shaft.

NOTE: Push rods can drop into oil pan if dislocated at lower end. Be careful not to lose push rods into engine when removing or installing rocker arms.

Right and left hand rocker arm units,

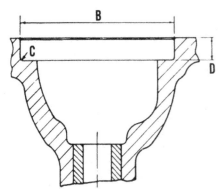

Fig. 28 — Bore (B) In cylinder head for valve seat should be 2.0165-2.0175 Inches for intake valves; 1.678-1.679 Inches for exhaust valves. Depth (D) should be 0.283-0.288 Inch for intake valves; 0.375-0.380 Inch for exhaust valves. Radius (C) should be 0.015 Inch for all valve seats.

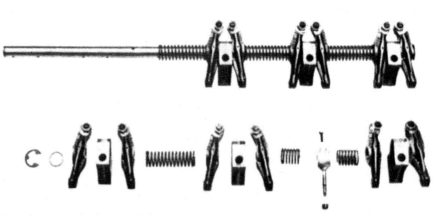

Fig. 29 — View of partially disassembled rocker arm shaft.

mounting brackets, and the oil feed tube must be positioned as shown. Oil feed holes in shaft must be installed toward valve stem side. Make sure sealing "O" ring does not roll out of retaining grooves in feed tube as shaft unit is installed.

Desired diametral clearance between rocker arms and shaft is 0.001–0.0035-inch. Renew shaft and/or rocker arms if clearance is excessive. Tighten rocker arm support cap screws to a torque of 55 ft.-lbs.

VALVE CLEARANCE

All Models

22. The recommended cold valve clearance (tappet gap) is 0.008 inch for intake valves and 0.018 inch for exhaust valves. Cold (static) setting of all valves can be made from just two crankshaft positions, using the procedure outlined in this paragraph and illustrated in Figs. 29A and 29B.

Remove timing plug from left front side of flywheel adapter housing and turn crankshaft until "TDC" timing mark on flywheel is aligned with timing pointer. Check the rocker arms for front and rear cylinders. If rear rocker arms are tight and front rocker arms have clearance, No. 1 piston is on compression stroke; adjust the six tappets shown in Fig. 29A. After adjusting the indicated tappets, turn crankshaft one complete turn until the "TDC" mark is again aligned and adjust the remaining tappets.

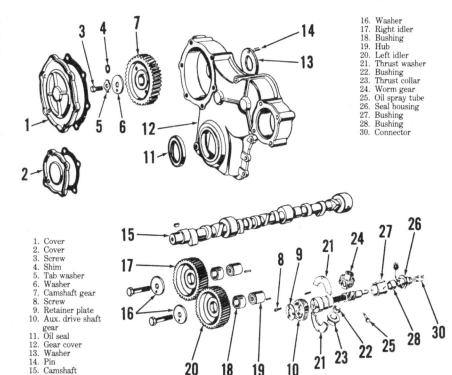

16. Washer
17. Right idler
18. Bushing
19. Hub
20. Left idler
21. Thrust washer
22. Bushing
23. Thrust collar
24. Worm gear
25. Oil spray tube
26. Seal housing
27. Bushing
28. Bushing
30. Connector

1. Cover
2. Cover
3. Screw
4. Shim
5. Tab washer
6. Washer
7. Camshaft gear
8. Screw
9. Retainer plate
10. Aux. drive shaft gear
11. Oil seal
12. Gear cover
13. Washer
14. Pin
15. Camshaft

Fig. 30 — Covers (1 and 2) and gears (7 and 10) must be removed before timing gear housing (12) can be removed for access to remainder of gears. Refer to text for details.

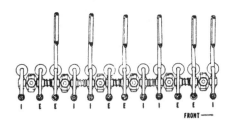

Fig. 29A — With TDC timing marks aligned and No. 1 piston on compression stroke, adjust indicated valves to specifications called out in text.

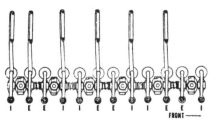

Fig. 29B — With TDC timing marks aligned and No. 6 piston on compression stroke, adjust indicated valves to specifications called out in text.

VALVE TIMING

All Models

23. Timing gears are keyed and valve timing will be correct if timing marks are properly aligned as outlined in paragraph 24.

NOTE: There is no adjustment provided for valve timing. Should timing be incorrect, gears can only be one or more teeth out of correct mesh.

COVER (HOUSING) AND GEARS

All Models

24. The timing gear cover housing (12 – Fig. 30), can only be removed after removing the camshaft gear (7), aux-iliary drive shaft gear (10) and the crankshaft pulley.

To remove the timing gear cover (12) or renew the oil seal (11), first support tractor and remove front end as a unit as outlined in paragraph 15. Turn crankshaft until "TDC" timing mark on flywheel is aligned (refer to Fig. 35), and No. 1 cylinder is on compression stroke.

To remove crankshaft pulley, remove cap screws and thrust block (1 – Fig. 31) with "O" ring (2).

NOTE: If pulley does not immediately become free, bump inner hub of pulley to separate tapered locking rings (3 and 4). Do not use a puller. Using a puller will cause tapered rings to lock and crankshaft or pulley may be damaged.

Remove outer tapered rings (3 and 4),

Fig. 31 — Crankshaft pulley and related parts. Refer to text for recommended removal procedure.

1. Thrust block
2. "O" ring
3. Outer taper ring
4. Inner taper ring
5. Spacer
6. Woodruff key

spacer (5), pulley and key (6). Remove the two covers (1 and 2 – Fig. 30) and gears (7 and 10), using a puller to remove camshaft gear (7). Cover can now be unbolted and removed.

The crankshaft gear and idler gears are now accessible for inspection or service. Allowable backlash between any two gears in timing gear train is 0.003 – 0.006-inch. Diametral clearance of idler gears to hubs is 0.003 – 0.006-inch and installed end play of idler gears is 0.002 – 0.012-inch. Bushings in idler gears must be reamed after installation to an inside diameter of 1.8750 – 1.8766-inches. Renew front crankshaft oil seal (11) in timing cover (12), if damaged or worn excessively. Press old seal out of cover. Press in new oil seal from front, with lip of seal towards inside of case to ¼-inch inside front face of timing case. Refer to Fig. 34.

Camshaft and auxiliary drive shaft can be withdrawn from cylinder block after timing gear housing is removed. Refer to paragraph 25 for details on auxiliary drive unit overhaul and to paragraph 26 for camshaft removal.

Before timing gear cover can be reinstalled, it is first necessary to remove the injection pump for proper retiming of auxiliary shaft. Remove right idler gear (17 – Fig. 32) and reinstall with timing punch marks on idler gears and crankshaft gear aligned as shown in Fig. 32.

NOTE: Left idler gear (20) may not have timing marks and marks may be ignored if present. Marks are not required for timing injection pump drive gear.

Because of the odd number of teeth in idler gears, timing marks on idler gears will only occasionally align. Remove idler gears and reinstall with all marks aligned.

Make sure the camshaft, auxiliary drive shaft and their thrust washers are properly positioned, then reinstall the timing gear cover and loosely install the retaining cap screws. Center the

crankshaft oil seal (or oil seal bore) in cover over the crankshaft, using the crankshaft pulley and tapered locking rings, then tighten the cover retaining cap screws to a torque of 65 ft.-lbs.

Install auxiliary drive shaft gear as outlined in paragraph 25 and camshaft gear as in paragraph 26.

AUXILIARY DRIVE SHAFTS AND GEARS

All Models

25. The auxiliary drive shaft drives the engine oil pump and injection pump. The auxiliary drive unit is shown exploded in Fig. 36 and in cross-section in Fig. 37.

The auxiliary drive shafts and gears are lubricated by engine oiling system. Renewable fitting (11) provides spray lubrication for drive worm.

To check the auxiliary shaft timing, remove injection pump as outlined in paragraph 50 and turn engine crankshaft until No. 1 piston is at "TDC" on compression stroke. Auxiliary shaft timing is correct if notches (N – Fig. 38) in pump drive shaft and adapter plate align as shown.

To remove pump drive gear, unbolt and remove adapter (A). Remove thrust sleeve (16 – Fig. 36) from top opening, then lift out gear (18).

Lower thrust collar (20) and bushing (19) may be either one or two pieces. Remove lower thrust collar/bushing using a suitable puller. Press new bushing into block until it bottoms. The one piece lower thrust collar/bushing requires no machining after installation. The two piece bushing/collar should be machined to an inside diameter of 1.6250 – 1.6266-inches after installation. The upper bushing (8) in the injection pump adapter should be finished to an inside diameter of 1.875-1.867-inches after installation.

Auxiliary shaft thrust washers (5) fit in machined slots in shaft (7). Outer portion of washers fit a machined recess in

front face of engine block and are retained by timing gear housing. Normal shaft end play of 0.0025 – 0.009-inch is adjusted by renewing the thrust washers and/or shaft.

Auxiliary drive shaft should have a diametral clearance of 0.0021 – 0.0053-inch in front bushing (12) and 0.001 – 0.0042-inch in rear bushing (13). Bushings are precision type and will not require reaming after installation. Make sure oil holes are aligned when bushings are installed.

Auxiliary shaft can be withdrawn from front after removing timing gear housing as outlined in paragraph 24. When reinstalling the auxiliary shaft, align engine timing marks as shown in Fig. 32 and pump drive shaft timing notch as shown in Fig. 38. Tighten cap screws (1 – Fig. 36) to a torque of 22 ft.-lbs.

Fig. 33 – Pressure oil passage (O) in idler gear hub and block.

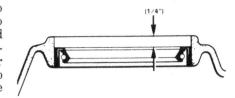

Fig. 34 – Press crankshaft seal ¼ inch inside front of case.

Fig. 32 – Front view of engine block with timing marks properly aligned. Timing gear train will never appear as shown, because gears (AS and 7) must be removed for access to timing gear housing.
AS. Accessory shaft gear
CR. Crankshaft gear
 7. Camshaft gear
17. Right idler
20. Left idler

Fig. 35 – View of timing marks with engine at Top Dead Center.

CAMSHAFT

All Models

26. To remove the camshaft, first remove the rocker arm assembly as outlined in paragraph 21 and timing gear housing as in paragraph 24. Unbolt and remove the fuel lift pump from right side of engine block. Remove engine oil pan. Remove tappets and identify for proper relocation during reassembly.

Remove camshaft and thrust washer. Normal camshaft bearing clearance is 0.0025 – 0.0045-inch for front bearing, 0.0025 – 0.0055-inch for other journals. Cylinder block bores are unbushed. Camshaft end play of 0.004 – 0.016-inch is controlled by thrust washer (13 – Fig. 30) located between front journal and camshaft gear. The thrust washer fits in a recess in front of engine block and is retained by timing gear housing and located by a dowel pin.

When reassembling the engine, reposition idler gear with timing marks aligned as outlined in paragraph 24 and shown in Fig. 32. With timing gear housing reinstalled, turn camshaft and gear until timing punch mark on idler gear is located between the two marks on camshaft gear and draw the gear into position using a suitable puller bolt. Tighten the camshaft gear retaining cap screw to a torque of 50 ft.-lbs. and lock in place by bending the tab washer.

ROD AND PISTON UNITS

All Models

27. Connecting rod and piston units are removed from above after removing the cylinder head as outlined in paragraph 16 and oil pan as outlined in paragraph 34. Cylinder numbers are stamped on the connecting rod and cap.

NOTE: When removing rod to piston units on turbocharged engines, be sure to turn connecting rod in a counter-clockwise direction to avoid damage to the cooling jets.

When reinstalling, make sure these numbers are aligned and face away from the camshaft side of the engine. Upper and lower big end rod bearing inserts are interchangeable. Bearings are available in standard size and undersizes of 0.010, 0.020 and 0.030-inch. Crankshafts used in turbocharged engines are "Tufftrided". Regrinding any of these crankshaft journals is recommended **ONLY** if the crankshaft can be retreated. **DO NOT** regrind shaft unless retreatment is possible. When installing connecting rod caps, tighten rod bolts/nuts to a torque of 75 ft.-lbs.

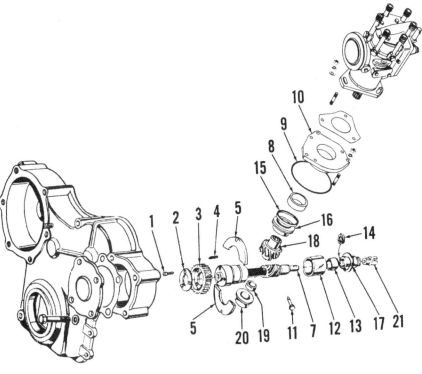

Fig. 36 — View of accessory shaft, gears, injection pump drive and oil pump.

1. Screw	7. Accessory shaft	12. Bushing	17. Seal housing
2. Retainer plate	8. Bushing	13. Bushing	18. Worm gear
3. Drive gear	9. "O" ring	14. Oil seal	19. Bushing
4. Dowel pin	10. Pump adapter	15. "O" ring	20. Thrust collar
5. Thrust washer	11. Oil spray tube	16. Thrust sleeve	21. Connector

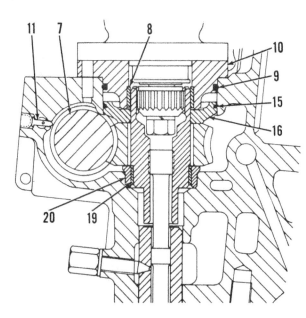

Fig. 37 — Cross-sectional view of accessory drive housing and injection pump drive. Oil spray tube (11) directs lubricant to worm on accessory shaft (7). Refer to Fig. 36 for parts identification.

Fig. 38 — Injection pump drive adapter with pump removed showing timing marks. Refer to text for shaft timing procedure.

A. Adapter
M. Pump timing mark
N. Worm gear timing notches
S. Master spline

PISTONS, SLEEVES AND RINGS

All Models

28. The aluminum alloy, cam ground pistons are supplied in a standard size only. The toroidal combustion chamber is offset in piston crown and piston is marked "FRONT" for proper assembly.

Specifications for the cast iron piston rings used in **turbocharged engines** are as follows:

End Gap—
Top Ring . . . . 0.016 – 0.026-inch
Second Ring . . . . . . 0.016 – 0.026-inch
Bottom Ring . . . . . 0.009 – 0.016-inch

Side Clearance in Groove—
Top Ring . . . . 0.0026 – 0.0044-inch
Second Ring . . . . 0.0024 – 0.0037-inch
Bottom Ring . . . . 0.0035 – 0.0048-inch

Refer to Fig. 40 for view of correct piston ring installation.

Specifications for the cast iron piston rings used in **naturally aspirated** engines are as follows:

End Gap—
Top Ring . . . . 0.010 – 0.016-inch
Second Ring . . . . . . 0.016 – 0.026-inch
Bottom Ring . . . . . . 0.009 – 0.016-inch

Side Clearance in Groove—
Top Ring . . . . . . 0.0035 – 0.0048-inch
Second Ring . . . . 0.0028 – 0.0040-inch
Bottom Ring . . . . 0.0020 – 0.0033-inch

Refer to Fig. 41 for view of correct piston ring installation.

The production cylinder sleeves are 0.001 – 0.003-inch press fit in cylinder block bores. Service sleeves are a transition fit (0.001-inch tight to 0.001-inch clearance). When installing new sleeves, make sure cylinder bore is absolutely clean, then apply clean oil to all except top two inches. Outside surface of cylinder liner must be clean and dry. Apply a band of locking type sealer one inch wide around top of liner directly under flange. Press liner into place using suitable tools.

NOTE: When correctly installed, top of liner must protrude 0.026 – 0.037-inch above top face of cylinder. Liner flange should be within 0.002-inch above or 0.004-inch below top face of cylinder block. Refer to Fig. 42.

PISTON PINS

All Models

29. The full floating piston pins are retained in piston bosses by snap rings.

Piston pin diameter is 1.4998 – 1.5000-inches for turbocharged engines and 1.3748 – 1.3750-inches for naturally aspirated engines. On all models, pins are available in standard sizes only and the renewable connecting rod bushing must be final sized to provide correct diametral clearance. Piston pin to bushing clearance should be 0.00075 – 0.0017-inch (turbocharged models) or 0.0008 – 0.0017-inch (naturally aspirated models). Be sure the pre-drilled oil hole (or holes) in bushings are aligned with holes in connecting rod. Hone new bushings to inside diameter of 1.50075 – 1.5015-inches (turbocharged models) or 1.3758 – 1.3765-inches (naturally aspirated models).

NOTE: Small end of connecting rod on turbocharged engines is wedge shaped, and ends of wrist pin bushing must be machined to match contour of wedge. Piston pin should be a thumb press fit in piston after piston is heated to 100-120 degrees F.

CONNECTING RODS AND BEARINGS

All Models

30. Connecting rod bearings are precision type, renewable from below after removing oil pan and oil pump/suction tube.

NOTE: Remove pipe from oil pressure relief valve to piston cooling feed connection on turbocharged engines.

Carbon and cylinder ridge must be removed before removing piston/rod assembly. Avoid damaging piston cooling jets on turbocharged engines. When renewing bearing shells, be sure that the projection engages milled slot in rod and cap and that correlation marks are in register and face away from camshaft side of engine.

Bearings are available in standard and undersizes of 0.010, 0.020 and 0.030-inch. Crankshafts used in turbocharged engines are "Tufftrided". **DO NOT** regrind any crankshaft journal on

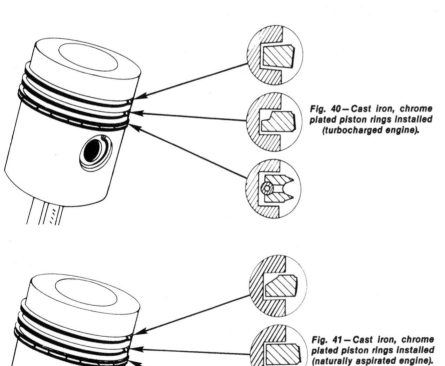

Fig. 40—Cast iron, chrome plated piston rings installed (turbocharged engine).

Fig. 41—Cast iron, chrome plated piston rings installed (naturally aspirated engine).

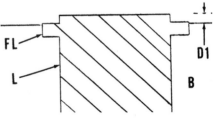

Fig. 42 — Cross-sectional view of cylinder block showing cylinder liner properly installed. Liner flange (FL) should be within 0.002 inch above to 0.004 inch below top surface of block.
B. Block
Dl. 0.026-0.037 inch   L. Cylinder liner

these engines unless retreatment is possible.

Connecting rod bearings should have a diametral clearance of 0.0019 – 0.0032-inch on the 2.499 – 2.4966-inch crankpin. Tighten rod cap retaining nuts to a torque of 75 ft.-lbs.

CRANKSHAFT AND BEARINGS

All Models

31. The crankshaft is supported in seven precision type main bearings. To remove the rear main bearing cap, it is first necessary to remove the engine, clutch, flywheel and rear oil seal. All other main bearing caps can be removed after removing oil pan and pump assembly.

Upper and lower halves of bearing inserts are not interchangeable. Upper (block) half is slotted and drilled to provide pressure lubrication to crankshaft and connecting rods. Inserts are interchangeable in pairs for all journals except center main bearing. When installing the thrust washers, make sure steel back is positioned next to block and cap and that grooved, bronze surface is next to crankshaft thrust faces.

Bearing inserts are available in undersizes of 0.010, 0.020 and 0.030-inch as well as standard. Thrust washers are available in standard thickness of 0.089 – 0.091-inch and in oversize thickness of 0.0965 – 0.0985-inch. **DO NOT** regrind any of the crankshaft journals on turbocharged engines unless "Tufftride" retreatment is possible.

Recommended main bearing diametral clearance is 0.0033 – 0.0046-inch and recommended crankshaft end play is 0.002 – 0.015-inch. Tighten main bearing cap retaining screws to a torque of 200 ft.-lbs. When renewing rear main bearing, refer to paragraph 32 for installation of rear seal and oil pan bridge pieces. Check crankshaft against the following specifications:

Main Journal Standard Diameter –
  Inches 2.9984-2.9992
Crankpin Journal Standard
  Diameter – Inches 2.4988-2.4996

CRANKSHAFT REAR OIL SEAL

All Models

32. Lip type rear oil seal is contained in a one-piece housing attached to rear face of engine block as shown in Fig. 44. The seal retainer can be removed after removing flywheel as outlined in paragraph 33. The lip seal is pressed into housing.

When installing cylinder block bridge piece at front or rear, lightly coat metal-to-metal contact faces with a non-hardening gasket cement and insert end seals as shown in Fig. 45, then use a straightedge as shown in Fig. 46 to make sure gasket faces are flush.

FLYWHEEL

All Models

33. To remove flywheel, first separate engine from transmission as outlined in paragraph 63 and remove the clutch as outlined in paragraph 64. Flywheel is secured to crankshaft mounting flange by six evenly spaced cap screws which also retain pto shaft drive hub.

The starter ring gear can be renewed after flywheel is removed. Heat ring gear evenly to approximately 475

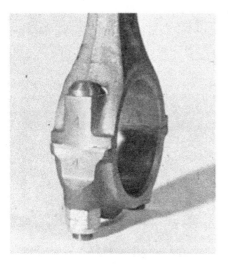

Fig. 43 – Assembled view of serrated connecting rod and cap.

Fig. 44 – Rear view of cylinder block and crankshaft showing rear oil seal housing.

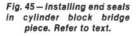

Fig. 45 – Installing end seals in cylinder block bridge piece. Refer to text.

Fig. 46 – Use a straight edge to align gasket faces of cylinder block and bridge piece.

degrees F. and install flywheel with beveled end of teeth facing front of engine.

NOTE: Flywheel must be installed so that with No. 1 piston at TDC (arrow on crankshaft flange straight up), "TDC1" mark on flywheel aligns with mark on inspection hole in flywheel housing; refer to Fig. 35.

Tighten flywheel cap screws to a torque of 77-115 ft.-lbs. Check flywheel runout with a dial gage after flywheel is installed. Maximum allowable flywheel runout is 0.012-inch.

ADAPTER PLATE

All Models

33A. Engine to clutch housing adapter can be removed after flywheel is removed. Be sure to reinstall adapter before installing flywheel. Tighten top two screws to 41-62 ft.-lbs. and six lower screws to 25-38 ft.-lbs. of torque.

OIL PAN (SUMP)

All Models

34. The heavy cast-iron oil pan serves as part of the tractor frame and attaching point for tractor front support. To remove oil pan, lift front end of tractor so all weight is just removed from front tires, then support in this position by blocking under adapter plate; refer to Fig. 48. Remove front engine mounts, then remove all fasteners that would not be accessible with mounts in place, then reinstall mounts. Tighten mounting bolts enough to secure mounts and still allow removal of oil pan fasteners. Remove cap screws that secure front axle support to oil pan and oil pan to adapter plate. Lift front end assembly to provide a slight separation between front axle support and oil pan. Drain engine oil, support pan with a jack, remove retaining screws, install guide studs, then lower oil pan.

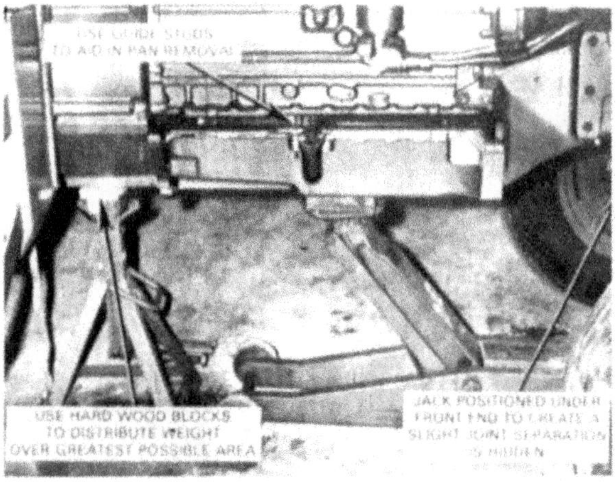

To reinstall oil pan, reverse removal procedure. Tighten fasteners to torques as follows:

Oil Pan To–
  Crankcase ....15 ft.-lbs.
  Adapter Plate
   (¾ inch) ....240-250 ft.-lbs.
  Adapter Plate
   (⅞ inch) ....300-400 ft.-lbs.
Tractor Front Axle Support To–
  Oil Pan (¾ inch) .....240-350 ft.-lbs.
  Oil Pan (1 inch) ......440-550 ft.-lbs.
  Front Engine
   Mounts ....130-200 ft.-lbs.

OIL PUMP

All Models

35. The rotary type oil pump is mounted on lower side of engine block and is driven by injection pump drive shaft. Refer to Fig. 49 for exploded view of oil pump components. After removing oil pan as described in paragraph 34, oil pump removal and reinstallation will be evident upon inspection of unit.

Service parts for oil pump are not available individually. Pump cover and rotors may be removed for inspection or cleaning.

Turbocharged engines use two piece rotors. Refer to Fig. 49. Oil pump clearances are as follows:

Inner Rotor Tip to Outer
  Rotor 0.003–0.005-inch
Outer Rotor and Pump
  Body Bore 0.006–0.013-inch
Outer Rotor End
  Clearance 0.0005–0.0025-inch
Inner Rotor End
  Clearance 0.0015–0.0035-inch

RELIEF VALVE

All Models

36. The relief valve (2–Fig. 50) is contained in valve body housing (1). The relief valve used in turbocharged en-

gines is a two stage unit with opening pressure for piston cooling jets opening at 30-37 psi. The relief valve on both engines is set to open at 50-60 psi. All relief valve parts (except valve body) are available individually.

DIESEL FUEL SYSTEM

The diesel fuel system consists of three basic units; fuel tank (and filters), injection pump and injector nozzles. When servicing any unit associated with the diesel fuel system, maintenance of absolute cleanliness is of utmost importance. Of equal importance is the avoidance of nicks and burrs on any of the working parts.

Probably the most important precaution that service personnel can impart to owners of diesel powered tractors is to urge them to use an approved fuel that is absolutely clean and free from foreign material. Extra precaution should be taken to be sure that no water enters the fuel storage tanks. Because of the high pressures and degree of control required of injection equipment, extremely high precision standards are necessary in the manufacture and servicing of diesel components. Extra care in daily maintenance will pay big dividends in extended

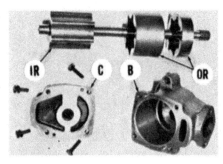

Fig. 49—Exploded view of engine oil pump. Turbocharged engine oil pump has two outer rotors.

B. Body   IR. Inner rotor
C. Cover   OR. Outer rotor

Fig. 48 – View showing blocking points when removing oil pan. Refer to text.

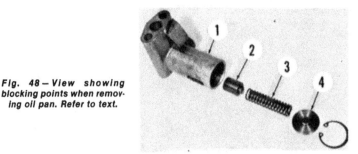

Fig. 50—Exploded view of oil pump pressure relief valve.

1. Valve body   3. Spring
2. Plunger   4. Seat

service life and the avoidance of costly repairs.

FILTERS AND BLEEDING

All Models

38. **MAINTENANCE.** The fuel filter head is fitted with two renewable type elements as shown in Fig. 51. Water drain plugs (D) should be opened daily and any water in the filter bases (FB) drained.

After each 200 hours of operation, the fuel filter elements (E) should be renewed. Close fuel shut-off valves at tank, thoroughly clean outside of filter housing, drain fuel from filters (loosen drain plugs), remove cap screws at top center of filter heads and remove the filter elements. Clean the filter bases and heads then reinstall with new elements and rubber sealing rings. Do not over tighten cap screws. Bleed the diesel fuel system as outlined in paragraph 39.

39. **BLEEDING.** To bleed the system, make sure tank shut-off valves are open, then have an assistant actuate the manual lever on fuel lift pump on right

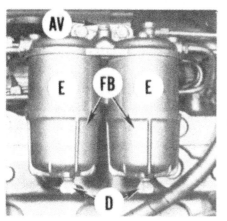

Fig. 51 — View of fuel filters used on both model tractors. Refer to text for filter changing procedure.

AV. Air vent	D. Drains
E. Filter elements	FB. Filter bases

side of engine block, and proceed as follows:

Loosen the air vent (AV – Fig. 51) on filter and continue to operate the lift pump until air-free fuel flows from vent plug hole. Tighten air vent, then loosen vent plugs (1 and 2 – Fig. 52) while continuing to operate the fuel lift pump. Tighten plug (1), then plug (2) as air is expelled. Loosen pressure lines at injector connections (C), move fuel shut-off control to run position, then crank engine with starting motor until air free fuel flows from connections. Tighten connections and start engine.

NOTE: Starter should not be operated for more than 30 seconds without allowing 30 seconds for cooling.

FUEL LIFT PUMP

All Models

40. The fuel lift pump (Fig. 53) is mounted on right side of engine block and is driven by the camshaft. Pump parts are available separately or in rebuilding kit. Output delivery pressure should be 5-7.4 psi at 800 rpm.

INJECTOR NOZZLES

All Models

All models are equipped with C.A.V. multi-hole nozzles which extend through the cylinder head to inject the fuel charge into combustion chamber machined in crown of piston.

WARNING: Fuel leaves the injector nozzle with sufficient force to penetrate the skin. Keep exposed portions of your body clear of nozzle spray when testing.

41. **TESTING AND LOCATING A FAULTY NOZZLE.** If rough or uneven engine operation or misfiring indicates a faulty injector, the defective unit can usually be located as follows:

With engine running at the speed where malfunction is most noticeable

Fig. 52 — View of injection pump and bleeders. Refer to text for bleeding procedure.

C. Injector connections
E. Filter element
1. Vent plug
2. Vent plug
4. Low idle speed set screw

(usually slow idle speed), loosen the compression nut on high pressure line for each injector nozzle in turn, and listen for a change in engine performance. As in checking spark plugs, the faulty unit is the one which, when its line is loosened, least affects the running of the engine.

If a faulty nozzle is found and considerable time has elapsed since the injectors have been serviced, it is recommended that all nozzles be removed and serviced or that new or reconditioned units be installed. Refer to the following paragraphs for removal and test procedure.

42. **REMOVE AND REINSTALL INJECTORS.** Before loosening any fuel lines, thoroughly clean the lines, connections, injectors and surrounding engine area with air pressure and solvent spray. Disconnect and remove the leak-off manifold and high pressure injection lines and cap all connections to prevent dirt entry into the system. Remove the two stud nuts and withdraw injector unit, with dust shield and copper washer from cylinder head.

Thoroughly clean the nozzle recess in cylinder head before reinstalling injector unit. It is important that seating

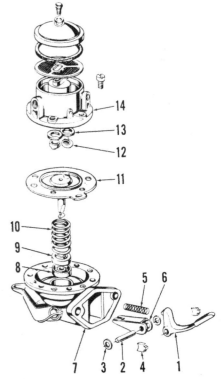

Fig. 53 — Exploded view of fuel lift pump.

1. Actuating lever	8. Seal
2. Shaft	9. Retainer
3. Washer	10. Spring
4. Retainer	11. Diaphragm
5. Spring	12. Valve
6. Link	13. Gasket
7. Body	14. Cover

surface be free of even the smallest particle of carbon or dirt which could cause the injector unit to be cocked and result in blow-by. No hard or sharp tools should be used in cleaning. Do not re-use the copper sealing washer (5–Fig. 55) located between injector nozzle and cylinder head, always install a new one.

NOTE: Thickness of sealing washers used on turbocharged engines is 0.028-inch and on naturally aspirated engines is 0.080-inch.

Each injector should slide freely into place in cylinder head without binding. Make sure that dust seal is reinstalled and tighten the retaining stud nuts to a torque of 15 ft.-lbs. After engine is started, examine injectors for blow-by, making the necessary corrections before releasing the tractor for service.

Fig. 54 — A suitable injector tester is required to completely test and adjust injector nozzles.

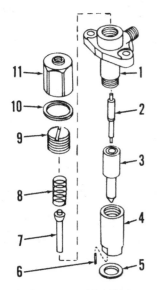

Fig. 55 — Exploded view of C.A.V. injector nozzle and holder assembly. Washer (5) is different thickness for turbocharged and naturally aspirated engines.

1. Nozzle holder
2. Nozzle valve
3. Nozzle body
4. Nozzle nut
5. Sealing washer
6. Dowel(s)
7. Valve spindle
8. Spring
9. Adjusting sleeve
10. Washer
11. Cap nut

43. TESTING. A complete job of testing and adjusting the injector requires the use of special test equipment. Only clean, approved testing oil should be used in the tester. The nozzle should be tested for opening pressure, seat leakage, back leakage and spray pattern. When tested, the nozzle should open with a sharp popping or buzzing sound, and cut off quickly at end of injection with a minimum of seat leakage and controlled amount of back leakage.

Before conducting the test, operate tester lever until fuel flows, then attach the injector. Close the valve to tester gage and pump tester lever a few quick strokes to be sure nozzle valve is not plugged, that fuel sprays emerge from nozzle tip, and that possibilities are good that injector can be returned to service without overhaul.

If adjustment is indicated in the preliminary tests, proceed as follows:

44. OPENING PRESSURE. Open the valve to tester gage and operate tester slowly while observing gage reading. Opening pressure should be 3086 psi on 2705 models; 2939 psi on 2675 models. If opening pressure is not as specified, remove the injector cap nut (11–Fig. 55), and turn adjusting sleeve (9) as required to obtain the recommended pressure.

NOTE: When adjusting a new injector or an overhauled injector with a new pressure spring (8) on 2675 models, set the pressure at 3160 psi. This additional pressure is to allow for initial pressure loss as the spring wears in.

45. SEAT LEAKAGE. The nozzle tip should not leak at pressure less than 2936 psi for 2705 models; 2789 psi for 2675 models. These pressures are 150 psi less than recommended opening pressure. To check for leakage, actuate tester lever slowly and observe the nozzle tip as gage needle approaches suggested test pressure. Hold pressure for 10 seconds. If drops appear or nozzle tip is wet, valve is not seating and injector must be disassembled and overhauled as outlined in paragraph 48.

46. BACK LEAKAGE. If nozzle seat as tested in paragraph 45 was satisfactory, check injector and connections for wetness which would indicate external leakage. If no leaks are noted, bring gage pressure to 2250 psi, release pump lever and observe time required for gage pressure to drop from 2200 to 1470 psi. For nozzle in good condition, this time should be 10-30 seconds. At higher temperatures (above 70 degrees F.), a period of time slightly less than 10 seconds may be considered satisfactory.

A faster drop would indicate a worn or scored nozzle valve or body and the nozzle assembly should be renewed.

NOTE: Leakage of the tester check valve or connections will cause a false reading, showing up in this test as excessively fast leakback. If all injectors tested fail to pass the test, tester rather than nozzles should be suspected as faulty.

47. SPRAY PATTERN. If leakage and pressure are as specified when tested as outlined in paragraphs 44 through 46, operate tester handle several times while observing spray pattern. Symmetrical sprays should come from tip of nozzle body.

If pattern is uneven, ragged or not finely atomized, overhaul the nozzle as outlined in paragraph 48.

48. OVERHAUL. Hard or sharp tools, emery cloth, grinding compound, or other than approved solvents or lapping compounds must never be used. An approved nozzle cleaning kit is available through C.A.V. Service Agency and other sources.

Wipe all dirt and loose carbon from exterior of nozzle and holder assembly. Refer to Fig. 55 and proceed as follows:

Secure nozzle in a soft-jawed vise or holding fixture and remove cap nut (11). Back off the adjusting sleeve (9) to completely unload pressure spring (8). Remove nozzle nut (4) and nozzle body (3). Nozzle valve (2) and body (3) are matched assemblies and **must never be**

Fig. 56 — Clean the pressure chamber in nozzle tip using the special reamer as shown.

Fig. 57 — Clean spray holes in nozzle tip using a pin vise and 0.011 inch (turbocharged models) or 0.012 inch (naturally aspirated models) wire probe.

intermixed. Place all parts in clean calibrating oil or diesel fuel as they are removed. Clean exterior surfaces with a soft wire brush, soaking in an approved carbon solvent if necessary, to loosen hard carbon deposits. Rinse parts in clean diesel fuel or calibrating oil immediately after cleaning, to neutralize carbon solvent and prevent etching of polished surfaces. Clean pressure chamber of nozzle tip using special reamer as shown in Fig. 56. Clean spray holes in nozzle with an 0.011-inch (turbocharged models) or 0.012-inch (naturally aspirated models) wire probe held in a pin vise as shown in Fig. 57. Wire probe should protrude from pin vise only far enough to pass through spray holes, to prevent bending and breakage. Rotate pin vise without applying undue pressure.

Clean valve seats by inserting small end of brass valve seat scraper into nozzle and rotating tool. Reverse the tool and clean upper chamfer, refer to Fig. 58. Use the hooked scraper to clean annular groove in top of nozzle body (Fig. 59). Use the same hooked tool to clean the internal fuel gallery.

With the above cleaning accomplished, back flush the nozzle by installing the reverse flusher adapter on injector tester and nozzle body in adapter, tip end first. Rotate the nozzle valve while flushing. After nozzle is back flushed, seat can be polished by applying a small amount of tallow on end of a polishing stick and rotating the stick as shown in Fig. 60.

Light scratches on valve piston and bore can be polished out by careful use of special injector lapping compound only. **DO NOT** use valve grinding compound or regular commercial polishing agents. **DO NOT** attempt to reseat a leaking valve using polishing compound. Clean thoroughly and back flush if lapping compound is used.

Reclean all parts by rinsing thoroughly in clean diesel fuel or calibrating oil and assemble valve to body while immersed in cleaning fluid. Reassemble injector while still wet. With adjusting sleeve (9 – Fig. 55) loose, reinstall nozzle body (3) to holder (1),

making sure valve (2) is installed and locating dowels are aligned as shown in Fig. 61. Tighten nozzle nut (4 – Fig. 55) to a torque of 50 ft.-lbs. **Do not overtighten.** Distortion may cause valve to stick and overtightening cannot stop a leak caused by scratches or dirt on lapped metal surfaces of valve body and nozzle holder.

Retest and adjust assembled injector assembly as outlined in paragraphs 41 through 47.

NOTE: If overhauled injector units are to be stored, it is recommended that a calibrating or preservative oil rather than diesel fuel be used for pre-storage testing. Storage of injectors containing diesel fuel for more than thirty days may result in necessity of recleaning prior to use.

INJECTION PUMP

All Models

The injection pump is a completely sealed unit. No service work of any kind should be attempted on the pump or governor unit without the use of special pump testing equipment and training. Inexperienced or unequipped service personnel should never attempt to overhaul a diesel injection pump.

49. **ADJUSTMENT.** Slow idle stop screw (1 – Fig. 62) should be adjusted with engine warm and running, to pro-

Fig. 59 – *Use the hooked scraper to clean annular groove in top of nozzle body.*

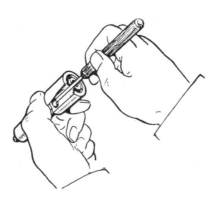

Fig. 60 – *Polish the nozzle seat using mutton tallow on a wood polishing stick.*

vide recommended slow idle speed of 825-875 rpm.

Linkage rod (L – Fig. 63) should be adjusted so that when throttle arm on injection pump is contacting low idle speed stop screw, front of bellcrank arm contacts mounting bracket (MB). Stud on overtravel rod end (R) must be fitted to rear hole of slot (S) in bellcrank arm for turbocharged models and in front hole for naturally aspirated models.

Adjust throttle cable at rear so that with hand throttle control lever set in low idle position, and injection pump throttle arm set in low idle position, there is 1/32-1/16-inch free play in cable.

To adjust fuel shut-off control, first pull fuel shut-off control knob 1/16-inch out from instrument panel. Adjust clevis (C – Fig. 62A) so that front of bellcrank arm is against mounting bracket (B). Adjust length of linkage rod (R – Fig. 62) until fuel shut-off arm (A) is against running position stop. Recheck adjustment to make sure shut-off arm contacts off position stop when control knob is pulled out, and contacts running position stop when control knob is pushed against in-

Fig. 61 – *Make sure locating dowels are carefully aligned when nozzle body is reinstalled.*

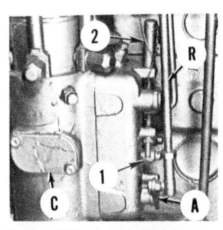

Fig. 62 – *Installed view of injection pump showing slow idle stop screw (1) and high idle speed sealing wire, cover sleeve (2), and timing cover (C).*

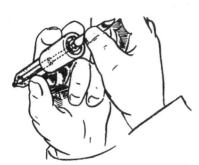

Fig. 58 – *Clean the valve seat using brass scraper as shown.*

strument panel.

Governed engine speed under load should be 2500 rpm, with a high idle (no-load) speed of 2675-2825 rpm. To adjust high idle speed, first remove wire seal and cover (2–Fig. 62) from high idle speed stop screw. Loosen high idle speed stop screw locknut, then adjust stop screw to proper rpm. Reinstall cover and wire seals.

50. **REMOVE AND REINSTALL.** Before attempting to remove injection pump, thoroughly wash pump and connections with clean diesel fuel or an approved solvent. Disconnect fuel shut-off and throttle linkage from injection pump. Disconnect high pressure fuel lines at both ends, unbolt bracket and lines as a unit. Cap all connections as lines are removed to prevent entrance of dirt. Disconnect and cap fuel inlet and return lines at pump, remove remaining attaching stud nuts and lift pump off.

Normal installation of injection pump can be accomplished without reference to crankshaft timing marks or internal timing marks on injection pump. Be sure timing scribe lines (T–Fig. 64) are aligned when pump is installed. The master spline on pump drive shaft aligns with splines in pump to assure proper timing if auxiliary shafts are correctly installed as outlined in paragraph 25. Bleed the fuel system after installation as outlined in paragraph 38. If necessary to check injection pump timing, refer to paragraph 51.

51. **PUMP TIMING TO ENGINE.** Static timing (end of injection) should occur at 28 degrees on both models. The top dead center mark for No. 1 cylinder is stamped on front face of flywheel as shown in Fig. 35.

NOTE: Static marks (26, 28 or 30 degrees) for No. 1 cylinder may or may not be visible through window.

Proceed as follows to check injection

pump timing without removing the pump.

Turn engine over in normal direction of operation (clockwise viewing crankshaft pulley), until No. 1 piston is on compression stroke and the indicated 28 degree mark is aligned in timing window. Remove the timing cover (C–Fig. 62) from injection pump body. The "F" mark on pump rotor should be aligned with square end of injection pump snap ring. Minor adjustment can be made by loosening injection pump mounting stud nuts and shifting the pump mounting flange. Major changes in timing necessitates removing and installing auxiliary drive shaft correctly as outlined in paragraph 25.

TURBOCHARGER

The engine used in the 2705 model tractor is equipped with an AiResearch turbocharger Model T-04B, which provides 11-13.5 psi boost pressure at 2500 rpm with the engine fully loaded.

TROUBLE SHOOTING

2705 Model

52. The following table can be used for locating difficulty with the turbocharger. Many of the probable causes are related to the engine air intake and exhaust systems which should be maintained in good condition at all times. Engine oil is used to cool and lubricate the turbocharger and damage can result from improper operation. The turbocharger should be allowed to cool by idling engine at approximately 1000 rpm for 2-5 minutes after engine has been operated under load. The engine should always be immediately restarted if it is killed while operating at full load. The oil

in the turbocharger will be hardened by the intense heat if not allowed to cool and damage is sure to result. Use extreme care to avoid damaging any of the turbocharger moving parts when servicing.

Symptoms	Probable Causes
Engine Lacks Power	1,4,5,6,7,8,9,10, 11,18,20,21,22,25,26,27,28,29,30
Black Exhaust Smoke	1,4,5,6,7,8,9,10, 11,18,20,21,22,25,26,27,28,29,30
Blue Exhaust Smoke	1,4,8,9,19,21,22, 32,33,34,36
Excessive Oil Consumption	2,8,17,19, 20,33,34,36
Excessive Oil in the Turbine End	2,7,8, 16,17,19,20,22,32,33,34,36
Excessive Oil in the Compressor End	1,2,4,5,6,8,9,16,19, 20,21,33,36
Insufficient Lubrication	12,15,16,23, 24,31,36
Oil In Exhaust Manifold	2,7,19,20,22, 28,29,30,33,34
Damaged Compressor Wheel	3,6,8, 20,21,23,24,36
Damaged Turbine Wheel	7,8,18,20, 21,22,34,36
Drag or Bind in Rotating Assy.	3,6,7,8, 13,14,15,16,20,21,22,31,34,36
Worn Bearings, Journals, Bearing Bores	6,7,8,12,13,14,15,16, 20,23,24,31,35,36
Noisy Operation	1,3,4,5,6,7,8,9,10, 11,18,20,21,22
Sludged or Coked Center Housing	2,15,17

Key To Probable Causes
1. Dirty air cleaner element.
2. Plugged crankcase breathers.
3. Air cleaner element missing, leaking, not sealing correctly, loose connections to turbocharger.
4. Collapsed or restricted air tube before turbocharger.
5. Restricted (damaged) crossover pipe from turbocharger to inlet manifold.

Fig. 62A – View of fuel shut-off adjustment for both models. Refer to text for details.
B. Bracket   C. Clevis

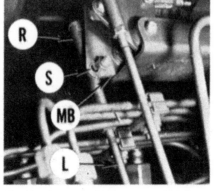

Fig. 63 – View of throttle linkage adjustment for both models.
L. Linkage rod   R. Overtravel rod end
MB. Mounting bracket   S. Slot

Fig. 64 – Pump should be properly timed when timing scribe lines (T) are aligned as shown.

6. Foreign object between air cleaner and turbocharger.
7. Foreign object in exhaust system from engine (check engine).
8. Turbocharger flanges, clamps or bolts loose.
9. Inlet manifold cracked, gaskets missing, connections loose.
10. Exhaust manifold cracked, burned, gaskets blown or missing.
11. Restricted exhaust system.
12. Oil lag (oil delay to turbocharger at start up).
13. Insufficient lubrication.
14. Lubricating oil contaminated.
15. Improper type lubricating oil used.
16. Restricted oil feed line.
17. Restricted oil drain line.
18. Turbine housing damaged or restricted.
19. Turbocharger seal leakage.
20. Worn journal bearings.
21. Excessive dirt buildup in compressor housing.
22. Excessive carbon buildup behind turbine wheel.
23. Too fast acceleration at initial start (oil lag).
24. Too little warm-up time.
25. Fuel pump malfunction.
26. Worn or damaged injectors.
27. Valve timing.
28. Burned valves.
29. Worn piston rings.
30. Burned pistons.
31. Leaking oil feed line.
32. Excessive engine pre-oil.
33. Excessive engine idle.
34. Coked or sludged center housing.
35. Oil pump malfunction.
36. Oil filter plugged.

TURBOCHARGER UNIT

2705 Model

53. **REMOVE AND REINSTALL.** Remove hood and disconnect air inlet housing from turbocharger. Remove turbocharger air intake pipe, outlet tube and exhaust outlet elbow. Disconnect oil inlet and return lines from turbocharger, then unbolt and remove unit from exhaust manifold.

To reinstall, reverse removal procedure. Pour 3-4 ounces of clean oil into oil inlet and spin turbocharger to lube bearing and journal surfaces. Tighten retaining nuts to a torque of 28-32 ft.-lbs. Leave oil return line disconnected and crank engine with starter for several 5 to 10 second bursts with fuel shut-off control in the "OFF" position, until oil begins to flow from return port, then tighten return line.

NOTE: Do not start engine until it is certain that turbocharger is receiving lubricating oil.

54. **OVERHAUL.** Remove turbocharger unit as outlined in paragraph 53. Mark across compressor housing, center housing and turbine housing (refer to Fig. 66), to aid alignment when assembling.

CAUTION: Do not rest weight of any parts of impeller on turbine blades. Weight of only the turbocharger unit is enough to damage the blades.

Keep turbine shaft from turning using appropriate type of wrench at center of turbine wheel (17) and remove nut (5).

NOTE: Use a "T" handle to remove nut in order to prevent bending turbine shaft.

On some turbine shafts, an Allen wrench must be used at turbine end while others are equipped with a hex and can be held with a standard socket. Lift compressor wheel (6) off, then remove center housing (22) from turbine shaft (17) while holding shroud (15) onto center housing. Remove back plate retaining screw (23), then remove back

plate (7), thrust bearing (11) and thrust collar (10). Carefully remove bearing retainers (13) from ends and withdraw bearings (14).

CAUTION: Be careful not to damage bearings or surface of center housing when removing retainers. The center two retainers do not have to be removed unless damaged or unseated. Always renew bearing retainers if removed from grooves in housing.

Clean all parts in a cleaning solution which is not harmful to aluminum. A stiff brush and plastic or wood scraper should be used after deposits have softened. When cleaning, use extreme caution to prevent parts from being nicked, scratched or bent.

Inspect bearing bores in center housing (22 – Fig. 66) for scored surface, out-of-round or excessive wear. Bearing bore diameter must not exceed 0.6223-inch. Make certain bore in center housing is not grooved in area where seal (16) rides. Bearings (14) should be

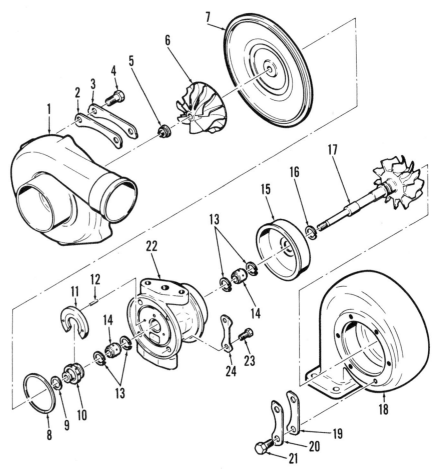

Fig. 66 — Exploded view of AiResearch turbocharger used on 2705 models.

1. Compressor housing	8. Seal ring	14. Bearing	19. Clamp
2. Clamp	9. Piston ring	15. Turbine wheel shroud	20. Lockplate
3. Lockplate	10. Thrust collar	16. Piston ring	21. Cap screw
4. Cap screw	11. Thrust bearing	17. Turbine wheel and	22. Center housing
5. Nut	12. Spring pin	shaft	23. Cap screw
6. Compressor wheel	13. Snap ring	18. Turbine housing	24. Lockplate
7. Backplate			

renewed each time unit is disassembled for service. Thrust bearing (11) should be measured at three locations around collar bore. Thickness should not be less than 0.1716-inch. Inside diameter of bore in backplate (7) should be 0.5005-inch. Thrust surface and seal contact area must be clean and smooth. Compressor wheel (6) must not show signs of rubbing with either the compressor housing (1) or the back plate (7). Make certain that impeller blades are not bent, chipped, cracked or eroded. Oil passage in thrust collar (10) must be clean and thrust faces must not be warped or scored. Ring groove shoulders must not have step wear. Bearing area width should not exceed 0.1748-inch. Width of groove for seal ring (8) should not exceed 0.0653-inch. Clearance between thrust bearing (11) and groove in collar (10) must be 0.001-0.004-inch, when checked at three locations. Inspect turbine wheel shroud (15) for evidence of turbine wheel rubbing. Turbine wheel (17) should not show evidence of rubbing and vanes must not be bent, cracked, nicked, or eroded. Turbine wheel shaft must not show signs of scoring, or overheating. Diameter of shaft journals should not be less than 0.3997-inch. Groove in shaft

for seal ring (16) must not be stepped and diameter of hub near seal ring must not be less than 0.6820-inch. Check shaft end play and radial clearance when assembling. If bearing inner retainers (13) were removed, they must be renewed.

Oil bearings (14) and install outer retainers using appropriate tools. Position wheel shroud (15) on the turbine shaft (17) and install seal ring (16) in groove. Apply a light, even coat of engine oil to shaft journals, compress seal ring (16) and install center housing (22). Install new seal ring (9) in groove of thrust collar (10), then install thrust bearing toward seal ring (9) end of collar. Install thrust bearing and collar assembly over shaft, making certain that pins in center housing engage holes in thrust bearing. Install new rubber seal ring (8), then install back plate making certain that seal ring (9) is not damaged. Install lock plates (24) and cap screws (23). Tighten screws (23) to a torque of 75-90 in.-lbs. and bend lock plates up around heads of screws. Install compressor wheel (6) and make certain that impeller is completely seated against thrust collar (10). Install nut (5) to 18-20 in.-lbs. torque, then use a "T" handle to turn nut an additional 90 degrees.

CAUTION: If "T" handle is not used, shaft may be bent when tightening nut (5).

NOTE: Coat threads of screws (21) with "FEL PRO" or similar high temperature compound before installing.

Assemble turbine housing (1) and compressor housing (18) to center housing (22) using clamps, lockplates and screws. Align housing locating marks made before disassembly, then tighten screws (4 and 21) to 100-130 in.-lbs. of torque.

Check shaft end play and radial play. If shaft end play (refer to Fig. 67) exceeds 0.003-inch, thrust collar (10 – Fig. 66) and thrust bearing (11) is worn excessively. End play of less than 0.001-inch indicates incomplete cleaning (carbon not completely removed) or dirty assembly and unit should be disassembled and re-cleaned. If turbine shaft radial play (refer to Fig. 68) exceeds 0.006-inch, unit should be disassembled and bearings, shaft and/or center housing should be renewed.

Fill reservoir with engine oil and protect all openings of turbocharger until unit is installed on tractor.

COOLING SYSTEM

RADIATOR

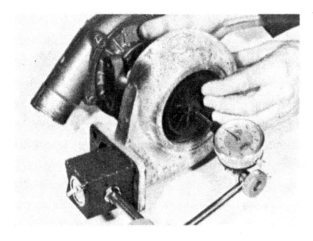

Fig. 67 – View showing method of checking turbine shaft end play. Shaft end play should be checked after unit is cleaned to prevent false reading caused by carbon build-up.

Fig. 68 – Turbine shaft radial play is checked with dial indicator through the oil outlet hole.

All Models

55. A 7 psi pressurized cooling system is used on all models. Cooling system capacity is approximately 31 U.S. quarts. Engine oil coolers are located next to the fuel lift pump on 2675 models and next to the radiator on 2705 models.

To remove radiator, first drain cooling system, then remove hood and side panels. Remove turbocharger heat shield from models so equipped. Disconnect windshield washer hose at reservoir. Disconnect two power steering hoses at control unit and pull hoses through rear support panel grommets. Remove oil cooler pipes attached to rear support bracket. Disconnect battery cables from batteries, then unplug main electrical harness from headlights, ether starting solenoid, air restriction indicator switch, water level indicator switch and air conditioning compressor connector. Work main harness free.

Disconnect air conditioner self sealing hose couplers and pull hoses through support bracket. Unbolt compressor and lay on top of air cleaner support. Disconnect ether starting air tube from air inlet tube, and remove air inlet tube. Disconnect air cleaner hose from bottom of muffler and pull hose out of front support panel. Remove radiator hoses and top radiator brace assembly. Remove engine oil cooler lines on 2705 models. Unbolt air conditioner condenser and hydraulic oil cooler from radiator. Loosen drive belts, unbolt fan shroud from radiator, then remove radiator.

THERMOSTAT

All Models

56. Two thermostats (4 – Fig. 69) with an opening temperature of 174-181 degrees F. are contained in water pump housing (3) and control coolant flow. Engine should not be operated without thermostats, which would allow coolant to bypass radiator and cause overheating.

WATER PUMP

All Models

57. **REMOVE AND REINSTALL.** To remove water pump, first remove radiator as outlined in paragraph 55. Remove exhaust system, muffler mounting bracket, air conditioning bracket, cooling fan, drive belts and idler pulley mounting bracket assembly.

Disconnect temperature sender wire, water hoses and engine oil cooler water line (2675 models only) from water pump, then unbolt and remove water pump. To reinstall water pump on all models, reverse removal procedure. Refer to Fig. 69 for exploded view of water pump.

57A. **OVERHAUL WATER PUMP.** To overhaul water pump, first remove pulley (7 – Fig. 69). Remove internal snap ring (8). Press shaft (13) and bearings (9) out front of impeller (16) and housing (3). Remove impeller (16), collar (17) and water seal (15). Remove flange (11), seal and retainer (12) from front of housing or rear of shaft (13). Press shaft out of bearings and spacer.

To reassemble water pump, reverse disassembly procedure noting the following points: Press bearings (9) and spacer (10) onto front of shaft (13) with shielded face of bearings facing away from each other. Pack bearings then half fill space between bearings with high temperature grease. Press complete bearing/shaft assembly into front of housing until it bottoms. Support pulley end of pump and press impeller onto rear of shaft (13) until impeller depth is 0.843-inch, refer to Fig. 69A.

ELECTRICAL SYSTEM

ALTERNATOR

All Models

58. A Motorola alternator with a built-in voltage regulator is used on both models. Alternator specifications are as follows:

Field Current (75°F.)
　Amperes 2.0-2.5
　Volts 12
Rated Hot Output (Amperes)
　at Maximum Operating Speed . . .55
Minimum Output (Amperes)
　at Minimum Operating Speed . . .48

To disassemble alternator, first scribe matching marks on frame halves (5 and 21 – Fig. 70). Remove voltage regulator (25), brush set (23), insulating cover (26),

then remove four through bolts. Carefully pry frame apart with a screwdriver between drive end frame (5) and stator frame (15) without allowing screwdriver to touch stator windings. Unsolder stator and voltage regulator leads from rectifier bridge (11). Remove rectifier bridge, capacitor (13) and stator (15). Remove "O" ring (16) from slip ring end frame (21), then remove D + terminal screw (18) and spade connector/insulator (24) from slip ring end frame (21).

Remove pulley, fan (1), washer (2), Woodruff key (4) and spacer (3), then tap rotor shaft on block of wood to push it out of drive end frame (5) bearing. Slide counterbored spacer (8) off of rotor (9) shaft. After removing bearing retainer (7), press bearing (6) out of drive end frame (5). Pull bearing (10) off rotor shaft. To reassemble alternator, reverse removal procedure.

NOTE: When reconnecting voltage

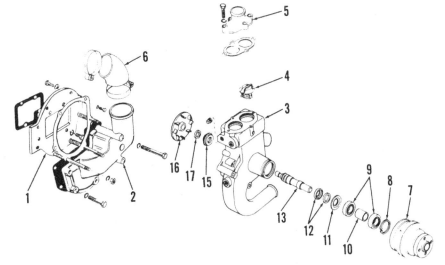

Fig. 69 – Exploded view of water pump, thermostats and associated parts.

1. Mounting plate	5. Water outlet	9. Bearing	13. Shaft
2. Water connection	6. By-pass hose	10. Spacer	15. Water seal
3. Water pump housing	7. Fan pulley	11. Seal flange	16. Impeller
4. Thermostats (2 used)	8. Snap ring	12. Seal and retainer	17. Collar

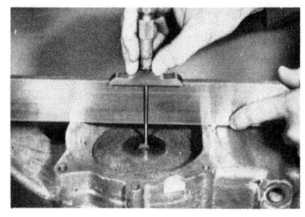

Fig. 69A – View of method used to measure impeller depth in water pump housing. Refer to text for specifications.

regulator (25) lead wires, connect yellow wire to D + terminal and green wire to brush assembly connector.

59. **TESTING.** Original brush set (23) may be reused if brushes protrude 3/16 inch or longer beyond bottom of holder, if brushes are not oil soaked or cracked and if brush set passes the following tests using an ohmmeter. Refer to Fig. 71. Check for continuity using an ohmmeter from brush 1 to terminal 3 and from brush 2 to mounting bracket (MB). Check for an open circuit between brushes (1 and 2) and from terminal (3) to mounting bracket (MB).

Rectifier bridge (11 – Fig. 70) may be tested with an ohmmeter after unsoldering stator and voltage regulator leads. Refer to Fig. 72 and following text for test procedures.

Connect one ohmmeter lead to ground (G – Fig. 72) and other lead to one of terminals (T), then note ohmmeter reading. Reverse ohmmeter leads and note reading. A good diode will read approximately 30 ohms resistance in one test mode and continuity with test leads reversed. Repeat test on remaining two terminals (T). Repeat test between (B +) and terminals (T), then repeat test between (R) and terminals (T). Renew rec-

tifier bridge if any one pair of readings is the same regardless of test lead polarity.

Check bearing surface of rotor shaft (9 – Fig. 70) for visible wear or scoring. Examine slip ring surface for scoring or wear and rotor winding for overheating or other damage. Check rotor for grounded, shorted or open circuits using an ohmmeter as follows:

Refer to Fig. 73 and touch ohmmeter probes to points (1-2) and (1-3). A reading near zero will indicate a ground. Touch ohmmeter probes to slip rings (2-3). Reading should be 4.0-5.2 ohms. If ohmmeter indicates high resistance, windings are open. If ohmmeter reads less than 4 ohms, windings are shorted or grounded and rotor will have to be renewed.

Clean slip rings (2 and 3) with fine crocus cloth. Spin rotor to avoid flat spots on slip rings.

Remove top link support of three-point linkage at top rear of tractor to release spring tension on pto shaft. Support front and rear of engine/front end assembly using suitable splitting stands or hydraulic floor jacks. Support tractor under front of transmission using suitable jacks. Insert axle wedges between front axle support to prevent engine and front end from tipping.

STARTER

All Models

60. A Lucas Model M-50 starter motor and solenoid is used on both models. Model and series numbers are located on main frame and solenoid. Specifications are as follows:

No Load Test

Volts 12
Amperes 115
Rpm 5500-8000
Pinion clearance 0.005-0.045-inch
Brush spring tension . . . . . . .42 ounces

Fig. 71 – Brush set (23 – Fig. 70) used in Motorola alternator. Refer to text for testing procedure.

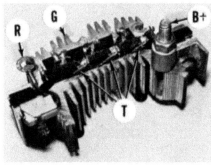

Fig. 72 – Test points for rectifier bridge (11 – Fig. 70). Refer to text for testing procedure.

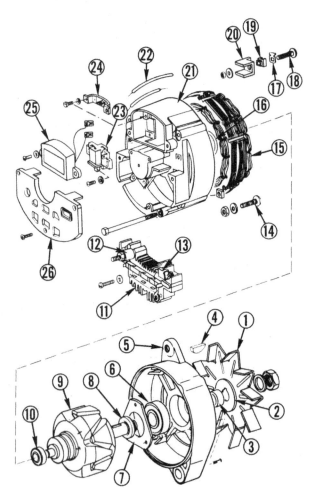

Fig. 70 – Exploded view of alternator with solid state voltage regulator (25).

1. Fan
2. Washer
3. Spacer
4. Woodruff key
5. Drive end frame
6. Bearing
7. Bearing retainer
8. Counterbored spacer
9. Rotor
10. Bearing
11. Rectifier bridge
12. Spacer
13. Capacitor
14. Ground terminal screw
15. Stator
16. "O" ring
17. Clip
18. D + terminal screw
19. Insulator
20. Insulator
21. Slip ring end frame
22. Wire
23. Brush set
24. Spade connector
25. Voltage regulator
26. Insulating cover

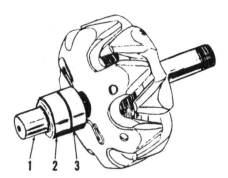

Fig. 73 – Removed rotor assembly showing test points to be used when checking for grounds (shorts) and opens.

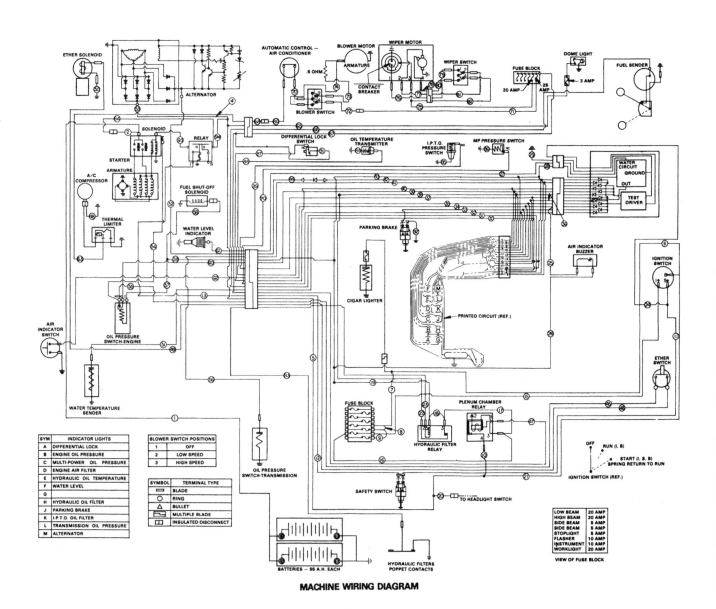

Fig. 74 — Wiring diagram typical of both models. Numbers refer to wire color(s).

1,3,4,5,6,7,8,9,10,21,53,57,58,59,64,78,80,94 . . . . Red
2,20,65 . . . . Brown
11,24 . . . . Orange/White
12,13 . . . . Orange/Black
14,30,47,51,67,74,91,93 . . . . Orange
15,16,17,18,19,25,28,37,43 . . . . Red/White
22,38 . . . . Lt. Blue/Black
23,83 . . . . Dk. Blue/White
26,34,62 . . . . Lt. Blue
27,61 . . . . Black/Red
29,48,50,76,79,84,85,86,92 . . . . Black
31,56 . . . . Pink

32 . . . . Dk. Green
37 . . . . Tan
35,54,73,75,82,88 . . . . White
36,45,55 . . . . Dk. Blue
39,60 . . . . Purple/White
40,87 . . . . Dk. Green/White
41 . . . . Gray
42,44,52,71,72 . . . . Yellow
46,49,77,81 . . . . Green
63,68,69,70 . . . . Purple
66 . . . . Black/White
89 . . . . Light Green
90 . . . . Dark Green

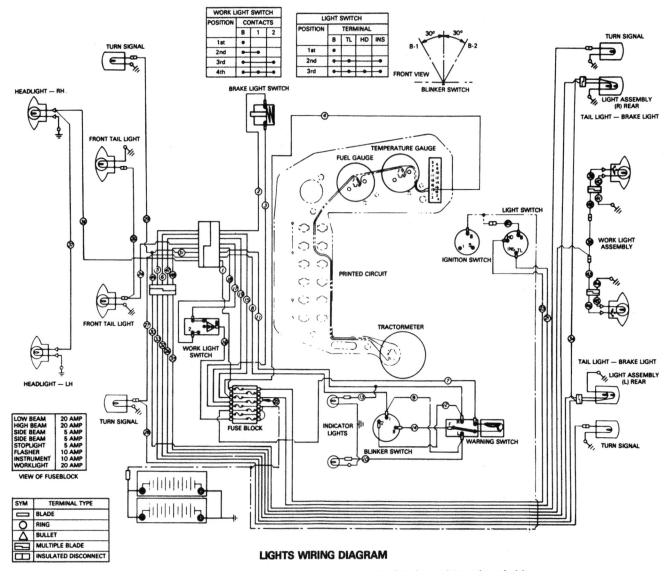

LIGHTS WIRING DIAGRAM

Fig. 75 — Light system wiring diagram typical of both models. Numbers refer to wire color(s).

1,23,46,37 . . . . .Orange	14,17,24,25 . . . . .Gray
2,5,33,34 . . . . .Yellow	18,35,48 . . . . .Pink
3 . . . . .Red/White	19,26,47 . . . . .Lt. Green/Black
4,6,15,21,22,32 . . . . .Dark Green	20 . . . . .Brown
7 . . . . .Red	38,39,42,43,46 . . . . .Blue
8,9,10,27,28 . . . . .White	40,41,44,45 . . . . .Black
11,12,13,29,30,31,49 . . . . .Light Blue	16 . . . . .Dark Blue

CIRCUIT DESCRIPTION

All Models

61. Refer to Figs. 74 and 75 for schematic wiring diagrams.

Two 95 Ampere Hour, 12 Volt batteries are connected in parallel and negative posts are grounded. Make sure both batteries are of equal capacity and condition. Quick disconnect couplings are used at all main connections.

ENGINE CLUTCH

Both models are equipped with a dry type single disc clutch mounted on the flywheel and released by a foot pedal. The clutch controls tractor travel, independent power take-off is controlled by a separate multiple disc wet clutch located in rear axle center housing. To avoid unnecessary wear on transmission

brake, clutch pedal should never be fully depressed before tractor is stopped.

ADJUSTMENT

All Models

62. If tractor is equipped with an eight-speed transmission, adjust clutch release cable until all free play is just removed. A force of approximately 5-10

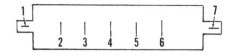

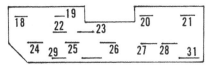

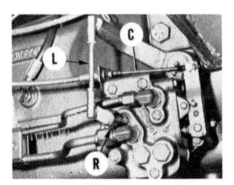

Fig. 75A — View of main wiring harness at fire-wall.

1. Lt. blue – turn signal
2. White – turn signal
3. Yellow – brake lights
4. Pink – rear work lights
5. Gray – front work lights
6. Tail light – left
7. Tail lights – right
8. Lt. green – pto
9. Black/white – fuel gage
10. Gray – trans. lube press.
11. Green/white – diff. lock
12. Blue/white – oil filters
13. Dk. green – Multi-Power press.
14. Purple – AC
15. White – trans. temp.
16. Brown – dome light
17. Orange – diff. lock and fuse block
18. Red
19. Yellow-alternator
20. Purple/white – oil press. switch
21. Dk. blue – starter
22. Black/red – water level indicator
23. Orange – water temp.
24. Lt. blue – air indicator
25. Red
26. Orange/black – starter relay
27. Red
28. Green – ether start
29. Purple – AC thermal limiter
31. Orange – head lights

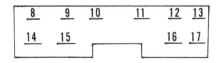

Fig. 76 — Adjust linkage rod (L) until roll pin (R) is vertical when clutch pedal is "UP" on models equipped with 24-speed transmission.

Fig. 77 — View of transmission brake adjustment point. Release fork has been removed for better view. Refer to text.

lbs. should be required to produce ¼-inch of pedal travel.

If equipped with a 24-speed transmission, pilot valve may be adjusted by turning linkage rod (L – Fig. 76) until roll pin (R) is vertical when clutch pedal is "UP".

To adjust the transmission brake on all models, remove inspection cover on bottom of transmission, loosen clutch release bearing sleeve locknut (LN – Fig. 77) with a punch and hammer.

NOTE: Always push clutch pedal down while sleeve locknut (LN) is being locked or unlocked.

With clutch pedal released (clutch engaged), use special tool No. 7282 or 7283 and turn nut (AS) until there is ⅜-7/16-inch clearance between contact surface of release bearing housing and front of transmission brake plate. Refer to Fig. 77. Be sure to tighten locknut.

TRACTOR SPLIT

All Models

63. Proceed as follows to detach (split) engine from transmission assembly: Remove hood and side panels. Disconnect engine end of tachometer drive cable, fuel shut-off cable, fuel shut-off spring and throttle cable. Pull throttle cable out of hole in support bracket.

Disconnect windshield washer hoses, main electrical plug and oil lines (from pipes) at firewall. Disconnect heater hoses and remove clamps from engine. Close fuel tank shut-off valves, then disconnect fuel lines (Fig. 21). Disconnect engine end of tachometer drive cable, fuel shut-off cable, fuel shut-off spring and throttle cable. Pull throttle cable out of hole in support bracket.

Remove clutch release fork after first disconnecting clutch release cable on eight-speed transmissions, or by disconnecting shift cable (C – Fig. 76), mounting bracket, links and linkage rod (L) from clutch shaft arm on 24-speed models; remove inspection cover on bottom of transmission, cut safety wires (W – Fig. 23) and remove release fork lock bolts (1). Remove release fork shafts from both sides of transmission housing, then remove release fork through inspection hole. Shift valve may have to be removed to facilitate removal of right shift fork shaft.

Remove top link support of three-point linkage at top rear of tractor to release spring tension on pto shaft. Support front and rear of engine/front end assembly using suitable splitting stands or hydraulic floor jacks. Support tractor under front of transmission using suitable jacks. Insert axle wedges between front axle support to prevent engine and front end from tipping.

NOTE: Remove front end weights if necessary to prevent front of tractor from tipping forward.

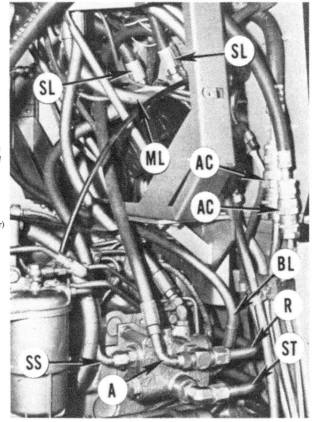

Fig. 78 — View of hoses which must be disconnected when splitting tractor.
A. Auxiliary
AC. Air conditioning
BL. Brake line
ML. Manifold line
R. Return tube
SL. Steering lines (to cylinder)
SS. Steering supply
ST. Supply tube (from pump)

Remove cap screws and bolts securing engine to transmission and insert two guide studs to aid in proper alignment during separation and assembly. Separate by rolling front end and engine forward.

NOTE: It may be necessary to remove floor panel inside cab to gain access to upper cap screws.

CAUTION: Carefully observe the joint between engine and transmission while loosening bolts to be sure that front end and mid-section of tractor is supported evenly. Change support as necessary if parting surfaces are not parallel.

When rejoining, carefully move engine back against transmission housing while aligning splines of clutch with transmission input shaft. If splines do not align, **DO NOT** force together, but turn flywheel slowly while sliding together.

NOTE: Make sure clutch release bearing is positioned squarely and that lube hose fitting is toward top.

Refer to paragraph 15A and tighten bolts to torques listed. Remainder of procedure is reverse of disassembly.

REMOVE AND REINSTALL CLUTCH ASSEMBLY

All Models

64. To remove clutch assembly for service, first split tractor as outlined in paragraph 63. Punch mark clutch cover (17 – Fig. 79), pressure plate (2) and flywheel so balance can be retained during reassembly. Install three 5/16-inch UNC x 2-inches long cap screws (C – Fig. 80) through holes in clutch cover and tighten until heads touch cover. Loosen cap screws securing clutch assembly to flywheel, remove clutch cover and friction disc.

To reinstall clutch, lube inside of pto coupler shield and install clutch disc in flywheel with hub **AWAY** from flywheel. While using a suitable clutch pilot tool, install clutch assembly and cap screws. Tighten cap screws alternately and evenly to a torque of 25-38 ft.-lbs. Remove three cap screws (C – Fig. 80) from clutch cover assembly.

Release lever height must be adjusted after clutch has been reassembled. Check lever height with a straight edge and machinists rule as shown in Fig. 81. Distance from straight edge to bottom of groove in back side of release lever should be one inch. Use special tool No. 7280 or equivalent, as shown in Fig. 82 to adjust lever to exact dimension. Locknut should be tightened to a torque of 60 ft.-lbs.

Screw release sleeve (13 – Fig. 79)

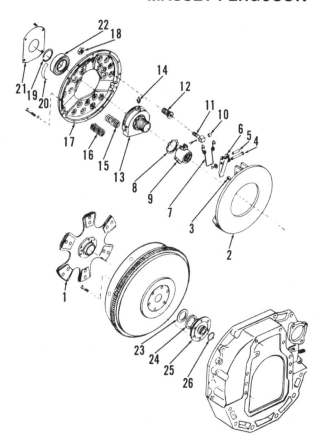

Fig. 79 – Exploded view of clutch disc and pressure plate assembly.

1. Clutch disc
2. Pressure plate
3. Bearing
4. Lever pivot pin
5. Lever to eyebolt pin
6. Release lever
7. Release lever spring
8. Spider locknut
9. Release spider
10. Eyebolt bearing
11. Eyebolt assy.
12. Eyebolt adjusting nut
13. Release bearing sleeve
14. Grease fitting
15. Dark maroon stripe spring
16. Light green stripe spring
17. Clutch cover
18. Locknut
19. Retaining ring
20. Bearing spring
21. Bearing cover
22. Release bearing
23. Coupler washer
24. Seal
25. Coupler
26. "O" ring

Fig. 80 – Install cap screws (C) before removing clutch cover assembly.

Fig. 81 – Using straightedge and machinists ruler to check release lever height.

with locknut (8) into release lever spider (9). Use special tool No. 6622 to measure distance from rear face of flywheel housing to rear face of release bearing housing. Refer to Fig. 83. Adjust to 5.408-5.418-inches by loosening locknut (8 – Fig. 84) and turning adjusting nut (A) to correct dimension. Tighten locknut after adjustment is correct. Adjust clutch as outlined in paragraph 62.

OVERHAUL CLUTCH

All Models

64A. Clutch cover (17 – Fig. 79), pressure plate (2) and flywheel should be punch marked in a line before removal so that balance can be maintained during reassembly. Remove clutch as outlined in paragraph 64, compress assembly in a suitable press and remove cap screws (C – Fig. 80) that were installed during the removal procedure. Use special tool No. 7280 or equivalent to remove release lever eyebolt locknuts (18 – Fig. 79), then release pressure on cover (17) and springs (15 and 16). Note spring location on pressure plate (2), then remove springs. Remove release levers (6).

Check springs (15 and 16) for overheating, breakage, wear or other damage and against the following pressure specifications:
Light Green Striped
Pressure at 1 27/32-inches 150-160 lbs.

Dark Maroon Striped
Pressure at 1 27/32-inches 130-140 lbs. Inspect remaining components for excessive wear or damage and renew as necessary.

To reassemble, lubricate release levers (6) lightly and assemble so that heads of pivot pins (4) are to the side shown in Fig. 85. Place springs (15 and 16 – Fig. 79) onto pressure plate.

NOTE: Three springs color coded dark maroon stripes (15) are installed on center inner pressure plate pins.

Place cover over springs aligning punch marks on cover and pressure plate made before disassembly. Apply pressure to cover until locknuts (18) can just be installed. Install release lever spider (9). Adjust release lever eyebolt nuts (12) until flush with end of bolt threads then tighten locknuts. Reinstall three cap screws (C – Fig. 80) into clutch assembly, then release pressure of hydraulic press.

Pads of clutch disc (1 – Fig. 79) are 0.410-0.448-inch thick when new. Renew disc as necessary. Reinstall clutch assembly as outlined in paragraph 64.

TRANSMISSION

All models are equipped with a basic eight-speed, sliding gear, constant mesh

transmission. Available as a factory installed option is a hydraulically actuated, three-range unit located on transmission input shaft which can be manually shifted when tractor is moving forward. Transmissions with the hydraulically activated three-range unit are referred to as "Multi-Power" transmissions and are capable of 24 forward and 6 reverse speeds.

A transmission brake is mounted on front cover of all models. Stator plates (4 – Fig. 100) are pinned to the case and discs (5) are splined to flats on the transmission input shaft. The brake is applied by clutch release bearing after clutch is fully released.

Massey-Ferguson Permatran lubricant is used in transmission and center housing. Oil capacity for transmission and rear axle is 30 gallons/with two filters. Fluid should be changed each 750 hours or annually.

TROUBLESHOOTING

24-Speed Models

65. Normal system pressure should be 290-310 psi. Pressure is controlled by

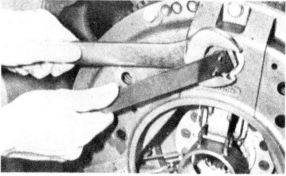

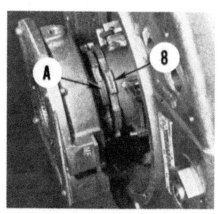

Fig. 82 – Using special tool No. 7280, adjust release lever to proper height. Refer to Fig. 81 and text.

Fig. 84 – View of locknut (8) and adjusting nut (A) that is used to change release bearing height. Refer to text for specifications.

Fig. 83 – Using tool No. 6622 to measure release bearing height. Refer to text.

Fig. 85 – Install lever pivot pins (4) in this direction only.

low pressure regulating valve. Refer to paragraph 112 for testing and adjustment procedures. System pressures below 275 psi can cause rapid wear of clutch friction components and pressures above 400 psi can cause immediate failure of piston "O" rings, sealing rings or retaining rings. Since extensive damage results from either of above conditions, it is recommended that complete input assembly be renewed if tractor has been operated with excessively low or high pressures for extended periods (usually over 50 hours) of time. Since oil is supplied to the input assembly from the hydraulic system, it is recommended that a complete hydraulic test be performed before starting input assembly diagnosis. Refer to paragraph 98 for hydraulic system testing and to paragraph 65A for shift valve testing.

Symptoms	Probable causes
External leaks	1,2,3,4,5
Power loss – High ratio	6,7,8
Intermediate ratio	8,9,10
Low ratio	6,7,9,10
Clutch slippage – High ratio	11,12,13
Intermediate ratio	15
Low ratio	11,14
Clutch slippage (Pressure OK) –	
High ratio	16,17,18,19,20
Intermediate ratio	21,22,23
Low ratio	20,21
Noise (Low Hours)	24,25,26,27
Noise (In low or intermediate only)	28
Noise (Over 100 hours)	28A

Key to Probable Causes

1. Loose lube lines or leakage from front cover tube(s) "O" ring(s).
2. Clutch flywheel seal or "O" ring worn (26 – Fig. 79).
3. Input shaft seals worn or damaged (7 or 12 – Fig. 100).
4. Square cut "O" ring on input housing defective (29 – Fig. 100).
5. Front cover "O" ring cut; leaks in low range only.
6. Intermediate piston "O" ring(s) failure (17 and 19 – Fig. 100).
7. Intermediate discs warped in excess of 0.030 inch (40 or 41 – Fig. 100).
8. Low discs warped in excess of 0.030 inch (37 or 39 – Fig. 100).
9. High discs warped in excess of 0.030 inch (4, 5, or 6 – Fig. 103).
10. Orifice plugged in high clutch pack (20 – Fig. 103).
11. Improper installation of timing pins in shift valve. Refer to paragraph 65A or Fig. 100C.
12. Worn teflon sealing rings on front bearing support (10 – Fig. 100).
13. Worn high piston sealing rings (11 and 22 – Fig. 103).
14. Low ratio piston "O" ring failure (30 and 32 – Fig. 100).
15. Clutch pack not pressurized in this mode; check adjustment of pilot valve as outlined in paragraph 62.
16. Incorrect disc and separator plates installed in high clutch pack (4 and 5 – Fig. 103).
17. Too much clearance in high clutch pack; refer to paragraph 71.
18. Planetary carrier not engaged in all discs (21 – Fig. 103).
19. Belleville springs too loose (44 – Fig. 100).
20. Worn low ratio discs (37 and 39 – Fig. 100).
21. Incorrect spacer(s) used (33 and 43 – Fig. 100).
22. Locating pin bent preventing proper assembly of discs (38 – Fig. 100).
23. Belleville springs upside down (44 – Fig. 100).
24. Thrust bearing on hub gear missing (46 – Fig. 100).
25. Thrust washer on gear sleeve missing (24 – Fig. 100).
26. Thrust bearing on bottom of planetary assembly missing (24 – Fig. 103).
27. Snap ring and spacer out of place in rear cover assembly (49 and 50 – Fig. 100).
28. Damaged planetary gears; renew input assembly.
28A. It is suggested that a factory new or rebuilt exchange input assembly be installed. The heat and internal damage for this length of operating time will probably warp and otherwise damage major parts of the unit beyond the usual rebuild limits.

CONTROL VALVE

24-Speed Models

65A. **TEST.** To perform pressure tests for shift valve and input assembly, connect gages as illustrated in Fig. 100A. Permatran fluid temperature should be between 110-120 degrees F.

Run engine at 1500 rpm, then move console lever to high. Gages 1 and 2 should pressurize. Intermediate gage (#2 – Fig. 100A) will hesitate then pressurize. Correct pressure is 290-310 psi.

Move lever to intermediate. Gages should drop to zero together.

Move console lever to low. Gages (#2 and #4 – Fig. 100A) should pressurize. Low gage (#4) should pressurize faster than intermediate gage (#2). Correct pressure is 290-310 psi.

Move console lever to intermediate. Gages should drop to zero together. Depress clutch pedal; gage (#2 – Fig. 100A) should pressurize. Correct pressure is 290-310 psi.

Symptoms	Probable causes
No pressure, drives only in intermediate	29
High pressure too low (50-60 psi), others ok	30
High pressure too low (200 psi), others ok	31
High pressure too low (200 psi), low pressure, low (100 psi)	32
Intermediate pressure drops slowly when shifting from low to intermediate or high to intermediate	33
Intermediate pressure increases too fast when shifting from intermediate to high or low	34
Low or high engages slower than intermediate, not consistent	35
Pressure in one circuit ok; low in another and lower in the other	36

Key to probable causes

29. Stuck unloading valve (7 – Fig. 100B).
30. High timing pin turned around (3 – Fig. 100C).
31. High and low timing pins in wrong holes (Fig. 100C).
32. High and low timing pins in wrong holes and reversed (Fig. 100C).
33. Intermediate orifices not installed properly (Fig. 100D).
34. Intermediate orifices reversed (Fig. 100D).

Fig. 86 – View of method used to determine number of plates required for transmission brake. Refer to text.

148.49-147.07 mm
(5.846-5.790")

35. Intermediate orifice stop not set to proper (1 5/32-inch) depth (Fig. 100D).

36. Leakage in circuit with lowest pressure.

65B. **R&R AND OVERHAUL.** Shift control valve (V – Fig. 22 or Fig. 100B) removal procedure is evident after inspection of unit. Extreme care must be taken to insure that all valve components are reassembled exactly as they came apart. Components should be identified during disassembly to aid in reassembly. To reinstall shift valve, reverse removal procedure. Tighten seal plate (9) retaining screws to a torque of 14-21 ft.-lbs.; tighten plug (11) to a torque of 41-62 ft.-lbs.

Overhaul procedure is evident after inspection of unit and reference to Fig. 100B and Fig. 100C. Clean all metal parts and valve body in a suitable solvent and air dry. Renew all seals, "O" rings and gaskets. Renew any spools, restrictors or valves that are scored or worn excessively or those that show evidence of sticking or binding.

Fig. 87 — Install new seal (S) and bearings (B) at front of pto housing before reinstallation of transmission.

Fig. 88 — Special tool No. 2801 (T) is used to aid in removal, refer to text.

TRANSMISSION BRAKE

All Models

66. **R&R AND OVERHAUL.** To remove transmission brake, first split tractor as described in paragraph 63. Transmission brake (4 and 5 – Fig. 100) is located in front of transmission front cover. Remove snap rings from anchor pins (8); remove plates (4 and 5) and keep in order. Four or five stationary and two rotating plates will be required; two stationary plates (4) must be used between rotating plates (5). Install snap rings (S) to anchor pins. Check distance from front surface of transmission housing to front of transmission brake plate using a straight edge as shown in Fig. 86. Add or deduct plates (4 and 5) as necessary to provide correct dimension of 5.790-5.846-inches. At least one stationary plate (5) should be at front and rear of stack. Reverse removal procedure to reassemble unit. Refer to paragraph 62 for clutch/transmission brake adjustment procedure.

TRANSMISSION FRONT COVER AND INPUT ASSEMBLY

All Models

68. **REMOVE AND REINSTALL.** Separate engine and transmission as outlined in paragraph 63. Remove shift valve, bracket and linkage as required. Remove transmission brake assembly as outlined in paragraph 66. Remove pto drive shaft, lube and clutch pack lines from front cover. Refer to paragraph 74 and remove shift cover, remove middle and lower shift shaft rails through rear, then move top rail to neutral position.

Remove cover retaining cap screws (C – Fig. 90). Attach Nuday special tool No. 2801 (T – Fig. 88), move front cover forward, rotate about 1/8 turn counterclockwise then remove from transmission.

To reinstall eight-speed transmission front cover, move shift rail (2 – Fig. 89) to raise interlock (3) so other shift rails will enter guides (4) in rear of front cover.

On all models, coat all cap screws with "Loctite". Use lockwashers in addition to "Loctite" on cap screws (C – Fig. 90), then tighten all screws to a torque of 25-38 ft.-lbs. Remainder of assembly is reverse of disassembly procedure. Check transmission brake dimension and adjustment as outlined in paragraph 66.

8-Speed Models

69. **OVERHAUL.** To overhaul front cover on 8-speed transmissions, first remove cover as outlined in paragraph 68.

Remove shift fork and rail, then remove interlock assembly in the sequence shown in Fig. 91A. Remove transmission brake and bearing support as outlined in paragraph 66 and remove inner seals if necessary.

Stand front cover (11 – Fig. 91) on input shaft (16) and drive against front gear (13) until the front bearing (5) has been moved approximately 1/4-inch. Move shift coupler (18) ahead and drive rubber plug from roll pin (15). Fabricate

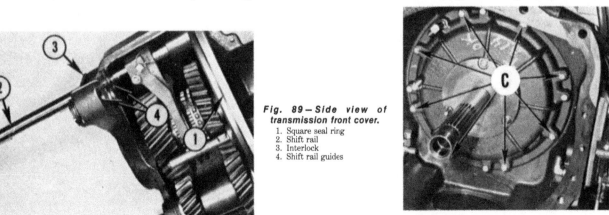

Fig. 89 — Side view of transmission front cover.
1. Square seal ring
2. Shift rail
3. Interlock
4. Shift rail guides

Fig. 90 — Use "Loctite" and lockwashers on cap screws (C) when reinstalling front cover.

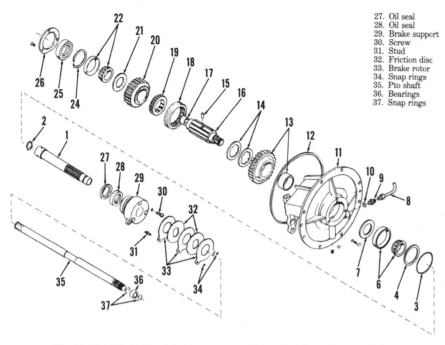

27. Oil seal
28. Oil seal
29. Brake support
30. Screw
31. Stud
32. Friction disc
33. Brake rotor
34. Snap rings
35. Pto shaft
36. Bearings
37. Snap rings

Fig. 91 – Exploded view of front cover assembly used on 8-speed transmissions.

1. Connecting shaft	8. Lube tube
2. Snap ring	9. Connector assembly
3. "O" ring	10. "O" ring
4. Shim (0.004-0.008-	11. Front cover
0.015 inch)	12. Seal
6. Bearing cup/cone	13. Reverse driving gear
7. Thrust washer	and bushing

14. Thrust washers	20. Input gear
15. Roll pin	21. Thrust washer
16. Input shaft	22. Bearing cone/cup
17. Bushing	24. Snap ring
18. Coupler	25. Main shaft bearing
19. Coupler hub	26. Bearing retainer plate

a tool similar to the one shown in Fig. 92 to remove roll pin as shown in Fig. 93. Remove rear cover bearing retainer plate (26 – Fig. 91) and push shaft (16) out of rear bearing cone. Withdraw shaft and parts from front cover, keeping parts in order to aid in reassembly.

Renew all components that are damaged or worn excessively. Input shaft (16) and connecting shaft (1) may be separated if necessary, by using a suitable puller. Snap ring (2) is used to align oil passage holes in shafts. Remove roller bearing (25) and snap ring (24) from rear case if necessary.

To reassemble front cover, first reassemble reverse idler assembly (Fig. 95). Install roller bearing (25 – Fig. 91) and snap ring (24) into shaft housing, if removal was necessary.

Install thrust washer (7) onto input shaft (16) with inside diameter chamfer toward shaft. Place snap ring (2) on inside of connecting shaft (1), align closed spline on connecting shaft with missing spline of input shaft (16). Push two sections together making sure oil holes are aligned.

Install transmission drive gear (20), one toothed thrust washer (14), shift coupler hub (19), with hole out, then install remaining toothed washer (14).

Install shift coupler (18) and reverse gear (13) into case and align thrust washer teeth, splines in hubs and splines in gears with holes aligned as shown in Fig. 96. Stand cover assembly on front shaft and drive rear bearing cone (22) cup onto shaft (16). Install bearing cup (6) cone onto front of shaft and tempo-

rarily install input shaft seal retainer leaving ¼ inch gap so roll pin hole is accessible. Install roll pin (15) with rubber plug (head of plug facing down) into input shaft. Install bearing cone/cup (6) the rest of the way.

Install brake support (29) on front cover (15) and tighten cap screw and studs (30 and 31) to 25-38 ft.-lbs. torque, then measure end play of input shaft (16). Remove brake support and insert shims (4) necessary to provide 0.001-inch end play to 0.006-inch preload on bearings.

Using proper tools, install oil seals in input shaft retainer. Coat rear bearing plate (26) retaining cap screws with "Loctite", then install and tighten to a torque of 14-21 ft.-lbs. Reinstall shift fork, shift rail and remaining components by reversing disassembly procedure.

24-Speed Models

70. **OVERHAUL.** To overhaul front cover on 24-speed transmissions, first remove cover as outlined in paragraph 68. Mark outer main sections (9, 15, 28 and 57 – Fig. 100) to aid in alignment during reassembly. Remove cap screws from rear cover (57). A shorter cap screw is located under interlock mechanism. Remove intermediate

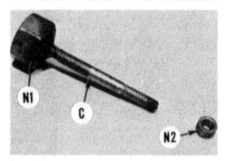

Fig. 92 – Fabricate from stock (listed below) to remove roll pin (15 – Fig. 91) from input shaft (16). One end of nut (N1) should fit slide hammer; other end should be welded to ¼ inch – 28 cap screw (C), approximately three inches long. Grind outside of nut (N2) to 13/32 inch outside diameter.

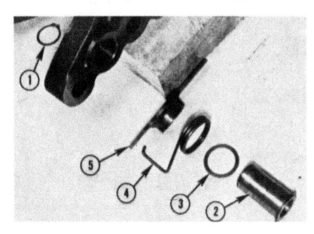

Fig. 91A – Disassembled view of interlock assembly. Refer to text.

1. Snap ring
2. Bushing
3. Washer
4. Torsion spring
5. Reverse lock assy.

Fig. 93 – After removing rubber plug from input shaft roll pin, pull roll pin using a slide hammer and tools as fabricated in Fig. 92.

clutch components which include: Belleville springs (44), three 9/16-inch thick spacers (43), locator plate (42), friction plate (41) and friction disc (40).

Unscrew six Allen head screws (62) then remove low clutch components including friction plate (39), friction disc (37), springs (36), locating pins (38), counter plate (60), round spacers (34) and three 7/16-inch spacers (33).

Remove support (28) and low piston (31). Remove thrust bearing (46), hub (45), gear sleeve (25), sun gear (22), then remove planetary carrier assembly (21). Remove high range clutch pack, thrust bearing (14) and races (13). Be sure front cover (15) and support (28) are marked for proper reassembly, then remove support. Remove intermediate piston (18) and springs (59) from front cover (15).

Clean, inspect and renew any components that are worn or damaged excessively.

Reassemble input by first placing springs (59 – Fig. 100) into front cover (15). Lubricate and install "O" ring (16) into front cover, install "O" rings (17) and (19) onto piston (18). Lubricate front cover piston bore, then install piston (18) into bore with small outer diameter of piston facing front of cover (15).

Place high clutch assembly (20) into front cover (15) with shaft toward front of cover (15). Install thrust races (13) and bearing (14) in recess of front cover. Coat outer edge of seal (7) with a suitable sealer then install into brake support (9). Install needle bearing (11) into transmission side of brake support 0.006-inch below end of support. Install seal rings (10) and "O" ring (12) on brake support (9). Install support (9) onto front

cover (15) using Nuday tool No. 7267 to protect seal (7). Coat threads with "Loctite" then install and tighten cap screw (6) and brake pins (8) to 25-38 ft.-lbs. of torque.

If bushing (23) in sun gear (22) requires renewal, inside diameter should be 2.866 – 2.872-inches. Refinishing is not normally required after installation. Gear sleeve bushing (26) must be finished after installation into gear sleeve (25) to a diameter of 1.8815 – 1.8890-inches. Inside diameter must not be more than 0.001 inch out-of-round. Press washer (24) onto gear sleeve with chamfered inside diameter toward large hub.

Carefully turn assembly over with shaft of high clutch assembly (20) down. Position planetary gears (21) so punch marks are straight out toward ring gear (20), then install planetary assembly (21) into ring gear (20). Install sun gear (22) with bushing away from hub, then install gear sleeve (25). Place "O" ring (61) into front cover (15) then install "O" rings (30 and 32) onto piston (31). Lube "O" rings and piston bore then install piston into brake support (28). Install brake support onto front cover aligning match marks made during disassembly. Temporarily secure with two cap screws.

NOTE: Apply "Loctite" to front cover cap screws and tighten to 25-38 ft.-lbs. of torque after front cover assembly is installed in transmission.

Place three thin, long spacers (33) between visibly threaded holes. Position

six round spacers (34) into support (28) as illustrated in Fig. 101. Install counter plate (60 – Fig. 100), low clutch disc (37) and friction plate (39) over dowel pin. Install six Allen head cap screws (62) into same holes as spacer (34 – Fig. 101) and tighten to a torque of 17-25 ft.-lbs.

Install nine pins (38 – Fig. 100) into counter plate (60) with a spring (36) over each pin. Install intermediate clutch disc (40) and friction plate (41), then place three thick, long spacers (43) over dowel pin and at 4 and 8 o'clock positions as shown in Fig. 102.

Place locator plate (42 – Fig. 100) over friction plate (41) with recesses over long thin spacers and narrow relief at rim toward bottom. Install Belleville springs (44), convex side up. Install planetary gear hub (45) and thrust bearing (46).

Align dowel pin and rear cover (57), then install rear cover. Install cap screws and tighten to a torque of 25-38 ft.-lbs.

NOTE: Short cap screw locates under shift rail guide.

71. HIGH CLUTCH PACK. To overhaul high clutch pack, refer to Fig. 103. Clutch pack and piston can be serviced without removing ring gear assembly from front cover.

Disassemble clutch pack by removing snap ring (1), counter plate (2), separating springs (3), friction plates (4), separating plates (5) and clutch plate (6). Remove flat head cap screws from piston retainer (7), remove retainer, then remove Belleville spring (8) and

Fig. 95 – Disassembled view of reverse idler assembly.
1. Screw
2. Reverse idler shaft
3. Reverse idler gear
4. Side washer
5. Roller bearings
6. Oil tube

Fig. 96 – Gear installation in shift cover.
H. Hole
5. Shift collar
13. Reverse gear
16. Input shaft
19. Coupler hub

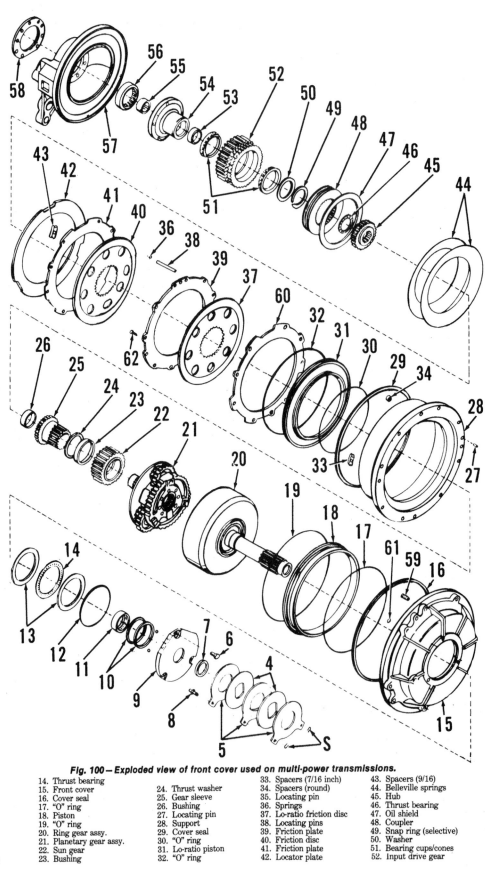

Fig. 100 — Exploded view of front cover used on multi-power transmissions.

4. Friction disc	14. Thrust bearing	33. Spacers (7/16 inch)
5. Brake rotor	15. Front cover	34. Spacers (round)
6. Cap screw	16. Cover seal	35. Locating pin
7. Oil seal	17. "O" ring	36. Springs
8. Stud	18. Piston	37. Lo-ratio friction disc
9. Brake support	19. "O" ring	38. Locating pins
10. Sealing rings	20. Ring gear assy.	39. Friction plate
11. Needle bearing	21. Planetary gear assy.	40. Friction disc
12. "O" ring	22. Sun gear	41. Friction plate
13. Thrust bearing race	23. Bushing	42. Locator plate
	24. Thrust washer	
	25. Gear sleeve	
	26. Bushing	
	27. Locating pin	
	28. Support	
	29. Cover seal	
	30. "O" ring	
	31. Lo-ratio piston	
	32. "O" ring	

43. Spacers (9/16)	53. Bushing
44. Belleville springs	54. Input gear bracket
45. Hub	55. Bushing
46. Thrust bearing	56. Bearing
47. Oil shield	57. Rear cover
48. Coupler	58. Retainer plate
49. Snap ring (selective)	59. Springs
50. Washer	60. Counter plate
51. Bearing cups/cones	61. "O" ring
52. Input drive gear	62. Allen screws

piston (19). Remove cap screws that retain input shaft (10) to planetary assembly and separate components. Clean, inspect and renew any components that are worn excessively or damaged.

To reassemble high clutch pack, first apply "Loctite" 222 or equivalent to surface between input shaft (10) and ring gear housing (20). Apply "Loctite" to cap screw threads then install and tighten to a torque of 14-21 ft.-lbs.

NOTE: Sealing ring area must be clean and free of "Loctite".

Apply "Permatran" lubricant to piston seals (11 and 22), then install seals on piston (19). After applying more "Permatran" to piston bore install piston into ring gear (20) using Nuday tools No's. 6783 and 6784.

Install Belleville spring (8) (concave to

piston) and retainer (7). Coat flat-head cap screws with "Loctite" 222 or equivalent, install and tighten them to a torque of 3-5 ft.-lbs.

Assemble parts (4, 5 and 6) into ring gear (20). Do not install separating spring rings (3) until after selection of snap ring (1). Install counter plate (2) with ridge up. Push parts firmly into ring gear. Snap ring (1–Fig. 103) is available in 1/8, 5/32 and 7/16-inch thicknesses. Select thickness of snap ring which will provide 0.070–0.100-inch clearance between snap ring (1) and counter plate (2) when measured as shown in Fig. 103A. Separating spring rings (3–Fig. 103) should not be installed when measuring clearance between snap ring and counter plate; however, clutch pack should be disassembled after snap ring (1) has been selected and a separating

spring ring (3) should be assembled outside each friction disc (4). Install counter

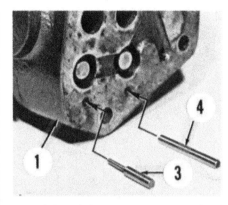

Fig. 100C — Install timing (restrictor) pins (3 and 4) correctly and in the proper holes.

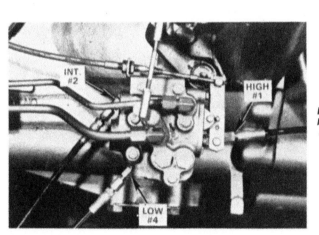

Fig. 100A — Connect hydraulic test gages to shift valve as illustrated. Refer to text.

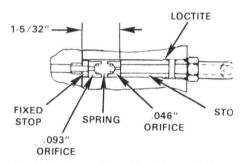

Fig. 100D — Intermediate orifice stop must be installed to correct depth of 1-5/32-inch from valve mounting surface. Refer to text for details.

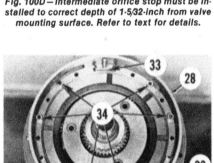

Fig. 101 — Install round spacers (31) and thin long spacers (33) into support (28).

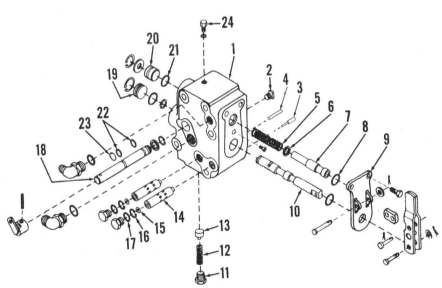

Fig. 100B — Exploded view of 24-speed shift valve. Install pilot washer (6) with chamfered edge towards spool land.

1. Valve body	6. Pilot washer	13. Valve plunger
2. Plug	7. Unloading valve	14. Pressure regulator
3. Stepped restrictor	8. "O" ring	valve
(timing) pin	9. Lever bracket assy.	15. Washer
4. Full length restrictor	10. Spool valve	16. Snap ring
(timing) pin	11. Plug	17. "O" ring
5. Spring	12. Plunger spring	18. Pilot valve

19. Spool plug
20. Unloading plug
21. "O" ring
22. "O" ring
23. Scraper seal
24. Plug

Fig. 102 — View of support (28) with thick spacers (43), intermediate clutch disc (40) and friction plate (41) properly installed.

plate (2) with ridge up, then install proper snap ring (1).

72. PLANETARY GEARS. To overhaul front cover planetary assembly, refer to Fig. 103 for exploded view and to following text: Remove needle bearing (25) and thrust bearing (24). Remove snap ring (26) and press shafts (17) out of planetary carrier (21).

NOTE: Each shaft has 50 needle bearings (16 and 13) and a bearing spacer (14) in each planetary gear (15).

Clean, inspect and renew any components that are damaged or worn excessively.

To reassemble unit, first install needle bearings (16 and 13) separated by spacer (14) into each gear (15), using Nuday tool No. 6618. Longer bearings (13) go into small end of gear (15).

Place thrust washer (12) on each side of gear, with flat side inward. Position gear (15) in carrier (21) using Nuday tool No. 6618 to locate gear, bearing needles and thrust washers, then press shafts (17) into carrier (21). Groove for snap ring (26) should be toward inside, and oil holes should be straight out. End of shafts (17) should be flush to 0.030-inch below surface of thin planet carrier flange. Reinstall snap ring (26) making sure that snap ring correctly engages groove of shafts (17). Bend tips of snap ring down approximately 30 degrees.

SHIFT COVER

All Models

74. REMOVE AND REINSTALL. To remove transmission shift cover, first drain transmission lubricant, (approximately nine gallons). Remove forward-reverse shift link (L – Fig. 104) then position gear shift lever in bottom of vertical neutral slot.

Remove plug from side cover and insert a cap screw into hole (C) tight enough to prevent shift shaft in cover from moving while upper portion of shift shaft is removed. Remove upper shift shaft out of way, remove shift cover cap screws then remove cover.

NOTE: Install front cover assembly, if removed, before installing shift cover.

Fig. 103 – Exploded view of planetary gear assembly and high clutch pack.

1. Snap ring	8. Belleville spring	14. Bearing spacer	20. Ring gear
2. Counter plate	9. Shim (0.020 inch)	15. Planetary gear	21. Planetary carrier
3. Separating spring	10. Input shaft	16. Short needle bearings	22. External seal
4. Friction disc	11. Internal seal	17. Planetary shaft	23. Planetary shaft
5. Separating plate	12. Thrust washer	18. Plug	24. Thrust bearing
6. Clutch plate	13. Long needle bearings	19. Hi-ratio piston	25. Needle bearing
7. Retainer			26. Snap ring

Fig. 103A – Procedure used to measure clearance between snap ring and counter plate.

Fig. 104 – To remove shift cover, remove forward-reverse shift link (L). Install cap screw (C) to prevent movement of shift shaft. Refer to text.

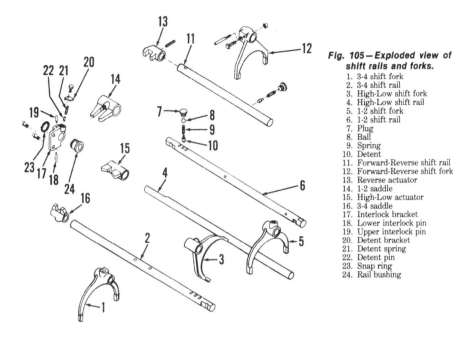

Fig. 105 – Exploded view of shift rails and forks.

1. 3-4 shift fork
2. 3-4 shift rail
3. High-Low shift fork
4. High-Low shift rail
5. 1-2 shift fork
6. 1-2 shift rail
7. Plug
8. Ball
9. Spring
10. Detent
11. Forward-Reverse shift rail
12. Forward-Reverse shift fork
13. Reverse actuator
14. 1-2 saddle
15. High-Low actuator
16. 3-4 saddle
17. Interlock bracket
18. Lower interlock pin
19. Upper interlock pin
20. Detent bracket
21. Detent spring
22. Detent pin
23. Snap ring
24. Rail bushing

Position rear gear in high range position and top and bottom shift rails in neutral. Center top and bottom shift fingers on shift cover. Rotate middle finger forward and push shift shaft down, keeping top and bottom fingers aligned.

Install short guide studs to aid in reassembly. Install cover, making sure all shift fingers are engaged. Remainder of procedure is reverse of disassembly procedure. Tighten shift cover cap screws to a torque of 25-38 ft.-lbs. Refill transmission as required with "Permatran" lubricant.

SHIFT RAILS AND FORKS

All Models

75. **R&R AND OVERHAUL.** After separating transmission and center housing as outlined in paragraph 79, remove detent retaining plugs (7 – Fig. 106), ball (8 – top detent only), springs (9) and detents (10).

Remove middle detent screws, bracket (20 – Fig. 105), spring (21) and detent pin (22). Remove interlock cap screws, bracket (17) and pins (19 – short pin) and (18 – long pin). Loosen shift fork set screws then remove rails (2), (4) and (6). Remove shift forks and keep in order to aid in reassembly.

Clean, inspect and renew components as necessary. To reinstall shift forks and rails, reverse disassembly procedure. Use "Loctite" 222 or equivalent and lockwires on all set screws. Tighten set screws to a torque of 21-32 ft.-lbs. Tighten interlock bracket (17) top retaining bolt to a torque of 63-95 ft.-lbs.;

use "Loctite" 222 or equivalent on detent plugs (7 – Fig. 106) and tighten to a torque of 41-62 ft.-lbs.

MAIN (TOP) SHAFT

All Models

76. **R&R AND OVERHAUL.** To remove main shaft, first separate transmission from center housing as outlined in paragraph 79, then remove shift rails and forks as outlined in paragraph 75. Use Nuday tool No. 6443 to remove locknut (1 – Fig. 107), then remove rear bearing cone (2). Slide bottom shaft rear gear forward, insert a bar or pto front section through main shaft assembly to aid in removal, then remove shaft and gear assembly.

Using Fig. 107 as a guide, remove components from main shaft. Renew any components that are damaged or worn excessively. Install bushings (10 and 15) 0.025-inch below edge of step in coupler side of gears (11 and 16) as required. After installation, bushings must be machined to an inside diameter size of 2.4852 – 2.5860-inches. Install large snap rings in gears (11 and 16).

To reinstall main shaft, reverse removal procedure. Apply "Loctite" 264 or equivalent to shaft threads, install nut (1) and tighten (using Nuday tool No. 6643) to a torque of 45 ft.-lbs., while rotating shaft to seat bearings; use Nuday tool No. 6621 to prevent bottom shaft from turning.

Adjust nut (1) until there is no end play or no pre-load. Stake nut to prevent turning; shaft end play must not be greater than 0.001-inch.

Fig. 106A – Pto line plate (P) must be removed to eliminate spring load on pto clutch pack standpipe.

Fig. 106 – Remove top and bottom rail detents (10), springs (9), and retaining plugs (7). Ball (8) is located in top rail only.

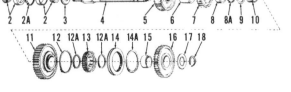

Fig. 107 – Exploded view of main shaft (top shaft).

1. Locknut
2. Bearing cone/cup
2A. Snap ring
3. Seal
4. Main shaft
5. 3rd driven gear
6. Spacer
7. 4th driven gear
8. Spacer
8A. Snap ring
9. Thrust washer
10. Bushing
11. 1st driven gear
12. Snap ring

12A. Snap ring
13. Carrier sleeve
14. Shift sleeve
14A. Snap ring

15. Bushing
16. 2nd driven gear
17. Thrust washer
18. Retaining ring

39

OUTPUT SHAFT

All Models

77. **R&R AND OVERHAUL.** To remove output shaft, first remove main shaft as outlined in paragraph 76.

Remove locknut (2—Fig. 108), using Nuday tool No. 6143; remove bearing cone (3) cup and bearing plate (1). Move shaft to rear, then out the side of transmission housing. Remove needle bearing (24) from inside rear of front shaft (25).

Remove snap ring (14) and remove components in order from shaft as shown in Fig. 108.

Clean, inspect and renew any components that are worn or damaged excessively.

To reassemble output shaft, reverse disassembly procedure. Select proper snap ring (14) to obtain roller bearing end play of 0.000–0.007-inch. Snap rings are available in 0.085, 0.090 and 0.0925-inch sizes. Install shaft assembly and bearing plate (1). Tighten large cap screws to a torque of 63-95 ft.-lbs. and small cap screw to 27-40 ft.-lbs.

Check shaft end play using a dial gage and install proper shim (7) so end play will be 0.032–0.055-inch. Shims are available in 0.015, 0.030 and 0.060-inch sizes.

Install bearing cone (3) and apply "Loctite" 264 or equivalent on shaft threads. Tighten nut (2) to a torque of 45 ft.-lbs. while rotating shaft to seat bearings. Back off nut (2) to obtain a 0.003-inch gap (measured with a feeler gage) between nut (2) and bearing cone (3). Tighten nut 1/16-turn which should position nut so that bearings are adjusted to no free play, no pre-load condition. End play should not exceed 0.001-inch. Stake nut to shaft after adjustment is complete.

COUNTERSHAFT

All Models

78. **R&R AND OVERHAUL.** To remove countershaft (25—Fig. 108), first remove main shaft as outlined in paragraph 76 and output shaft as outlined in paragraph 77. Remove snap ring (15—Fig. 108), shift hub (16), collar (17), gear (19), counterplate (20) and shim(s) (21). At front of shaft (25) remove snap ring (30), gear (29), spacer (28) then remove shaft.

Clean, inspect and renew any components that are damaged or worn excessively. To reassemble countershaft reverse disassembly procedure and note the following additional steps: Bushing (18) has an inside diameter of 2.5025–2.5033-inches and should be installed 0.025 inch from small edge of hub.

Assemble components (15) through (22) onto shaft (25) outside transmission

0.004–0.024-inch end play in gear (19). Shim (12) may not be necessary if equipped with wider gear (19) which is identified by groove (G–Fig. 109).

NOTE: On multi-power transmission, spacer (28) is 25/64-inch wide and similar spacer located between snap ring (30) and gear (29) is 7/32-inch.

Tighten bearing retainer (20) cap screws to 14-21 ft.-lbs. of torque. Install or remove shims (21) necessary to obtain 0.001–0.006-inch bearing pre-load on countershaft (25).

TRACTOR SPLIT

All Models

79. To detach (split) tractor between transmission and rear axle center housing, first drain transmission and center housing. Split engine from transmission as outlined in paragraph 63. Remove front cover if necessary as outlined in paragraph 68. Place suitable jacks under drawbar and transmission, remove cab front bolts and place jack stands on each side to front of cab. Raise cab approximately one inch. Remove batteries, battery tray and step bracket from left side of transmission.

Remove two center housing to transmission cap screws and install guide studs. Remove remaining cap screws, bolts and nuts, then carefully split tractor.

To rejoin center housing to transmission, reverse removal procedure and perform the following additional steps: Renew seal (S–Fig. 87) and bearings (B) at front of center housing as outlined in paragraph 93. Secure drive pinion shaft coupler with cotter pin (6–Fig. 114). Use guide studs to aid in alignment during reassembly. Rotate transmission bottom shaft as necessary to align coupler splines. Tighten bottom bolts to 173-260 ft.-lbs. of torque, six point cap screws to 130-200 ft.-lbs. and 12 point cap screws to 200-290 ft.-lbs. Coat threads of front cab cap screws with

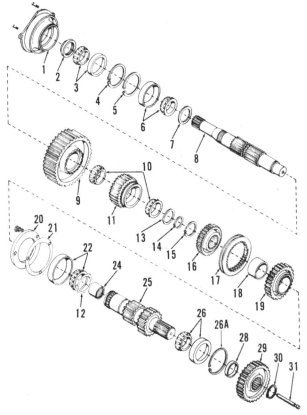

Fig. 108—Exploded view of counter shaft and output shaft.

1. Bearing plate
2. Locknut
3. Bearing cone/cup
4. Snap ring
5. Snap ring
6. Bearing cone/cup
7. Selective shim
8. Output shaft
9. Output gear
10. Bearing cones
11. Output gear
12. Shim (0.20 inch)
13. Washer
14. Selective snap ring
15. Snap ring
16. Shift hub
17. Shift collar
18. Bushing
19. 4th driving gear
20. Counter plate
21. Selective shim
22. Bearing cone/cup
24. Output shaft bearing
25. Counter shaft
26. Bearing cone/cup
26A. Snap ring
28. Spacer
29. Input gear
30. Snap ring
31. Oil tube

Fig. 109—Shim (12–Fig. 108) between bearing cone (22) and counter shaft (25) may not be required if groove (G) is present. Refer to text.

"Loctite" and tighten to a torque of 240-350 ft.-lbs.

CENTER HOUSING, MAIN DRIVE BEVEL GEARS AND DIFFERENTIAL

The rear axle center housing contains the main drive bevel gears and differential, power take off gears, pto shaft and pto clutch.

CENTER HOUSING

All Models

81. **REMOVE AND REINSTALL.** To remove center housing assembly, first remove both axle housings as outlined in paragraph 85. Use suitable stands to support rear of cab and rear of transmission. Remove differential as outlined in paragraph 82 and remove pto clutch pack as outlined in paragraph 93. Detach necessary hydraulic lines from rear of hydraulic filters. Detach and support filter bracket, leaving front lines attached. Remove bolts and cap screws from transmission center housing, install guide studs and carefully roll center housing away from transmission.

To rejoin center housing to transmission, first inspect and renew bearings (B – Fig. 87) and oil seal (S), if necessary. Position new gasket on center housing. Place coupler (5 – Fig. 114), with hole to rear onto pinion gear shaft (9), install cotter pin (6) then lubricate seal (S – Fig. 87).

Carefully move center housing into position aligning guide studs. Rotate bottom shaft as needed to align coupler with splines. Install bolts and cap screws, tightening to torques specified in Fig. 111. Remainder of procedure is reverse of disassembly.

DIFFERENTIAL AND BEVEL GEAR

All Models

82. **REMOVE AND REINSTALL.** To remove differential bevel ring gear, first remove axle housings as outlined in paragraph 85 and fuel tank, park brake and lift cover as outlined in paragraph 134.

Remove top link bracket (1 – Fig. 112), gear (6), washer (10) and shaft (11). Adequately support tractor front axle, cab, transmission and center housing. Attach "C" clamp to bevel ring gear, lift slightly

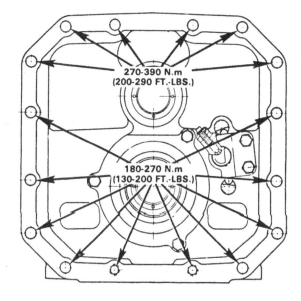

Fig. 111 – Tighten center housing bolts and cap screws to the above specifications.

270-390 N.m (200-290 FT.-LBS.)

180-270 N.m (130-200 FT.-LBS.)

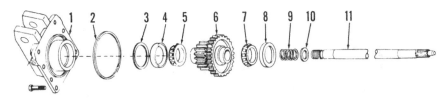

Fig. 112 – Exploded view of rear pto assembly. Refer to Figs. 123 and 126 for views of front gears and clutch pack components.

1. Coupler link	4. Cup	7. Cone
2. Seal	5. Cone	8. Cup
3. Shim (0.008 inch)	6. Cluster gear	9. Spring
		10. Washer
		11. Pto secondary shaft

Fig. 112A – Checking end play of secondary shaft. Refer to Fig. 112 for legend.

Fig. 113 – Disconnect control rod end (E) from ball. Remove "E" lock and unhook link (L) from pin (P) to avoid internal damage when removing lift cover.

and remove differential standpipe (S–Fig. 113), if equipped. Remove differential carrier plates (1 and 27–Fig. 114), then lift differential from center housing.

NOTE: Drive pinion (9) must be installed as outlined in paragraph 84 (if removed), before installing differential.

To reinstall differential and bevel gear, reverse removal procedure and perform the following additional steps: Check and adjust side carriers (1 and 27–Fig. 114) by referring to following procedures. Bevel gear backlash should be checked and adjusted by following procedure outlined in paragraph 82A.

With differential assembly suspended into center housing, install left side carrier (27) with drain hole facing down and without shims (26). Tighten cap screws to 63-95 ft.-lbs. of torque. Install right side carrier (1) and tighten cap screws to 35 in.-lbs. of torque. Rotate differential several turns then retighten cap screws to 75 in.-lbs. Check clearance between flange of right side carrier (1) and center housing to be sure that clearance is the same all the way around. If clearance is not equal, loosen screws, then retighten equally. Rotate differential again and measure gap (next to cap screws) between center housing and flange of carrier (1). Select proper thickness of shims (26R) as determined by chart in Fig. 115. Shims are available in thicknesses of 0.003, 0.006, 0.012 and 0.024-inch.

82A. Bevel ring and pinion gear backlash should be 0.003–0.025-inch.

NOTE: Check bearing pre-load and backlash after any shim change or repositioning.

To increase backlash, remove shim(s) (26L–Fig. 114) from left side carrier (27) and add that same shim to right side under carrier (1). To decrease backlash, remove shims (26R) from right side carrier (1) and install same shims at (26L) behind left side carrier (27). If necessary to further decrease backlash after all shims (26R) have been removed from under right side, a shim (33) can be made from MF part No. 892-120 M1. The special made 0.010 inch thick shim can be installed between bearing cone (2) and carrier (1); however, a shim of equal thickness must be installed at (26L) under left carrier (27) to maintain correct bearing pre-load.

83. **OVERHAUL.** To overhaul differential assembly, first scribe alignment marks on major components to aid in reassembly, then remove case retaining nuts (30–Fig. 114), differential case half (23), clutch discs (19) and plates (18), side gear (16) and thrust washer (13).

Procedure for non-locking differential is basically the same.

Clean, inspect and renew any components that are damaged or worn excessively.

NOTE: During reassembly lube differential components with "Permatran" fluid; apply a suitable grease to inside bore of differential gears and to outer hubs of side gears.

Install thrust washer (13) and side

gear (16) into flanged case half (4). Install differential pinion gears (15) and thrust washers (14), then insert cross shaft (17) through pinion gears. Align hole in cross shaft with hole in flanged case half (4). Drive pin (32) into cross shaft hole, install set screw (3) into hole in case half (4), then tighten set screw to a torque of 63-95 ft.-lbs.

Install case studs (31) if required using "Loctite" 271 equivalent. Install side gear (16), thrust washer (13), clutch

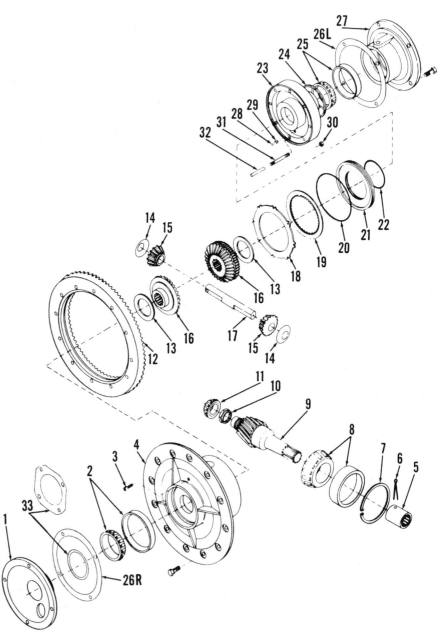

Fig. 114—Exploded view of differential lock clutch, main drive bevel gears, differential and associated parts used on models with differential lock. Models without differential lock option are similar.

1. Carrier plate	10. Adjusting nut	17. Cross shaft	25. Bearing cone/cup
2. Bearing cone/cup	11. Bearing cone	18. Clutch plate	26. Shim (see Fig. 115)
3. Set screw	12. Ring gear	19. Clutch disc	27. Carrier plate
4. Case half	13. Side gear thrust	20. External "O" ring	28. Ball (3/16 inch
5. Pinion coupler	washer	21. Differential lock	diameter)
6. Cotter key	14. Pinion thrust washer	piston	29. Check valve
7. Snap ring	15. Differential pinion	22. Internal "O" ring	30. Case retaining nuts
8. Bearing cone/cup	gears	23. Case half	31. Case studs
9. Pinion gear	16. Side gears	24. Seal ring	32. Retaining pin

plate (18), and clutch disc (19). Alternate plates and discs.

NOTE: A disc (19) should be last on the stack.

Inspect check valve (29) and oil inlet hole in case half (23). Install "O" rings (20) and (22) onto piston (21), apply petroleum jelly to rings and install piston into case half with raised edge of piston facing last disc (19).

Install case half on differential, apply "Loctite" 271 or equivalent to studs, then install nuts and tighten to a torque of 27-40 ft.-lbs. Install new seal rings (24) and bearing cone (25) when necessary. Install ring gear, using "Loctite" 271 or equivalent on cap screw threads. Tighten screws to a torque of 180-230 ft.-lbs.

DRIVE PINION

All Models

84. **R&R AND OVERHAUL.** To remove differential drive pinion, first separate center housing as outlined in paragraph 81 and remove differential as outlined in paragraph 83.

Install Nuday wrench No. 6628 on to pinion adjusting nut (10 – Fig. 114), then loosen nut by turning pinion shaft with wrench and Nuday tool 6621. Remove snap ring (7), then remove pinion assembly from front of housing using a suitable puller and a 5/8-inch-11 UNC threaded rod as illustrated in Fig. 116. Remove bearing cones/cups (8 and 11 – Fig. 114) as required.

Clean, inspect and renew any components that are worn excessively or damaged.

To reassemble unit, first install bearing cone (8) onto pinion shaft (9), then install nut (10) (almost all the way) and

bearing cone (11). Install snap ring (7) and bearing cup (8) into housing. Install pinion gear assembly into housing, tap assembly forward to seat bearing cup (8) against snap ring (7). Lubricate bearing cones (8 and 11) and tighten nut (10) using Nuday tools No's. 6628 and 6621, until torque required to turn pinion shaft is 5 ft.-lbs.

Loosen nut and tap shaft lightly into rear bearing cone (11) until 0.010-inch shaft end play is obtained.

Tighten nut (10) gradually until a rolling resistance of 24-30 in.-lbs. is obtained (with new bearings) or 12-15 in.-lbs. (with used bearings). Stake nut (10) then reinstall remaining components by reversing removal procedure.

REAR AXLE AND FINAL DRIVE

All models use a planetary final drive reduction unit (Fig. 117) located in inner end of axle housing next to center housing. Final drive lubricant reservoir is common with transmission, differential and hydraulic reservoir. Axle stub shafts (21) are free floating, with inner ends supported by differential side gears and outer ends serving as planetary sun gears. Service to any part of axle or final drive requires removal of unit as outlined in paragraph 85.

AXLE ASSEMBLY

All Models

85. **REMOVE AND REINSTALL.** To remove axle housing, first drain oil from center housing. Remove bottom connection of lift link on side(s) requiring service. Install front axle wedges to

prevent front end from tipping and securely support rear of tractor.

Remove wheel on side requiring service. Disconnect shift coupler, remove bolts and nuts securing rear of cab and carefully raise rear of cab high enough for axle housing clearance.

NOTE: Care should be taken to avoid cab binding or glass breakage.

Attach a hoist or place a suitable jack under housing, remove retaining cap screws and carefully slide the complete final drive assembly away from rear axle center housing.

To reinstall unit, first remove old sealer and oil from mating surfaces. If removed, brake plate (23 – Fig. 117) must be positioned on ring gear dowel pins (29).

NOTE: Shorter stub axle goes on right side of tractor.

Apply a bead of a suitable silicone sealer in recess of center housing. Install dowel (28) into center housing to aid alignment during reassembly. Move axle housing into position making sure planetary meshes properly, then slide assembly completely into position. Coat retaining cap screws with hydraulic grade "Loctite" or equivalent then install and tighten to a torque of 130-200 ft.-lbs.

NOTE: Hubs are marked for right and left side and are not interchangeable.

Remainder of procedure is reverse of disassembly. Tighten cab retaining bolts to a torque of 20-24 ft.-lbs. and cab locknuts to 100-150 ft.-lbs. Apply grease to hub retaining bolts and tighten to a torque of 300-420 ft.-lbs.

AXLE AND BEARINGS

All Models

86. **OVERHAUL.** With final drive unit removed as outlined in paragraph 85, remove brake plate (23 – Fig. 117), cap screw retainer (20), cap screw (29),

GAP — mm (in.)	SHIMS USED
0.025-0.13 (.001-.004)	0 No Shims
0.13-0.330 (.005-.013)	0.24 (.009)
0.331-0.558 (.014-.022)	0.46 (.018)
0.559-0.787 (.023-.031)	0.68 (.027)
0.788-1.016 (.032-.040)	0.91 (.036)
1.017-1.244 (.041-.049)	1.13 (.045)
1.245-1.475 (.050-.058)	1.36 (.054)

Fig. 115 — Chart showing differential spacer shims (26 — Fig. 114).

Fig. 116 — Remove snap ring from front bearing cup (8). Use a 5/8 inch — 11 UNC threaded rod (R) and suitable puller to remove pinion assembly.

plate (19) and shims (2). Remove planetary assembly (8), thrust washer (10) and bearing cone (12).

If ring gear (22) requires renewal, new gear must be installed with 0.002 – 0.009-inch interference fit, then dowel pin holes drilled as illustrated in Fig. 118. Brake plate (23 – Fig. 117) must fit freely over dowel pins (29).

Remove old sealer from mating surfaces, apply "Loctite" to outer edge of inner oil seal (14), then insert seal into housing (15), ½-inch from bearing cup seat. Install inner and outer bearing cups (13 and 16) and fill area behind cup (16) with a suitable multi-purpose Lithium base grease. Press bearing cone (17) onto axle shaft (1), then insert shaft into housing. Block shaft so it cannot move when planetary assembly is installed. Install bearing cone (12) and thrust washer (10), then install planetary assembly into housing. Assemble shims (2), cap screw (29) and retainer plate (19), then insert cap screw and tighten to a torque of 200-325 ft.-lbs.

Check axle end play using a dial gage, then add or remove shims (2) if necessary to obtain 0.001 – 0.005-inch pre-load on bearings. Reinstall and tighten cap screws to a torque of 200-325 ft.-lbs.

Apply multi-purpose Lithium base grease to area between bearing (17) and bottom of seal recess then install oil seal (18) flush to 0.030-inch below end of housing. Install brake plate (23) over ring gear pins. Apply grease to cap screw retainer (20) then install over cap screw.

87. PLANET CARRIER. Disassembly and reassembly is evident after inspection of carrier assembly and referral to Fig. 117. There are 56 needle bearings (6) per planet.

BRAKES

All Models

88. ADJUSTMENT. Normal brake pedal free play at pedal is 1 – 1¾-inches. Engine should be running when adjustments are made.

CAUTION: Before adjusting brakes, make sure park brake is applied and that transmission is in neutral.

To adjust brakes, loosen locknut (L – Fig. 119), then turn brake rod adjusting nuts (A) until correct pedal travel is obtained. Equalize adjustment so gap (G) doesn't vary more than 0.015 inch between poppets and vertical links.

88A. ADJUST PARKING BRAKE. To adjust park brake, pull brake handle out approximately 2½ inches, push lever (L – Fig. 168) to engaged position, screw adjusting nut up to lever, then tighten locknut.

NOTE: Lever should be installed at a right angle to cable. Adjustment is possible with fuel tank in place.

89. FLUID AND BLEEDING. Lubricant (Permatran), from rear axle housing is utilized as brake hydraulic fluid. Pressurized oil is supplied to master cylinder from flow control valve. Overflow from motor cylinder to pressure relief valve returns to sump at side of transmission. Refer to Fig. 144 for view of steering and brake system

hydraulic lines and connections. Refer to paragraph 104 for brake hydraulic system diagnosis. Whenever servicing brake master cylinder, or if cylinder has been drained, refill with proper lubricant, then bleed brakes as follows:

Run engine at idle speed for several minutes. Adjust engine speed to 1500 rpm, then have an assistant depress brake pedals, while bleed screws (B – Fig. 120) are opened. Bleed screws should remain open until no air appears, then tighten bleed screws. Bleeder fittings are located on both sides of center housing above rear axle housings.

90. R&R AND OVERHAUL MASTER CYLINDER (Brake valve). Brake valve (3 – Fig. 144) is located next to power steering valve and removal

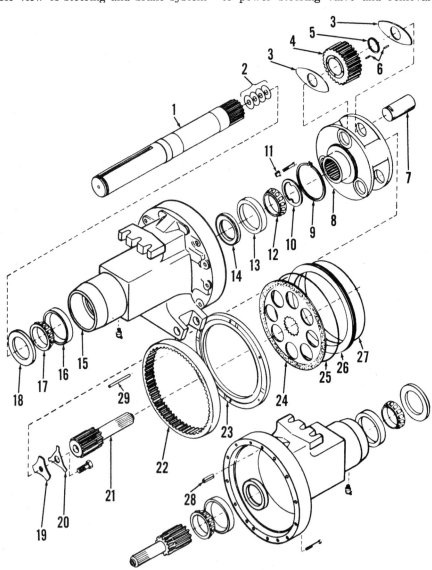

Fig. 117 – Exploded view of right final drive assembly and brake disc.

1. Axle shaft	7. Planetary shaft	15. R.H. housing	23. Brake plate
2. Shims (0.002-0.005-	8. Planetary carrier	16. Outer bearing cup	24. Brake disc
0.010 and 0.018 inch)	9. Snap ring	17. Outer bearing cone	25. Outer seal
3. Thrust plate	10. Washer	18. Oil seal	26. Piston
4. Planetary gear	11. Spacer	19. Plate	27. Inner seal
5. Spacer washer	12. Inner bearing cone	20. Retainer	28. Dowel pin
6. Needle bearings	13. Inner bearing cone	21. Stub axle	29. Ring gear dowel pins
(56 per planet)	14. Oil seal	22. Ring gear	(6)

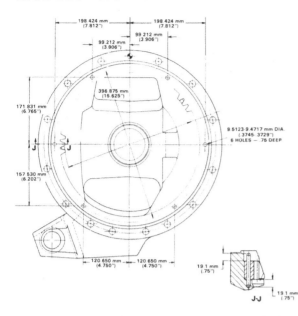

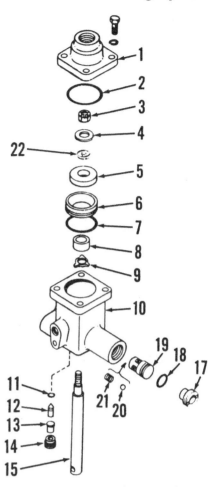

Fig. 118—Drill dowel pin holes as illustrated when renewing ring gear (22—Fig. 117).

procedure is evident after inspection of unit. Be sure to cover all lines and ports to prevent entrance of dirt.

Before disassembly, thoroughly clean exterior of valve. Remove cap (1—Fig. 121), piston assembly (15) and "O" ring (2).

Unscrew retainer (17) and using a wire hook, remove check seat (19) from body (10). Remove "O" ring (18) from check seat. Tip inlet port down and remove remaining components.

Unscrew metering valve retainers (14) and remove plunger needles (13), directional valve needles (12) with "O" rings (11). Remove "O" ring (9) from center of body (10). Separate parts of piston rod assembly. Clean and dry all metal components.

Renew all seals and inspect metal components for excessive wear or damage. Check needle valve (12) and seat for pitting, excessive wear, scoring or other damage.

Brake valve reassembly is evident after inspection of unit and referral to Fig. 121. Bleed and adjust brakes as outlined in paragraph 89.

CAUTION: Use only Massey-Ferguson Permatran fluid in system.

91. BRAKE PISTONS AND DISCS. Before any service can be performed on brake discs or actuating pistons, final drive unit must be removed as outlined in paragraph 85.

With final drive assembly off, brake disc (24—Fig. 117) and plate (23) can be removed. After plate and disc are removed, piston (26) can be pulled from housing bore.

NOTE: Distance between center housing and brake plate measured with feeler gages (F—Fig. 122) should be 0.004—0.007-inch. Brake plates are available in four sizes. Standard size diameter is 16.367—16.370-inches; other sizes are: 16.368—16.369, 16.372—16.373 and 16.373—16.375-inches.

Renew "O" rings (25 and 27—Fig. 117) after first coating with petroleum jelly. Align holes in back of piston (26) with pins and push piston into bore. Remainder of assembly is reverse of removal procedure. Bleed brakes as outlined in paragraph 89.

Fig. 121—Exploded view of brake valve (master cylinder).

1. End cap		12. Directional valve needle
2. "O" ring		13. Plunger needle
3. Locknut		14. Plunger retainer
4. Washer		15. Piston rod
5. Piston washer		17. Retainer
6. Piston		18. "O" ring
7. "O" ring		19. Check seat
8. Guide sleeve		20. Ball
9. Washer		21. Spring
10. Valve body		22. Sealing washer
11. "O" ring		

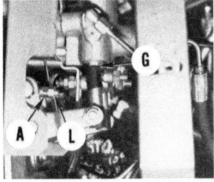

Fig. 119—Normal brake pedal free play is 1—1¾ inches. Refer to text for adjustment procedure.

A. Adjusting nuts
G. Gap
L. Locknut

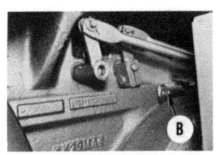

Fig. 120—Bleed screws (B) are located on rear axle center housing at top of rear axle housing on both sides of tractor. Refer to text for bleeding procedure.

Fig. 122—When renewing center housing or brake plate, measure clearance between piston and housing using three feeler gages (F) as shown.

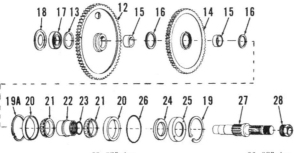

Fig. 123 — Exploded view of pto gears. Refer to Fig. 126 for exploded view of clutch pack and components and to Fig. 112 for exploded view of pto rear section.

12. 540 rpm driven gear
13. Thrust washer
14. 1000 rpm driven gear
15. Bushing
16. Thrust washer
17. Roller bearing
18. End cap
19. Snap ring (0.132, 0.140 and 0.150 inch)
19A. Snap ring (0.156 inch only)
20. Bearing cup
21. Bearing cone
22. Hub/gear

23. "O" ring	26. "O" ring
24. Cap seal	27. Shaft
25. Cap	28. Nut

Fig. 123A — Use hooks (H and J) to support and remove pto gears (12 and 14 — Fig. 123). Refer to text for details.

Fig. 124 — Pto clutch installed in center housing. Refer to text for removal procedure.

C. Clip
F. Inlet tube fitting
P. Locating pin
2. Cap screws
3. Cap screws

POWER TAKE OFF

The independent type power take off is driven by a hydraulically operated, multiple disc clutch located in the rear axle center housing. Output shaft speed can be changed by installing either the 540 rpm shaft or the 1000 rpm shaft (27 – Fig. 123) after using Nuday tool No. 6105 to unscrew nut (28).

CLUTCH PACK

All Models

93. **REMOVE AND REINSTALL.** To remove clutch pack, first drain center housing. Remove top link plate (1 – Fig. 112), drive gear (6), spring (9), washer (10) and shaft (11). Do not

disconnect flexible lines, but unbolt valve body (Fig. 145) from left side cover plate. Remove oil pick up tube from side cover, remove cover cap screws, then move cover away from center housing.

Remove clutch pack alignment pin (P – Fig. 124). Remove two cap screws (2), loosen third cap screw (3), rotate assembly down and remove clip (C) from standpipe.

Slide coupler (1 – Fig. 125) onto long standpipe (5), rotate clutch pack slightly, withdraw standpipe (5), then remove clutch pack assembly.

If bearings (B – Fig. 87) or seal (S) require renewal, first separate tractor as outlined in paragraph 81. Use a suitable grade of Bearing Mount Loctite in bearing bore. Front of rear bearing should be 2⅜-inches from front surface; front of forward bearing should be 1¼ inches from front surface. Use Nuday tools (6678, 6684, 6609) or equivalent and press against lettered end of bearings. Renew seal (S) after first coating outside diameter of seal with a suitable sealer. Use Nuday tools (6688, 6705, 6609) or equivalent to install seal.

To reinstall clutch pack, renew "O" rings and reverse removal procedure. Tighten clutch pack retaining cap screws to a torque of 8-10 ft.-lbs., side cover retaining cap screws to 45-50 ft.-lbs., and link plate retaining cap screws to 130-200 ft.-lbs. torque.

94. **OVERHAUL.** To overhaul clutch pack, first remove unit from tractor as

Fig. 126 — Exploded view of pto clutch pack and related components. Refer to Fig. 123 for exploded view of pto gear assembly and to Fig. 112 for exploded view of pto rear section.

F1. Front needle bearing
G. Hole
R1. Rear needle bearing
S. Slot
2. Clutch arm/hub
3. Seal rings
4. Clutch cover
5. Check valve
6. Ball (3/16 inch diameter)
7. "O" ring
8. "O" ring
9. Piston
10 & 10A. Thrust race
11 & 11A. Thrust bearing
12. Clutch brake cup
13. Brake actuator
14. Inner hub
15. Separating plate
16. Clutch plate
17. Friction disc
18. Bushing
19. Snap ring
20. Spring
21. Pin
22. Drive/gear cover

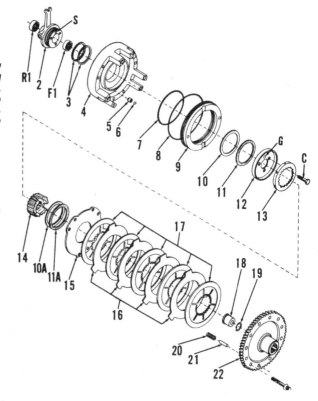

Fig. 125 — To remove clutch, it is necessary to slide coupler (1) onto standpipe (5).

outlined in paragraph 93. Remove screws which attach pump drive gear (22–Fig. 126) to cover (4). Carefully remove springs (20) and pins (21), clutch plates (16), friction discs (17), separating plate (15), thrust bearing (11A), bearing race (10A), inner hub (14) and brake actuator (13).

Remove clutch piston (9) from cover (4) using three ¼-inch-20 puller cap screws. Inspect oil passage and check valve (5 and 6) in cover. Remove clutch brake cap screws (C), clutch brake cup (12), thrust bearing (11) and thrust washer (10). Separate clutch arm (2) from cover. Clean, inspect and renew any components that are worn excessively or damaged.

Renew bearings (F1 and R1) in clutch arm (2) if required, by referring to Fig. 127 and following text: Bearing (F1) should be installed 2 inches from surface; bearing (R1) should be installed 0.012-inch from rear surface.

Renew seal rings (3–Fig. 126) onto clutch arm (2) hub. Install clutch arm (2) into clutch cover (4). Lubricate piston "O" rings (7 and 8) with petroleum jelly, install onto piston (9). Install piston into clutch cover (4) with puller holes out. Install thrust race (10) and bearing (11). Install clutch brake cup (12) making sure to align hole (G) in cup with slot (S) in

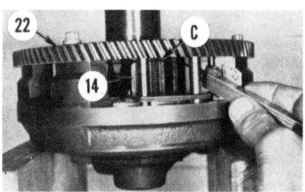
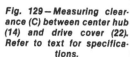

Fig. 129 — Measuring clearance (C) between center hub (14) and drive cover (22). Refer to text for specifications.

clutch arm/hub (2). Apply a suitable grade of "Loctite" onto clutch brake cup cap screws (C), install and tighten to a torque of 6-10 ft.-lbs.

Clearance between hub and clutch cover housing should be 0.004-0.010-inch when measured with two feeler gages as shown in Fig. 128. If clearance is incorrect, inspect condition of thrust bearing and washer (10 and 11–Fig. 126). Install inner hub (14) (recessed end out), thrust race (10A) and bearing (11A). Install pins (21) in cover (4), then temporarily install separating plate (15) and drive gear/cover (22). Install cover retaining screws to a torque of 25-38 ft.-lbs. Check clearance (C) between center hub (14) and drive cover

(22) as shown in Fig. 129. If clearance is not 0.004-inch inspect condition of thrust bearing and washer (10A and 11A–Fig. 126).

Remove cover (22) and install springs (20). Install clutch discs (17) and plates (16), beginning with one clutch disc (17) and alternating clutch discs and plates until all five discs (17) and all four plates (16) are installed.

Renew bushing (18) and snap ring (19) in drive gear cover (22). Inside diameter of bushing must be machined to 0.626-0.628-inch after installation. Inside diameter must be concentric with pitch diameter of splines within 0.002-inch. Flange end of bushing must be square with axis of splines within 0.001-inch total indicator reading.

Install cover and tighten retaining cap screws to a torque of 25-38 ft.-lbs. Measure clearance between plate (15) and cover (4) as shown in Fig. 130. Clearance (C) must be 0.099-0.139-inch for proper brake operation. Inspect thrust washer (10A–Fig. 126), thrust bearing (11A) and cover (22) for excessive wear if clearance is not correct.

Measure clearance (C–Fig. 131) between top clutch disc (17) and cover (22) as shown. Inspect plates (16–Fig. 126) and discs (17) for excessive wear if clearance is less than 0.070-0.188-inch.

Attach the magnetic "V" mount of dial

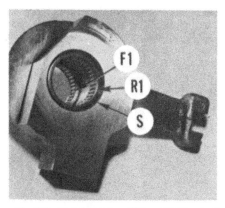

Fig. 127 — Install bearings (F1) and (R1) to depth specified in text.

Fig. 130 — Measuring clearance (C) between plate (15) and cover (4). Refer to text for specifications.

Fig. 128 — Measuring clearance between center hub (14) and drive cover (22). Refer to text for specifications.

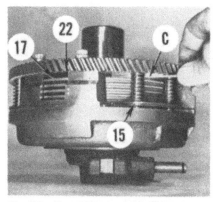

Fig. 131 — Measuring clearance (C) between top clutch disc (17) and cover (22). Refer to text for specifications.

Fig. 132 — Method used to check housing runout. Refer to text for procedure.

S. Surface
4. Cover
22. Drive cover

indicator base to the protruding machined hub of cover (22) and position dial indicator so that plunger contacts machined surface of cover (4) as shown in Fig. 132. Slide magnetic mount around cover hub and observe runout. If runout exceeds 0.004-inch, loosen cover retaining cap screws, tap edge of gear (22) lightly with a soft faced hammer, then retighten retaining screws to 25-35 ft.-lbs. torque and recheck runout. Repeat procedure until runout does not exceed 0.004-inch.

Test brake release by applying 9 psi of air pressure to inlet port. Inner hub (14) should be able to be turned. To reinstall clutch refer to paragraph 93.

OUTPUT SHAFT SEALS

All Models

95. **RENEW.** To renew output shaft seals, first drain center housing and remove output shaft retaining nut (28–Fig. 123) using (Nuday tool No. 610S) or equivalent. Remove output shaft (27), then renew "O" ring (23).

Install pto shaft, remove snap ring (19) and seal (24) with seal retainer (25). Renew "O" ring (26). "O" ring prevents bearing cup (20) from sliding out.

Install seal (24) into retainer (25) with lip forward after applying a suitable sealer to outer edge of retainer. Nuday tools No's. 6610, 6688 or 6705 may be used to draw seal into retainer.

Install retainer (25) into tractor being careful not to cut "O" ring (26). Install pto shaft (27). Attach dial gage and measure pto shaft end play, which should be 0.001-0.009-inch. Thickness of snap ring (19) should be changed to obtain required end play. Snap ring is available in 0.150, 0.140 and 0.132-inch thicknesses. Refill center housing with Permatran fluid.

PTO DRIVEN GEARS

All Models

96. **R&R AND OVERHAUL.** To remove pto driven gears, first remove fuel tank as outlined in paragraph 134 (leave tank strapped to bottom plate), drain center housing, remove lift control valve cover as outlined in paragraph 131 and lift cover as outlined in paragraph 134. Place a piece of wood under pto gears (12 and 14–Fig. 123) to prevent damage to gear teeth. Remove oil filler neck (D–Fig. 134) from center housing. Remove pto shaft (27–Fig. 123), seal assembly (24 and 25) and "O" ring (26) as outlined in paragraph 95. Push bearing cup (20) and cone (21) assembly towards rear of tractor far enough to allow gear (14) hub to move toward rear of tractor.

Fabricate two hooks from ⅛ x ½-inch strap steel. Hook (H–Fig. 123A) is used to support gear (14–Fig. 123) while removing gear (12) with hook (J–Fig. 123A). After removing gear (12–Fig. 123), remove gear (14) with hook (J–Fig. 123A). Make sure that all three thrust washers (13 and 16–Fig. 123) are removed.

Clean, inspect and renew any components that are worn excessively or damaged.

To reassemble pto gears, reverse disassembly procedure. Install top shaft (11–Fig. 112), gears and seals. Check end play with a dial gage as illustrated in Fig. 112A and outlined in paragraph 95. Allowable end play is 0.001-0.009-inch.

HYDRAULIC SYSTEM

All models are equipped with a dual element main hydraulic pump and a single element auxiliary pump. The dual

element pump is mounted to the inside of the right side cover and the auxiliary pump is mounted on the inside of the left side cover. Both pumps are connected by an intake manifold tube and are driven by the ipto clutch housing gear (22–Fig. 123).

High volume (16 gpm)-high pressure side of dual element pump supplies high pressure circuits including three point linkage control valve (29–Fig. 160) and flow control valve (FV–Fig. 144). Flow control valve directs oil to power steering and brake systems. Refer to Fig. 144. Excess oil is sent back to relief plate assembly through line (E–Fig. 136). Relief valve (2) controls maximum operating pressure to three point linkage control valve and right side auxiliary valve (2–Fig. 137), if equipped.

Low volume (6 gpm)-low pressure side of dual element pump supplies oil to pto valve/differential lock valve (4–Fig. 137), if equipped and 24-speed shift valve, if equipped. Pressure valve (6) controls pressures in low volume circuits. The auxiliary hydraulic pump also supplies oil to the left auxiliary control valve assembly.

LUBRICANT

All Models

97. The hydraulic system fluid serves as a lubricant for the transmission, differential, final drive and pto gear train. The transmission and rear axle serve as reservoir for system.

Recommended fluid is Massey-Ferguson Permatran lubricant. System capacity including filter change is 30 gallons. To check fluid level, be sure tractor is on level ground and remove dipstick (D–Fig. 134) from rear of center housing. Fill system through dipstick opening.

System filters (F–Fig. 135) should be

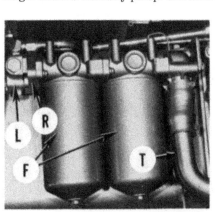

Fig. 134—Hydraulic fluid level dipstick filler tube (D) is located at rear of tractor.

Fig. 135—Hydraulic filters (F) should be renewed after first 25 hours of operation, then renew filters each 750 hours or annually.
F. Filters
L. Lubrication control   R. Oil cooler relief valve
   valve   T. Tubes

Fig. 136—View of right side cover.
E. Excess oil   4. Ipto/differential lock
F. Flow valve line   valve
L. Elbow fitting   6. Low pressure
PS. Pressure switch   regulating valve
2. Relief valve   8. Side cover

renewed after first 25 hours of operation and every 750 hours or annually. If fluid is badly contaminated, system should be disassembled for cleaning.

TEST AND ADJUST

All Models

98. All tests should be conducted with tractor hydraulic fluid at normal operating temperature of 110-120 degrees F. Be sure reservoir is filled to proper level with Permatran fluid at all times during tests.

Always cover openings in lines and ports as soon as possible to prevent entrance of dirt and unnecessary loss of fluid.

CAUTION: To prevent injury, engine should be stopped, all hydraulically actuated equipment should be resting on the ground and all hydraulic pressure within circuits should be relieved of pressure before removing any test port plug or disconnecting any hydraulic line.

99. **MAIN PUMP—HIGH VOLUME SECTION.** To test high volume section of dual element main hydraulic pump, connect flow meter using proper fittings from test kit No. MFN 2110 (2040) or equivalent to location (A–Fig. 144). Plug line to flow control valve with suitable fitting and put flow meter outlet hose into reservoir filler inlet (D–Fig. 134). Run engine at 2000 rpm and partially close flow meter load valve to create a pressure of 2000 psi on flow meter pressure gage. If flow reading is less than 11.8 gpm, pump repair or renewal is required. Refer to paragraph 122 for pump overhaul procedure.

100. **MAIN PUMP—LOW VOLUME SECTION.** To test low volume side of dual element main hydraulic pump, connect flow meter inlet using elbow (E–Fig. 138) and route flow meter outlet into reservoir filler inlet (D–Fig. 134). Run engine at 2000 rpm and close load valve until 200 psi is obtained on flow meter gage. If low pressure circuits are not engaged, flow reading should be

4.5 gpm. If pressure is too low, repair or renew pump. Refer to paragraph 122 for pump overhaul procedure.

101. **AUXILILIARY HYDRAULIC PUMP.** To test auxiliary hydraulic pump, first connect flow meter inlet to one of the left outboard remote couplers and connect outlet from flow meter to reservoir filler inlet (D–Fig. 134). Run engine at 2000 rpm, operate appropriate auxiliary control (O–Fig. 144A) in cab, then adjust flow meter until 2000 psi is indicated on flow pressure gage.

NOTE: It may be necessary to hold auxiliary lever against detent.

If flow is less than 11.4 gallons per minute, pump repair or renewal is required. Refer to paragraph 124 for overhaul procedure.

102. **LUBRICATION CONTROL VALVE.** The lubrication control valve (L–Fig. 135) receives return oil from the three point linkage and relieved oil from the low pressure regulating valve (6–Fig. 137). The lubrication control valve regulates oil flow to the transmission at a minimum of 15 psi–4 gpm to a maximum of 32 psi–12 gpm. Elbow (E1–Fig. 139) supplies cooler oil to the tractor brakes and contains a 0.078-inch diameter orifice. Elbow (E2) contains a 0.030-inch diameter orifice and is connected to transmission lubrication pressure sensing line to prevent overheating and erratic behavior of the lift cover valve. Inspect orifice in elbow (E2) if draft control operates erratically or if hydraulic system warning light flashes on and off intermittently.

To test lubrication control valve, connect a 100 psi gage into line (3–Fig. 140). Pressure should be 15 psi at 1000 rpm and 32 psi at 2500 rpm. Removal, overhaul and installation procedures will be evident after inspection of unit and referral to Fig. 139. Plug (7) should be tightened to 25-35 ft.-lbs. torque.

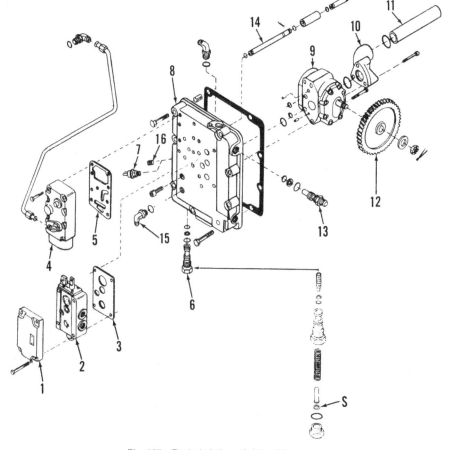

Fig. 137—Exploded view of right side cover.

S. Shims
1. Valve plate cover
2. Auxiliary valve
3. Intermediate plate
4. Ipto/differential lock valve

5. Gasket
6. Low pressure regulator valve (300 psi)
7. Multi-power pressure switch

8. Side cover
9. Main hydraulic pump
10. Inlet manifold
11. Tube
12. Pump drive gear (67 teeth)

13. Adapter
14. Standpipe
15. 45° elbow
16. Plug

Fig. 138—To test low volume side of main pump and connect test hose to fitting (E). Refer to text for details.

102A. OIL COOLER RELIEF VALVE. The oil cooler relief valve (R–Fig. 135) is used to prevent circuit pressure from exceeding 80 psi. Relieved oil is dumped back into hydraulic system filters.

To test oil cooler relief valve, connect flow meter inlet to "Tee" fitting (2–Fig. 140) and outlet to disconnected line (1). Run engine at 1000 rpm and slowly close flow meter load valve. Record pressure indicated on flow pressure gage. Relief valve setting should be 80 psi.

Overhaul procedure is evident after inspection of unit and referral to Fig. 141.

CAUTION: Use care when removing snap ring (1). Washer (2) is spring loaded and may come out with a great deal of force.

103. BRAKE AND STEERING SYSTEMS. Refer to Fig. 144 for view of brake and steering hydraulic system. Testing should be performed using a pressure gage and flow meter. Before testing brake or steering system, first make sure flow valve (FV) is receiving an adequate supply of oil by performing a flow supply test as outlined in paragraph 104. All tests should be conducted after Permatran fluid has reached 110-120 degrees F. Test the brake system first. Oil flows through brake circuit before steering circuit.

104. FLOW VALVE. To test oil supply to flow valve (FV–Fig. 144), connect flow meter inlet to flow valve supply tube (ST) and flow meter outlet tube to hydraulic system filler tube (D–Fig. 134). With flow meter load valve fully open, adjust and record rpm necessary to obtain 6 and 10 gpm flow, this is used in further tests.

Adjust flow meter load valve to 1800 psi then adjust and record rpm necessary to obtain 6 and 10 gpm flow. If unable to obtain proper flow, performance test high pressure/high volume side of dual element main hydraulic pump as outlined in paragraph 99.

105. BRAKE CIRCUIT PRESSURE. To perform brake circuit pressure test,

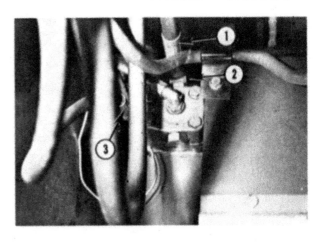

Fig. 140–Connect test gages to proper port when testing lube control valve or oil cooler relief valve. Refer to text for details.

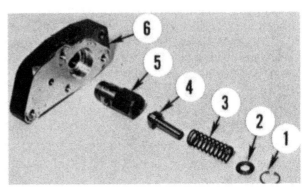

Fig. 141–Exploded view of oil cooler relief valve.
1. Snap ring
2. Washer
3. Spring
4. Poppet
5. Housing
6. Adapter block

connect a 0-2000 psi gage at fitting (F–Fig. 144) or at brake valve inlet (I). Some models will be equipped with a cartridge type external relief valve (4) which opens at 800 psi; other models will be equipped with a combination relief/flow control valve (BC) which opens at 600 psi.

Run engine at 1500 rpm and fully depress brake pedals. Brake circuit should maintain pressure of 800 psi if equipped with external relief valve (4) or 600 psi if equipped with combination valve (BC). Malfunction of brake cartridge (BC) or external relief valve (4) is indicated if pressure is not maintained. Refer to paragraphs 10 or 11 for overhaul procedure.

Shut off engine and apply 75 lbs. pressure to brake pedal. A reading less than 250 psi indicates master cylinder or piston malfunction. Refer to paragraph 90 for master cylinder overhaul and to paragraph 91 for brake piston repair procedure.

106. BRAKE CIRCUIT FLOW. To perform brake circuit flow test, connect flow meter inlet to line (I–Fig. 144) at master cylinder and connect flow meter outlet to hydraulic system filler tube (D–Fig. 134).

With load valve fully open, run engine fast enough to obtain exactly 6 gpm flow and record engine speed necessary to maintain this gpm flow. Adjust flow meter load valve to obtain the correct

brake circuit relief pressure depending upon valve. Correct brake circuit relief pressure is 800 psi with external cartridge (4–Fig. 144) or 600 psi if equipped with combination valve (BC). Record flow reading while maintaining rpm that was necessary to produce 6 gpm flow at no load. Repeat this test with 10 gpm delivery to flow valve. Malfunction of brake cartridge (BC) or external relief valve (4) is indicated if flow in both tests is not at least 1.3 gpm. Refer to paragraphs 10 or 11 for overhaul procedure.

107. STEERING CIRCUIT PRESSURE. To perform steering circuit pressure test, "Tee" a 0-2000 psi gage in-

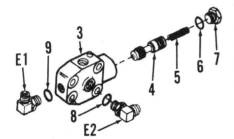

Fig. 139–Exploded view of lube control valve.
E1. Elbow 1 (0.078 inch orifice)
E2. Elbow 2 (0.030 inch orifice)
3. Valve body
4. Spool
5. Spring
6. "O" ring
7. Plug
8. "O" ring
9. "O" ring

Fig. 143–View of relief pressure adjustment procedure. Refer to text for details.

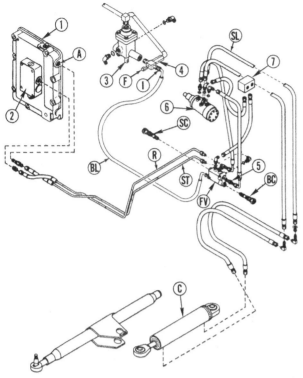

Fig. 144 — View showing external hydraulic system components and hoses. External relief valve (4) is not supplied on all models.

- A. Test port
- BC. Brake cartridge
- BL. Brake line
- C. Cylinder
- F. Fitting
- FV. Flow valve assy.
- I. Brake valve inlet
- R. Return line
- SC. Steering cartridge
- SL. Steering lines (to cylinder)
- ST. Supply tube
- 1. Rear cover
- 2. Rear plate
- 3. Brake valve
- 4. External relief valve
- 5. Check valve (port sv)
- 6. Steering control unit
- 7. Manifold block

port adjusting screw (T – Fig. 158A), run engine at 1500 rpm and turn adjusting screw counter-clockwise to fully raise lift links. If pressure is not 2050-2300 psi on both gages, adjust relief valve as outlined in paragraph 111.

NOTE: After tests and adjustments are made adjust transport stop as outlined in paragraph 130.

111. To adjust relief pressure, refer to Fig. 143 and remove cap and washer covering adjusting screw (S) and loosen jam nut (J). Start engine and turn adjusting screw until pressure is 2050-2300 psi. Reinstall components that were removed or detached for tests.

112. LOW PRESSURE REGULATING VALVE. Valve (6 – Fig. 136 or 137) controls pressure of fluid delivered to pto valve/differential lock valve (4 – Fig. 137) and 24-speed transmission shift valve (V – Fig. 22), if so equipped.

To test valve, "Tee" a 600 psi gage into elbow (15 – Fig. 137) on side cover that supplies oil to 24-speed shift valve (Fig. 100B). To test 8-speed transmission, remove plug (16 – Fig. 137) and install 600 psi gage.

Pressure should be 290-310 psi with either pto or 24-speed transmission engaged. To adjust pressure turn off engine and remove valve (6).

Overhaul procedure is evident after inspection of unit and reference to Fig. 137. To increase system pressure add shims (S), to decrease pressure remove shims (S). Tighten valve to a torque of 30-40 ft.-lbs. then retest system pressure.

112A. **AUXILIARY CONTROL VALVES.** The left auxiliary control valve is attached to left side cover and controls oil delivery to left rear couplers.

Either one or two auxiliary control valves may be located between relief side plate and right side cover. Couplers controlled by these valves should be located on right side.

to line (SS – Fig. 78). Adjust engine speed to 1500 rpm and turn steering wheel from lock to lock. A reading other than 1800 psi indicates malfunction of flow control/relief cartridge (SC – Fig. 144). Refer to paragraph 9 for overhaul procedure.

108. STEERING CIRCUIT FLOW. To perform flow valve supply test, first connect flow meter inlet into "SV" port of flow control valve (FV – Fig. 144) and connect outlet line from flow meter to hydraulic system reservoir filler tube (D – Fig. 134). Open load valve and run engine at rpm necessary to obtain exactly 6 gpm delivery. Close load valve to obtain exactly 1600 psi on flow meter pressure gage and record flow reading while maintaining same rpm that was

necessary to produce a 6 gpm flow at no load. Repeat this test with 10 gpm delivery to flow valve. Malfunction of flow control cartridge (SC – Fig. 144) is indicated if result is less than 5 gpm. Refer to paragraph 9 for overhaul procedure.

110. **MAIN HYDRAULIC SYSTEM.** If system is equipped with right auxiliary valve connect a 5000 psi gage to right coupler set, run engine and operate auxiliary valve lever (R – Fig. 144A) in direction necessary to obtain pressure reading on gage. Adjust relief valve as outlined in paragraph 111 if pressure is not 2050-2300 psi.

If system is not equipped with a right auxiliary valve, connect a 5000 psi gage at (C – Fig. 143) using a suitable "Tee" fitting and "Tee" a gage into one lift cylinder. Remove plug covering trans-

Fig. 144A — View of auxiliary valve control levers.
- I. Inboard (left couplers)
- O. Outboard (left couplers)
- P. Plate
- R. Right side (optional)

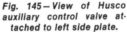

Fig. 145 — View of Husco auxiliary control valve attached to left side plate.
- A. Adjustable relief valve
- 1. Working section
- 2. Working section
- 3. Valve outlet plate

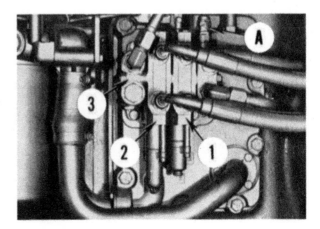

Control valves may be manufactured by either Bosch or Husco. Husco is used only on left side cover while Bosch can be used on either side, but will be the only type used on right side cover. Both sections (1 and 2 – Fig. 145) of Husco valves are detented and contain kickout mechanisms, however only section (2) has a float position. Adjustable relief valve (4) is installed in valves inlet port plate.

113. **HUSCO VALVES.** To test valve relief pressure, connect a 5000 psi gage to coupler and run engine to 1500 rpm. Operate auxiliary control (O – Fig. 144A) to rear and record gage reading. If pressure of 2050-2300 psi is not obtained, adjust main relief in left auxiliary valve as outlined in paragraph 115. If kickout is working properly, it may be necessary to hold lever in position to obtain reading.

115. To adjust relief pressure, remove acorn nut (A – Fig. 145) then loosen jam nut on adjusting screw. Turn screw clockwise to increase or counterclockwise to decrease pressure setting. Tighten jam nut, recheck pressure and repeat procedure if necessary.

116. To test spool kickout, connect flow meter inlet to bottom coupler and flow meter outlet to top coupler. Open flow meter, run engine to 2000 rpm, move appropriate lever (O or I – Fig. 144A) forward to first position. Slowly close flow meter load valve until lever kicks out (returns to neutral). Adjustment is necessary if kickout does not occur between 1900-2100 psi.

117. To adjust spool kickout, remove rubber plug (P or P1 – Fig. 152), then turn screw (40 or 34) clockwise to increase or counter-clockwise to decrease spool kickout pressure.

119. **BOSCH VALVES.** To test relief pressure of Bosch auxiliary valves (Fig.

153), connect 5000 psi gage to coupler, run engine at 1500 rpm, move auxiliary lever (O – Fig. 144A) to pressurize gage port. Gage should indicate 2050-2300 psi.

To adjust relief pressure remove lower cap and washer covering relief valve, loosen locking nut and turn adjusting screw clockwise to increase (or counterclockwise to decrease) pressure.

120. To test detent kickout, connect flow meter inlet to bottom left outboard coupler and connect outlet to top outboard coupler. With engine running at 2000 rpm and lever (O – Fig. 144A) in forward position, gradually close flow meter until valve lever returns to neutral. Adjust valve if kickout does not occur between 1900-2040 psi.

To adjust kickout pressure, first turn off engine; remove end cap (13 – Fig. 153) from bottom of spool valve requiring adjustment. Push inward on plug (8) and while preventing detent sleeve (11) from moving, remove internal snap ring (9). Withdraw spring (7) and detent cam (6). Turn adjusting screw (5) clockwise to increase (or counter-clockwise to decrease) kickout pressure. Reinstall components.

NOTE: If detent balls (2) drop out, remove detent sleeve (11) from end of valve. Install detent cam (6) into housing (1), then install detent balls (2) into holes around housing. Slide detent sleeve (11) over balls. Make sure detent balls enter center recess of detent cam (6) and detent sleeve (11) is positioned to its internal detent notch located nearest end of sleeve faces away from retainer sleeve (10).

120A. **PTO VALVE MODULATION TESTING.** To test pto "feathering" capability, connect a 600 psi gage to pressure switch (PS – Fig. 136) outlet port. With engine running, it should be possible to modulate pressure between

0-310 psi by operating pto valve lever slowly.

MAIN PUMP AND RIGHT SIDE COVER

All Models

121. **REMOVE AND REINSTALL.** To remove main hydraulic pump and right side cover, drain center housing, thoroughly clean side cover area, then mark control linkages and hydraulic lines.

If so equipped, disconnect control linkage and lines from right auxiliary control valves (2 – Fig. 137) and differential lock control (4). Identify and disconnect any electrical wiring from switches, then remove side cover cap screws and remove side cover and pump as a unit.

Procedure for detaching pump from side cover is evident after inspection of unit. Refer to paragraph 122 for pump overhaul procedure. Reinstall side cover and hydraulic pump by reversing removal procedure and by performing the following additional steps: Nut retaining gear (12 – Fig. 137) should be tightened to 35-50 ft.-lbs. torque. Check to be sure that clearance between back of gear and protrusion of pump casting around shaft seal is at least 0.010-inch. Install new "O" rings to manifold (10) and standpipe (14). Install guide studs into center housing to aid installation procedure. Make sure all components and drive gears match properly.

122. **OVERHAUL.** To overhaul main hydraulic pump, first remove pump and right side cover as outlined in paragraph 121. Separate pump and side cover while noting position of "O" rings and dowels. Clean pump thoroughly then remove drive gear (12 – Fig. 137), inlet manifold (10) and "O" ring.

Scribe alignment marks on pump rear cover (1 – Fig. 147), center body (15) and front cover (10) to aid in reassembly. Remove Woodruff key from shaft (4),

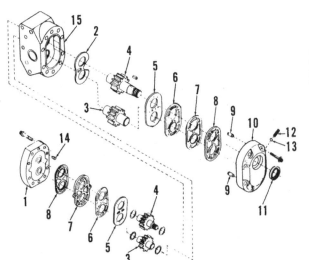

Fig. 147 – Exploded view of dual element main hydraulic pump. Front cover (10) cap screws are longer than rear cover cap screws.

1. Rear cover
2. Thrust plate
3. Idler gear
4. Drive gear
5. Diaphragm plate
6. Back-up gasket
7. Protector gasket
8. Diaphragm seal
9. Dowel
10. Front cover
11. Seal
12. Spring
13. Check ball
14. Dowel
15. Center body

Fig. 148 – Install large gears in pump. Painted end (P) of idler gear (3) faces front cover (10 – Fig. 147).

then remove cover bolts. Tap drive shaft (4) to remove front cover (10) from center body (15). Remove seal (11) from front cover then carefully remove diaphragm plate (5), gasket (6), protector (7), seal (8), steel ball (13) and spring (12). Remove drive gear (4), idler gear (3) and thrust plate (2). Repeat procedure for rear section, items (12 and 13) are in front cover only.

Clean, inspect and renew any components that are worn, damaged or those that do not meet the following specifications: Minimum diameter of shafts for gear (3 and 4) is 0.8094-inch for those in front cover (10); 0.6845-inch for those in rear cover (1). Minimum width of front (large) gears is 0.7051-inch; 0.3716-inch for rear (small) gears. Inside diameter of gear pockets in center body (15) should not be greater than 2.002-inch for front pump and 1.7185-inch for rear pump. Maximum inside diameter of shaft bearings for front cover (10) is 0.8156-inch and for rear cover (1) 0.6906-inch. Wear groove patterns in front or rear covers should not exceed 0.0015-inch.

To reassemble pump, first lubricate all parts with clean oil. Use new gaskets and seals during reassembly. Install thrust plate (2) (with bronze side towards gear faces), into center body (15). Install large drive gear (4) and large idler gear (3) as shown in Fig. 148.

Install diaphragm seal (8 – Fig. 147) into front cover (10) with groove facing into cover, then install protector gasket (7) and back-up gasket (6). Insert check ball (13) and spring (12) into front cover (10) bore containing angled drilling.

Install diaphragm plate (5) on top of seals and spring (12) with bronze side facing gears. Install dowel pins (9) into center body (15), apply petroleum jelly to facing surfaces, then join center body and front cover using scribe marks to insure proper alignment. Install front

cover cap screws finger tight. Install shaft seal (11) using special tool No. 1145 or equivalent.

Install small drive gear (4) and small idler gear (3) as shown in Fig. 149. Install diaphragm seal (8 – Fig. 147) into rear plate (1) with open side of "V" facing into groove. Install protector gasket (7) and back-up gasket (6) into rear cover, then install diaphragm plate (5) with bronze side facing gears. Install dowel pins (14) and make sure roll pin in rear plate (1) is properly installed. Apply petroleum jelly to facing surfaces then join rear cover to center body. Install and tighten cover bolts (rear first – then front) to a torque of 23-26 ft.-lbs.

Install drive gear (12 – Fig. 137), washer and retaining nut. Tighten nut to a torque of 35-50 ft.-lbs., then install new cotter key. Pour clean hydraulic fluid into pump ports and rotate pump drive gear to check for binding or excessive drag.

122A. After installing pump, fill hydraulic system reservoir to proper level with Permatran fluid, start engine and operate at about half speed for three minutes without operating hydraulic controls. Move controls to load system intermittently for the next three minutes, then increase engine speed to full operating rpm and operate controls to load system intermittently for an additional three minutes. Recheck fluid level.

AUXILIARY PUMP AND LEFT SIDE COVER

123. **REMOVE AND REINSTALL.** Left side cover may be equipped with either Bosch or Husco auxiliary control valves. Removal procedure for either type of valve assembly will be obvious after inspection of unit.

To remove left side cover and auxiliary hydraulic pump, first drain center housing, thoroughly clean side cover area, then mark control linkages, hydraulic lines and tubes to aid in reassembly. Remove oil tube (T – Fig. 135), auxiliary control valve, side cover retaining cap screws, then remove side cover and auxiliary pump as a unit.

Pump removal is evident after inspection of unit. Refer to paragraph 124 for

pump overhaul procedure. Reinstall side cover/auxiliary pump by reversing removal procedure and by performing the following additional steps: Nut retaining pump drive gear should be tightened to 35-50 ft.-lbs. torque. Check to be sure that clearance between back of gear and protrusion of pump casting around shaft seal is at least 0.010-inch. Install new "O" rings and make sure tubes (11 and 14 – Fig. 137) are installed properly into main hydraulic pump. Install guide studs into center housing to aid in installation. Remainder of procedure is reverse of disassembly procedure. Fill hydraulic system to proper level with "Permatran" fluid.

124. **OVERHAUL.** To overhaul auxiliary hydraulic pump, first remove left side cover and pump assembly as outlined in paragraph 123. Separate pump from side cover, then scribe alignment marks across front cover (10 – Fig. 150) and main body (15) to aid in alignment during reassembly. Remove pump cap screws then tap end of drive gear to separate front cover from main body. Remove drive gear (4), idler gear (3) and diaphragm plate (5).

NOTE: Before removing spring (12) or ball (13) note their location for proper reassembly.

Remove spring and ball, back-up gasket (6), protector gasket (7) and diaphragm seal (8). Remove shaft seal (11) from front cover and remove thrust plate (2) from main body (15).

Clean, inspect and renew any components that are worn excessively, damaged or do not meet the following specifications: Minimum diameter of shaft for gear (3 and 4) is 0.8094-inch; minimum width of gear face is 0.7051-inch; maximum inner diameter of shaft bearings in main body and front cover is 0.8156-inch. Renew all seals and gaskets.

To reassemble pump, first lubricate all parts with clean hydraulic fluid. Install thrust plate (2) in main body (15) with bronze side toward gears. Install drive (4) and idler (3) gears in main body. Install diaphragm seal (8) in front cover with open side of "V" facing into groove, then install protector gasket (7) and

Fig. 149 – Small gears installed in pump. Painted end (P) of idler gear (3) faces rear cover (1 – Fig. 147).

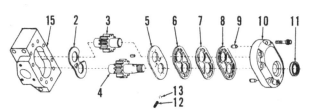

Fig. 150 – Exploded view of auxiliary hydraulic pump.
2. Thrust plate
3. Idler gear
4. Drive gear
5. Diaphragm plate
6. Back-up gasket
7. Protector gasket
8. Diaphragm seal
9. Dowel
10. Front cover

11. Seal
12. Spring

13. Check ball
15. Center body

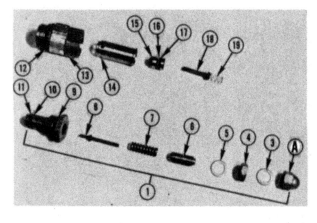

Fig. 151 — Exploded view of Husco auxiliary valve main relief. Acorn nut (and valve) is also shown at (A — Fig. 152).

A. Acorn nut
1. Pilot section
3. Washer
4. Jam nut
5. Copper washer
6. Adjusting screw
7. Pilot spring
8. Pilot poppet
9. "O" ring
10. Back-up washer
11. "O" ring
12. "O" ring
13. Main housing
14. Check poppet
15. Relief poppet
16. Back-up washer
17. Detent sleeve
18. Piston poppet
19. Spring

back-up gasket (6). Insert ball (13) and spring (12) in bore of front cover (10) which contains the angled drilling. Install diaphragm plate (5) with bronze side toward gears and coat gear faces with petroleum jelly. Install dowel pins (9). Join main body and front cover using scribe marks to insure proper alignment. Install and finger tighten front cover cap screws, then install shaft seal (11) using tool No. 1145 or equivalent. Tighten cap screws to a torque of 25-28 ft.-lbs. Install drive gear, washer and retaining nut. Tighten retaining nut to a torque of 35-50 ft.-lbs. and secure with new cotter key. Pour hydraulic fluid into pump ports and rotate pump drive gear to check for binding or excessive drag. Refer to paragraph 122A for pump break-in procedure.

HUSCO AUXILIARY CONTROL VALVES

125. **R&R AND OVERHAUL.** Procedure for removal of auxiliary control valve is evident after inspection of unit. Hydraulic lines and linkage should be identified to aid in reassembly. To reinstall valve, reverse removal procedure.

Main relief valve assembly (Fig. 151) may be disassembled for inspection. Only seals and washers are available separately. If other parts are required, valve must be renewed.

Carefully disassemble, clean and inspect relief valve using Fig. 151 as a guide. Make sure all parts slide freely, but are not worn excessively or otherwise damaged. Renew seals and washers if valve does not require renewal. Reinstall relief into housing (7 – Fig. 152) and adjust relief pressure as outlined in paragraph 115.

NOTE: Valve section Fig. 152 components should be clearly identified and placed in a specific order as they are disassembled to facilitate reassembly. Do not switch components and use care

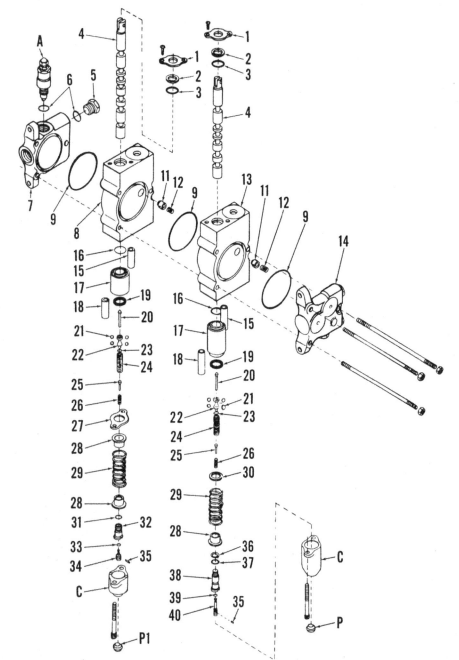

A. Acorn nut/main relief valve	20. Valve poppet
C. End cap	21. Detent balls (3/16 inch)
P. Plug	22. Detent cam
Pl. Plug	23. Steel spacer
1. Retainer plate	24. Spring
2. Wiper rings	25. Spring guide
3. "O" ring	26. Spring
4. Spool	27. Seal plate
5. Plug	28. Spring seat
6. "O" ring	29. Centering spring
7. Housing	30. Washer
8. Housing	31. "O" ring
9. "O" ring	32. Screw
11. Valve poppet	33. "O" ring
12. Poppet spring	34. Screw
13. Housing	35. Screw insert
14. Outlet section	36. Snap ring
15. Sleeve spacer	37. "O" ring
16. "O" ring	38. Cap screw
17. Spool sleeve	39. "O" ring
18. Spacer	40. Adjusting screw
19. "O" ring	

Fig. 152 — Exploded view of Husco auxiliary control valve. Main relief valve (A) is shown exploded in Fig. 151.

when handling to prevent damage to machined surfaces.

To overhaul valve sections, remove retainer plates (1 – Fig. 152), wiper rings (2) and "O" rings (3) from valve spools (4). Remove bolts, spacers (18) and caps (C). Withdraw spools (4) from bores, compress centering spring (29) and remove cap screws (32 or 38). Carefully release compression tool and separate retainers (28), centering spring (29) and plate (27) from end of spool.

Tip spool (4) downward and catch detent parts (20, 22, 25 and 26), then carefully slide detent sleeve (17) from spool while catching detent balls (21).

Clean and inspect all components carefully. Renew components that are worn excessively or otherwise damaged. Renew all seals, lubricate components with clean hydraulic fluid, then reassemble by reversing disassembly procedure. Reinstall spool assemblies in their respective bores.

Assemble valve sections as shown in Fig. 152. Make sure machined surfaces are clean and free of burrs. Renew "O" rings (9) and any components that are worn excessively or otherwise damaged. Tighten 3/8-inch retaining nut to a torque of 33 ft.-lbs. and 5/16-inch nuts to 14 ft.-lbs. of torque.

Reinstall unit on tractor, check and add "Permatran" fluid to hydraulic system as required. Test and adjust systems pressures by following instructions given in paragraph 113.

BOSCH AUXILIARY CONTROL VALVES

126. R&R AND OVERHAUL. Procedure for removal of auxiliary control valve is evident after inspection of unit. Hydraulic lines and linkages should be identified to aid in reassembly. To reinstall valve, reverse removal procedure.

Different types of Bosch auxiliary valves are used and a plate attached to the main body valve will list number. The following procedures apply to Bosch 0521-601-164 and 0521-601-165 valves; however, No. 0521-601-163 is similar and Fig. 153 can be used as a guide.

NOTE: Before beginning overhaul, check for parts availability. Valve components should be clearly identified and placed in order to facilitate proper reassembly. Do not interchange components between valves and use care when handling to prevent damage to machined surfaces.

Clean exterior thoroughly with a suitable solvent. Remove end cap (13 – Fig. 153) then unscrew retainer sleeve (10) from body (14). Withdraw spool assembly (1 through 12, 15 through 19, 23 through 26 and 35) through bottom of casting. Remove wiper seal retainer plate (38), wiper seal (37) and spool seal (36). Remove mounting plate, seals (27) and back-up rings (28) from valve body (14).

Thread a suitable machine screw into retainer plug (33), press plug in and remove retaining ring (34). Withdraw retainer plug (33), washer (32), "O" ring (31) and spring (30). Tip valve body (14) and extract steel ball (29). If valve spool assembly requires disassembly, insert spool back into valve body (14) and thread retainer sleeve (10) back into body. Push in on end plug (8), release retaining ring (9), then remove end plug and internal spring (7). Lift detent sleeve (11) from end of housing (1) and remove detent balls (2). Remove internal detent cam (6) from inside housing (1). Unscrew retainer sleeve (10) from end of body (14) and withdraw spool assembly with housing (1). Compress spring assembly (23 through 25), then unscrew housing (1) from end of spool. Carefully release spring tension, then remove spring (24) and retainers (23 and 25).

Remove housing (1) from retainer sleeve (10). Count number of turns required to unscrew kickout adjustment plug (5), then remove spring (4) and piston (3) from housing (1). Remove pin (16) from spool bore. Carefully apply air pressure to hole (H) in valve spool (35) to force washers (17), seal (18 and 19) and cone piston (15) from bore.

Clean, inspect and renew any components that are worn excessively. Renew all seals. There should not be excessive side clearance between spool (35) and its bore. Lubricate components with clean hydraulic fluid, then assemble unit by reversing disassembly procedure. Test and adjust detent and kickout pressures as outlined in paragraphs 119 and 120.

IPTO/DIFFERENTIAL LOCK CONTROL VALVE

127. R&R AND OVERHAUL. Removal, overhaul and installation of valve is evident after inspection of unit and reference to Fig. 154. All components should be cleaned, inspected and renewed if there is evidence of excessive wear or damage. Unloaded height of spring (8) should be 1.540-inches. Height should be 1.44-inches with a 7-8 lb. load applied. Shims (S) are 0.006-inch thick and should be placed between spool (7) and spring (8) when necessary to correct spring free length.

The following points should be noted when overhauling differential lock portion of valve: Remove cap (17) and "O"

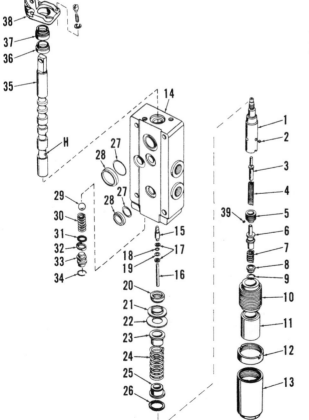

Fig. 153 – Exploded view of Bosch (part No. 0521-601-164 or 0521-601-165) auxiliary control valve.

1. Housing
2. Detent balls (3)
3. Piston
4. Spring
5. Kickout adjustment plug
6. Detent cam
7. Spring
8. End plug
9. Retainer
10. Retainer sleeve
11. Detent sleeve
12. Nut
13. End cap
14. Body
15. Cone piston
16. Thrust pin
17. Steel washers
18. "O" ring
19. Back-up washer
20. Seal cup
21. Support ring
22. Washer
23. Retainer
24. Centering spring
25. Retainer
26. Washer
27. "O" ring
28. Support ring
29. Ball
30. Spring
31. "O" ring
32. Back-up washer
33. Retainer plug
34. Retainer
35. Spool
36. Spool seal
37. Wiper seal
38. Retainer plate

ring (18), then remove shaft (19) and spool (12). To assemble unit, reverse disassembly procedure.

NOTE: Overhaul procedure is basically the same if unit is not equipped with differential lock option.

HYDRAULIC LIFT SYSTEM

The high volume/high pressure side of the dual element hydraulic pump sup-

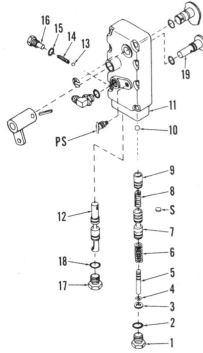

Fig. 154—Exploded view of ipto/differential lock control valve. Refer to Fig. 136 for location of valve assembly.

PS. Pressure switch
S. Shim
1. Plug
2. "O" ring
3. Washer
4. Retainer
5. Lower plunger
6. Spring (large coils)
7. Spool
8. Spring
9. Upper spool
10. Steel ball
11. Housing
12. Spool
13. Detent ball
14. Spring
15. "O" ring
16. Plug
17. Cap
18. "O" ring
19. Shaft

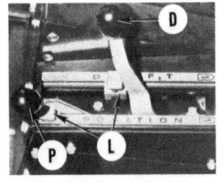

Fig. 155—View of draft (D) and position (P) controls and locators (L).

plies fluid for system operation. Refer to paragraph 110 for hydraulic system tests. Three controls regulate the lift system operation, they are: Draft (depth selection of the implement), position (to raise or lower the implement) and intermix control (adjustable for smooth draft operation). Refer to Fig. 155 for view of draft and position controls and to Fig. 156 for view of intermix control knob.

QUICK CHECKS

128. Attach 1000 lbs. to lift links, warm system fluid temperature to at least 68 degrees F., then raise links to the full transport position. A noticeable

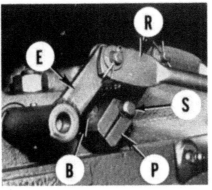

Fig. 156—View of intermix control knob (I). Refer to text for adjustment procedure.

Fig. 157—View of draft and position control linkage.

B. Clamp bolt
E. External arm
P. Position control arm
R. Draft control rod
S. Position control shaft

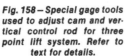

Fig. 158—Special gage tools used to adjust cam and vertical control rod for three point lift system. Refer to text for details.

E. Extension arm
G. Gage tool No. 6204
T. Tool No. 6623

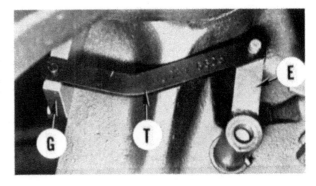

change in sound will indicate proper functioning of unloading valve. Measure position of links using scribe marks across lift cover and lift arm, shut off engine, then move control levers to full down position. Links should not move for five minutes (use scribe marks for reference). If links move, check adjustments as outlined in paragraph 130 or for internal leakage within system.

Move draft control lever (D-Fig. 155) to transport position, run engine at 1000 rpm then slowly move draft control to down position while observing lift linkage. Components should show no signs of excessive binding or looseness throughout travel.

Run engine at 2000 rpm and move depth control to full transport position. Links should raise from full down position to transport position within 2½ seconds, indicating proper oil flow to linkage control valve. Repeat with draft control moved to full down position. Links should fully lower within 2 seconds. Repeat last two tests using position control lever, results should be the same.

ADJUSTMENTS

129. **INTERMIX LINKAGE.** To adjust intermix linkage, first remove lift control valve cover as outlined in paragraph 131. Lower lift arms to full down position, turn intermix control knob (I–Fig. 156) fully clockwise, then turn knob counter-clockwise approximately eight full turns to locate follower arm (F–Fig. 166) at mid-position on intermix cam (C).

Disconnect draft control rod (R–Fig. 157) from external arm (E). Loosen bolt (B) on position control arm (P). Make sure position control link (15–Fig. 169) is connected to intermix cam (C). Install gage tool No. 6204 (G–Fig. 158) to rear surface of lift cover then install special tool No. 6623 (T) to extension arm (E).

Disconnect vertical sensing rod from ball joint (J–Fig. 167) on draft feed back arm. Adjust socket of vertical rod until ball on draft feed back arm can be connected without binding or lifting vertical sensing rod and breakout plunger (12–Fig. 169) must be flush with outer surface of tool No. 6204 (G–Fig. 158).

Remove special tools and tighten clamp bolt (B–Fig. 157). Operate arms (E and P) at the same time in same direction and check linkage for freedom of movement without binding. Moving both arms in opposite directions should result in some free movement.

Reconnect linkage, make sure transport adjusting screw and draft range screws are partially threaded. Refer to paragraph 130 for draft and transport adjustment procedures.

130. DRAFT AND TRANSPORT POSITION ADJUSTMENTS. To adjust draft and transport controls, attach 1000 lbs. of weight to lift links, start and run engine at 1500 rpm then operate lift links up and down several times to purge air from system. Remove plugs from control valve cover, (refer to Fig. 159).

To adjust transport position, move draft control lever completely to the top and move position control lever all the way down. Lift cylinders should fully extend and relief valve should be releasing as indicated by the chattering sound. If relief valve is not releasing pressure, turn upper adjusting screw (T) counterclockwise until pressure relief valve starts releasing. Scribe reference lines across lift cover casting and lift arm hub. Turn upper adjusting screw (T) clockwise until scribe lines separate 1/8-inch or until relief valve just stops blowing.

To adjust draft range, lower lift links completely then move draft control lever 4-11/16-inches back from full down position. Turn adjusting screw (D) clockwise until links just stop raising. When adjustment is correct, weighted links will retain their position.

To adjust position control, run engine at 1500 rpm and place position control lever in full down position. Loosen clamp screw (B–Fig. 157) and rotate position control shaft (S) until lift links just start to raise. Tighten clamp bolt to a torque of 12-18 ft.-lbs.

Fig. 160 — View of lift control valve and valve cover. Refer to Fig. 161 for exploded view of linkage.

CS. Cap screws
E. Elbows
6. Valve cover
11. Standpipe
18. Bellcrank arm
27. Standpipe
29. Valve body
34. Standpipe
35. Spool valve assy.

Fig. 161 — Exploded view of lift control linkage. Refer also to Fig. 160.

C. Clip
S. Spring
2. Transport arm
3. Transport sensing lever
13. Bellcrank spacer
15. Breather shield
21. Draft adjustment bellcrank
22. Link
23. Tie link rod
38. Cap screw
39. Adjusting screw

LIFT CONTROL VALVE AND VALVE COVER

131. REMOVE AND REINSTALL. To remove lift control valve and valve cover, first lower lift links to relieve all hydraulic system pressure. Disconnect hoses from elbow fittings (E–Fig. 160). Remove cover retaining cap screws then remove valve cover.

To reinstall valve cover, renew "O" rings and valve cover gasket and use guide studs to aid in installation. Remainder of procedure is reverse of removal procedure. Perform control adjustments as outlined in paragraphs 129 and 130.

132. OVERHAUL LINKAGE. To overhaul lift control valve linkage, first remove valve cover (6–Fig. 160) as outlined in paragraph 131. Remove bolt screw (38–Fig. 161), "E" locks and washers that retain arms (2 and 21) to valve cover.

Clean, inspect and renew any components that are worn excessively or damaged. Reinstall linkage components by reversing disassembly procedure. Apply "Loctite" hydraulic sealant to cap screw (38), renew all "O" rings and check for freedom of linkage movement.

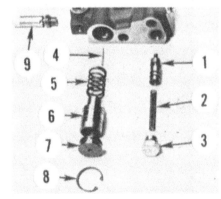

Fig. 163 — Partial exploded view of lift control valve. Refer to Fig. 164 for view of other half of valve.

1. Inlet throttle valve
2. Spring
3. Plug and "O" ring
4. Discharge valve pin
5. Large spring
6. Servo piston
7. Plug
8. Snap ring
9. Safety relief valve

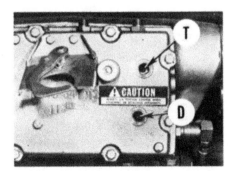

Fig. 159 — Draft (D) and transport (T) position adjustment locations.

Fig. 162 — Pilot orifice (P) should be 0.060 inch. Notice positions of "O" rings.

133. **OVERHAUL LIFT CONTROL VALVE.** Remove valve as outlined in paragraph 131. Remove "O" rings and 0.060-inch diameter pilot orifice (P – Fig. 162). Remove plug (11 – Fig. 164) with "O" ring then remove 0.030 – 0.032-inch diameter dampening orifice (10). Remove main spool (21). Remainder of disassembly procedure is evident after inspection of unit and reference to Figs. 162, 163 and 164.

Clean, inspect and renew any components that are damaged or worn excessively. Renew all "O" rings and reassemble unit by reversing disassembly procedure.

LIFT COVER

134. **REMOVE AND REINSTALL.** To remove lift cover, first drain fuel tank, disconnect fuel supply lines (L – Fig. 165) under tank and disconnect wires to sending unit (8). Remove cover plates (12), then remove four bottom braces securing fuel tank (10) and shield in place. Carefully remove tank and shield.

Disconnect position control rod from arm (P – Fig. 157) and draft control rod (R) from arm (E). Remove control valve cover as outlined in paragraph 131. Remove snap ring (S – Fig. 166) and disconnect position control link (L) from cross shaft cam (C).

Remove standpipe (S – Fig. 167) and disconnect vertical sensing rod (R) from ball joint (J). Disconnect lines from left side of lift cover then remove top two rear cap screws (CS).

Disconnect park brake cable (C – Fig. 168) from lever (L) and remove snap ring (S). Remove bolts securing cable and bracket to lift cover. Raise links slightly and place a bar across rear of lift cover. Disconnect lift rods and cylinder from lift arms. Remove cover retaining cap screws, then carefully remove cover using a suitable hoist. Make sure components do not bind or catch.

To reinstall lift cover, first install guide studs to aid alignment and install new gasket onto top of center housing. Make sure draft sensing arm is adjusted as outlined in paragraph 129. If equipped with differential lock, make sure standpipe (27 – Fig. 169) with new "O" rings (28) is installed properly in center housing.

Carefully lower cover into position making sure that binding does not occur. Remainder of installation is reverse of removal procedure. Adjust park brake as outlined in paragraph 88A and intermix linkage as outlined in paragraph 129.

Fig. 166 — Remove snap ring (S) and disconnect position control link (L) from cross shaft cam (C).

Fig. 167 — Remove standpipe (S) and disconnect sensing rod (R) from ball joint (J). Disconnect lines from top left side of lift cover and remove top two rear cap screws (CS).

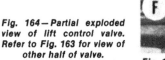

Fig. 164 — Partial exploded view of lift control valve. Refer to Fig. 163 for view of other half of valve.

10. Dampening orifice (small hole)
11. Plug and "O" ring
12. Steel ball
13. Spring
14. Cap and "O" ring
15. Unloading valve spool
16. Spring
17. Steel plug and "O" ring
18. Steel ball
19. Spring
20. Plug and "O" ring
21. Valve spool

Fig. 165 — View of fuel tank and related components.

L. Fuel lines
1. Fuel tank cap
2. Filler neck
3. "O" ring
4. Adapter
5. Sending unit cap
6. Spacer
7. Screw
8. Sending unit
9. Pad
10. Fuel tank
11. Platform
12. Cover plates

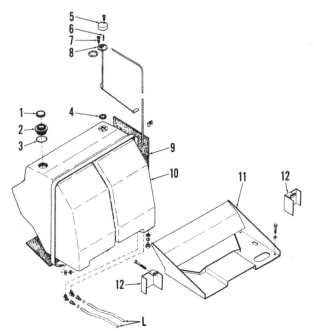

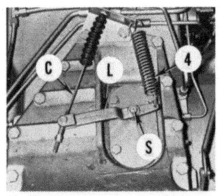

Fig. 168 — Disconnect park brake cable (C) from lever (L) and remove snap ring (S). Remove bolts securing cable and spring bracket to lift cover.

135. **REMOVE AND REINSTALL LIFT ARMS.** If necessary to remove both lift arms first remove lift control valve and valve cover (1 – Fig. 169) as outlined in paragraph 131. Disconnect feed back link from intermix cam (C).

NOTE: If only one lift arm is to be removed, disregard preceding paragraph.

Disconnect lift rods and cylinders, remove cap screw and lock from end of cross shaft (8) and carefully pull lift arm(s) from tractor. To reinstall lift arms, reverse removal procedure.

136. **REMOVE AND REINSTALL CROSS SHAFT, BUSHINGS AND SEALS.** To remove cross shaft (8 – Fig. 169) and bushings (7), first remove lift cover as outlined in paragraph 134. Remove lift arms as outlined in paragraph 135. Remove intermix cam (C) then remove shaft from either side. To install cross shaft, reverse removal procedure.

Apply heat to area surrounding bushings to loosen "Loctite" seal, then remove bushing (7). Apply "Loctite" to outside diameter of new bushings, then install new bushings until they are positioned slightly below chamfer in housing (2).

To renew cross shaft seals, remove both lift arms as outlined in paragraph 135. Remove intermix cam (C – Fig. 169). Remove seals (6) by moving cross shaft to left and right.

Install special tool No. 7072 (T – Fig. 170) over right side of shaft with flared end of tool squarely into bushing chamfer. Lubricate seal then install over shaft (8 – Fig. 169) and into its groove. Push shaft into lift cover while holding special tool. Remove tool and repeat procedure for opposite side. Assemble unit by reversing disassembly procedure.

LIFT CYLINDERS

137. **R&R AND OVERHAUL.** Before disconnecting lift cylinders, raise and block lift arms to avoid damage to internal controls of three point linkage. Disconnect hydraulic line at cylinder

then disconnect cylinder from lift arm. Remove bolt securing lower collar to lower pin. Use a puller bolt, nut, plate and a piece of pipe to pull pin enough to permit collar removal. Bolt hole is slightly offset toward lift cylinder. Pull pin and remove lift cylinder.

To disassemble the cylinder, refer to Fig. 171. Remove hose fitting and work through open port to move retainer ring (5) into deep groove in piston rod (4). Refer to views (A, B and C – Fig. 171A). Piston rod can now be withdrawn. Assemble by reversing disassembly sequence as shown in views (D, E and F).

SENSING SHAFT

138. **R&R AND OVERHAUL.** To remove sensing shaft and linkage, first block lift arms to prevent damage to internal linkage when lift arms are removed. Drain center housing then remove lower links from sensing shaft. Remove lift cylinders following procedure outlined in paragraph 137 and fuel tank as outlined in paragraph 134.

Loosen bolts (B and B1 – Fig. 172), remove washers, square keys and plates (11 and 20). Remove spring (16) and sensing rod (17). Slide sensing shaft (1) toward right side of tractor and push right bushing (23) and thrust washer (24) out of case. Continue withdrawing until guide plate (5) is free of pin (6).

Lift assembly from left end of shaft and remove left bushing (8) and thrust washer (7). Remove sensing arm through top of center housing. Slide sensing shaft (1) out towards left side of housing.

Clean, inspect and renew any components that appear to be damaged or worn excessively. Renew snap rings (9 and 22) and seals (10 and 21). Make sure guide pin (6) is installed at left side of center housing. If pin was removed, use "Loctite" on threads when reinstalling.

To reassemble sensing shaft unit, first install sensing arm (12) and clamp (15) inside center housing.

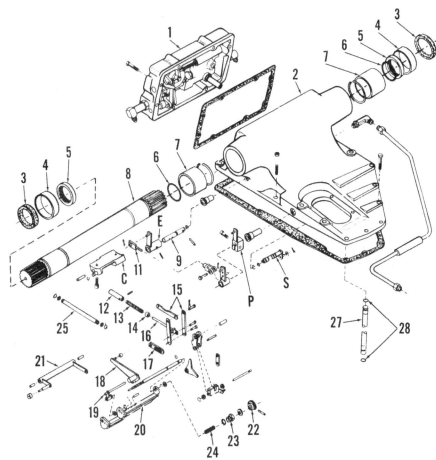

Fig. 170 – Using special tool No. 7072 (T) to install seal (6). Refer also to Fig. 169.

Fig. 169 – Exploded view of typical hydraulic lift cover and related components.

C. Intermix cam	5. Oil seal	14. Spacer	22. Control knob
E. Draft control shaft	6. "O" ring	15. Position feed back link	23. Friction disc
P. Position control arm	7. Sleeve bearing	16. Break out rod	24. Intermix spring
S. Position control shaft	8. Cross shaft	17. Position control link	25. Supply standpipe
1. Control valve cover	9. Draft control shaft	18. Position follower assy.	26. Position input shaft
2. Lift cam assy.	11. Connecting rod	19. Position relay assy.	27. Differential lock
3. Dust seal	12. Break out plunger	20. Intermix cradle	standpipe
4. Sleeve	13. Spring	21. Draft feed back assy.	28. "O" rings

NOTE: Roller end of arm faces guide plate pin (6) and bolt (14) head should be up and towards front of tractor.

Install retaining ring (2) into its groove on sensing shaft then insert shaft through left side of center housing with retaining ring end of shaft towards left side of tractor. Slide washer (3) over left

end of shaft with chamfered side toward snap ring (2).

NOTE: New washer (3) should be 0.012-inch thick and has larger inside diameter than thrust washer (7 and 24).

Attach vertical rod (17) to sensing cam (4). Slide cam over left end of shaft (1) against washer (3), with cam arm facing away from washer (3). Install guide plate (5) over left end of shaft with narrowest side up. Slide sensing shaft (1) in-

to bore of center housing and install guide plate (5) over pin (6). Roller (13) on sensing cam must enter slot in guide plate (5).

Install a 0.156-inch thrust washer (7) on left side of shaft, then install bushing (8) and snap ring (9). Move sensing shaft (1) toward left side of center housing until snap ring (2) is against chamfered washer (3).

Install bushing (23) on right side and temporarily secure with old snap ring

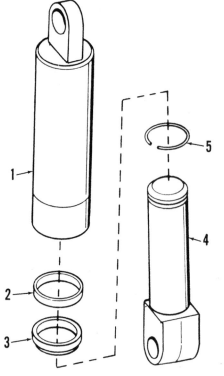

Fig. 171 — Exploded view of hydraulic lift cylinder.

1. Barrel
2. Barrel seal
3. Wiper seal
4. Rod assy.
5. Retainer ring

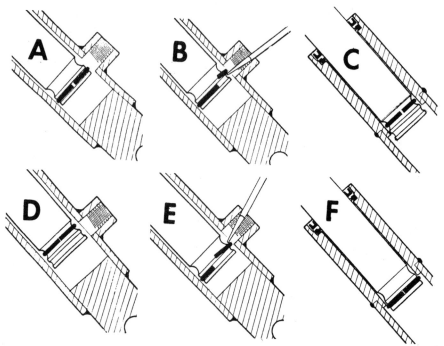

Fig. 171A — Cylinder is disassembled by moving retainer ring to deeper groove as shown in views (A, B and C). Reassemble with retainer ring in deep groove, then move ring to shallow groove at end as shown in views (D, E and F). Retainer ring can be moved from one groove to the other using screwdriver or similar tool through port opening as shown.

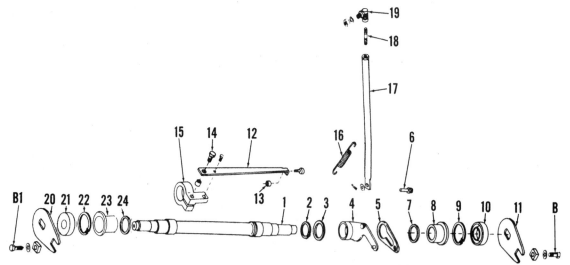

Fig. 172 — Exploded view of typical draft sensing linkage.

1. Sensing shaft
2. Snap ring
3. Variable thickness thrust washer
4. Draft signal cam
5. Guide plate
6. Threaded pin
7. Selective thrust washer
8. Bushing
9. Snap ring
10. Seal
11. Plate
12. Arm
13. Rollers
14. Bolt
15. Clamp arm
16. Spring
17. Sensing rod
18. Rod stud
19. Ball joint link
20. Plate
21. Seal
22. Snap ring
23. Bushing
24. Thrust washer

22). Attach dial indicator as shown in Fig. 173 and check end play of sensing shaft. Remove snap ring (22–Fig. 172) and bushing (23), then install thrust washer (24) (on right side) of sufficient thickness to limit shaft end play to 0.002–0.005-inch. Thrust washers are available in 0.125, 0.156 and 0.188-inch sizes. Chamfer on inner diameter of thrust washer (24) should be toward center housing. Reinstall bushing (23) and snap ring (22).

Coat outer edge of seal (10) with a

Fig. 173 — Check sensing shaft (1) end play using a dial gage as illustrated. Refer to text for details.

suitable sealer, place special tool No. 6620 over end of shaft (1), slide new oil seal over special tool until it is 0.010 inch below housing surface. Repeat procedure with seal (21) on opposite side. Install plates (11 and 20), square keys, washers and retaining cap screws (B and B1). Install spring (16) between vertical control rod (17) and draft signal cam (4).

Remainder of procedure is reverse of disassembly. Adjust linkage as outlined in paragraphs 129 and 130.

MASSEY-FERGUSON

Models ■ MF 2745 ■ MF 2775 ■ MF 2805

Previously contained in I & T Shop Manual No. MF-39

I&T

SHOP MANUAL

SHOP MANUAL

MASSEY-FERGUSON

MODELS

MF2745   MF2775   MF2805

Tractor serial number is stamped on left side of instrument console. Engine serial number is stamped on right front cylinder block valley. Transmission serial number is stamped on right side of transmission housing. Axle serial number is stamped on upper right side of center housing.

INDEX (By Starting Paragraph)

BRAKES
Adjust parking brake 130
Adjustment 126
Brake pistons & discs 129
Fluid & bleeding 127
Overhaul master cylinder 128

CLUTCH
Adjustment 96
Overhaul 99, 100
Tractor split 97

COOLING SYSTEM
Thermostats 88
Water pump 90

ELECTRICAL SYSTEM
Alternator 92
Circuit description 95
Starter 94
Wiring diagram . . . . .Fig. 93

ENGINE
Assembly R&R 20
Cam followers 27, 28
Camshaft 45, 46
Connecting rods & bearings 51
Crankshaft & bearings 52
Cylinder head 21
Flywheel .54
Flywheel housing 55
Oil cooler 60, 61
Oil pan .56

ENGINE CONT.
Oil pump 57, 58
Oil spray jets 62
Pistons, sleeves & rings 48, 49
Piston pins 50
Rear oil seal 53
Rocker arms 29
Timing gear cover 32
Timing gears 33
Valves & seats 23, 24
Valve clearance 30
Valve guides 25
Valve springs 26

FINAL DRIVE
Differential & bevel gear 120
Differential lock valve 146
Rear axle & final drive 123

FRONT SYSTEM
Axle 1
Steering cylinder 17
Steering valve 15
Test & adjustment 8

FUEL SYSTEM
Bleeding .65
Fuel filters 64
Fuel lift pump 82, 83
Injection pump 74, 78
Injector nozzles 66

HYDRAULIC SYSTEM
Auxiliary hydraulic pump 151
Lift system 159
Lubrication 136
Lubrication control valve 157
Main pump 141
Multi-Power control valve 113
Oil cooler relief valve 158
Pto/differential lock valve 145
Remote control valves 147, 155
Test & adjust 137, 149

POWER TAKE-OFF
Clutch assembly 132
Drive gear assembly 134
Driven gear assembly 135
Pto control valve 145

TRANSMISSION (8-speed)
Front cover 103
Shift cover 102
Main transmission overhaul 106
Main transmission R&R 105
Transmission brake 101

TRANSMISSION (Multi-Power)
Front cover 114
Shift control valve 113
Test & adjustment 111
Troubleshooting 110

TURBOCHARGER
Troubleshooting 84
Remove & reinstall 85
Overhaul .86

CONDENSED SERVICE DATA

	Models	
2745	**2775**	**2805**

GENERAL

	2745	2775	2805
Engine Make	——————————— Perkins ———————————		
Engine Model	AV8.540	AV8.640	ATV8.640
Number of Cylinders	——————————— 8 ———————————		
Bore – Inches	4.25	——————— 4.63	
Stroke – Inches	——————— 4.75 ———————		
Displacement – Cu.-In.	539.1	——————— 640	
Compression Ratio	16.5:1	16:1	15:1
Main Bearings, No. of	——————————— 5 ———————————		
Cylinder Sleeves	——————————— Dry ———————————		
Forward Speeds	——————————— 8 or 24 ———————————		
Reverse Speeds	——————————— 6 ———————————		

TUNE-UP

	2745	2775	2805
Firing Order	——————————— 1-8-7-5-4-3-6-2 ———————————		
Valve tappet gap – Cold			
Intake (inch)	0.012	——————— 0.010 ———————	
Exhaust (inch)	0.012	——————— 0.025 ———————	
Valve Face And Seat Angle-			
Intake	45°	——————— See paragraph 24 ———————	
Exhaust	——————————— 45° ———————————		
Bosch Fuel Injection-			
Injection Timing			
Crankshaft degrees	20	22	21
Injectors			
Opening Pressure, psi	3530	3450	2870
Spray Hole Dia., (inch)	0.0118-0.0122	0.0140-0.0144	0.0136-0.0140
C.A.V. Fuel Injection-			
Injection Timing			
Crankshaft degrees	——————————— NA ———————————		
Injectors			
Opening Pressure, psi	2940	——————— 2870 ———————	
Spray Hole Dia., (inch)	0.0112-0.0116	0.0120-0.0124	0.0140-0.0142
Governed Speeds –			
Engine rpm			
Low Idle	——————————— 725-775 ———————————		
High Idle	——————— 2730-2860 ———————		2625-2750
Loaded	——————— 2600 ———————		2500
540 rpm pto			
Low Idle	196-210	——————— NA ———————	
High Idle	740-775	——————— NA ———————	
Loaded	705	——————— NA ———————	
1000 rpm pto			
Low Idle	——————————— 347-371 ———————————		
High Idle	——————— 1306-1368 ———————		1255-1315
Loaded	——————— 1243 ———————		1196
Horsepower at pto Shaft	140	160	190
Battery-			
Volts	——————————— 12 ———————————		
Ground Polarity	——————————— Negative ———————————		
Capacity Amp/hr	——————————— 95 ———————————		

SIZES — CAPACITIES — CLEARANCES	2745	Models 2775	2805
Crankshaft Main Journal—			
Diameter, inch		3.9967-3.9972	
Bearing Clearance, inch	0.004-0.0063	0.0026-0.0049	
Crankshaft Crankpin—			
Diameter, inch		2.9980-2.9985	
Bearing Clearance, inch	0.0018-0.0038	0.0013-0.0045	
Crankshaft End Play, Inch	0.002-0.017	0.003-0.015	
Camshaft Journal Diameters—			
Front, inch	2.3715-2.3725	2.4965-2.4975	
Second, inch	2.2415-2.2425	2.3665-2.3675	
Third, inch	2.2315-2.2325	2.3565-2.3575	
Fourth, inch	2.2215-2.2225	2.3465-2.3475	
Fifth, inch	2.2115-2.2125	2.2115-2.2125	
Camshaft Bearing Clearance—			
Front, inch		0.0023-0.0055	
Remaining, inch	0.0025-0.0053	0.0033-0.0059	
Camshaft End Play, inch		0.004-0.014	
Cooling System—Quarts	42	46	
Crankcase Oil—Quarts			
Including filters	22	24.5	
Transmission, Differential and Hydraulic Lift—			
Gallons		30	

FRONT SYSTEM

AXLE ASSEMBLY

Refer to Fig. 1 for view of both Western style (non-adjustable) axle (17) and rowcrop (wide-adjustable) axle (17A). Hub assemblies are also available in Heavy Duty (28) and Extra Heavy Duty (28A).

NOTE: Front end parts may vary in size and method of attachment according to tractor serial number.

1. **R&R AND OVERHAUL AXLE ASSEMBLY.** Follow procedure in paragraph 17 and remove steering cylinder cover to obtain access to the top of steering arm. Factory locating marks for spline alignment on steering yoke (8) and steering arm (4) should be visible, if not, scribe locating marks to aid in reassembly at this time. Disconnect tie rod ends from spindle arms (20 – Fig. 1), then remove snap ring (9) from top of steering arm (4) and lower assembly from tractor. Remove side panel lower mounting brackets. Unbolt block weight, located in front frame rail on rowcrop models only, and lower from frame.

CAUTION: Block weight is approximately 600 lbs.

Detach removable weights and mount-

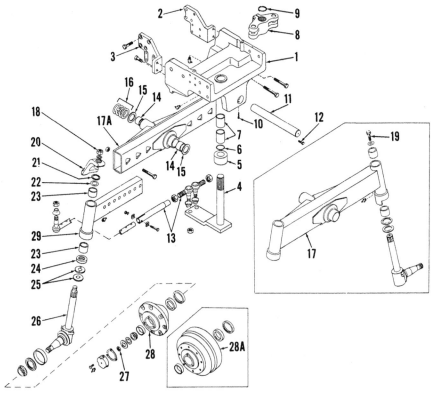

Fig. 1—Exploded view of Western style and rowcrop axle units with associated parts.

1. Front axle support	8. Steering yoke	17. Front axle (Western style)
2. Left engine support	9. Snap ring	17A. Front axle (rowcrop)
3. Right engine support	10. Lock screw	18. Spindle nut
4. Steering arm	11. Axle pivot shaft	19. Spindle bolt
5. Spacer	12. Bolt	20. Spindle arm
6. "O" ring	13. Tie rod assembly	21. Seal
7. Steering arm bushing	14. Pivot pin bushing	22. Shim
	15. Thrust washers	23. Spindle bushing
	16. Shims	
24. Thrust bearing		
25. Thrust washers		
26. Spindle		
27. Hub nut		
28. Heavy duty hub		
28A. Extra heavy duty hub		
29. Axle extension		

ing bracket. Support tractor under oil pan, remove wheel weights and front wheels. With front axle on a floor jack, remove pivot pin lock screw (10) and using a slide hammer, pull out pivot pin (11). Lower axle assembly.

When installing pivot pin bushings (14), bushing split should be to one side and bushing must be 0.020-0.040 inch below surface of axle. After installing bushings, slide pivot pin (11) into housing and check for clearance of 0.004-0.007 inch. Install front axle with thrust washers (15) placed on both sides of axle. Install correct thickness of shims (16) at rear of axle only to obtain a maximum end play of 0.010 inch.

Set axle assembly in place and align holes in axle support and pivot pin for lock screw (10), then tighten lock screw to 85-95 ft.-lbs. torque.

Steering arm bushings (7) require the same specifications as the pivot pin bushings. Align marks on steering arm (4) with steering yoke (8) and install snap ring (9).

2. R&R FRONT WHEEL BEAR-INGS. If wheel bearings were repacked or renewed, upon reassembly tighten hub nut (27 – Fig. 1) to 60 ft.-lbs. while rotating hub. Back off nut one flat. Wheel hub end play should be 0.002-0.005 inch. Install wheel and tighten bolts to 125 ft.-lbs. torque.

3. TIE RODS AND TOE-IN. Automotive type tie rod ends are used; however, the right tie rod on Western style axle has ends integral with connecting tube. On all models, recommended toe-in is 0-9/16 inch and is adjusted by screwing tie rod ends into threaded tube. Adjust tie rods evenly on models equipped with adjustable tie rods on both sides. Be sure to tighten locknut to torque of 125-150 ft.-lbs. after adjustment is correct. Nut attaching tie rod ends to steering arms should be tightened to 125-150 ft.-lbs. torque for ⁵⁄₈-inch nut or 200-230 ft.-lbs. torque if ⁷⁄₈-inch nut is used.

4. SPINDLE AND SPINDLE BUSH-INGS. Two spindle bushings (23 – Fig. 1) are used at each end of axle extension (29) or axle (17). Remove spindle (26) noting location of thrust washers (25) or thrust bearings (24) and shims (22). Heat housing to ease removal of bushings. Before installing new bushings, clean inside of housing and outside of bushings. Coat outside of bushing with "Loctite" Stud N' Bearing Mount or equivalent. Inside diameter of installed bushing should be 1.8750-1.8785 inch. Install spindle and align "X" marks on spindle with spindle arm (20). Tighten bolt (19) or nut (18) to 180-200 ft.-lbs. of torque. Measure gap between spindle arm and

axle. Install shims (22) as needed to obtain a gap of 0.005-0.030 inch.

STEERING SYSTEM

Refer to Fig. 2 and Fig. 3 for views of early and late model steering hydraulics. The hydraulic system on early production tractors receives a constant flow of oil to the steering reservoir (1 – Fig. 2), through a special orificed "Y" fitting connected in the line between the transmission lube control valve and the main system oil cooler. A line on top of reservoir sends excess oil back to transmission sump.

The steering pump (2), driven by the engine timing gears, receives oil from reservoir and pumps it to the steering control valve (3). The high pressure hose has a "T" fitting attached at the pump to supply oil to a resonator (4). The resonator dampens fluctuations (vibra-

tions) in pressure.

The pump is gear type, positive displacement with an integral flow divider and steering system main relief valve. Hose (7) attaches to elbow at top of pump and returns excess oil back to reservoir. When the steering control valve is not in operation (not turning), oil passes through control valve to the power brake master cylinder (5). This provides continuous fluid to the brake system. Return hose (9) allows excess oil from brake master cylinder to return to steering reservoir. Operating the control valve, routes oil to the selected steering cylinder (6).

Late production tractors differ from early production previously described as follows. The power steering pump, resonator, and reservoir are eliminated. Hydraulic fluid is supplied from the main hydraulic pump (1 – Fig. 3) to a priority flow control valve (2), then to steering control valve (3). Priority flow control valve housing contains steering pressure and brake relief valves.

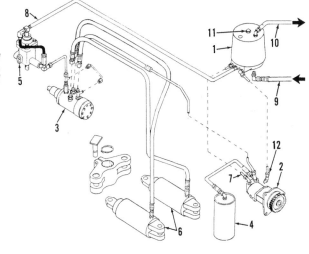

Fig. 2 – Exploded view of early model hydraulic steering system and associated parts.

1. Reservoir
2. Pump
3. Steering control valve
4. Resonator
5. Brake master cylinder
6. Steering cylinder
7. Return hose from pump
8. Return hose from brake master cylinder
9. Supply hose
10. Return hose
11. Acorn nut
12. Pump supply hose

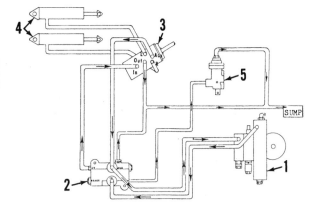

Fig. 3 – Schematic view of late model hydraulic steering system and associated parts.

1. Main hydraulic pump
2. Priority flow control valve
3. Steering control valve
4. Steering cylinder
5. Brake master cylinder

RESERVOIR FILTER

Early Models

5. **REMOVAL.** Refer to Fig. 2 for view of steering reservoir. To remove filter, first loosen and raise or completely remove hood. Drain fluid from reservoir (1) through drain plug or a pump line connection. Disconnect return hose (10); then, remove acorn nut (11) and "O" ring. Lift filter cover from base. Upon reassembly, filter cover aligning key must index between the two notches on base. Remove snap ring, washer and spring. Remove sealing cup and filter.

NOTE: Do not replace filter, manufacturer states sufficient filtration is provided by main hydraulic system filters.

Reassembly is the reverse of disassembly. Tighten acorn nut to 10-15 ft.-lbs. torque. Before starting engine see FILLING AND BLEEDING section paragraph 6.

FILLING AND BLEEDING

All Models

6. Any operation that required draining or repair of hydraulic steering parts, will make it necessary to fill and bleed steering system. Check main hydraulic fluid level before proceeding. Dipstick and fill point are located at right side of differential housing on rear of tractor. Use Massey-Ferguson Permatran Oil or equivalent in hydraulic system.

On early production models with external pump, fill steering reservoir, then loosen pump supply line (12–Fig. 2) at pump to remove any air before starting engine. On late models without external pump, bleeding of lines is not needed as it receives a constant flow of fluid from the main transmission pump. On all models, run engine at 800-1000 rpm and turn steering wheel from lock to lock until steering feels smooth. This will remove air in steering control assembly. Turn off engine and check for possible leaks. Restart engine and run at 1200 rpm, turn steering and hold at stops in both directions until relief valve blows for approximately 10 seconds. Repeat procedure at least 3 times, then check for leaks.

TROUBLESHOOTING

7. If malfunction of the steering system exists, first make an operational check of some other hydraulic unit to eliminate the hydraulic system as the cause of trouble, then proceed as follows:

Hard Or No Steering Movement. Check for improper oil or low fluid level.

Check steering pump, if applicable, for proper operation and renew if faulty. Check steering relief valve; clean and adjust as needed. Check steering control assembly; renew steering control if spool is damaged.

Erratic Wheel Movement. Check for improper oil or low fluid level. Check steering pump, if applicable, for proper operation and renew if faulty. Check steering control assembly; renew steering control if spool is damaged.

TEST AND ADJUSTMENT

Models With External Pump

8. **SYSTEM OPERATING PRESSURE.** The power steering operating pressure is controlled by a relief valve in the power steering pump. Relief pressure should be 2500-2600 psi. To test, install a "T" fitting on high pressure hose at resonator and attach a 5000 psi

gage. With engine running, turn steering wheel to either lock and hold long enough to read relief pressure. Adjustment is made by adding or removing shims in relief valve.

Refer to Fig. 4 for cross-section of relief valve. To adjust, remove relief valve from pump then clean and air dry. Unscrew cartridge (2) from plug (1). Adjustment can be accomplished with varying thickness of shims (3). Upon reassembly, renew all "O" rings (9) and back-up washers (10).

9. **SYSTEM FLOW TEST.** Refer to Fig. 5 for flow meter connections. Check steering pump output by attaching flow meter inlet hose to pressure side (3) of pump and flow meter outlet hose to line going to steering control valve (4). Open flow meter load valve and run engine at 2000 rpm until fluid temperature is 110-150 degrees F., then increase engine speed to 2500 rpm. Slowly close flow meter load valve until a pressure of

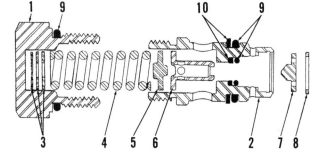

Fig. 4—Cross-section of steering pump relief valve.
1. Plug
2. Cartridge
3. Shims
4. Spring
5. Guide
6. Poppet seat
7. Poppet
8. Snap ring
9. "O" ring
10. Back-up washer

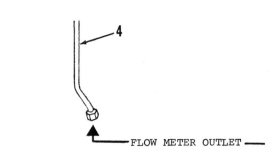

FLOW METER OUTLET

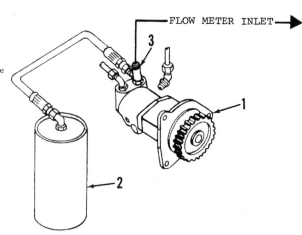

FLOW METER INLET

Fig. 5—View showing flow meter connections.
1. Steering pump
2. Resonator
3. Pressure side of pump
4. Line to steering control valve

2500 psi is obtained and observe rate of flow. If meter reads less than minimum of 4.0 gpm, repair pump.

Models Without External Pump

10. **SYSTEM OPERATING PRESSURE.** The power steering operating pressure is controlled by a relief valve cartridge in the priority flow control valve. Priority flow control valve is located under the right side of the cab on transmission housing. Relief pressure should be 2500-2600 psi. To test, install a "T" fitting on supply line to steering control valve at "SV" port of priority flow control valve (2 – Fig. 3) and attach a 5000 psi gage. With engine running, turn steering wheel to either lock and hold long enough to read relief pressure. Adjustment is made by turning in or out relief plug inside relief valve cartridge.

Refer to Fig. 6 for disassembled view of relief valve cartridge. To adjust, remove relief valve from flow control valve then clean and air dry cartridge. Record positions of "O" rings and back-up washers on cartridge before disassembly. Depress spool (9) and remove snap ring (4) from cartridge housing (1). Remove spool assembly and spring (2). Mark relief plug (5) and spool (9) then count the number of turns when removing relief plug for reference when reassembling spool. Thoroughly clean all parts and inspect for damage or wear. Renew all "O" rings and back-up washers prior to reassembly.

Reassemble spool and make adjustments as follows: Turning relief plug into spool will increase relief pressure while turning out plug decreases pressure. Reinstall spool (9) with spring (2) in cartridge housing and retain with snap ring (4). Install relief valve cartridge in flow control valve and retest.

11. **SYSTEM FLOW TEST.** To test steering circuit flow, first it is necessary to check main pump oil output to flow control valve (2 – Fig. 3) so that pump flow rate and engine rpm relationship can be recorded. Disconnect flow control supply line and attach inlet hose on flow meter to supply line. Connect flow meter outlet to sump at rear of tractor. Run engine to heat oil to 110-120 degrees F. Fully open flow meter load valve and run engine at rpm required to obtain a 6 US gpm reading, then record engine rpm. Increase engine speed to obtain a 10 US gpm reading and record engine rpm.

After testing and recording engine speeds, remove inlet hose of flow meter and reconnect supply line to flow control valve. Disconnect line to steering control valve at "SV" port of flow control valve and attach inlet hose of flow meter to "SV" port. Open flow meter load valve

and start engine. When oil reaches operating temperature (110-120 degrees F.) set engine speed at rpm recorded to obtain 6 US gpm flow to control valve. Record gpm flow reading from "SV" port then adjust flow meter load valve to obtain 1600 psi and record flow reading. Repeat test, setting engine speed for 10 US gpm delivery to flow control valve and record flow readings from "SV" port. A malfunction in flow control valve or cartridge is indicated if flow of 5 US gpm or less is recorded.

EXTERNAL PUMP

All Models So Equipped

12. **REMOVE AND REINSTALL.** Clean pump and surrounding area and drain fluid from reservoir. Disconnect pump pressure and return lines. Cap all openings to prevent dirt from entering pump or lines. Unbolt and remove pump with drive assembly. When reinstalling pump, renew "O" ring between pump drive assembly and engine. Reconnect lines, fill and bleed system as outlined in paragraph 6.

13. **OVERHAUL DRIVE ASSEMBLY.** Remove pump and drive assembly as in paragraph 12. Detach drive gear assembly from pump. Remove snap ring (1 – Fig. 7) and drive gear (2) from shaft (4). Remove inner snap ring (3) and shaft

with bearing (4) from pump drive housing. Reassembly is the reverse of disassembly while noting the following: Renew gasket against pump mounting flange, then coat pump splines with 60% Molybdenum Disulfide Grease. Install pump drive on pump and tighten cap screws to 25-38 ft.-lbs. torque. Backlash on pump drive assembly gear with engine timing gears is 0.002-0.007 inch.

14. **OVERHAUL PUMP.** Remove pump and drive gear assembly as in paragraph 12. Detach gear assembly from pump as in paragraph 13. Remove line connections from pump and mark the pump housings front, center and rear sections for reference upon reassembly. Remove plug (8 – Fig. 8), spring (7), spool (5) and relief valve (1). Remove pump housing bolts, then tap shaft on a wood block to separate front, center and rear housings.

NOTE: Do not pry housings apart, permanent damage to machined surfaces or pump shafts may result.

Refer to Fig. 9 for disassembled view of pump. Remove diaphragm plate (9) from front housing (5), then remove springs (7). Carefully turn front housing over to allow steel balls (8) to be removed. Remove back-up gasket (10), white Teflon protector gasket (11), and molded "V" diaphragm seal (12) from front housing. Remove from front hous-

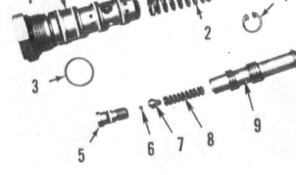

Fig. 6 – View of disassembled flow/relief valve cartridge.

1. Cartridge housing
2. Spring
3. Filter spring
4. Snap ring
5. Relief plug
6. Ball
7. Poppet
8. Spring
9. Spool

Fig. 7 – Disassembled power steering pump drive assembly.

1. Snap ring
2. Drive gear
3. Inner snap ring
4. Shaft and bearing
5. Pump drive housing

ing, drive shaft outer seal. Remove "O" ring (6) from groove in rear housing (1). Thoroughly clean and air dry all metal parts and check for wear or damage that will require replacement of parts. Inspect widths of gear faces and diameter of gear shafts in bearings of front and rear housings. Check for wear on gear pockets of center housing. Wear or scoring of rear housing gear facing surface should not exceed 0.0015 inch. Renew shaft seal, rear housing seal, diaphragm and diaphragm seals.

Lubricate all metal parts with hydraulic oil. Insert diaphragm seal (12) in front housing with open side of "V" facing into groove. Install white Teflon protector gasket (11) and back-up gasket (10) into diaphragm seal. Set the 2 steel balls (8) and springs (7) in place. With bronze side towards rear of pump, set diaphragm plate (9) in place over springs and into diaphragm seal. Use a thin coat of petroleum jelly to seal machined faces of housings. Mate center housing with half moon cavities toward rear, to front housing, aligning marks made previously on outside of housings. Insert gear shafts into center and front housing. Insert seal (6) into rear housing, then set rear housing onto center housing. Align marks made previously on outside of housing. Seat rear housing on center housing by tapping lightly with a plastic mallet.

Install and alternately tighten bolts to 23-26 ft.-lbs. torque. Insert spool (5 – Fig. 8) and spring (7) into pump with slot end of spool (6) toward spring, then install plug (8). Renew "O" rings (2 and 4) and back-up washer (3) on relief valve (1) before installing relief valve into pump. Install line fittings then pour hydraulic fluid into pump and turn pump shaft by hand. Pump should turn freely after a few revolutions. Install pump drive on pump as in paragraph 13. After installation of pump assembly to engine, fill and bleed system as outlined in paragraph 6.

Start engine and run at approximately 1400 rpm without operating the steering for the first 3 minutes. For the next 3 minutes, intermittently operate steering back and forth. For an additional 3 minutes, run engine at 2600 rpm and turn steering both directions lock to lock until relief valve blows. Shut off engine and check for leaks.

STEERING CONTROL VALVE

All Models

15. **R&R AND OVERHAUL.** Remove hood. Identify hoses to simplify reinstallation, then detach hoses from valve. Remove panel from top of instrument cluster. Remove bolt securing steering "U" connection to control valve extension tube. Remove bolts securing steering control mounting bracket, then remove bracket and steering control valve as a unit. Separate mounting bracket from steering control for disassembly. To reinstall, reverse removal procedure. Refer to paragraph 6 to fill and bleed system.

16. Refer to Fig. 10 for exploded view of steering control valve assembly. Plug all ports and clean surface area before disassembly. It is extremely important to prevent any foreign matter from entering control valve. Install a fitting in one of the four ports in housing (11). Clamp fitting in a vise so input shaft (24) is pointing downward.

NOTE: Protect valve parts from scratching, burring or any other damage as operation and sealing of parts depends on their finish and flatness.

Remove bolts and end cover (1) from retainer (8). Thrust washer (2) may adhere to end cover (1) when removed. If thrust washer (2) was not removed with end cover, remove it from top of commutator (3). Remove retainer (8) and rotor seal (7). Remove commutator assembly (3 and 4) and manifold (5). Remove metering element (6), spacer plate (9) and drive link (10), then remove drive link (10) and spacer plate (9) from metering element (6).

Reposition housing (11) so input shaft points upward. Mark upper cover (34) and housing (11) for reference when assembling. Remove cap screws retaining upper cover to housing, then using a smooth upward pull, remove input shaft (24), upper cover (34) and spool (23) as an assembly. Remove spacer (20) from housing. Slide spool (23) and input shaft (24) assembly from upper cover. Remove shims (32) from cavity in upper cover or from face of thrust washer (30) on input shaft assembly. Note number and size of shims for ease of reassembly. Remove seal ring (33) from upper cover. Remove dirt seal (39) from upper cover. If internal shaft seal kit is to be installed, remove snap ring (38), bronze stepped washer (37), seal ring (36) and seal (35) from upper cover. Remove snap ring (31), thrust washer (30), thrust bearing (29), thrust washer (28) and wave washer (27) from input shaft. Drive out torsion bar retaining pin (26), then remove torsion bar (18) and spacer (21). With input shaft pointing up, hold spool and rotate shaft back and forth until drive ring (22) slides free, then turn input shaft until actuator ball (25) disengages from groove in shaft. Separate shaft from spool and remove actuator ball.

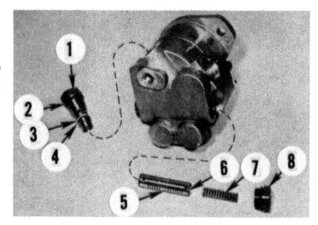

Fig. 8 – View showing rear of pump with relief valve cartridge and flow divider.

1. Relief valve
2. "O" ring
3. Back-up washer
4. "O" ring
5. Spool
6. Slot end of spool
7. Spring
8. Plug

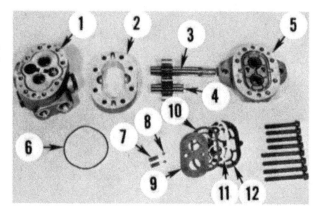

Fig. 9 – Exploded view of external power steering pump.

1. Rear housing
2. Center housing
3. Drive gear
4. Driven gear
5. Front housing
6. "O" ring
7. Spring
8. Steel ball
9. Diaphragm plate
10. Back-up gasket
11. White Teflon gasket
12. Diaphragm seal

Clean all parts in a suitable solvent. Visually inspect parts and renew any showing excessive wear, scoring or other damage. Renew all seals. Inspect the following parts closely.

Check drive link (10) "pin slot" width; slot must not exceed 0.001 inch difference at any point on slot length. Check torsion bar and pin, pin diameter must not exceed 0.001 inch difference at any point. Measure thickness of commutator ring (4) and commutator (3). If commutator ring is 0.0015 inch or more thicker than commutator, renew the matched set. Measure the thickness of metering element assembly (6); the difference in thickness of rotor and stator must not exceed 0.002 inch. Check clearance between rotor and stator of metering element assembly (6). The distance between "lobe" of rotor and nearest point to stator should not exceed 0.006 inch. Renew complete assembly if necessary.

After inspection and renewal of

necessary parts, begin reassembly of control valve. Install wave washer (27), thrust washer (28), thrust bearing (29), thrust washer (30) and snap ring (31) on input shaft. Insert actuator ball (25) into spool and align groove on input shaft with actuator ball. Turn input shaft counterclockwise into spool. Use the mid-section of torsion bar (18) as a gage between end of spool and thrust washer as shown in Fig. 11 to insure the neutral position of the ball on ramp. Place the assembly in a vertical position with end of input shaft resting on bench. Insert drive ring (22 – Fig. 10) into spool until drive ring is fully engaged on input shaft spline.

Remove torsion bar used as gage. Install spacer (21) in spool and on top of drive ring (22). Install torsion bar into spool. Align cross-holes in torsion bar and input shaft, then drive torsion bar retaining pin (26) through shaft and into torsion bar. Pin must be pressed into shaft until end of pin is about 1/32-inch

below surface of shaft. Place spacer (20) over spool and carefully install spool assembly into housing.

Position original shims (32) on thrust washer (30) if the original input shaft and cover are to be used. Renew seal (35), seal ring (36), bronze stepped washer (37) and install snap ring (38), but do not install outer dirt seal (39) at this time. Lubricate new upper cover seal (33), place seal in upper cover (34) and install upper cover assembly on housing (11). Align the reference marks on outside of upper cover and housing and install cap screws finger tight. Place a hose clamp around mating area of upper cover and housing and tighten to align parts. Tighten cap screws to 18-20 ft.-lbs. torque.

NOTE: If either input shaft or upper cover or both have been renewed, the following procedure for shimming must be used.

With upper cover installed using original shims as previously outlined, invert unit in a vise so input shaft is pointing downward and prevent shaft from rotating. Insert drive link (10) splines in spool and turn drive link until end of spool is flush with end of housing. Remove drive link and line up slot in drive link with torsion bar pin. Install drive link back into spool with torsion bar pin engaged in drive link slot. Check relationship of spool end to housing end. Spool end must be within 0.0025 inch of flush in either direction. If not within 0.0025 inch of flush, remove cover and add or remove shims (32) as necessary. Shims are available in 0.0028, 0.005, 0.010 and 0.030 inch sizes.

With drive link installed, place spacer plate (9) on housing with plain side up. Install metering element (6) over drive link and onto spacer plate. Install

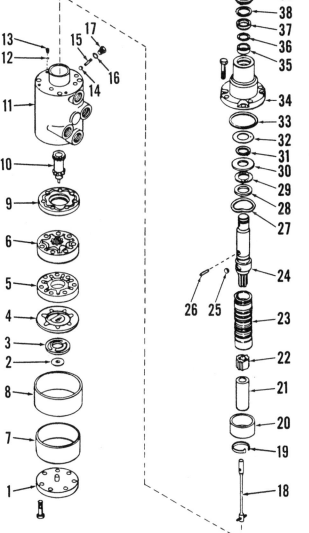

Fig. 10 — Exploded view of steering control valve assembly.

1. End plate
2. Thrust washer
3. Commutator
4. Commutator ring
5. Manifold
6. Metering element
7. Rotor seal
8. Retainer
9. Spacer plate
10. Drive link
11. Housing
12. "O" ring
13. Plug
14. Recirculating ball
15. Roll pin
16. "O" ring
17. Plug
18. Torsion bar
19. Preload spring
20. Spacer
21. Input shaft spacer
22. Drive ring
23. Spool
24. Input shaft
25. Actuator ball
26. Pin
27. Wave washer
28. Thrust washer
29. Thrust bearing
30. Thrust washer
31. Snap ring
32. Shim
33. Upper cover seal
34. Upper cover
35. Seal
36. Seal ring
37. Bronze stepped washer
38. Snap ring
39. Dirt seal

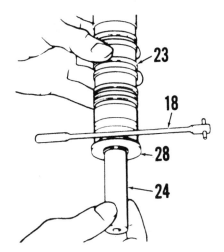

Fig. 11 — With spool (23) assembled on shaft (24), use torsion bar (18) as shown between thrust washer (28) and end of spool to establish neutral position.

manifold (5) with circular slotted side up and align cap screw holes with metering element, spacer and housing. Install commutator ring (4) with single groove side toward manifold, then set commutator (3) over drive link end with counterbore for thrust washer (2) facing out. Make certain drive link end is engaged in elongated hole in commutator.

Install rotor seal (7) and retainer (8). Place thrust washer (2) on pin of end cover (1) using a small amount of grease to hold washer in position. Install end cover making certain pin enters center hole in commutator. Align holes and install cap screws. Alternately and progressively tighten cap screws to a torque of 15-19 ft.-lbs. while turning input shaft to check assembly for binding. Once assembled, turn steering control valve over in vise and install new dirt seal (39) over input shaft.

STEERING CYLINDERS

All Models

17. **REMOVE AND REINSTALL.** Remove grill and side panels. Slide oil cooler and air conditioning condenser to left side out of way. Unbolt air conditioning dehydrator from mounting bracket. Disconnect hoses from steering cylinders. Lift up steering cylinder cover and remove from right side of tractor. Remove cylinder pin retaining bolts, then remove pins from cylinders. Raise front of tractor until wheels clear ground. Turn front axle to free piston rod-end of cylinder, then remove cylinder.

Installation is the reverse of removal. Make certain grease fittings are accessible when installed. Refer to paragraph 6 to fill and bleed system.

18. **OVERHAUL.** Clean and dry outside of cylinder before disassembly. Refer to Fig. 12 for exploded view of steering cylinder. Pull piston rod (1)

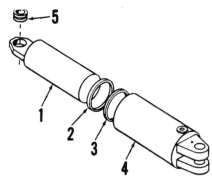

Fig. 12—Exploded view of steering cylinder.
1. Piston rod
2. Scraper ring seal
3. Oil seal
4. Cylinder barrel
5. Bushing

from cylinder barrel (4). Remove inner barrel groove. Check parts for wear and damage. Renew scraper ring seal and oil seal before reassembly. Lubricate all parts with clean hydraulic fluid for reassembly.

Install oil seal (3) in barrel groove, with "O" ring side of seal toward the piston rod end of barrel. Slide scraper ring seal (2) onto piston rod so bronze lip side faces rod eye, then position seal approximately 2 inches from eye end of piston rod. Install piston rod in barrel until scraper ring seal contacts lip on barrel. Tap scraper ring seal into counterbore seat in barrel. If bushing (5) is to be renewed, installed bushing should be 0.010 inch below flush and have an inside diameter of 1.1255-1.1265 inch.

ENGINE AND COMPONENTS

Model 2745 is equipped with a Perkins AV8.540 naturally aspirated, direct injection, V-8 diesel engine. Cylinder bore diameter of 4.25 inches and a stroke of 4.75 inches provide 539.1 cubic inch displacement.

Models 2775 (naturally aspirated) and 2805 (turbocharged) are equipped with AV8.640 or ATV8.640, direct injection, V-8 diesel engines respectively. Cylinder bore of 4.63 inches and a stroke of 4.75 inches provides 640 cubic inch displacement for both models.

FRONT END/ENGINE SPLIT

All Models

19. Remove air intake cap, muffler, hood and side panels. Disconnect ether starting aid tube from air inlet tube and disconnect electrical wires from air restriction indicator switch. Disconnect ground wire from air inlet tube. Remove air intake tube and braces, muffler support and exhaust pipe on models with left side muffler. Disconnect radiator braces from engine. Disconnect windshield washer hose at cab and pull hose free so it can be removed with front end assembly.

Remove oil line clamp in front of support panel. Disconnect power steering hoses that go to steering cylinders at steering control valve. Disconnect oil cooler lines at support panel. Disconnect battery cables from batteries, then disconnect wiring from engine oil pressure switch, alternator, coolant temperature indicator, coolant level indicator, fuel shut-off solenoid, starter motor and auxiliary relay located under a cover behind alternator. Disconnect main wiring harness plug at cab. Move

alternator belt shield down from adjusting bracket, then loosen alternator belts and remove them from alternator. Drain coolant and disconnect coolant hoses between engine and radiator from engine. Remove coolant expansion bottle. Unbolt fan and lay in fan shroud. Disconnect air conditioning self-sealing hoses at cab. Unbolt air conditioning compressor and lay it on top of radiator.

NOTE: If engine is to be separated from transmission, it is necessary to remove clutch release fork so a floor jack can be placed under front of transmission housing to support tractor when front end is split from engine.

Disconnect clutch release cable from release arm on side of transmission. On models with Multi-Power transmission, disconnect shift cable with mounting bracket and links from shift valve. Disconnect linkage rod from clutch shaft arm to shift valve. Remove inspection cover on bottom of transmission. Cut lock wires on release fork lock bolts, then remove lock bolts. Pull release fork shafts out of each side of transmission housing. If necessary, remove Multi-Power shift valve to pull out right-side release fork shaft. Remove release fork through inspection hole in bottom of transmission.

Place a floor jack under transmission and another under front end assembly. Insert wooden blocks between front axle and axle support to prevent tipping. Remove eight cap screws that secure front end to engine and move front end assembly away from engine. Be sure to guide power steering hoses through grommets in front support panel when rolling front end forward.

Installation is the reverse of removal procedure noting the following: Cap screws should be tightened to the following torques: ⅝-inch—130-200 ft.-lbs.; ¾-inch—240-350 ft.-lbs.; 1 inch—440-580 ft.-lbs. Make sure release bearing lubrication hose is routed to rear of release shaft when installing clutch release fork. Tighten release fork lock bolts to 130-200 ft.-lbs. torque and secure with lock wire. Adjust clutch and Multi-Power shift valve as outlined in paragraph 96.

R&R ENGINE

All Models

20. To remove the engine and clutch as a unit, first separate the front end assembly as outlined in paragraph 19. If equipped with external power steering pump, drain power steering reservoir and remove all power steering fluid lines from pump, resonator and reservoir.

Remove reservoir and mounting bracket.

Remove coolant lines from support panel to water pump, from support panel to expansion bottle and from expansion bottle to water pump. Disconnect oil lines from fittings in support panel and remove windshield hose from "T" fitting behind support panel. Disconnect tachometer cable, fuel shut-off cable, throttle cable and excess fuel cable from engine, then pull cables free of support panel or lower heat shield. With all cables and oil lines removed from support panel, unbolt and lift out support panel with mounting brackets. On models with right side muffler, remove lower heat shield.

Attach rear engine lifting hooks to cylinder head. On models with right side muffler, remove exhaust heat shields and exhaust pipe bracket. Close fuel tank shut-off valves and disconnect fuel return line behind battery tray and fuel inlet line from sedimenter. Remove necessary fuel line retaining clamps. Remove fuel line and securing clamp above left cylinder head, then install engine lift hooks onto front of cylinder heads. Remove pto top link support (Fig. 13) at rear of tractor, to release tension on pto shaft. Support engine as shown in Fig. 14 then remove starter for access to cap screw inside flywheel housing. Remove all cap screws retaining engine to transmission. Install two 7-inch guide studs to aid alignment during removal and installation.

NOTE: Floor panel inside cab may require removal to gain access to upper cap screws.

Check all wires and tubes to be sure they are disconnected and positioned so they will not be damaged. Separate engine from transmission.

When reinstalling engine observe the following: Lube input shaft splines and clutch release bearing. Make sure clutch release bearing is positioned with lube hose fitting at top. Carefully move engine back against transmission housing while aligning splines of clutch with transmission input shaft. Engine must be flush against transmission before tightening any cap screws. **DO NOT** force together. Tighten cap screws as follows: ½-inch — 63-95 ft.-lbs.; ⅝-inch — 130-300 ft.-lbs.; ¾-inch — 240-350 ft.-lbs.; and ⅞-inch — 300-400 ft.-lbs. torque. Remainder of procedure for joining engine to transmission housing is reverse of disassembly.

CYLINDER HEADS

All Models

21. To remove cylinder heads, first

remove air intake cap, muffler, hood and side panels. Disconnect electrical wires from air restriction switch and ether starting aid tube from air inlet tube. Remove air intake tube with braces and disconnect radiator braces from engine. On models with right side muffler, remove exhaust pipe bracket and heat shield.

Drain power steering oil reservoir on models with external power steering pump, then disconnect oil lines and remove reservoir with mounting bracket. Drain engine coolant, remove thermostat housings, expansion bottle and coolant line from support panel to water pump. Remove intake crossover pipes, high pressure fuel injection lines, intake manifolds, injector fuel return manifolds, injector fuel return lines, exhaust manifold shields, exhaust manifolds and injectors.

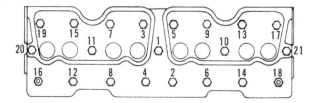

NOTE: Be sure to remove old injector copper sealing washers from heads.

Remove lower heat shield struts and cap screws securing support panel brackets to rear of cylinder heads. Raise and secure support panel. Remove breather tube clamp bolt from rear of right cylinder head and fuel line clamp bolt from rear of left cylinder head. Remove valve covers, rocker arm shaft assemblies and push rods. On MF 2745 models disconnect rocker shaft oil feed pipes from cylinder head. Remove cylinder head bolts in the reverse order of tightening sequence shown in Fig. 15, then remove cylinder heads.

Cylinder heads can be resurfaced providing injector nozzle protrusion does not exceed 0.143-inch as shown in Fig. 16. Original thickness of cylinder head is 3.985-4.015 inch and normally surfac-

Fig. 13 — Remove pto top link support to release tension on pto shaft.

Fig. 14 — Support engine evenly to prevent binding when removing or replacing engine.

Fig. 15 — Tighten cylinder head bolts using the sequence shown.

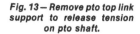

ing of less than 0.015 inch will be possible.

NOTE: After resurfacing cylinder head, it will be necessary to renew valve seat inserts and reseat valves as described in paragraph 23 for MF 2745 models and paragraph 24 for MF 2775 and MF 2805 models.

Injector nozzle holes must be radiused after resurfacing. Manufacturer recommends that a warm water pressure test at 30 psi should follow machining of cylinder head.

To install heads, reverse removal procedure and perform the following additional steps: Thoroughly clean all mating surfaces. Install guide studs into block and tighten to 30-35 ft.-lbs. torque. Apply a light coat of non-hardening sealer to both sides of head gaskets and position head gaskets on cylinder block. If sealing rings are used in the upper corners of head gaskets, be sure rubber faces of sealing rings are free of sealing compound when installed. Install cylinder heads and head bolts, tighten head bolts in 15 ft.-lbs. increments in the sequence shown on Fig. 15, until reaching a torque of 120-125 ft.-lbs. on MF 2745 models and 150-155 ft.-lbs. on MF 2775 and MF 2805 models. Tighten rocker arm shaft assemblies on all models to 34-36 ft.-lbs. torque. Adjust

valve lash by following procedure outlined in paragraph 30. Use new sealing rings on injectors and tighten injector retainers to a torque of 10-12 ft.-lbs. on MF 2745 models and 28-30 ft.-lbs. on MF 2775 and MF 2805 models.

Retorque cylinder heads "hot" after engine reaches normal operating temperature and again retorque "hot" after 10 hours of operation. Readjust valve clearance as outlined in paragraph 30 after retorquing heads.

INJECTOR SLEEVES

All Models So Equipped

22. **REMOVE AND REINSTALL.** To replace injector sleeves, it will be necessary to use MF special tools #2807 and #2808. Tap old injector sleeve with a ⅞-inch x 14 UNF thread tap then use MF special tool #2808 to remove injector sleeve. Lubricate outer "O" ring seal with petroleum jelly and install on new injector sleeve. Using MF special tool #2807 pull injector sleeve into cylinder head.

VALVES AND SEATS

MF 2745 Models

23. Exhaust valves seat on renewable

valve seat inserts while intake valves seat directly in cylinder head. Valve face and seat angles are 45 degrees for all valves. Seat runout relative to valve guide bore should not exceed 0.002-inch.

Protrusion of valve above cylinder head can be checked using a straight edge and feeler gage as shown in Fig. 17. Desired protrusion is 0.042-0.053 inch for intake and 0.0485-0.0585 inch for exhaust. Minimum allowable protrusion is 0.001 inch for intake and 0.0085 inch for exhaust. Valves and/or inserts should be renewed when specifications are exceeded.

Exhaust valve seats are held in place by rolling metal of cylinder head face over a 0.015-0.020 inch x 45 degree chamfer on insert as shown in Fig. 18. Seats can be renewed and head rerolled only once. A letter "R" stamped next to valve seat indicates seat has been renewed and head rerolled.

New inserts should be installed if cylinder head is resurfaced. Valve seat insert height should be reduced before installing in resurfaced heads to allow for decreased depth of seat bore in head. Cut valve face side of insert to reduce height. Outer circumference of insert must be rechamfered at (C) after cutting insert, to allow rolling of cylinder head. Bore diameter in head for valve seat insert must not exceed 1.770 inch. Always stamp a letter "R" next to renewed seat inserts.

NOTE: Intake valve seat inserts are cataloged for field installation if desired.

MF 2775 And MF 2805 Models

24. Intake and exhaust valves seat on renewable seat inserts. Valve face and

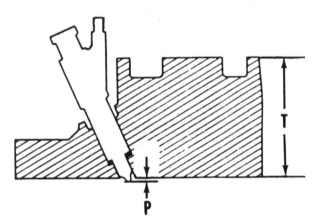

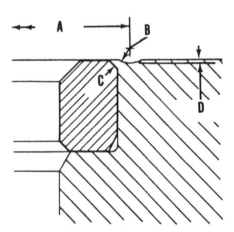

Fig. 16 — Cylinder head may be resurfaced up to 0.015 inch providing overall thickness (T) is not less than 3.970 inch and injector nozzle protrusion (P) does not exceed 0.143 inch.

Fig. 17 — Use a straightedge and feeler gage to check valve head protrusion. Refer to text.

Fig. 18 — Cross-section of exhaust valve seat insert in cylinder head used on MF 2745 models. When installing insert use the following dimensions:

A. Roller circumference
(1.834 – 1.854 inch)
B. Radius of roll
(0.125 inch)

C. Insert chamfer
(0.015 – 0.020 inch x 45°)
D. Rolled metal depth
(0.004 – 0.005 inch)

seat angle may be either 30 or 45 degrees for intake and 45 degrees for exhaust. Seat runout relative to valve guide bore should not exceed 0.002 inch.

Depth of valve below cylinder head can be checked using a straightedge and feeler gage as shown in Fig. 19. Desired clearance is 0.051-0.067 inch for intake and 0.041-0.057 inch for exhaust. Maximum allowable depth is 0.102 inch for intake and 0.097 inch for exhaust. Valves and/or inserts should be installed or renewed when specifications are exceeded.

Inserts are pressed fit and should be replaced as follows: Drive out old valve seat insert then clean and inspect cylinder head bore. Cylinder head bore for valve seat insert (A – Fig. 20) must not exceed 2.1795 inch for intake and 1.9095-inch for exhaust. Bore depth (B) for all inserts is 0.405-0.410 inch. To insure proper bottoming of insert in bore, maximum radius of counterbore (C) is 0.015 inch. Inserts (D) are installed with chamfered end toward cylinder head.

New inserts should be installed if cylinder head is resurfaced. Valve seat insert height should be reduced the same amount cylinder head was resurfaced

before installing in head, to allow for decreased depth of cylinder head bore. Cut valve face side of insert to reduce height.

VALVE GUIDES

All Models

25. Intake and exhaust valve guides are interchangeable. Intake and exhaust valve stem to guide wear limit with new valve is 0.0055 inch. Press new guide into head until distance between guide and cylinder head spring seat (Fig. 21) is 0.783-0.800 inch for MF 2745 models and 0.516-0.536 inch for MF 2775 and MF 2805 models.

The following specifications are the same for all models and both intake and exhaust valve guides. Inside diameter of new guides should be 0.375-0.376 inch while desired diametral clearance with valve stem is 0.0015-0.0035 inch. Outside diameter of new guides should be 0.626-0.6265 inch with an interference fit in cylinder head of 0.0003-0.0018 inch. Seal diameter of guides should be 0.555-0.569 inch. Be sure to resurface valve seats after new guides are installed.

VALVE SPRINGS

All Models

26. Springs, retainers, locks and stem seals are interchangeable for intake and exhaust valves. A single spring is used on MF 2745 models and dual springs are used on MF 2775 and MF 2805 models. Renew springs if they are distorted, heat discolored or fail to meet the test specifications which follow. Renew other parts if they show signs of wear or damage. Close coils of all springs should be toward cylinder head.

Model MF 2745
  SINGLE SPRING
    Test length (closed)....1.833 inches
      Lbs. test ....68.8-76.0 lbs.
    Test length (open).....1.362 inches
      Lbs. test ....160.4-177.4 lbs.

Models MF 2775 and MF 2805
  INNER SPRING
    Test length (closed)....1.273 inches
      Lbs. test ....31.0-34.4 lbs.
    Test length (open).....0.924 inches
      Lbs. test ....78.1 lbs.
  OUTER SPRING
    Test length (closed)....1.495 inches
      Lbs. test ....61.3-67.7 lbs.
    Test length (open).....1.025 inches
      Lbs. test ....156 lbs.

CAM FOLLOWERS

MF 2745 Models

27. Mushroom type cam followers (tappets) are contained in removable guide blocks (B – Fig. 22). The guide block (containing the two cam followers for one cylinder) can be removed without major engine disassembly. Remove auxiliary drive housing as outlined in paragraph 43, and parts necessary to remove block valley cover. Remove block valley cover, valve covers, rocker arm shaft assemblies, and push rods. Mark and remove guide blocks with cam followers.

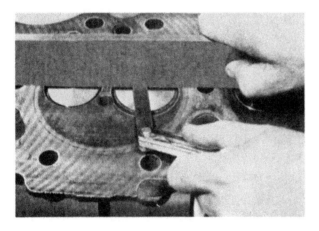

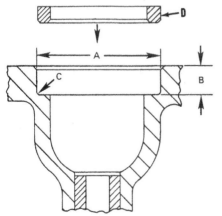

Fig. 19 – On MF2775 and MF2805 models, use straightedge and feeler gage to check valve head depth as shown.

Fig. 21 – Measure valve guide height as shown. Refer to text.

Fig. 20 – Cross-section of valve seat insert (D) and valve seat insert bore in cylinder head used on MF 2775 and MF 2805 models. Refer to text for specifications.

The 0.7475-0.7485 inch diameter tappets should have 0.006-0.0085 inch clearance in tappet bores. Installation is the reverse of removal. Refer to paragraph 43 to reinstall auxiliary drive housing and retime engine.

MF 2775 And MF 2805 Models

28. The mushroom type cam followers (tappets) operate in machined bores in the cylinder block and can be renewed after removing camshaft as outlined in paragraph 46. The 0.7475-0.7485 inch diameter tappets should have 0.0015-0.00375 inch diametral clearance in crankcase bores.

ROCKER ARMS

All Models

29. Rocker arm covers can be removed without any additional disassembly of engine. Components of rocker shaft units are interchangeable in all positions. Refer to Fig. 23 for exploded view of rocker arm assembly. Two different diameter rocker shafts (1) have been used, 0.7485-0.7495 inch or 0.8722-0.8734 inch. Shaft diameter should be checked before renewing parts.

The cast iron rocker arms (7) are equipped with renewable bushings (10). For small diameter rocker arm (B – Fig. 24), inside diameter of fitted bushing should be 0.7515-0.7537 inch with a 0.0020-0.0052 inch diametral clearance to shaft. If large diameter rocker arm (A) is used, inside diameter of fitted bushing should be 0.8750-0.8772 inch with a 0.0016-0.0050 inch diametral clearance to shaft. Check size of rocker shaft, then install new bushing with split and oil hole positioned as shown in Fig. 24.

When reassembling rocker shaft units, make certain boss side of bracket (8 – Fig. 23) points to right and set screw

hole (2) in rocker shaft (1) is towards left shaft end when facing exhaust side of cylinder head. Locating set screw (3) must enter large hole (2) in rocker shaft (1). Oil restrictor pin (13) should be installed in rocker arm, as required, until flush with top of rocker arm. Adjust tappet gap as outlined in paragraph 30.

VALVE CLEARANCE

All Models

30. The recommended valve clearance (tappet gap) when engine is cold is 0.012 inch for intake and exhaust valves on MF 2745 models and 0.010 inch for intake and 0.025 inch for exhaust valves on MF 2775 and MF 2805 models. A "hot" valve clearance of 0.010 inch is recommended for intake and exhaust valves on MF 2745 models only.

On all models, adjustment is made by turning crankshaft clockwise (facing crankshaft pulley) until intake and exhaust valves are rocking on a cylinder (intake valve just opening and exhaust valve just closing). Adjust both valves on corresponding cylinder as shown in Fig. 25. If valves are rocking on number four cylinder, adjust valves on number one cylinder and vice versa, until completed.

VALVE TIMING

All Models

31. To check valve timing, first remove valve covers on number one and number four cylinders. Rotate crank-

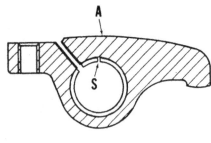
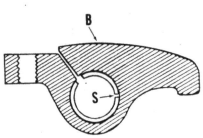

Fig. 24 – Install bushing in large diameter rocker arm (A) and small diameter rocker arm (B) with bushing split (S) and oil hole positioned as shown. Refer to text for bushing and shaft specifications.

Fig. 22 – Cam follower guide blocks (B) are attached to cylinder block with cap screws and are removable in pairs as shown.

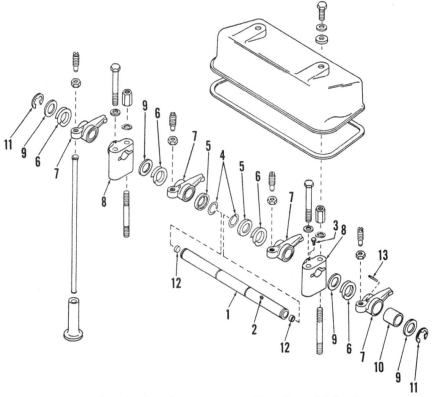

Fig. 23 – Exploded view of rocker arm assembly and associated parts.

1. Rocker shaft	4. Wire ring	7. Rocker arm	10. Bushing
2. Hole for setscrew	5. Collar	8. Bracket	11. "E" Ring
3. Setscrew	6. Spring	9. Washer	12. Plug

shaft until number four cylinder intake valve is completely open, then adjust intake valve gap on number one cylinder to 0.045 inch on MF 2745 models and 0.046 inch on MF 2775 and MF 2805 models. Rotate crankshaft until number four cylinder exhaust valve is completely open, then adjust exhaust valve gap on number one cylinder to 0.038 inch on MF 2745 models and 0.049 inch on MF 2775 and MF 2805 models. On all models, rotate crankshaft clockwise until intake and exhaust valves of number one cylinder are rocking (intake just opening and exhaust just closing) and a 0.015 inch feeler gage will just fit in both valves.

TDC mark on crankshaft and pointer on timing cover should be within plus or minus 2½ degrees. If marks are not within specifications then timing gears are at least one tooth out of time. Refer to paragraph 33 for information on timing gears.

TIMING GEAR COVER

All Models

32. The timing gear cover has an upper and lower section. To remove complete cover, first separate front end from engine as outlined in paragraph 19. Remove drive belts, muffler support, air conditioning compressor, muffler mounts, fan drive assembly, thermostat housings, right side cylinder block coolant adapter, water pump and fan belt idler pulley. Remove oil pan as outlined in paragraph 56. Disconnect tachometer drive cable from angle drive on upper timing gear cover.

NOTE: Crankshaft pulley differs in attachment on early and late production tractors. Removal procedure for each is as follows:

On early production models (Fig. 26), remove crankshaft pulley cap screw (1), washer (2) and clamping ring halves (3), then remove pulley (P).

On late production models (Fig. 27), remove three cap screws (1) and thrust

block (2) with "O" ring (3). If pulley does not loosen, place a block of wood against inner hub of pulley (Fig. 28) and use a sharp blow to separate outer taper ring (4 – Fig. 27) and inner taper ring (5). Remove taper rings (4 and 5), spacer (6), pulley and key (7).

Remove oil pan deflector shield then unbolt and remove upper and lower timing gear covers. Renew crankshaft oil seal if worn excessively. Press new seal in front of timing gear cover until seal is 0.290-0.300 inch below flush of surface.

To install timing gear cover, first install lower portion of timing gear cover and oil pan deflector shield. Tighten cap screws to a torque of 26-30 ft.-lbs. and retaining nut to 35-40 ft.-lbs. Install upper cover making sure tachometer drive is properly positioned. Do not tighten cap screws until water pump is installed. Install water pump and tighten cap screws securing timing gear cover top half to cylinder block and auxiliary drive housing to 12-15 ft.-lbs. of torque. Tighten bolts securing timing gear cover top half to cylinder block to 21-24 ft.-lbs. of torque. Install crankshaft pulley.

On early production models (Fig. 26), tighten pulley retaining cap screw (1) to 275-300 ft.-lbs. of torque.

On late production models (Fig. 27), tighten three pulley retaining cap screws to 80 ft.-lbs. of torque.

Remainder of procedure is reverse of disassembly.

TIMING GEARS

All Models

33. Fig. 29 shows a view of timing gear train with cover removed. Engine is properly timed when number one piston is at TDC on compression stroke and timing marks (TM) are aligned as shown.

NOTE: Because of the odd number of teeth on idler gear (4), timing marks will not align each time number one piston is on compression stroke. Failure of marks to align does not necessarily indicate engine is mistimed.

Fuel injection pump drive gear (2) is properly timed when number one piston is at TDC on compression stroke and pump is located at correct timing position. On models with C.A.V. pump, pointer in injection pump drive housing should align with mark on injection pump drive hub as shown at (T – Fig. 30). On models with Bosch pump, groove on governor spring spindle should line up with "O" mark on timing circlip as shown at (T – Fig. 30A).

Timing gear backlash should be 0.006-0.009 inch between the following gears: Injection pump drive gear (2 – Fig. 29) and camshaft/injection pump idler gear (8); camshaft gear (3) and crankshaft/camshaft idler gear (4); crankshaft/camshaft idler gear (4) and crankshaft gear (5). Backlash should be 0.004-0.017 inch between camshaft/injection pump idler gear (8) and camshaft gear (3).

Oil pump idler gear (6) and crankshaft gear (5) backlash should be 0.009-0.012 inch. Oil pump gear (7) and oil pump idler gear (6) should have a backlash of 0.012-0.015 inch. Minimum backlash between all gears is 0.003 inch. Refer to paragraphs 57 or 58 for information on oil pump and related parts.

Check backlash by pushing rearward on gears being measured to prevent false readings because of shaft end play. If backlash is not within limits renew

Fig. 26 – Early production models use a single cap screw (1) to retain crankshaft pulley. Cap screw torque is 275-300 ft.-lbs.

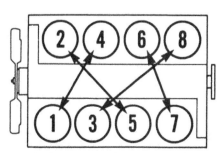

Fig. 25 – Adjustment of valves is made by rocking valves of a cylinder and adjusting valves on corresponding cylinder and vice versa as shown. Refer to text for tappet gap.

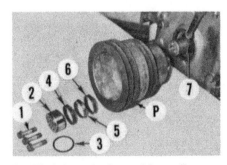

Fig. 27 – Late production models use three cap screws (1) with locking taper rings as shown, to retain crankshaft pulley (P). Cap screw torque is 80 ft.-lbs.

Fig. 28 – After removing cap screws and thrust block, use a sharp blow rearward to free crankshaft pulley (P) used on late production models.

gears, idler shafts, bushings or other items concerned. Before attempting to remove any gears, first remove rocker arm covers and rocker arms as outlined in paragraph 29 to avoid possible piston or valve train damage if camshaft or crankshaft should be turned independently of the other. Refer to the appropriate following paragraph to service timing gears.

34. **INJECTION PUMP DRIVE GEAR.** Injection pump drive gear (2 – Fig. 29) is a transition fit and is retained by three cap screws. Gear has elongated holes for retaining cap screws to allow injection pump timing adjustment.

To install gear, first set number one piston at TDC on compression stroke then release a valve on number one cylinder so that it rests on top of the piston. Place a dial indicator on top of valve stem and carefully set number one piston at exactly TDC. On models equipped with C.A.V. pump, align pointer in injection pump drive housing with mark on injection pump drive hub as shown at (T – Fig. 30). On models equipped with Bosch pump, line up groove on governor spring spindle with "O" mark on timing circlip as shown at (T – Fig. 30A). On all models, install gear on shaft with tachometer drive adapter and finger tighten cap screws. Ensure number one piston is exactly at TDC on compresssion stroke and injection pump marks are properly aligned, then turn gear counterclockwise just enough to remove backlash and tighten cap screws to 30-35 ft.-lbs. torque. Refer to paragraph 33 for backlash specifications.

35. **CAMSHAFT/INJECTION PUMP IDLER GEAR ASSEMBLY.** Camshaft/injection pump idler gear assembly (8 – Fig. 29) consists of two retaining cap screws, front thrust plate, hub, idler gear and rear thrust washer. Idler gear has a renewable bushing that should not require machining after installation. Inside diameter of fitted bushing should be 2.2500-2.2522 inch. Gear should have a diametral clearance of 0.0010-0.0039 inch on the 2.2483-2.2490 inch diameter hub.

When installing gear assembly, position oil hole in hub to the right side of engine and tighten cap screws to 27-30 ft.-lbs. torque. End play on idler gear should be 0.0215-0.0250 inch with a wear limit of 0.035 inch. Refer to paragraph 33 to time engine and for backlash specifications.

36. **CAMSHAFT GEAR.** Camshaft gear (3 – Fig. 29) is a transition fit and is retained to camshaft by three cap screws. A dowel pin positively locates gear to shaft.

When installing camshaft gear, do not drive gear onto shaft as expansion plug in cylinder block at rear of camshaft may become dislodged. Tighten cap screws to 38-40 ft.-lbs. torque. Refer to paragraph 33 to time engine and for backlash specifications.

37. **CRANKSHAFT/CAMSHAFT IDLER GEAR ASSEMBLY.** On MF 2745 models, crankshaft/camshaft idler gear assembly (4 – Fig. 29) consists of two retaining nuts, front thrust plate, hub and idler gear. Idler gear has renewable half bushings that will require machining after installation.

Specifications for bushings are shown in Fig. 31. Gear should have a diametral

Fig. 30 – Align pointer and mark on injection pump drive hub (T) to time C.A.V. injection pump with engine. Refer to text.

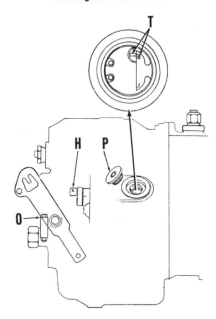

Fig. 30A – Align groove on governor spring spindle with "O" mark on timing circlip (T) when timing Bosch injection pump.

Fig. 29 – View of timing gear train with cover removed. Plate (A) covers hole for external hydraulic steering pump drive gear used on early production models.

2. Injection pump drive gear
3. Camshaft gear
4. Crankshaft/camshaft idler gear
5. Crankshaft gear
6. Oil pump idler gear
7. Oil pump gear
8. Camshaft/injection pump idler gear
TM. Engine timing marks

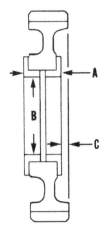

Fig. 31 – Machine crankshaft/camshaft idler bushings for MF 2745 models as follows:

A. Total bushing width – 1.008-1.010 inch.
B. Bushing diameter – 2.2500-2.2522 inch.
C. Bushing depth – 0.272-0.276 inch.

clearance of 0.0012-0.0037 inch on the 2.2483-2.2490 inch diameter hub.

When installing gear assembly, position hub with oil hole toward crankshaft gear and tighten retaining nuts to 21-24 ft.-lbs. torque. End play on idler gear should be 0.009-0.020 inch with a wear limit of 0.030 inch. Refer to paragraph 33 to time engine and for backlash specifications.

38. On MF 2775 and MF 2805 models, crankshaft/camshaft idler gear assembly (4 – Fig. 29) consists of two retaining cap screws, front thrust plate, hub, idler gear and rear thrust washer. Idler gear has a renewable bushing that should not require machining after installation. Inside diameter of fitted bushing should be 2.2500-2.2522 inch. Gear should have a diametral clearance of 0.0010-0.0039 inch on the 2.2483-2.2490 inch diameter hub.

When installing gear assembly, position hub so oil hole is towards crankshaft and tighten retaining cap screws to 27-30 ft.-lbs. torque. End play on idler gear should be 0.0215-0.0250 inch with a wear limit of 0.035 inch. Refer to paragraph 33 to time engine and for backlash specifications.

39. **CRANKSHAFT GEAR.** Crankshaft gear (5 – Fig. 29) can be removed using a suitable puller. On early production models, it will be necessary to remove crankshaft pulley locating pin before removing gear.

When installing crankshaft gear, first remove crankshaft/camshaft idler gear as outlined in paragraphs 37 or 38, then retime engine and check backlash as outlined in paragraph 33.

40. **CRANKSHAFT/OIL PUMP IDLER GEAR ASSEMBLY.** Crankshaft oil pump idler gear assembly (6 – Fig. 29), consists of two retaining cap screws, front thrust plate, hub and idler gear. Idler gear has a renewable

bushing and should not require machining after installation. Inside diameter of fitted bushing should be 2.000-2.0022 inch. Gear should have a diametral clearance of 0.0012-0.0041 inch on the 1.9981-1.9988 inch diameter hub.

Install gear assembly and tighten retaining cap screws to 17-19 ft.-lbs. torque. Idler gear end play should be 0.015-0.022 inch with a wear limit of 0.032 inch. Refer to paragraph 33 for backlash specifications.

41. **OIL PUMP GEAR.** Oil pump gear (7 – Fig. 29) is a press fit and disassembly of oil pump is necessary to renew gear. Refer to paragraphs 57 or 58 for information on oil pump gear and oil pump.

42. **AUXILIARY DRIVE HOUSING.** The auxiliary drive housing mounts on top of the engine block and houses the injection pump drive and on early production models, the external power steering pump drive. The auxiliary drive housing can be removed independently of the timing gears by following the procedures outlined in paragraph 43; or during timing gear overhaul by following only those portions which apply.

43. REMOVE AND REINSTALL. To remove the auxiliary drive housing assembly, first remove timing gear cover as outlined in paragraph 32 and injection pump as outlined in paragraph 75 or paragraph 79. On early production models, remove external power steering pump with drive, as outlined in paragraph 12. Remove fuel lift pump and push rod. Remove auxiliary housing retaining nuts and lift housing (H – Fig. 32) from engine. On models equipped with C.A.V. fuel injection pump, refer to paragraph 44 for overhaul of auxiliary housing assembly.

Reinstall auxiliary drive housing by reversing removal procedure and performing the following additional steps: Front faces of auxiliary drive housing and cylinder block should be flush. Tighten 5/16-inch retaining nuts to 12-15 ft.-lbs. torque and ⅜-inch retaining nuts to 21-24 ft.-lbs. torque. Retime injection pump to engine as outlined in paragraph 34.

44. OVERHAUL. (Models With C.A.V. Injection Pump.) Refer to Fig. 33 for exploded view of auxiliary housing and associated parts. Remove camshaft/injection pump idler gear assembly

Fig. 32 – Lift auxiliary drive housing (H) off cylinder block.

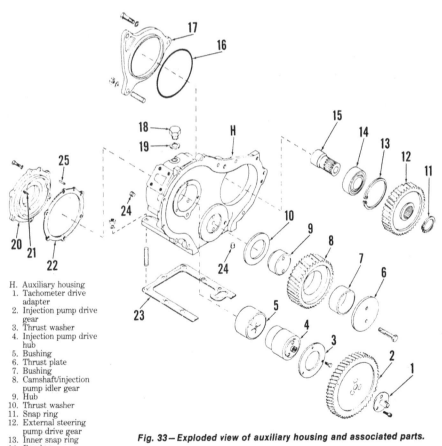

H. Auxiliary housing
1. Tachometer drive adapter
2. Injection pump drive gear
3. Thrust washer
4. Injection pump drive hub
5. Bushing
6. Thrust plate
7. Bushing
8. Camshaft/injection pump idler gear
9. Hub
10. Thrust washer
11. Snap ring
12. External steering pump drive gear
13. Inner snap ring
14. Bearing
15. Shaft
16. "O" ring
17. Steering pump drive housing
18. Injection pump timing hole plug
19. Washer
20. Injection pump adapter plate
21. Injection pump timing pointer
22. Gasket
23. Gasket
24. Oil plugs
25. Guide dowel

Fig. 33 – Exploded view of auxiliary housing and associated parts.

(6 through 10) as outlined in paragraph 35 and injection pump drive gear (2) as outlined in paragraph 34. Unbolt injection pump drive thrust washer (3) then withdraw injection pump drive shaft (4) and remove injection pump adapter plate (20).

Renew injection pump drive shaft bushing (5) if damaged or worn excessively. Bushing will require machining after installation. Inside diameter of finished bushing should be 2.3750-2.3768 inch. The 2.3722-2.3735 inch journal diameter of shaft should have a diametral clearance of 0.0015-0.0046 inch in bushing. Install bushing, aligning oil holes, until flush with front face of housing bore. End play of injection pump drive shaft (4) should be 0.006-0.010 inch. Renew thrust washer (3) or shaft (4) if specifications are exceeded.

If equipped with external power steering pump drive (11 through 17), refer to paragraph 13 for servicing.

Renew injection pump timing pointer (21) in injection pump adapter plate (20) if bent or damaged, then reinstall adapter plate. Reinstall injection pump drive gear (2) as outlined in paragraph 34 and camshaft/injection pump idler gear assembly (6 through 10) as outlined in paragraph 35.

CAMSHAFT

MF 2745 Models

45. To remove camshaft, first remove cam followers as outlined in paragraph 27 and camshaft gear as outlined in paragraph 36. Unbolt camshaft thrust washer and withdraw camshaft as shown in Fig. 34.

Front camshaft bearing is renewable and should not require machining when installed. The remaining four bearings are machined directly in cylinder block. Insert bearing with front marking on bearing toward front of cylinder block and align oil passage hole. Bearing should be pressed in until flush with front face of cylinder block.

Normal camshaft bearing clearance is 0.0023-0.0055 inch for front journal and

0.0025-0.0053 inch for all others. Maximum running clearance for all five journals is 0.0095 inch. Camshaft journal diameters are as follows:

    Front journal . . . . . 2.3715-2.3725 inch
    No. 2 journal . . . . . 2.2415-2.2425 inch
    No. 3 journal . . . . . 2.2315-2.2325 inch
    No. 4 journal . . . . . 2.2215-2.2225 inch
    No. 5 journal . . . . . 2.2115-2.2125 inch

Maximum wear of journals is 0.0015 inch below lower production limit.

Cam lift for both intake and exhaust is 0.3325-0.3355 inch. End play on camshaft should be 0.004-0.014 inch with a wear limit of 0.020 inch. Renew thrust washer and/or camshaft if specifications are exceeded.

Pressure lubrication for rocker arms is provided by third camshaft bearing journal. When installing camshaft, tighten thrust washer retaining cap screws to 10-12 ft.-lbs. torque. The remainder of reassembly is the reverse of disassembly.

MF 2775 And MF 2805 Models

46. To remove camshaft, first remove auxiliary drive housing as outlined in paragraph 43 and on MF 2805 models, remove turbocharger as outlined in paragraph 85. On both models remove camshaft gear as outlined in paragraph 36 and parts necessary to remove block valley cover. Secure the cam followers (valve tappets) in their uppermost position as shown in Fig. 35, then unbolt camshaft thrust washer and withdraw camshaft as shown in Fig. 34.

Front camshaft bearing is renewable and should not require machining when installed. The remaining four bearings are machined directly in cylinder block. Insert bearing with front marking on bearing toward front of cylinder block and align all three oil passage holes. Bearing should be pressed in until flush with front face of cylinder block.

Normal camshaft bearing clearance is 0.0023-0.0055 inch for front journal and 0.0033-0.0059 inch for all others. Maximum running clearance for all five journals is 0.009 inch. Camshaft journal diameters are as follows:

    Front journal . . . . . 2.4965-2.4975 inch
    No. 2 journal . . . . . 2.3665-2.3675 inch
    No. 3 journal . . . . . 2.3565-2.3575 inch
    No. 4 journal . . . . . 2.3465-2.3475 inch
    No. 5 journal . . . . . 2.2115-2.2125 inch

Maximum wear of journals is 0.001 inch below lower production limit.

Cam lift for intake is 0.3386-0.3416 inch and for exhaust is 0.3456-0.3486 inch. End play of camshaft should be 0.004-0.014 inch with a wear limit of 0.020 inch. Renew thrust washer and/or camshaft if specifications are exceeded.

Pressure lubrication for rocker arms is provided by front and rear camshaft bearing journals. When installing camshaft, tighten thrust washer retaining cap screws to 10-12 ft.-lbs. torque. The remainder of reassembly is the reverse of disassembly.

ROD AND PISTON UNITS

All Models

47. Connecting rod and piston units can be removed from above after removing cylinder heads as outlined in paragraph 21 and oil pan as outlined in paragraph 56. Cylinder numbers are stamped on connecting rod and cap. Connecting rod bearings are available in undersizes of 0.010, 0.020 and 0.030 inch as well as standard size. Piston is installed in cylinder with valve relief pockets up (toward camshaft).

When assembling, tighten connecting rod cap screws to 100-105 ft.-lbs. torque.

PISTONS, SLEEVES AND RINGS

MF 2745 Models

48. The aluminum alloy, cam-ground pistons are supplied in standard size only. Piston top has a toroidal combustion chamber and relief pockets for valve heads. Cylinder numbers are stamped on the crown of each piston. New pistons should be stamped with corresponding cylinder number.

Maximum piston skirt diameter measured at right angles to piston pin and just above bottom ring groove should be 4.2425-4.2435 inch. Piston pin

Fig. 34 — Remove thrust plate and withdraw camshaft.

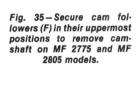

Fig. 35 — Secure cam followers (F) in their uppermost positions to remove camshaft on MF 2775 and MF 2805 models.

bore diameter should be 1.7500-1.7503 inch. Ring groove widths are as follows:

Top ring groove . . . . . 0.097-0.098 inch
Second ring groove 0.0957-0.0967 inch
Third and fourth ring
grooves . . . . 0.1895-0.1905 inch

New pistons have a diametral clearance in new sleeves of 0.0065-0.0085 inch when measured between piston skirt and sleeve at 90 degrees to piston pin.

Crown of installed piston may require machining to provide a piston height of 0.000-0.0085 inch below top face of cylinder block (D – Fig. 37).

NOTE: Piston height is measured from top face of cylinder block and not from cylinder sleeve.

Each piston is fitted with two compression rings and two oil control rings with one of the oil control rings located below piston pin. The barrel faced chrome top compression ring may be installed either side up in groove. The second chrome compression ring is internally stepped and installed with stepped side toward top of piston. Upper and lower chrome oil control rings are spring loaded and can be installed either side up. Position oil ring spring in groove and be sure latch pin enters both ends of spring, then install the chrome oil control ring over spring with gap of ring opposite latch pin in spring. Stagger ring end gaps around piston. Ring specifications are as follows:

End gap (measured in 4.250-inch diameter bore)–

Top ring . . . . 0.008-0.021 inch
Second ring . . . . 0.017-0.027 inch
Third and fourth oil
rings . . . . 0.017-0.024 inch
Side clearance in groove–
Top ring . . . . 0.003-0.005 inch
Second ring and oil
rings . . . . 0.002-0.004 inch
Maximum wear limit . . . . . . 0.008 inch
Ring width–
Top and second
rings . . . . 0.0928-0.0938 inch
Third and fourth oil
rings . . . . 0.1865-0.1875 inch

Standard cylinder sleeve bore diameter is 4.250-4.251 inches and sleeves may not be machined oversize. Cylinder sleeves should be renewed if scored, if wear exceeds 4.255 inch, or if bore is more than 0.002 inch out-of-round. The production cylinder sleeves are 0.001-0.003 inch press fit in cylinder block bores. Service sleeves have 0.001 inch interference to 0.001 inch loose fit in block bores and should not require machining after installation. Cylinder block bore diameter should be 4.4565-4.4575 inches.

Use a heavy duty sleeve puller and installer to renew sleeves. After removal of sleeve, clean cylinder bore, giving particular attention to counterbore. Carefully examine sleeve contact area in cylinder block and remove any burrs,

rust spots or other foreign material.

Thoroughly clean new service sleeves in solvent and air dry. Use a spray can to lubricate cylinder block bore and outside of sleeve then press sleeve fully into place. When correctly installed, sleeve should protrude above cylinder block 0.020-0.026 inch.

Once sleeves are installed and reach ambient temperature of block, check bores for distortion. New service sleeves should have an installed diameter of 4.252-4.253 inch with a distortion limit of 0.002 inch.

MF 2775 And MF 2805 Models

49. The aluminum alloy, cam-ground pistons are supplied in standard size only. Piston has a toroidal combustion chamber in crown. Cylinder numbers are stamped on the crown of each piston. New pistons should be stamped with corresponding cylinder number. Pistons are non-directional and can be installed in either bank of cylinder block. Pistons are cooled by oil jet tubes which spray oil on bottom of piston crowns. If piston failure occurs or excessive

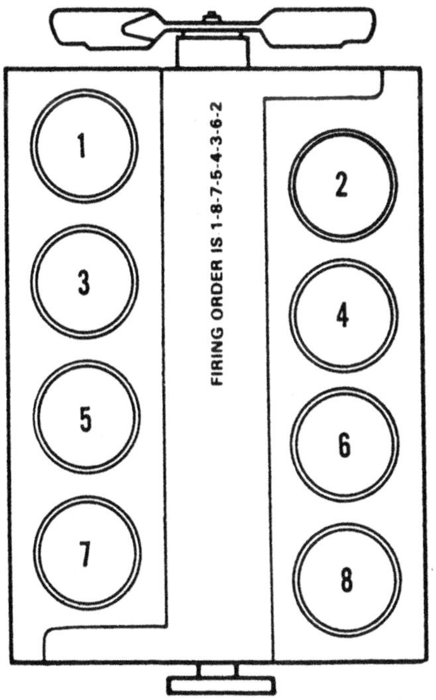

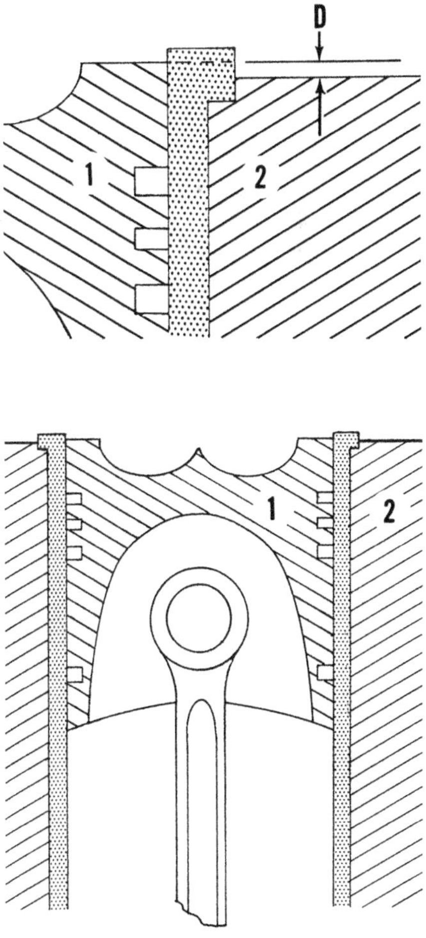

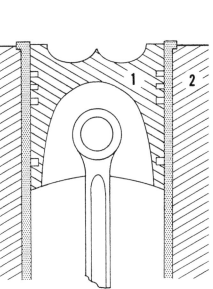

Fig. 38 – On MF 2775 and MF 2805 models, piston crown (1) should be 0.0015-0.010 inch (D) above block deck (2).

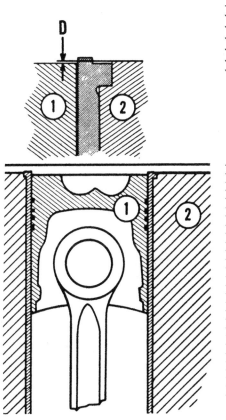

Fig. 37 – Piston crown (1) should be 0.000-0.0085 inch (D) below block deck (2).

FIRING ORDER IS 1-8-7-5-4-3-6-2

Fig. 36 – Schematic view of block showing cylinder numbering sequence.

heating of piston is evident, check operation of oil jet tubes as outlined in paragraph 62.

Maximum piston skirt diameter measured at right angles to piston pin and just above bottom ring groove should be 4.621-4.622 inch. Piston pin bore diameter should be 1.7499-1.7501 inch. Ring groove widths are as follows:

Top ring groove . . . 0.0973-0.0983 inch
Second ring groove 0.0963-0.0973 inch
Third and fourth ring
   grooves . . . . 0.1895-0.1905 inch

New pistons have a diametral clearance in new sleeves of 0.008-0.010 inch when measured between piston skirt and sleeve at 90 degrees to piston pin.

Crown of installed piston may require machining to provide a piston height of 0.0015-0.010 inch above top face of cylinder block (D – Fig. 38).

NOTE: Piston height is measured from top face of cylinder block and not from cylinder sleeve.

Each piston is fitted with two compression rings and two oil control rings with one of the oil control rings located below piston pin. The barrel faced chrome top compression ring may be installed either side up in groove. The second chrome compression ring can be either internally stepped or internally beveled and installed with stepped or beveled side toward top of piston. Upper and lower chrome oil control rings are spring loaded and can be installed either side up. Position oil ring spring in groove and be sure latch pin enters both ends of spring, then install the chrome oil control ring over spring with gap of ring opposite latch pin in spring. Stagger ring end gaps around piston. Ring specifications are as follows:

End gap (measured in 4.630 inch bore) –
Top ring . . . . 0.019-0.026 inch
Second ring and oil
   rings . . . . 0.019-0.029 inch
Side clearance in groove –
Top ring . . . . 0.0035-0.0055 inch
Second ring . . . . . . 0.0025-0.0045 inch
Third and fourth oil
   rings . . . . 0.002-0.004 inch
Maximum wear limit . . . . . . 0.008 inch
Ring width –
Top and second
   rings . . . . 0.0928-0.0938 inch
Third and fourth oil
   rings . . . . 0.1865-0.1875 inch

Cylinder sleeves should be renewed if scored, or if wear exceeds 4.634 inch on the original 4.630-4.631 inch diameter bore and if bore is more than 0.002 inch out-of-round. The production cylinder sleeves are 0.001-0.003 inch press fit in cylinder block bores. Service sleeves have 0.000-0.002 inch clearance in block bores and should not require machining

after installation.

Use a heavy duty sleeve puller and installer to renew sleeves. After removing piston oil cooling jets, pull sleeve from top of cylinder block. Clean parent bore giving particular attention to counterbore. Carefully examine sleeve contact area in cylinder block and remove any burrs, rust spots or other foreign material. Cylinder block bore diameter for cylinder sleeve should be 4.8065-4.8075 inch.

Thoroughly clean new service sleeves in solvent and air dry. Use a spray can to lubricate parent bore in cylinder block and outside of sleeve then press sleeve fully into place. When correctly installed, sleeve should protrude above cylinder block 0.024-0.030 inch.

Once sleeves are installed and reach ambient temperature of block, check bores for distortion. New service sleeves should have an installed diameter of 4.631-4.632 inch with a distortion limit of 0.002 inch. Install piston oil cooling spray jets and tighten cap screws to 12-15 ft.-lbs. torque.

PISTON PINS

All Models

50. The 1.7498-1.7500 inch diameter,

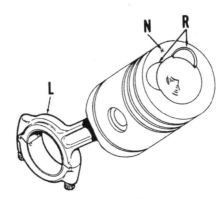

Fig. 39 – Valve relief pockets (R) and long side of rod must be on same side when rod and piston unit is assembled. Install piston and rod so valve pockets (R) are towards camshaft.

Fig. 40 – Main bearing caps are also secured with cross-bolts (B).

floating type piston pins are retained in piston bores by snap rings and are available in standard size only. To ease piston pin removal and installation, heat piston to 100-120 degrees F.

The renewable connecting rod bushing must be final sized after installation to provide the recommended 0.0007-0.0018 inch diametral clearance of piston pin. Be sure the pre-drilled oil hole in bushing is properly aligned with hole in connecting rod.

On MF 2745 models, piston pin clearance in piston bosses should be 0.000-0.0005 inch. Valve relief pockets in piston crown (R – Fig. 39) and long leg of rod (L) must be on the same side of assembly when pin is installed.

On MF 2775 and MF 2805 models, piston pin fit in piston should be 0.0001 inch tight to 0.0003 inch loose. Pistons are non-directional, but if old pistons are used, the manufacturer recommends reassembly in original directions.

CONNECTING RODS AND BEARINGS

All Models

51. Connecting rod bearings are precision type, renewable from below after removing oil pan and rod bearing caps. Rods and caps are marked with corresponding cylinder numbers. When renewing bearings make sure that projection in bearing engages milled slots in rod and cap and correlation marks on rod and cap are in register. Bearings are available in undersizes of 0.010, 0.020 and 0.030 inch as well as standard.

On MF 2745 models, connecting rod bearings should have a diametral clearance of 0.0018-0.0038 inch on the 2.9980-2.9985 inch diameter crankpin. Recommended connecting rod side clearance is 0.015-0.022 inch with both rods fitted on crankpin journal. Crankshaft is induction hardened and will not require rehardening if journal is reground undersize.

On MF 2775 and MF 2805 models, connecting rod bearings should have a diametral clearance of 0.0013-0.0045 inch on the 2.9980-2.9985 inch diameter crankpin. Recommended connecting rod side clearance is 0.014-0.021 inch with both rods fitted on crankpin journal. Crankshaft is induction hardened on the naturally aspirated MF 2775 models and will not require rehardening if journal is reground undersize. The crankshaft on turbocharged MF 2805 models must be "RENITRIDED" if journal is reground undersize.

NOTE: Manufacturer replacement crankshafts for MF 2775 and MF 2805 models are "NITRIDED".

On all models tighten connecting rod cap screws to a torque of 100-105 ft.-lbs. when units are installed.

CRANKSHAFT AND BEARINGS

All Models

52. The crankshaft is supported in five precision type main bearings and end play is controlled by thrust washer halves located adjacent to center main bearing in block portion only. The first four main bearings and thrust washer halves can be renewed from below after removing starter motor, oil pan, oil pump, oil cooler and lines, and main bearing cross bolts (B—Fig. 40). To remove the rear main bearing cap, it is first necessary to remove engine, clutch, flywheel and rear oil pan bridge piece. If removal of crankshaft is required, remove crankshaft pulley, timing gear cover, flywheel housing, rear oil seal housing, crankshaft counterweights and rod bearing caps.

NOTE: To avoid the possibility of damage to pistons or valve train if camshaft or crankshaft should be turned independently of the other, remove rocker arm covers and rocker arms as outlined in paragraph 29.

Main bearing caps are numbered front to rear (1–Fig. 41) and are also stamped with a cylinder block identification number (2). Crankshaft counterweights are stamped for proper location on crankshaft corresponding to web numbers (1–Fig. 42). To install counterweights, oil weights and retaining screws lightly, then tighten weight retaining screws to 80-85 ft.-lbs. torque on MF 2745 models and 115-120 ft.-lbs. torque on MF 2775 and MF 2805 models. Loosen retaining screws and retorque to the same specifications. This procedure will properly seat weight and set the screws.

Recommended main bearing diametral clearance is 0.0040-0.0063 inch on MF 2745 models and 0.0026-0.0049 inch on MF 2775 and MF 2805 models. On all models standard main journal diameter is 3.9967-3.9972 inch and standard crankpin journal diameter is 2.9980-2.9985 inch.

Bearing inserts are available in undersizes of 0.010, 0.020 and 0.030 inch as well as standard size. Crankshaft is induction hardened on naturally aspirated MF 2745 and MF 2775 models and will not require rehardening if journal is reground undersize. The crankshaft on turbocharged MF 2805 models must be "RENITRIDED" if journal is reground undersize.

NOTE: Manufacturer replacement crankshafts for MF 2775 models and MF 2805 are "NITRIDED" and must be "RENITRIDED" if journal is reground undersize.

Recommended crankshaft end play is 0.002-0.017 inch on MF 2745 models and 0.003-0.015 inch on MF 2775 and MF 2805 models with a maximum end play of 0.020 inch on all models. Thrust washer halves are in standard thicknesses of 0.122-0.125 inch and oversize of 0.1295-0.1325 inch.

On all models tighten main bearing retaining cap screws to a torque of 190-210 ft.-lbs. and rod bearing cap bolts to a torque of 100-105 ft.-lbs.

Tighten cross-bolts on MF 2745 models to 95-100 ft.-lbs. torque. On MF 2775 and MF 2805 models, tighten ½-inch cross-bolts to 105-110 ft.-lbs. torque and 9/16-inch cross-bolts to 155-160 ft.-lbs. torque.

Refer to paragraph 53 for installation of rear housing and oil pan bridge piece and to paragraphs 57 or 58 for installation of oil pump and idler. Retime engine as outlined in paragraph 33. The remainder of reassembly is the reverse of disassembly.

CRANKSHAFT REAR OIL SEAL

All Models

53. A spring loaded type lip seal is located in a housing bolted to rear of engine (1–Fig. 43). The housing and seal can be removed after splitting tractor, removing clutch and flywheel. During production, the oil seal is installed with its rear face flush with rear face of seal housing. Check condition of seal contact surface of crankshaft before installing new seal in retainer. The seal can be pressed further into retainer to provide another contact surface if crankshaft is grooved. The recommended locations are: Flush, ⅛-inch below flush and ¼-inch below flush of housing rear face. Coat lip of seal and crankshaft surfaces with oil before assembling, then install housing while carefully sliding lip of seal over crankshaft end. Tighten cap screws to 12-15 ft.-lbs. torque.

If cylinder block bridge piece is removed, fit the two small cylindrical rubber seals (S–Fig. 44) into recesses in ends of bridge piece. Use a straightedge (Fig. 45) to make sure bottom (oil pan) surface of bridge piece is aligned with gasket surface of block before tightening retaining screws.

Fig. 42—Counterweights are stamped (1) for proper location on crankshaft.

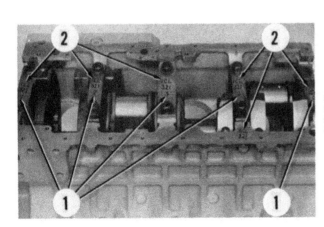

Fig. 41—Main bearing caps are numbered front to rear (1) and are stamped with cylinder block identification number (2).

Fig. 43—View of rear oil seal housing (1). Use a guide dowel to aid in removal and installation of flywheel.

FLYWHEEL

All Models

54. The flywheel can be removed with the coupler after separating engine from clutch housing and removing clutch. Install a guide dowel as shown in Fig. 43 to aid in removal and installation of flywheel.

The starter ring gear can be renewed after flywheel is removed. Remove old ring gear either by cutting or by immersing flywheel in cold water up to edge of ring gear and heating ring gear. Heat new ring gear evenly to approximately 480 degrees F. and install on flywheel with beveled end of teeth facing front of engine.

Lubricate "O" ring seal on front side of PTO coupler with lithium base grease, then install PTO coupler and flywheel. Tighten the twelve retaining cap screws to 77-115 ft.-lbs. torque. Maximum allowable face runout is 0.001 inch for each inch from flywheel center line to point of measurement.

FLYWHEEL HOUSING

All Models

55. The cast flywheel housing is located by engine block dowels and secured with cap screws. The housing can be removed after splitting tractor and removing flywheel. When installing flywheel housing, tighten cap screws to 48-52 ft.-lbs. torque on MF 2745 models and 61-92 ft.-lbs. torque on MF 2775 and MF 2805 models.

OIL PAN

All Models

56. The heavy, cast iron oil pan serves as part of the tractor frame and as an attaching point for tractor front support. If oil pan must be removed with engine installed in tractor, the following pro-

cedure is suggested, however, extreme care must be observed to prevent personal injury or damage to tractor.

Drain oil pan and on MF 2745 models, drain and remove oil cooler with oil transfer pipes as outlined in paragraph 60. Raise front of tractor until weight is just off of front tires and block up solidly under front of transmission housing. Unbolt front engine mounts and remove oil pan retaining cap screws that are not accessible with engine mounts in place. If front main bearing cap is to be removed, remove cross-bolts (C – Fig. 46) at this time. Reinstall front engine mounts but finger tighten engine mount cap screws only. Remove cap screws securing front and rear of oil pan to tractor. Position a jack under front end and raise slightly to separate oil pan from front end. Position a rolling floor jack under oil pan and remove remaining oil pan retaining cap screws, then lower oil pan away from tractor.

Install oil pan by reversing removal procedure while noting the following: Tighten oil pan retaining cap screws to 12-15 ft.-lbs. torque. Tighten cap screws securing front and rear of oil pan to tractor as follows: ¾-inch – 240-350 ft.-lbs. torque; ⅞-inch – 300-400 ft.-lbs. torque; 1-inch – 440-580 ft.-lbs. torque. Tighten front engine mount cap screws to 130-200 ft.-lbs. torque.

OIL PUMP

MF 2745 Models

57. The gear type oil pump is mounted underneath the front main bearing cap and is driven by the crankshaft gear through an idler gear. Oil is drawn up through a strainer and oil inlet tube mounted on the pump, and is delivered from the pump through a tube to the oil cooler mounted on side of oil pan. Oil enters cooler through the front tube and exits through rear tube to the oil pressure relief valve body. Regulated oil pressure then flows through the double

filters before lubricating the engine bearings.

The oil pump can be removed after removing the oil cooler, withdrawing the two tubes used to direct oil to and from the oil cooler, then removing the oil pan. Refer to paragraph 56. Detach the oil inlet strainer brackets, then unbolt and remove the oil pump from bottom face of crankcase.

Refer to Fig. 47 for drawing of oil pan, pump, cooler, relief valve and associated parts. Refer to Fig. 48 for exploded view of the pump assembly. Normal pump gear end play is 0.002-0.006 inch and maximum permissible wear limit is 0.010 inch. Radial clearance between gear teeth and body should be 0.002-0.008 inch and wear limit is 0.010 inch. Backlash between drive and driven gears should be 0.028-0.032 inch.

Inside diameter of bushings (3 and 5) should be machined to 0.7505-0.7520 inch if renewed to provide a diametral clearance of 0.001-0.003 inch on shaft (4). Bushings (7) for driven gears (6) will require machining if renewed. Machine bushings to an inside diameter of

Fig. 45 — Align bridge piece with block using a straight edge before tightening cap screws 12-15 ft.-lbs. torque.

Fig. 44 — Be sure to fit the two small cylindrical rubber seals (S) into recesses at ends of bridge piece as shown.

Fig. 46 — Unbolt engine mounts to obtain access to front oil pan retaining cap screws. Remove cross-bolts (C) on both sides of engine, if front main bearing cap is to be removed.

0.6875-0.6885 inch to provide a diametral clearance of 0.0010-0.0025 inch on shaft (8).

Oil pump drive gear (2) is a press fit and can be renewed using a suitable puller. When installing, press gear onto shaft (4) until front face of gear is flush with shaft as shown at D in inset. Remove cover (9) before installing gear to prevent damage to shaft and/or end cover.

Prime oil pump before installing then tighten oil pump and mounting bracket retaining cap screws to 21-24 ft.-lbs. torque.

Refer to paragraphs 33 and 40 for information on oil pump idler gear and for backlash specifications.

MF 2775 And MF 2805 Models

58. The gear type oil pump (P – Fig. 49) is mounted underneath the front main bearing cap and driven by the crankshaft gear through an idler. Oil is drawn up through a strainer and oil inlet tube (T) mounted on the pump, and is delivered from the pump through a passage in the block to the oil cooler mounted at C on side of block. Oil enters cooler then exits to the oil pressure relief valve (R) where oil pressure is regulated; then oil flows through the double filters before lubricating the engine bearings. Oil pump is equipped with a dump valve (D) attached directly to pump to prevent extreme oil pressure.

The oil pump can be removed after removing the oil pan as outlined in paragraph 56. Detach the oil inlet strainer brackets, then unbolt and remove the oil pump from bottom face of crankcase.

Refer to Fig. 50 for exploded view of oil pump assembly. Normal pump gear end play is 0.002-0.006 inch with a maximum wear limit of 0.010 inch. Radial clearance between gear teeth and body should be 0.004-0.006 inch and wear limit is 0.010 inch. Backlash between drive and driven gears should be 0.028-0.032 inch.

Inside diameter of new bushings (4) should be machined to 0.7505-0.7520 inch to provide a diametral clearance of 0.001-0.003 inch on shaft (5). Bushings (7) for driven gears (6) will require machining if renewed. Machine bushings to an inside diameter of 0.6875-0.6885 inch to provide a diametral clearance of 0.0010-0.0025 inch on shaft (8).

Oil pump drive gear (2) is a press fit and can be renewed using a suitable puller. With end cover (9) removed, press gear onto shaft (5) until front face of gear is flush with end of shaft as shown at D in inset. Disassemble and inspect dump valve (13 through 17) for wear or damage and renew parts as necessary.

Prime oil pump before installing then tighten oil pump and mounting bracket retaining cap screws to 21-24 ft.-lbs. torque.

Refer to paragraphs 33 and 40 for information on oil pump idler gear and for backlash specifications.

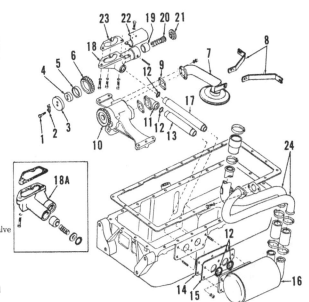

Fig. 47 – Drawing of oil pan, pump, cooler, relief valve and related parts. Inset shows relief valve housing (18A) used on late production models.

1. Cap screw
2. Lock plate
3. Plate
4. Hub
5. Bushing
6. Idler gear
7. Strainer
8. Brackets
9. Gasket
10. Oil pump
11. Elbow
12. "O" rings
13. Pipe from pump to cooler
14. Gasket
15. Plate
16. Oil cooler
17. Pipe from cooler to relief valve
18. Relief valve housing
19. Relief valve plunger
20. Spring
21. Cap
22. Oil deflector plate
23. Gasket
24. Water tubes to and from oil cooler

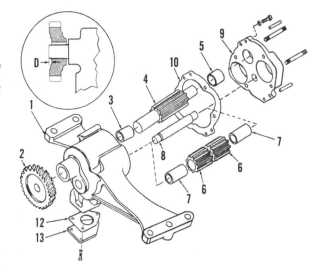

Fig. 48 – Exploded view of oil pump. Inset shows distance (D) gear (2) should be pressed onto shaft (4).

1. Oil pump housing
2. Gear
3. Bushing
4. Drive shaft and gear
5. Bushing
6. Driven gears
7. Bushings
8. Driven shaft
9. Cover
10. Gasket
11. Body
12. Gasket
13. Plate

Fig. 49 – Location of oil pump (P), inlet tube (T), dump valve (D) and oil pressure relief valve (R). Oil cooler/filter assembly mounts at C.

RELIEF VALVE

All Models

59. The oil pressure relief valve is shown in Fig. 47 for MF 2745 models and Fig. 51 for MF 2775 and MF 2805 models. Relief valve can be removed from engine block after removing oil pan as outlined in paragraph 56. The valve is preset to 60-65 psi and no attempt should be made to adjust the pressure except by renewal of worn or damaged parts. Hollow end of plunger should be away from spring when assembling valve. Tighten retaining cap screws to 12-15 ft.-lbs. torque.

OIL COOLER

MF 2745 Models

60. An engine oil cooler (16-Fig. 47) is attached to left side of the engine oil pan. Oil enters cooler through tube (13) from oil pump (10) and leaves through longer tube (17). Coolant pipes (24) are used to circulate water from cylinder head through cooler and back to water pump.

To remove oil cooler, first drain coolant by removing plug at bottom of tube stack (25-Fig. 52). Disconnect hoses, then unbolt and remove cooler from side of oil pan. Use care to prevent damage to tubes (13 and 17), plate (15) and "O" rings (12) at both ends of tubes. The tubes must be withdrawn before removing the oil pan.

No attempt should be made to disassemble oil cooler unless the unit can be properly pressure tested after reassembly. Mark body (26) and tube stack (25) before disassembly so parts can be correctly aligned when assembling. Remove screws and washers locating tube stack in body, then slide the tube stack out towards the closed end until "O" ring (27) can be removed. Slide tube stack the other direction and out of body.

CAUTION: Do not pound on either end of the oil cooler. Slide the tube stack by supporting the stack and knocking the body mounting flange gently but firmly on a soft but firm surface such as a stout wooden bench.

Clean all parts thoroughly and use new "O" rings (27 and 27A). Oil or grease "O" rings, grooves in tube stack and bores in body before assembling. Install "O" ring (27A) at connection end of tube stack and carefully insert closed end of stack through body (26) being sure that previously affixed marks will align. Slide tube stack through body until "O" ring groove at closed end of tube stack is exposed and install "O" ring (27). Slide the tube stack back into body aligning the marks. Install locating screws and washers and test for leakage. Fabricate a suitable adapter to cover and seal the oil ports with provision to pressurize the oil passages. Fill coolant side of cooler with water, pressurize the oil side with air pressure of 90-150 psi, then submerge the unit in water. The tube stack should be renewed if air escapes from coolant connections; new "O" rings should be installed if air escapes from ends.

MF 2775 And MF 2805 Models

61. An engine oil cooler (Fig. 53) is mounted on top of the double oil filter head (16) and is attached to left side of engine block. Oil enters cooler through a passage in the block from the oil pump, then oil exits to the oil pressure relief valve. After oil pressure is regulated at relief valve, oil flows back to the double filter head (16) where it passes through the oil filters (20) before lubricating engine bearings. Coolant pipes (12 and 13) are used to circulate water from cylinder head through cooler and back to water pump.

No attempt should be made to disassemble oil cooler unless the unit can

be properly pressure tested after reassembly.

To remove oil cooler, first drain coolant by removing plug (8) in end of tube stack. Remove oil pressure line connected to filter head (16) and oil pressure switch (14). Disconnect upper support bracket and coolant hoses then unbolt and remove oil cooler/filter head assembly from block. Separate cooler housing (5) from filter head (16). Mark cooler housing (5) and one end of tube stack (7) before disassembly so parts can be correctly aligned when assembling. Disassemble oil cooler by-pass valve by removing plug (1) with fiber washer (2), spring (3) and plunger (4). Remove screws and washers locating tube stack in housing, then slide the tube stack out toward the closed end until "O" ring can be removed. Slide tube stack the other direction and out of housing.

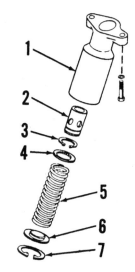

Fig. 51 — Exploded view of relief valve used on MF 2775 and MF 2805 models.

1. Housing	
2. Plunger	5. Spring
3. Snap ring	6. Spring seat
4. Washer	7. Snap ring

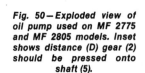

Fig. 50 — Exploded view of oil pump used on MF 2775 and MF 2805 models. Inset shows distance (D) gear (2) should be pressed onto shaft (5).

1. Oil pump housing
2. Gear
3. Gasket
4. Bushings
5. Drive shaft and gear
6. Driven gears
7. Bushings
8. Driven shaft
9. End cover
10. Gasket
11. Plate
12. Gaskets
13. Dump valve housing
14. Cotter pin
15. Ball
16. Spring
17. Spring seat
18. Inlet tube and strainer
19. Brackets

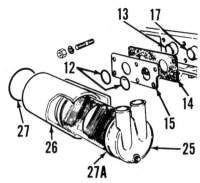

Fig. 52 — Exploded view of engine oil cooler used on MF 2745 models.

12. "O" rings	17. Exit tube
13. Delivery tube	25. Tube stack
14. Gasket	26. Cooler body
15. Plate	27 & 27A. "O" rings

CAUTION: Do not pound on either end of the oil cooler. Slide the tube stack by supporting the stack and knocking the housing mounting flange gently but firmly on a soft but firm surface such as a stout wooden bench.

Clean all parts thoroughly and use new "O" rings (6). Oil or grease "O" rings, grooves in tube stack and bores in housing before assembling. Install "O" ring at connection end of tube stack and carefully insert closed end of stack through housing being sure that previously affixed marks will align. Slide tube stack through housing so "O" ring (6) groove at closed end of tube stack is exposed and install "O" ring. Slide the tube stack back into housing aligning the marks. Install locating screws and washers and test for leakage. Fabricate a suitable adapter to cover and seal the oil ports with provision to pressurize the oil passages. Fill coolant side of cooler with water, pressurize the oil side with air pressure of 60 psi, then submerge the unit in water. The tube stack should be renewed if air escapes from coolant connections; new "O" rings should be installed if air escapes from ends.

Inspect oil cooler by-pass valve and renew worn or damaged parts. Assemble valve with hollow end of plunger (4) toward spring (3). Remainder of assembly is the reverse of disassembly. Tighten cooler to filter head retaining nuts to 12-15 ft.-lbs. torque and filter head to block retaining cap screws 21-24 ft.-lbs. torque.

OIL SPRAY JETS

MF 2775 And MF 2805 Models

62. Engines are equipped with four oil spray jet tube assemblies located at bottom of cylinders which spray oil on the bottom of piston crowns to aid in cooling pistons. Oil pressure to spray jets is regulated by a control valve under block valley cover. Control valve is pre-set to 40 psi and is unadjustable. If piston damage is indicated due to overheating of piston, check piston oil spray control valve or oil spray jets and renew or adjust as required.

To inspect or renew oil spray control valve, remove injection pump as outlined in paragraph 75 or paragraph 79, external power steering pump, if equipped, as outlined in paragraph 12 and parts necessary to remove block valley cover. On MF 2805 models, remove turbocharger as outlined in paragraph 85. Remove control valve, then drive out pin (4-Fig. 54) and withdraw plunger (3) and spring (2). Reinstall control valve and tighten to 12-15 ft.-lbs. torque.

Adjustment of oil jet tubes can be accomplished after removal of rod and piston units as outlined in paragraph 47. Insert a rod (such as welding rod) long enough to protrude above cylinder liner when inserted into jet pipe (Fig. 55). Rod should be 1½ inches inboard and ⅝-inch to right of center when viewed from side looking toward center of engine. Oil

spray jet tube retaining cap screws should be tightened to 12-15 ft.-lbs. torque.

DIESEL FUEL SYSTEM

The diesel fuel system consists of three basic units; fuel tank (and filters), injection pump and injector nozzles. Early production models were equipped with a C.A.V. fuel system, where as current production models are equipped with a Bosch fuel system. Be sure to refer to the appropriate paragraph or sections when servicing the fuel system. Maintenance of absolute cleanliness is of utmost importance when servicing any unit associated with the diesel fuel system. Of equal importance is the avoidance of nicks, burrs or handling damage on any of the working parts.

Probably the most important precaution that service personnel can impart to owners of diesel powered tractors is to urge them to use an approved fuel that is absolutely clean and free from foreign materials. Extra precaution should be taken to make certain that no water enters the fuel storage tanks. Because of the high pressures and degree of control required of injection equipment, extremely high precision standards are necessary in the manufacture and servicing of diesel components. Extra care

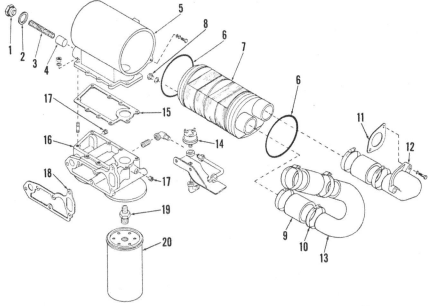

Fig. 53 — Exploded view of oil cooler/filter assembly and associated parts.

1. Plug	6. "O" rings	11. Gasket	16. Filter head
2. Fiber washer	7. Tube stack	12. Inlet pipe	17. Plugs
3. Spring	8. Coolant drain plug	13. Outlet pipe	18. Gasket
4. Plunger	9. Hose	14. Oil pressure switch	19. Oil filter adapters
5. Cooler housing	10. Clamp	15. Gasket	20. Oil filter

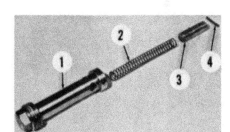

Fig. 54 — Exploded view of oil spray jet control valve.

1. Cartridge	3. Plunger
2. Spring	4. Pin

Fig. 55 — Adjust direction of oil spray jet by inserting a rod (R) through cylinder bore into jet tube (J). Refer to text for specifications.

in daily maintenance will pay big dividends in long service life and avoidance of costly repairs.

FUEL SEDIMENTER

All Models

63. Fuel system is equipped with a sedimenter between fuel lift pump and gas tank. Sedimenter is mounted on the left side of flywheel housing as shown in Fig. 56. Water should be drained from sedimenter daily or every 10 hours. To drain, loosen drain cock (1) at bottom of sedimenter. If servicing of sedimenter is required, renew "O" rings and seal rings before assembling.

FILTERS

All Models

64. **MAINTENANCE.** The fuel filter head is fitted with two renewable type elements as shown in Fig. 56 or Fig. 57. C.A.V. fuel filters (Fig. 56) are equipped with water drain cocks at bottom of filters. Drain cocks should be opened daily or every 10 hours to remove any water in filter bases. After each 200 hours of operation, the fuel filter elements should be renewed.

Renew fuel filter elements on C.A.V. fuel systems as follows: Close fuel shut-off valves at tank, thoroughly clean outside of filter housing, drain fuel from filters (loosen drain plugs), remove cap screws at top center of filter heads and remove the filter elements. Clean the filter bases and heads then reinstall with new elements and rubber sealing rings. Do not overtighten cap screws.

Bosch fuel system has two spin-on fuel filter elements (Fig. 57). When renewing, tighten by hand until sealing surface just touches then tighten another quarter turn. Bleed the diesel fuel system as outlined in paragraph 65.

BLEEDING

All Models

65. To bleed the system, open bleed plug(s) on top of fuel filter head (2-Fig. 56 or Fig. 57), then actuate priming plunger (3) above filter head on both fuel systems until air-free fuel flows from plug opening(s). Close bleed plug(s) while fuel is flowing. Loosen fuel return line (4-Fig. 58 or Fig. 59) on injection pump and actuate priming plunger until air-free fuel flows from fitting then tighten fuel return line. If engine was run out of fuel or fuel tank completely

drained, it may be necessary on C.A.V. fuel systems to remove bleed plug (5-Fig. 58) on right side of injection pump, move fuel shut-off control to run position, then crank engine with starter motor until air-free fuel flows from port. Close bleed plug while fuel is flowing.

NOTE: Starter should not be operated for more than 30 seconds without allowing 30 seconds for cooling.

Partially open the throttle and attempt to start the engine. If engine does not start, loosen compression nut at all injectors and turn engine over with starter until fuel escapes from all loosened connections. Tighten compression nuts and start engine.

INJECTOR NOZZLES

All Models

WARNING: Fuel leaves the injector nozzle with sufficient force to penetrate the skin. Keep exposed portions of your body clear of nozzle spray when testing.

66. **TESTING AND LOCATING A FAULTY NOZZLE.** If rough or uneven engine operation or misfiring indicates a faulty injector, the defective unit can usually be located as follows:

With engine running at the speed where malfunction is most noticeable (usually slow idle speed), loosen the compression nut on high pressure line for each injector in turn and listen for a change in engine performance. As in checking spark plugs, the faulty unit is the one which, when its line is loosened, least affects the running of the engine.

If a faulty nozzle is found and considerable time has elapsed since the injectors have been serviced, it is recommended that all nozzles be removed and serviced or that new or reconditioned units be installed. Refer to the following paragraphs for removal and test procedure.

Fig. 56—Drain water accumulated in sedimenter at drain cock (1). When bleeding fuel system, open bleed plug (2) and actuate primer pump (3). Refer to text.

Fig. 57—Bosch fuel system uses two spin-on fuel filters. When bleeding, open bleed plugs (2) and actuate priming plunger (3).

Fig. 58—C.A.V. fuel injection pump showing location of fuel return line (4) and injection pump fuel bleed plug (5).

67. REMOVE AND REINSTALL. Before loosening any fuel lines, thoroughly clean the lines, connections, injectors and surrounding area with air pressure and solvent spray. Disconnect and remove the leak-off line; disconnect pressure line and cap all connections as they are loosened to prevent dirt from entering the system.

On MF 2745 models, remove the stud nuts and withdraw injector unit from cylinder head. On MF 2775 and MF 2805 models, unbolt injector retaining clamp and if necessary, invert the clamp and slide it under injector, then strike clamp with a hammer as shown in Fig. 60 to remove injector unit.

Thoroughly clean the nozzle recess in cylinder head before reinstalling injector unit. It is important that seating surface be free from even the smallest particle of carbon or dirt which could cause the injector unit to fail to seat and result in blow-by. No hard or sharp tools should be used in cleaning. Do not re-use copper sealing washer located between injector nozzle and cylinder head, always install a new washer. Each injector should slide freely into place in cylinder head bore without binding. Make sure that high pressure fuel line fits squarely at both ends and that leak-off line compression seals are in good condition. On MF 2745 models, tighten retaining stud nuts to 10-12 ft.-lbs. torque. On MF 2775 and MF 2805 models, tighten injector retaining clamp nut to 28-30 ft.-lbs. torque. After engine is started, examine injectors for blow-by, making the necessary corrections before releasing tractor for service.

68. TESTING INJECTOR UNIT. A complete job of testing and adjusting the injector requires the use of special test equipment. Only clean, approved testing oil should be used in tester tank. The nozzle should be tested for opening pressure, seal leakage, back leakage and spray pattern. When tested, the nozzle should open with a sharp popping or buzzing sound, and cut off quickly at end of injection with a minimum of seat leakage and a controlled amount of back leakage.

Before conducting the test, operate tester lever until fuel flows, then attach the injector. Close the valve to tester gage and pump tester lever a few quick strokes to be sure nozzle valve is not plugged, that four sprays emerge from nozzle tip, and that possibilities are good that injector can be returned to service without overhaul. If adjustment is indicated in the preliminary tests, proceed as follows:

69. OPENING PRESSURE. Open the valve to tester gage and operate tester lever slowly while observing gage reading.

Opening pressure for C.A.V. fuel systems should be as follows:

MF 2745 models . . . . .2940 psi
MF 2775 and MF 2805 models. . 2870 psi

Opening pressure for Bosch fuel systems should be as follows:

MF 2745 models . . . . .3530 psi
MF 2775 models . . . . .3450 psi
MF 2805 models . . . . .2870 psi

If not as specified, remove cap nut (1-Fig.61) and turn adjusting sleeve (2) as required to obtain the recommended pressure.

NOTE: When adjusting a new injector or an overhauled unit with a new pressure spring (5), set the pressure at 3090 psi for all models with C.A.V. fuel systems and 3670 psi on MF 2745 and MF 2775 models and 3090 psi on MF 2805 models with Bosch fuel systems, to allow for initial pressure loss as the spring "settles in".

70. SEAT LEAKAGE. The nozzle tip should not leak at a pressure 150 psi less than opening pressure. To check for leakage, actuate tester lever slowly and as the gage needle approaches the suggested test pressure, observe the nozzle tip. Hold the pressure for 10 seconds; if drops appear or nozzle tip is wet, the valve is not seating and injector must be disassembled and overhauled as outlined in paragraph 73.

71. BACK LEAKAGE. If nozzle seat as tested in paragraph 70 was satisfactory, check the injector and connections for wetness which would indicate external leakage. If no leaks are found, bring gage pressure to just below opening pressure and observe the time required for gage pressure to drop from 2200 psi to 1470 psi. For nozzle in good condition, this time should not be less than 10 seconds. A faster drop would indicate a worn or scored nozzle valve piston or body, and the nozzle assembly should be renewed.

NOTE: Leakage of the tester check valve or connections will cause a false reading, showing up in this test as excessively fast leakback. If all injectors tested fail to pass the test, the tester rather than the units should be suspected.

72. SPRAY PATTERN. If leakage and pressure are as specified when tested as outlined in paragraphs 69 through 71, operate the tester handle several times while observing the spray pattern. Four finely atomized, equally spaced, conical sprays should emerge from nozzle tip, with equal penetration into the surrounding atmosphere.

Fig. 61 — Exploded view of an injector unit of the type used on both C.A.V. and Bosch fuel systems.

1. Cap nut	8. Inlet filter
2. Adjusting sleeve	9. Gasket
3. Spring seat	10. Inlet adapter
4. Gasket	11. Dowel
5. Spring	12. Nozzle valve
6. Valve spindle	13. Nozzle body
7. Nozzle holder	14. Nozzle nut

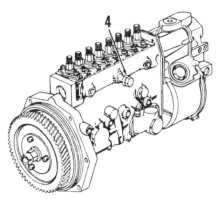

Fig. 59 — Bosch fuel injection pump showing location of fuel return line (4).

Fig. 60 — Use an inverted injector retaining clamp (C) to remove stuck injector units.

If patter is uneven, ragged or not finely atomized, overhaul the nozzle as outlined in paragraph 73.

73. **OVERHAUL.** Hard or sharp tools, emery cloth, grinding compound, or other than approved solvents or lapping compounds must never be used. An approved nozzle cleaning kit is available through a number of specialized sources.

Wipe all dirt and loose carbon from exterior of nozzle and holder assembly, refer to Fig. 61 and proceed as follows:

Secure the nozzle in a soft-jawed vise or holding fixture and remove cap nut (1). Back off the adjusting sleeve (2) to completely unload pressure spring (5). Remove nozzle retaining cap nut (14) and nozzle body (13). Nozzle valve (12) and body (13) are matched and must never be intermixed. Place all parts in clean calibrating oil or diesel fuel as they are removed. Clean the exterior surfaces with a soft wire brush, soaking in an approved carbon solvent if necessary, to loosen hard carbon deposits. Rinse the parts in clean diesel fuel or calibrating oil immediately after cleaning, to neutralize the carbon surfaces. Clean the pressure chamber of nozzle tip using the special reamer as shown in Fig. 62. Clean the spray holes in nozzle body with the proper diameter wire probe held in a pin vise as shown in Fig. 63. Nozzle spray hole diameters are as follows:

C.A.V. fuel system:
MF 2745 models . 0.0118-0.0122 inch
MF 2775 models . 0.0140-0.0144 inch
MF 2805 models . 0.0136-0.0140 inch

Bosch fuel system:
MF 2745 models . 0.0112-0.0116 inch
MF 2775 models . 0.0120-0.0124 inch
MF 2805 models . 0.0140-0.0142 inch

Wire probe should protrude from pin vise only far enough to pass through spray holes to prevent bending and breakage. Rotate pin vise without applying undue pressure.

Clean valve seats by inserting small end of valve seat scraper into nozzle and rotating tool as shown in Fig. 64. Use the proper size drill bit or wire to clean small feed channel bore as shown in Fig. 65. Clean carbon deposits from fuel gallery side of nozzle body with special groove scraper as shown in Fig. 66.

With cleaning accomplished, back flush the nozzle by installing the reverse flusher adapter on injector tester and nozzle body in adapter, tip-end first. Secure with the knurled adapter nut and insert and rotate the nozzle valve while flushing. After nozzle is back-flushed, seat can be polished using a small amount of tallow on end of polishing stick and rotating the stick while moderate pressure is applied.

Light scratches on valve piston and bore can be polished out by careful use of special injector lapping compound only. **DO NOT** use valve grinding compound or regular commercial polishing agents. **DO NOT** attempt to reseat a leaking valve using polishing compound. Clean thoroughly and back-flush if lapping compound is used.

Reclean all parts by rinsing thoroughly in clean diesel fuel or calibration oil and assemble valve to body while immersed in the cleaning fluid. Reassemble the injector while still wet. With adjusting sleeve (2-Fig. 61) loose, reinstall nozzle body (13) to holder (7) making sure valve (12) is installed and locating dowel (11) is properly aligned. Tighten nozzle retaining nut (14) to a torque of 50 ft.-lbs. Do not overtighten. Distortion may cause valve to stick and no amount of overtightening can stop a leak caused by scratches or dirt.

Retest the assembled injector assembly as outlined in paragraphs 68 through 72.

INJECTION PUMP

The injection pump is a completely sealed unit. No service work of any kind should be attempted on the pump or governor unit without the use of special pump testing equipment and proper

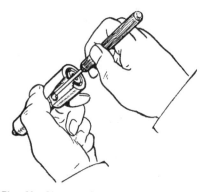

Fig. 66—Clean carbon deposits from nozzle body using a suitable tool.

Fig. 62—Clean the pressure chamber in nozzle tip using the special reamer as shown.

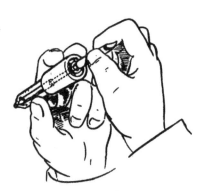

Fig. 64—Clean the valve seat using scraper as shown.

Fig. 63—Clean spray holes in nozzle tip using a pin vise and wire probe.

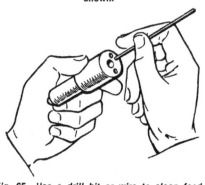

Fig. 65—Use a drill bit or wire to clean feed channel bore.

Fig. 67—Locations of slow idle speed stop screw (S) and high speed stop screw (H) on C.A.V. injection pump are as shown.

training. Inexperienced or unequipped service personnel should never attempt to overhaul an injection pump.

C.A.V. Fuel System

74. **ADJUSTMENT.** The slow idle stop screw (S-Fig. 67) should be adjusted with engine warm and running, to provide the recommended slow idle speed of 725-775 rpm.

The high speed stop screw (H) is factory set and sealed. High idle (no-load) speed should be 2625-2750 rpm for MF 2805 models and 2730-2860 rpm for MF 2745 and MF 2775 models.

There are three types of fuel shut-off controls on the C.A.V. injection pump. Two are cable operated and one is an electric solenoid. The cable actuated fuel shut-off control models either have the cable attached directly to fuel shut-off valve or the cable is attached to a mechanical linkage that actuates fuel shut-off valve. To adjust models with cable attached directly to fuel shut-off control valve, loosen cable retaining set screw on control valve and set control knob 1/16-inch away from instrument panel. Making certain control valve is in the "ON" position (vertical), tighten cable retaining set screw. Check operation and readjust if necessary.

To adjust models with cable attached to a mechanical linkage (Fig. 68), disconnect cable (C) from lever and set control knob 1/16-inch away from instrument panel. Push lever forward until it stops on bracket, then adjust cable (C) so it can be reattached without moving lever. Adjust rod (R) so fuel shut-off valve (F) is in the "ON" position (vertical). Check operation and readjust if necessary.

No adjustment is required if electrical solenoid is used.

An excess fuel control, located on right side of injection pump (Fig. 69) is used to aid in cold starting. To adjust, disconnect cable from excess fuel control lever (E) and set control knob 1/16-inch away from instrument panel.

Push control lever (E) forward until it rests on stop, then adjust cable so it can be reattached without moving lever. Check operation and readjust if necessary.

75. **REMOVE AND REINSTALL.** Before attempting to remove the injection pump, thoroughly wash pump and connections with clean diesel fuel or approved solvent.

CAUTION: Allow engine to cool before washing or seizure of parts may result.

Remove hood, then disconnect throttle cable, excess fuel cable and stop cable or electrical connection on injection pump. Remove high pressure lines and cap all connections as lines are removed. Disconnect low pressure fuel lines from injection pump, then unbolt and withdraw injection pump from engine.

Install injection pump by engaging master spline on pump shaft in pump drive shaft, then line up timing marks on pump and adapter plate flange (T-Fig. 70). Tighten pump retaining cap screws to 12-15 ft.-lbs. torque. The remainder of reassembly is the reverse of disassembly. Bleed fuel system as outlined in paragraph 65.

76. **PUMP TIMING.** Injection pump timing should be correct when timing marks (T-Fig. 70) on pump flange and adapter plate flange are aligned, and when pump drive hub mark and pointer are aligned as shown in Fig. 72 with number one piston at TDC on compression stroke.

If drive hub marks do not align, proceed as follows: Turn crankshaft in normal direction of rotation until notch (N-Fig. 71) on crankshaft pulley lines up with indicator (I) on timing gear cover and number one piston is at TDC on compression stroke. Release a valve on number one cylinder so that it rests on top of piston.

NOTE: It is recommended that an "O"

Fig. 69—Excess fuel control is located at E.

Fig. 70—When installing injection pump, align timing marks (T).

Fig. 71—Align notch (N) on crankshaft pulley with indicator (I) to check or adjust pump timing. Refer to text.

Fig. 68—View of fuel shut-off control using mechanical linkage.

Fig. 72—Remove plug (P) to check pump timing in auxiliary drive housing.

ring be placed on valve stem to prevent it from falling into the cylinder if crankshaft is turned.

Place a dial indicator on top of valve stem and carefully set number one piston at exactly TDC. Remove plug (P-Fig. 72) from auxiliary drive housing and check alignment of pointer with mark on injection pump drive hub. If minor adjustment is required, disconnect tachometer drive cable and remove injection pump drive gear cover (Fig. 73). Loosen cap screws (S) and align timing marks (Fig. 72) in auxiliary drive housing. Ensure number one piston is exactly at TDC on compression stroke and marks (Fig. 72) are aligned in auxiliary drive housing, then tighten cap screws to 30-35 ft.-lbs. torque.

77. If the validity of injection pump timing marks (T-Fig. 70) are questioned, verify marks using Timing Tool No. 8106 as follows: Remove injection pump as outlined in paragraph 75. Set number one piston at TDC as outlined in paragraph 76. Position timing tool as shown in Fig. 74. Loosen set screw (2) and slide shaft (1) into injection pump shaft engaging master spline. Lock shaft (1) in place. Set chamfered edge of bracket (3) at engine checking angle of 312 degrees. Release gage (5) and slide it against pump mounting flange then lock

Fig. 73—Loosen cap screws (S) to make minor adjustments of injection pump drive to engine timing.

Fig. 74—Static time injection pump using special timing tool as shown. Refer to text.

in place. Turn timing tool as shown in Fig. 75 to remove backlash and check alignment of mark on flange with gage. Adjust pump drive shaft as outlined in paragraph 76.

Bosch Fuel System

78. **ADJUSTMENT.** Slow idle speed

Fig. 75—Remove backlash then check alignment of timing mark (T) with timing tool.

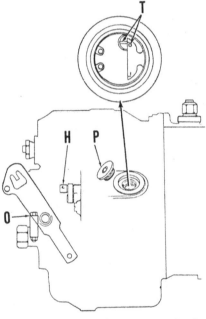

Fig. 76—Bosch fuel injection pump showing location of fuel shut-off control adjustment screw (O), high speed adjustment screw (H) and injection pump timing marks (T).

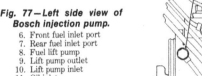

Fig. 77—Left side view of Bosch injection pump.
6. Front fuel inlet port
7. Rear fuel inlet port
8. Fuel lift pump
9. Lift pump outlet
10. Lift pump inlet
11. Oil inlet
12. Oil outlet
13. Pump oil filler plug
14. Governor oil filler plug
15. Maximum fuel adjustment

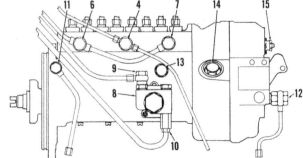

adjustment is made with engine warm and running. Move hand throttle to slow run position against detent, then adjust idle speed to 725-775 rpm by turning adjusting nuts at pedal end of throttle cable.

High speed adjustment screw (H-Fig. 76) is factory set and sealed. High idle (no-load) speed is given in the last four digits of pump setting code. If pump setting code is JA99E 1000/2750, then high idle speed should be 2750 rpm.

Fuel shut-off control adjustment screw (O) on injection pump, is factory set and should not be disturbed.

Excess fuel control, used to aid in cold starting, is activated by moving hand throttle to maximum speed position with engine off. When engine starts, excess fuel control automatically shuts off. Excess fuel control is non-adjustable.

79. **REMOVE AND REINSTALL.** Before attempting to remove the injection pump, thoroughly wash pump and connections with clean diesel fuel or approved solvent.

CAUTION: Allow engine to cool before washing or seizure of parts may result.

Remove hood, then disconnect throttle cable. Remove high pressure lines and cap all connections as lines are removed. Disconnect oil lubrication lines from front and rear of pump. Disconnect low pressure fuel lines from injection pump and fuel lift pump. Disconnect tachometer drive cable, then remove injection pump drive gear cover. Disconnect support bracket from bottom side of pump. Unbolt and withdraw pump complete with drive gear.

To install pump, turn crankshaft in normal direction of rotation until notch (N – Fig. 71) on crankshaft pulley lines up with indicator (I) on timing gear cover and number one piston is at TDC on compression stroke. Remove plug (P-Fig. 76) on governor housing and locate timing mark groove in governor spring spindle. Slide injection pump into place engaging injection pump drive gear with idler gear. If groove in governor spring spindle is not visible, with-draw pump and re-engage gear, then secure pump to engine. Adjust static timing as outlined in paragraph 80. Remove oil filler plugs (13 and 14-Fig. 77) and add 0.5 pt. of engine oil in governor housing and 0.6 pt. in pump body. Connect injection pump lines and bleed fuel system as outlined in paragraph 65. Remainder of reassembly is the reverse of disassembly.

80. **STATIC TIMING.** To check or adjust static timing, turn crankshaft in normal direction of rotation until notch (N – Fig. 71) on crankshaft pulley lines up with indicator (I) on timing gear cover and number one piston is at TDC on compression stroke. Release a valve on number one cylinder so that it rests on top of piston.

NOTE: It is recommended that an "O" ring be placed on valve stem to prevent it from falling into the cylinder if crankshaft is turned.

Place a dial indicator on top of valve stem and carefully set number one piston at exactly TDC. Remove plug (P – Fig. 76) from right side of governor housing and check alignment of groove on governor spring spindle with "O" mark on timing circlip as shown at (T). If minor adjustment is required, disconnect tachometer drive cable, then remove injection pump drive gear cover shown in Fig. 73. Loosen drive gear retaining cap screws (S) and align timing marks in governor housing. Ensure number one piston is exactly on TDC compression stroke and timing marks are properly aligned in governor housing, then turn gear counterclockwise just enough to remove backlash and

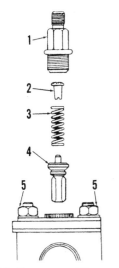

Fig. 78 – Injection pump delivery valve components must be removed when spill timing the injection pump.

1. Outlet
2. Peg
3. Spring
4. Delivery valve
5. Nuts

tighten drive gear cap screws to 37 ft.-lbs. torque. If the validity of timing marks is questioned, spill timing of injection pump will be required as outlined in paragraph 81.

81. **SPILL TIMING.** Refer to Fig. 77 for location of fuel ports. Set number one piston at TDC on compression stroke as outlined in paragraph 80. Remove number one fuel outlet (1 – Fig. 78) and withdraw peg (2), spring (3) and delivery valve (4), then fit a spill line on the outlet as shown in Fig. 79.

CAUTION: Do not loosen the nuts (5 – Fig. 78) on either side of outlet port or internal phasing of pump may be changed.

Install an eight-inch long line with a fuel reservoir (7 and 8 – Fig. 79) on the front fuel inlet port of injection pump and cap off the rear inlet port to prevent drainage.

Move throttle from idle position to full run position. With engine set at TDC, zero the dial indicator, then turn crankshaft against normal rotation approximately 45 degrees. Open fuel reservoir (8 – Fig. 79), fuel should flow freely from spill line (6). With fuel flowing, turn crankshaft in normal direction of rotation until fuel is a fast drip from spill line. Dial indicator reading should be at the correct static timing point as given in test specifications which follow:

Model	Static Timing B.T.D.C.	Piston Position B.T.D.C.
MF 2745	20	0.182 inch
MF 2775	22	0.219 inch
MF 2805	21	0.200 inch

If minor adjustment is required, disconnect tachometer drive cable, then remove injection pump drive gear cover shown in Fig. 73. Loosen drive gear retaining cap screws (S) and reposition pump shaft in the required direction. Tighten drive gear retaining cap screws to 37 ft.-lbs. torque. Remove fuel reservoir and spill line. Install fuel outlet, tightening outlet fitting (1 – Fig. 78) to 44 ft.-lbs. torque. The remainder of reassembly is the reverse of disassembly.

FUEL LIFT PUMP

C.A.V. Fuel System

82. The fuel lift pump is mounted on the right side of auxiliary drive housing and is driven by a lobe on injection pump drive hub. Pump parts are available separately or in rebuilding kit. Output delivery pressure should be 10 psi.

Bosch Fuel System

83. The fuel lift pump (8 – Fig. 77) is mounted on the left side of injection pump and is driven by injection pump camshaft. Pump parts are not available separately and complete assembly should be renewed if defective. Output delivery pressure should be 10 psi.

TURBOCHARGER

MF 2805 Models

The Perkins ATV8.640 engines used on MF 2805 models are equipped with exhaust driven turbochargers. The turbocharger used is an AiResearch TO4B which provides 9-11 psi boost pressure at 2500 rpm with engine fully loaded.

TROUBLESHOOTING

84. The following table can be used for locating difficulty with turbocharger. Many of the probable causes are related to the engine air intake and exhaust systems which should be maintained in good condition at all times. Engine oil is used to cool and lubricate the turbocharger and damage can result from improper operation. A minimum oil

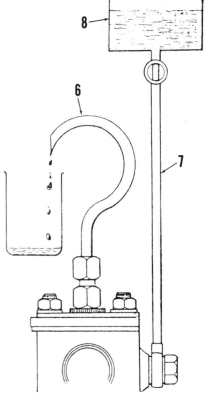

Fig. 79 – When correctly timed, fuel flow will be a fast drip from spill line.

6. Spill line
7. 8 inch line
8. Fuel reservoir

pressure of 10 psi at turbocharger oil inlet port with engine at idle speed is required for sufficient lubrication of turbocharger bearings and no less than 30 psi at all other times. The turbocharger should be allowed to cool by idling engine for 2-5 minutes after engine has been operated under load. Never shut down engine at high operating speed, this will cause turbocharger to continue to spin in excess of one hundred thousand rpm without lubrication and damage turbocharger bearings. Use extreme care to avoid damaging any of the turbocharger moving parts when servicing.

SYMPTOMS PROBABLE CAUSES

Engine Lacks Power 1,4,5,6,7,8,
    9,10,11,18,20,21,22,25,26,27,28,29,30
Black Exhaust Smoke 1,4,5,6,7,8,
    9,10,11,18,20,21,22,25,26,27
Blue Exhaust Smoke 1,2,4,8,9,17,
    19,21,22,29,32,33,34
Excessive Oil Consumption 2,8,19,
    20,33,34
Excessive Oil in the
    Turbine End 2,7,8,16,17,19,20,
    22,32,34,36
Excessive Oil in the
    Compressor End 1,2,4,5,6,8,9,
    16,19,20,21,33,36
Insufficient Lubrication 12,15,16
    23,24,31,36
Oil In Exhaust Manifold 2,7,19,20,
    22,28,29,30,33,34
Damaged Compressor Wheel 3,6,
    8,20,21,23,24,36
Damaged Turbine Wheel 7,8,18,
    20,21,22,34,36
Drag or Bind in Rotating
    Assembly . .3,6,7,8,13,14,15,16,20,21,
    22,31,34,36
Worn Bearings, Journals, Bearing
    Bores 6,7,8,12,13,14,15,16,20,
    23,24,31,35,36
Noisy Operation 1,3,4,5,6,7,8,9,10,
    11,18,20,21,22
Sludged or Coked Center
    Housing 2,15,17

KEY TO PROBABLE CAUSES

1. Dirty air cleaner element.
2. Plugged crankcase breathers.
3. Air cleaner element missing, leaking, not sealing correctly, loose connections to turbocharger.
4. Collapsed or restricted air tube before turbocharger.
5. Restricted (damaged) crossover pipe from turbocharger to inlet manifold.
6. Foreign object between air cleaner and turbocharger.
7. Foreign object in exhaust system from engine (check engine).
8. Turbocharger flanges, clamps or bolts loose.
9. Inlet manifold cracked, gaskets missing, connections loose.

10. Exhaust manifold cracked, burned, gaskets blown or missing.
11. Restricted exhaust system.
12. Oil lag (oil delay to turbocharger at start up).
13. Insufficient lubrication.
14. Lubricating oil contaminated.
15. Improper type lubricating oil used.
16. Restricted oil feed line.
17. Restricted oil drain line.
18. Turbine housing damaged or restricted.
19. Turbocharger seal leakage.
20. Worn journal bearings.
21. Excessive dirt buildup in compressor housing.
22. Excessive carbon buildup behind turbine wheel.
23. Too fast acceleration at initial start (oil lag).
24. Too little warm-up time.
25. Fuel pump malfunction.
26. Worn or damaged injectors.
27. Valve timing.
28. Burned valves.

29. Worn piston rings.
30. Burned pistons.
31. Leaking oil feed line.
32. Excessive engine pre-oil.
33. Excessive engine idle.
34. Coked or sludged center housing.
35. Oil pump malfunction.
36. Oil filter plugged.

TURBOCHARGER UNIT

85. **REMOVE AND REINSTALL.** Remove muffler, air cleaner cap and hood. Disconnect braces and remove shields necessary to remove intake crossover pipes. Remove exhaust pipe, exhaust outlet elbow and air intake pipe on turbocharger. Disconnect oil inlet and return lines from turbocharger, then unbolt and remove unit from exhaust manifold.

To reinstall, reverse removal procedure. Tighten turbocharger retaining nuts to 21-24 ft.-lbs. torque. Pour 3-4 ounces of clean oil into oil inlet and spin

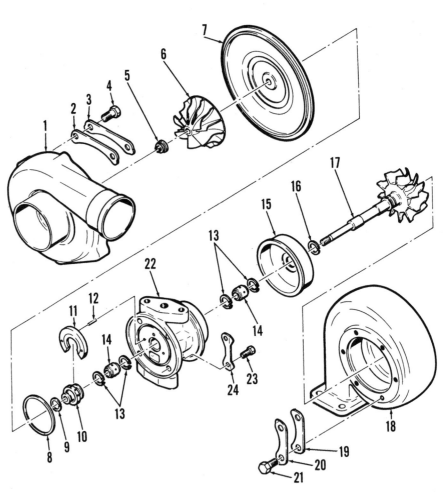

Fig. 80 — Exploded view of AiResearch turbocharger used on MF 2805 models.

1. Compressor housing		
2. Clamp	8. Seal ring	14. Bearing	19. Clamp
3. Lockplate	9. Piston ring	15. Turbine wheel shroud	20. Lockplate
4. Cap screw	10. Thrust collar	16. Piston ring	21. Cap screw
5. Nut	11. Thrust bearing	17. Turbine wheel and	22. Center housing
6. Compressor wheel	12. Roll pin	shaft	23. Cap screw
7. Backplate	13. Snap ring	18. Turbine housing	24. Lockplate

turbocharger to lube bearings and journals. Leave oil return line disconnected and crank engine with starter until oil begins to flow from return port, then tighten return line.

NOTE: Do not start engine until it is certain that turbocharger is receiving lubricating oil.

86. **OVERHAUL.** Remove turbocharger unit as outlined in paragraph 85. Mark across compressor housing (1 – Fig. 80), center housing (22) and turbine housing (18) to aid alignment when assembling. Unbolt and remove compressor housing and turbine housing from center housing. Use a box end wrench clamped in a vise to hold center of turbine wheel (17) and remove nut (5).

NOTE: A "T" handle wrench is recommended to remove nut so bending of turbine shaft is avoided.

Lift off compressor wheel (6), then remove center housing (22) and shroud (15) from turbine shaft (17). Remove cap screws (23) and separate center housing (22) from backplate (7). Remove seal ring (8) from center housing. Roll pins (12) in center housing should not be removed unless renewal is required. Withdraw thrust collar (10) with thrust bearing (11) from backplate. Slide thrust bearing (11) and piston ring (9) from thrust collar (10). No attempt should be made to remove star spring on center housing side of backplate (7), if renewal is required, spring and backplate can only be renewed as a unit. Remove outer snap rings (13), then withdraw bearings (14) from each side of center housing.

Clean all parts in a cleaning solution which is not harmful to aluminum. A stiff brush and plastic or wood scraper should be used after deposits have softened. When cleaning, use extreme caution to prevent parts from being nicked, scratched or bent. Thoroughly clean oil cavity (C – Fig. 81) and oil squirt hole (H) in center housing. Small passage in oil squirt hole (H) can be cleaned with a 0.053-0.060 inch diameter wire.

Inspect bearing bores in center housing (22 – Fig. 80) for scored or scratched surfaces. Bearing bore diameter must not exceed 0.6228 inch. Turbine end seal bore may be either stepped or standard depending on production. Turbine end seal bore diameter must not exceed 0.713 inch for stepped and 0.703 inch for standard. Inside diameter of bearings (14) should be 0.4010-0.4014 inch and outside diameter should be 0.6182-0.6187 inch. These bearings and their retaining rings (13) should be renewed each time unit is disassembled for service.

Thrust bearing (11) should be carefully inspected for wear or damage. Thickness should not be less than 0.1716 inch. Faces of thrust bearing must be flat within 0.0003 inch and diameter of bore for thrust collar must not exceed 0.430 inch. Make certain oil passages in thrust bearing are clean and free of obstructions. Oil passage in thrust collar (10) must be clean and thrust faces must not be warped or scored. Piston ring groove shoulders must not have step wear. Width of groove for piston ring (9) should be 0.064-0.065 inch and should not exceed 0.066 inch. Thrust bearing groove width in thrust collar (10) should be 0.1740-0.1748 inch and should not exceed 0.1752 inch; diameter at bottom of groove should not be less than 0.370 inch. Inside diameter of backplate (7) bore should be 0.4995-0.5005 inch and must not exceed 0.5010 inch. Thrust surface and seal contact area must be clean and smooth. Compressor wheel (6) must not show signs of rubbing with either compressor housing (1) or the backplate (7). Make certain that impeller blades are not bent, chipped, cracked or eroded. Inspect turbine shroud (15) for evidence of turbine wheel rubbing. Turbine wheel (17) should not show evidence of rubbing and vanes must not be bent, cracked, nicked or eroded. Vane tips must not be less than 0.025 inch thick.

Turbine wheel shaft must not show signs of scoring, scratching or overheating. Diameter of shaft journals should be 0.3997-0.4000 inch and should not be less than 0.3994 inch. Groove in shaft for piston ring (16) must not be stepped and width of groove must not exceed 0.0735 inch. Diameter of hub near piston ring should be 0.682-0.683 inch and must not be less than 0.681 inch.

Upon reassembly, the following parts should be renewed: Snap rings (13), bearings (14), piston rings (9 and 16), seal ring (8), lockplates (3, 20 and 24) and cap screws (4, 21 and 23).

Install inner snap rings (13) into center housing (22) with rounded edge toward bearings. Oil bearings (14), then insert bearings in each side of center housing. Install outer snap rings with rounded edge toward bearings. Fill piston ring groove with high vacuum silicon grease manufactured by Dow-Corning or equivalent. Install piston ring (16) and shroud (15) on turbine wheel assembly (17), then guide wheel assembly shaft through bearings (14). Slide shaft into center housing (22) as far as it will go. Install new piston ring (9) in groove of thrust collar (10), then install thrust bearing so smooth side of bearing (11) is toward piston ring end of collar. Install thrust bearing and collar assembly over shaft, making certain that pins in center housing engage holes in thrust bearing. Install new seal ring (8), then install backplate (7) making certain that piston ring (9) is not damaged. Install lock plates (24) and cap screws (23). Tighten cap screws (23) to 75-90 in.-lbs. torque and bend lock plates up around heads of screws. Place serrated end of turbine wheel assembly in a box end wrench clamped in a vise, then install compressor wheel (6) on turbine wheel shaft. After oiling washer face and threads of nut (5), install nut on shaft and tighten to a torque of 18-20 in.-lbs., then use a "T" handle to turn nut an additional 90 degrees.

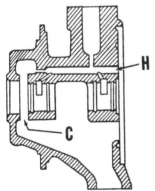

Fig. 81 — Sectional view of turbocharger center housing.

C. Turbine end oil cavity   H. Oil squirt hole

Fig. 82 — View showing method of checking turbine shaft end play. Shaft end play should be checked after unit is cleaned to prevent false reading due to carbon build-up.

CAUTION: If "T" handle is not used, shaft may be bent when tightening nut (5).

Install turbine housing (18) and compressor housing (1) to center housing (22) while aligning locating marks made before disassembly. Secure turbocharger assembly with clamp plates (2 and 19), lock plates (3 and 20) and cap screws (4 and 21)—coat threads with Fel Pro high temperature compound. Tighten cap screws to 100-130 in.-lbs. torque, then bend lock plates up around screw heads.

Check shaft end play and radial play. If shaft end play (Fig. 82) exceeds 0.004 inch, thrust collar (10–Fig. 80) and/or thrust bearing (11) is worn excessively. End play of less than 0.001 inch indicates incomplete cleaning (carbon not all removed) or dirty assembly and unit should be disassembled and cleaned. If turbine shaft radial play (Fig. 83) exceeds 0.006 inch, unit should be disassembled and bearings, shaft and/or center housing should be renewed. Maximum permissible limits of all of these parts may result in radial play which is not acceptable. Fill reservoir with engine oil and protect all openings of turbocharger until unit is installed on tractor.

COOLING SYSTEM

A 7 psi radiator cap along with a 6 psi relief valve assembly on coolant expansion bottle, controls cooling system pressure. Capacity is approximately 50 U.S. quarts on MF 2745 models and 55 U.S. quarts on MF 2775 and MF 2805 models.

THERMOSTATS

All Models

88. Thermostats are located in housings on front of each cylinder head (4 and 5–Fig. 84). MF 2745 models use one thermostat in each housing while MF 2775 and MF 2805 models use two thermostats in each housing. The thermostat should begin to open at 174-181 degrees F. and should be fully open at 199-205 degrees F. Engine should not be

Fig. 83—Turbine shaft radial play is checked with dial indicator through the oil outlet hole.

AIR INTAKE SYSTEM

All Models

87. An air filter restriction gage located in the instrument panel is used to determine when air cleaner service is required. Advise customers to observe instrument and to service air cleaner only when the needle remains in the red zone with engine running. Proper maintenance is important and of particular importance is avoidance of leaks which would permit unfiltered air to enter the engine.

Never wash filter in gasoline, fuel oil or solvent. Use only solution of MF part number 1900 726 M1, mixed according to directions on the carton, or lukewarm water and non-foaming detergent to clean the outer element. Install new outer element after six washings or once each year whichever occurs first. Be sure outer element is dry before installing. The inner filter element should not be cleaned, but new unit should be installed yearly or when restriction gage needle stays in red zone after outer element has been renewed.

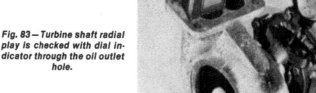

Fig. 84—View of installed water pump with hoses connected.

1. Water pump pulley
2. Fan drive
3. Idler pulley
4. Right thermostat housing
5. Left thermostat housing
6. Oil cooler

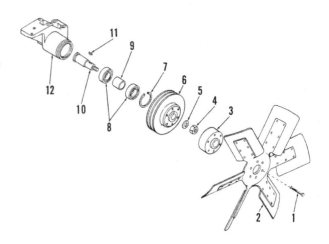

Fig. 85—Exploded view of fan drive assembly.

1. Cap screw
2. Fan assembly
3. Fan spacer
4. Nut
5. Washer
6. Pulley
7. Snap ring
8. Bearings
9. Spacer
10. Fan drive shaft
11. Woodruff key
12. Housing

operated without thermostats, which would allow coolant to by-pass radiator and cause overheating.

FAN DRIVE

All Models

89. Refer to Fig. 84 for installed view of fan drive and related parts. Fig. 85 shows an exploded view of fan drive. To remove fan drive, first remove air intake cap, muffler, hood, fan, fan spacer, belts and muffler support. Remove air conditioning compressor with mounting bracket if equipped. Unbolt and withdraw fan drive.

Installation is the reverse of removal. Tighten fan drive to water pump cap screws 70-75 ft.-lbs. torque and fan drive to timing case cap screw 35-40 ft.-lbs. torque.

If fan drive was disassembled for repair, upon reassembly tighten pulley retaining nut (4 – Fig. 85) 65-70 ft.-lbs. torque.

WATER PUMP

All Models

90. To remove water pump (1 – Fig. 84), first remove fan drive as outlined in paragraph 89. Drain cooling system, then remove thermostat housings (4 and 5) with hoses. Loosen hose clamps and slide remaining hoses down on connections away from water pump. Remove retaining nuts and cap screws, then slide water pump forward off studs.

When disassembling the pump, remove plug (1 – Fig. 86), insert a suitable mandrel into main shaft bore at pulley end and press out impeller shaft (11) with impeller (15). After removing snap ring (5), shaft (8) and bearings (6) can be pressed out from impeller end.

When reassembling, pack area between the two bearings (6) half full of high melting-point grease. If bearings are removed, install with open sides together next to spacer (7). Tighten nut (2) to 90-95 ft.-lbs. torque. Rear face of impeller center should be 0.015-0.020 inch beyond flush with rear surface of housing as shown in Fig. 87.

IDLER PULLEY

All Models

91. Fig. 88 shows an exploded view of idler pulley assembly. After removal, separate pulley from mounting bracket by driving out pin (13) and unscrewing

adjusting bolt (10), then remove nut (16), washer (15) and plate (14).

To renew bearings (4), remove bearing retainer (2) and snap ring (3). Press shaft (8) out rear of pulley (6), then press bearings (4) and spacer (5) out front of pulley.

When reassembling, install bearings (4) with open sides together next to spacer (5). Pack area between bearings half full of high melting-point grease. Temporarily install bearing retainer (2) to press shaft (8) with stepped washer (7) into pulley. Remove bearing retainer to install snap ring (3), then reinstall retainer.

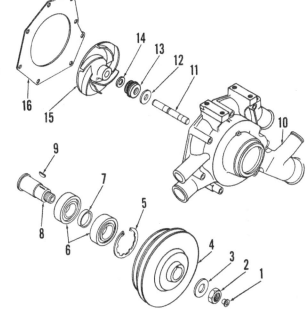

Fig. 86 — Exploded view of water pump assembly.
1. Plug
2. Nut
3. Washer
4. Pulley
5. Snap ring
6. Bearings
7. Spacer
8. Front shaft
9. Woodruff key
10. Pump housing
11. Impeller shaft
12. Water slinger
13. Seal
14. Ceramic collar
15. Impeller
16. Gasket

Fig. 87 — Use a straightedge and feeler gage as shown to measure impeller protrusion above water pump housing. Refer to text for specifications.

Fig. 88 — Exploded view of idler pulley assembly and associated parts.
1. Cap screw
2. Bearing retainer
3. Snap ring
4. Bearing
5. Spacer
6. Pulley
7. Stepped washer
8. Pulley shaft
9. Bracket
10. Adjusting screw
11. Washer
12. Collar
13. Pin
14. Plate
15. Washer
16. Nut
17. Spacer

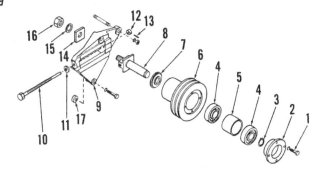

ELECTRICAL SYSTEM

ALTERNATOR

All Models

92. A Motorola alternator with a built-in voltage regulator is used on both models. Alternator specifications are as follows:

Field Current (75° F.)

Amperes 2.0-2.5
Volts 12
Rated Hot Output (Amperes) at
Maximum Operating Speed 55
Minimum Output (Amperes) at
Maximum Operating Speed 48

To disassemble alternator, first scribe matching marks on frame halves (5 and 21 – Fig. 89). Remove voltage regulator (25), brush set (23), insulating cover (26), then remove four through bolts. Carefully pry frame apart with a screwdriver between drive end frame (5) and stator frame (15) without allowing screwdriver to touch stator windings. Unsolder stator and voltage regulator leads from rectifier bridge (11). Remove rectifier bridge, capacitor (13) and stator (15). Remove "O" ring (16) from slip ring end frame (21), then remove D+ terminal screw (18) and spade connector/insulator (24) from slip ring end frame (21).

Remove pulley, fan (1), washer (2), Woodruff key (4) and spacer (3), then tap rotor shaft on block of wood to push it out of drive end frame bearing (6). Slide counterbored spacer (8) off of rotor (9) shaft. After removing bearing retainer (7), press bearing (6) out of drive end frame (5). Pull bearing (10) off rotor shaft. To reassemble alternator, reverse removal procedure.

NOTE: When reconnecting voltage regulator (25) lead wires, connect yellow wire to D+ terminal and green wire to brush assembly connector.

93. TESTING. Original brush set (23) may be reused if brushes protrude 3/16-inch or longer beyond bottom of holder, if brushes are not oil soaked or cracked and if brush set passes the following tests using an ohmmeter. Refer to Fig. 90. Brush set is good if ohmmeter shows continuity when checking brush 1 to terminal 3 and from brush 2 to mounting bracket (MB). An open circuit is required between brushes 1 and 2 and from terminal 3 to mounting bracket (MB).

Rectifier bridge (11 – Fig. 89) may be tested with an ohmmeter after unsoldering stator and voltage regulator leads. Refer to Fig. 91 and following text for test procedures.

Connect one ohmmeter lead to ground (G – Fig. 91) and other lead to one of terminals (T), then note ohmmeter reading. Reverse ohmmeter leads and note reading. A good diode will read approximately 30 ohms resistance in one test mode and continuity with test leads reversed. Repeat test on remaining two terminals (T). Repeat test between (B+) and terminals (T), then repeat test between (R)

Fig. 90 — Refer to text for testing procedure of brush set.

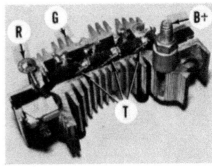

Fig. 91 — Test points for rectifier bridge. Refer to text.

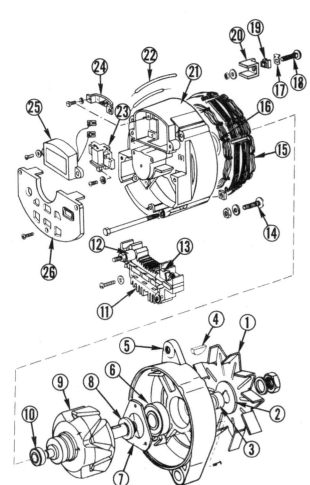

Fig. 89 — Exploded view of alternator with solid-state voltage regulator.

1. Fan
2. Washer
3. Spacer
4. Woodruff key
5. Drive end frame
6. Bearing
7. Bearing retainer
8. Counterbored spacer
9. Rotor
10. Bearing
11. Rectifier bridge
12. Spacer
13. Capacitor
14. Ground terminal screw
15. Stator
16. "O" ring
17. Clip
18. D+ terminal screw
19. Insulator
20. Insulator
21. Slip ring end frame
22. Wire
23. Brush set
24. Spade connector
25. Voltage regulator
26. Insulating cover

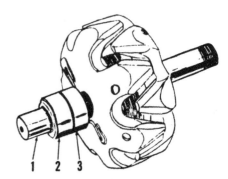

Fig. 92 — Removed rotor assembly showing test points to be used when checking for grounds (shorts) and opens.

and terminals (T). Renew rectifier bridge if any one pair of readings is the same regardless of test lead polarity.

Check bearing surface of rotor shaft (9 – Fig. 89) for visible wear or scoring. Examine slip ring surface for scoring or wear and rotor winding for overheating or other damage. Check rotor for grounded, shorted or open circuits using an ohmmeter as follows:

Refer to Fig. 92 and touch ohmmeter probes to points (1-2) and (1-3). A reading near zero will indicate a ground. Touch ohmmeter probes to slip rings (2-3). Reading should be 4.0-5.2 ohms. If ohmmeter indicates high resistance, windings are open. If ohmmeter reads less than 4 ohms, windings are shorted or grounded and rotor will have to be renewed.

Clean slip rings (2 and 3) with fine crocus cloth. Spin rotor to avoid flat spots on slip rings.

Stator (15 – Fig. 89) should be checked for shorted circuit between stator leads and stator frame. With one ohmmeter lead on stator frame, touch the other to each stator lead. Stator is defective and should be renewed if any reading shows continuity.

STARTER

All Models

94. A Delco-Remy starter motor and solenoid, model number 1114190 is used on all production models. Service replacement starters are model number 1114175. Model, series, type and serial numbers are located on main frame. Specifications for each model are as follows:

Starter No. 1114190
No Load Test-
  Volts 9
  Amperes (w/solenoid) 140-190
  RPM 4000-7000
Starter No. 1114175
No Load Test-
  Volts 9
  Amperes (w/solenoid) 140-215
  RPM 4000-7000
Pinion clearance for both models is 21/64-25/64 inch.

CIRCUIT DESCRIPTION

All Models

95. The two 95 ampere hour, 12 volt batteries are connected in parallel and negative posts are grounded. Make sure both batteries are of equal capacity and condition. Refer to Fig. 93 through Fig. 98 for schematic wiring diagrams.

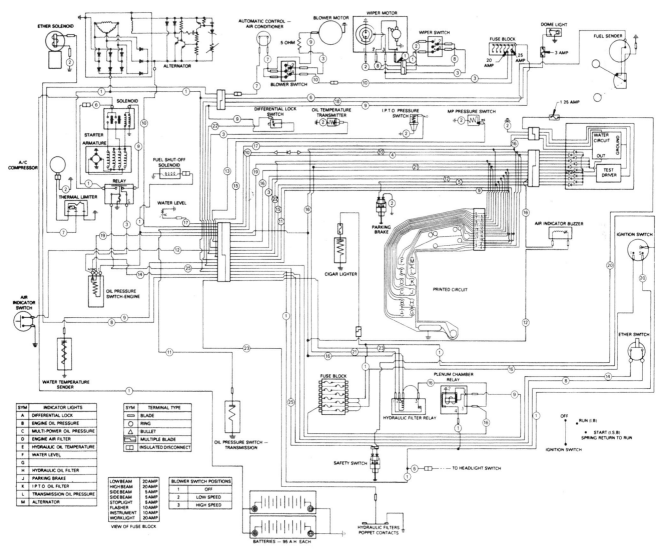

Fig. 93 — Wiring diagram typical of all models equipped with 12 instrument panel warning lights and a printed circuit board.

1.............Red	6.............Brown	11.............Pink	16.............Red/White	21.......Lt. Blue/Black	26.............Dk. Brown
2.............Black	7.............Purple	12.............Lt. Blue	17.............Black/Red	22.......Dk. Green/White	27.............Lt. Brown
3.............White	8.............Green	13.............Lt. Green	18.............Black/White	23.......Dk. Blue/White	28.............Blue
4.............Gray	9.............Orange	14.............Dk. Blue	19.............Purple/White	24.............Tan/Black	29.............Red/Black
5.............Tan	10.............Yellow	15.............Dk. Green	20.............Orange/White	25.............Orange/Black	30.............Lt. Green/Black

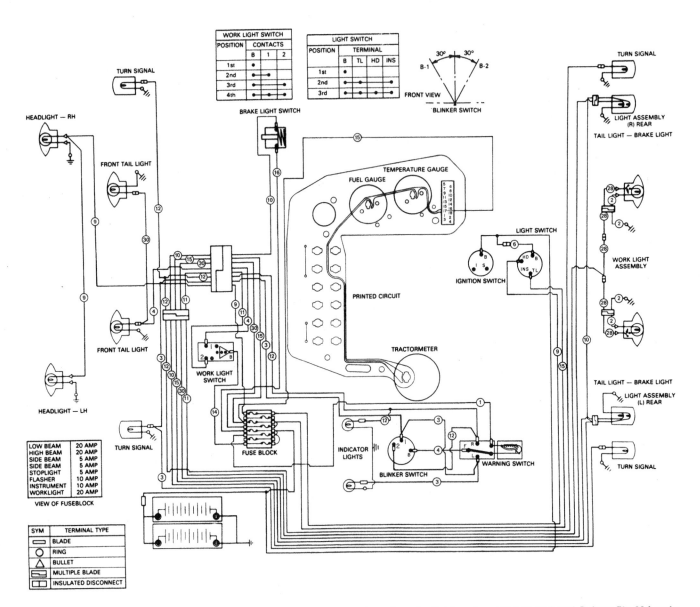

Fig. 94 — Light wiring diagram typical of all models equipped with 12 instrument panel warning lights and a printed circuit board. Refer to Fig. 93 for wire color code.

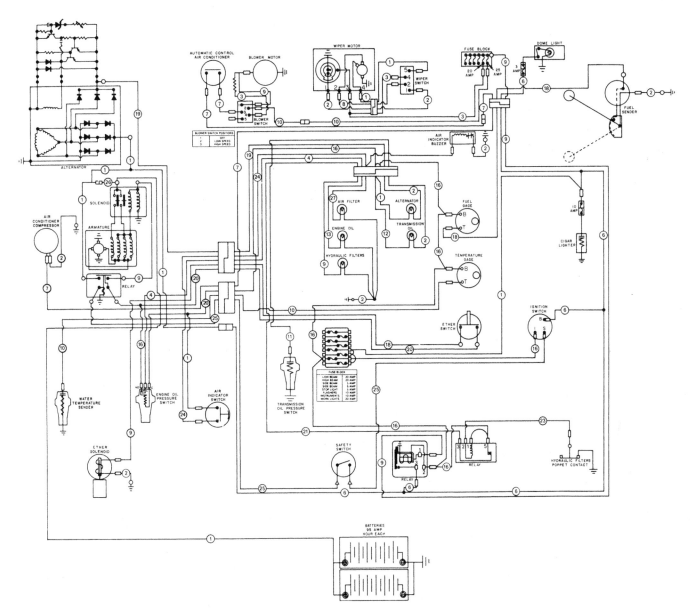

Fig. 95 — Wiring diagram typical of all models with right side muffler and five or six instrument panel warning lights. Refer to Fig. 93 for wire color code.

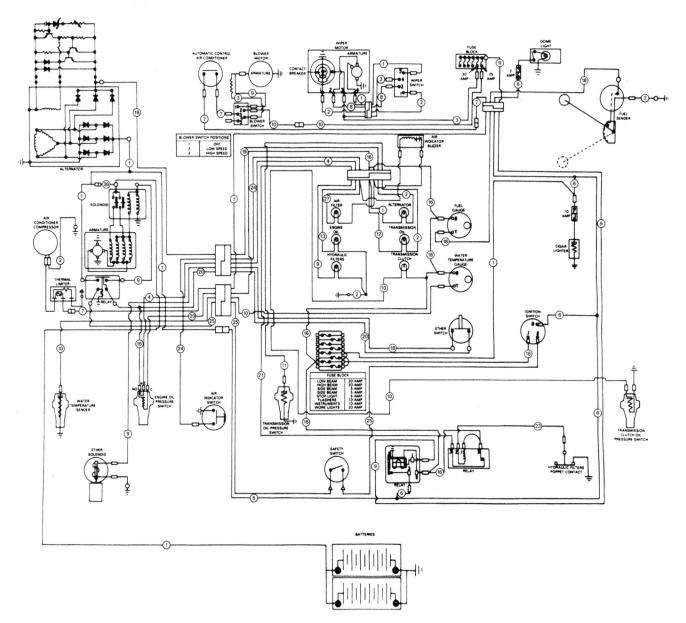

Fig. 96 — Wiring diagram typical of all models with left side muffler and five or six instrument panel warning lights. Refer to Fig. 93 for wire color code.

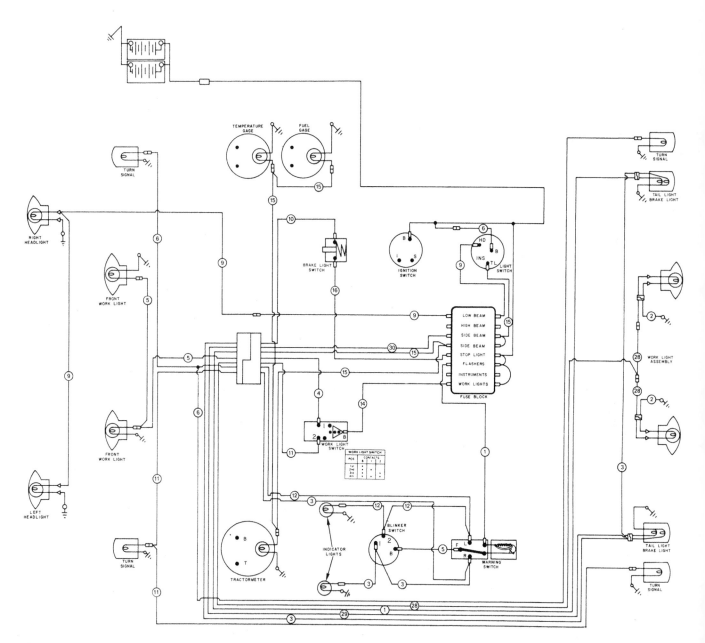

Fig. 97—Light wiring diagram typical of all models using five or six instrument panel warning lights. Refer to Fig. 93 for wire color code.

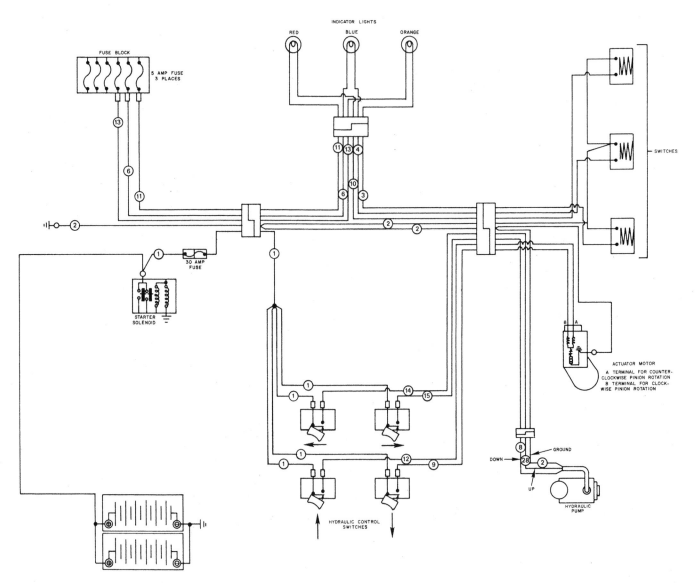

Fig. 98 — Remote control wiring diagram typical of all models. Refer to Fig. 93 for wire color code.

ENGINE CLUTCH

All models are equipped with a dry-type, double-disc clutch mounted on the flywheel and actuated by a foot pedal. Two different types of clutch cover assemblies are used. Early models are equipped with a conventional clutch cover assembly and late models are equipped with an angle spring clutch cover assembly. Refer to the appropriate paragraph for servicing of each.

The clutch controls tractor travel while the independent power take-off is controlled by a separate multiple-disc, wet-type clutch located in rear axle center housing. Fully depressing clutch pedal engages transmission brake which is used to aid in shifting of gears. To avoid unnecessary wear on transmission brake, clutch pedal should never be fully depressed before tractor is stopped.

ADJUSTMENT

All Models

96. If tractor is equipped with an eight-speed transmission, adjust clutch release cable until all free play is just removed. A force of approximately 5-10 lbs. should be required to produce

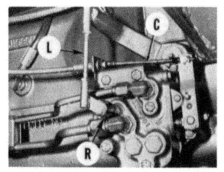

Fig. 99 — Adjust linkage rod (L) so roll pin (R) is vertical when clutch pedal is up on models equipped with Multi-Power transmission.

Fig. 100 — View of transmission brake adjustment point on early model tractors. Clearance (C) should be ⅜-7/16-inch. Refer to text.

¼-inch of pedal travel.

If equipped with a Multi-Power transmission, pilot valve may be adjusted by turning linkage rod (L – Fig. 99) until roll pin (R) is vertical when clutch pedal is up.

Transmission brake is properly adjusted when there is 3/8-7/16 inch clearance between front transmission brake plate and contact surface of release bearing housing. To adjust the transmission brake on early models, remove inspection cover on bottom of transmission, loosen clutch release bearing sleeve locknut (N – Fig. 100) with a punch and hammer.

NOTE: Always push clutch pedal down while sleeve locknut (N) is being locked or unlocked.

With clutch pedal released, use special tool No. 7282 or 7283 and turn adjusting sleeve nut (A) until adjustment is correct then tighten locknut (N). Readjust clutch cable if required.

To adjust the transmission brake on late models, first remove inspection cover on bottom of transmission. Rotate flywheel until adjusting slot on clutch cover assembly is at bottom, then remove adjusting ring lockstrap. Depress clutch pedal and turn adjusting ring as shown in Fig. 101. Each notch on adjusting ring is equal to 0.020 inch movement of release bearing.

NOTE: Clutch pedal must be depressed in order to turn adjusting ring.

Release clutch pedal and check clearance. Once clearance is correct, reinstall adjusting ring lockstrap. Readjust clutch cable if required.

All models are equipped with a safety start switch designed to prevent starting unless clutch pedal is fully depressed. Two types of safety start switches are used. Probe type (1 – Fig. 102) should not require adjustment. Plunger type (2 – Fig. 103) is properly adjusted when switch is closed during last ½-inch of clutch pedal travel.

Fig. 101 — Hold clutch pedal down then turn adjusting ring (31).

ENGINE/TRANSMISSION SPLIT

All Models

97. Proceed as follows to detach (split) the engine and front end assembly from transmission. Disconnect batteries, then remove hood and side panels. Disconnect main wiring harness plug at cab and electrical connector located behind right cylinder head. Disconnect air conditioning self-sealing hoses, heater hoses and oil line to external power steering reservoir, if equipped, at cab. Remove oil line at top of brake master cylinder. Disconnect necessary power steering hoses at steering control valve. Disconnect oil lines from fittings on cab side of support panel and remove windshield hoses from fittings on cab. Disconnect tachometer cable, fuel shut-off cable, throttle cable and excess fuel cable from engine, then pull cables free of support panel or lower heat shield. Close fuel tank shut-off valves and disconnect fuel return line behind battery tray and fuel inlet line from sedimenter. Remove necessary fuel line retaining clamps.

Disconnect clutch release cable from release arm side of transmission. On models with Multi-Power transmission, disconnect shift cable with mounting bracket and links from shift valve.

Fig. 102 — Probe type safety start switches (1) should not require adjustment.

Fig. 103 — Plunger type safety start switches (2) are adjustable. Refer to text.

Disconnect linkage rod from clutch shaft arm to shift valve. Remove inspection cover on bottom of transmission. Cut lockwires on release fork lock bolts, then remove lock bolts. Pull release fork shafts out of each side of transmission housing. If necessary, remove Multi-Power shift valve to pull out right-side release fork shaft. Remove release fork through inspection hole on bottom of transmission. Remove pto top link support at rear of tractor (Fig. 104) to release tension on pto shaft. Place a rolling floor jack under engine oil pan and another under front of transmission. Insert wooden blocks between front axle and axle support to prevent tipping.

NOTE: Remove front end weights if necessary to prevent front of tractor from tipping forward.

Remove starter for access to cap screw inside flywheel housing. Remove all cap screws retaining engine to transmission. Install two 7-inch guide studs to aid alignment during removal and installation.

NOTE: Floor panel inside cab may require removal to gain access to upper cap screws.

Check all wires and tubes to be sure that they are disconnected and positioned so they will not be damaged. Separate engine and front end assembly from transmission.

When reinstalling engine observe the following: Lube input shaft splines and clutch release bearing. Make sure clutch release bearing is positioned with lube hose fitting at top. Carefully move engine back against transmission housing while aligning splines of clutch with transmission input shaft. Engine must be flush against transmission before tightening any cap screws. **DO NOT** force together. Tightening torques for cap screws are as follows: ½-inch–63-95 ft.-lbs.; ⅝-inch–130-300 ft.-lbs.; ¾-inch–240-350 ft.-lbs.; ⅞-inch–300-

400 ft.-lbs. Remainder of procedure for joining engine and front end assembly to transmission is reverse of disassembly.

REMOVE AND REINSTALL CLUTCH

All Models

98. To remove clutch assembly for service, first split tractor as outlined in paragraph 97. Punch mark clutch cover (24 or 37 – Fig. 105), floater plate (8) and flywheel (6) so balance can be retained during reassembly.

On early models (prior to serial number 9R6433), install three 5/16-inch UNC x 2-inch long cap screws (C – Fig. 106) through holes in clutch cover and tighten until heads touch cover.

On late models (serial number 9R6433 and up), insert two ½-inch square blocks of wood between clutch cover and release bearing housing as shown in Fig. 107. Position blocks so mounting bolt holes (H) are accessible. Unbolt and withdraw clutch assembly.

NOTE: If clutch parts are to be reused, keep them in their original order for reassembly.

To reinstall clutch on all models, lube inside of pto seal, then install front clutch disc (7 – Fig. 105) into flywheel with disc hub facing flywheel. Install floater plate (8) aligning punch marks, then install rear clutch disc (7A) with disc hub facing transmission. While using a suitable clutch pilot tool, install clutch cover assembly. Evenly tighten clutch cover retaining cap screws to 25-38 ft.-lbs. of torque.

On early models, remove the three cap screws (C – Fig. 106) and proceed as follows: If clutch assembly was overhauled as outlined in paragraph 99, release lever height must be adjusted after clutch has been reassembled. Check lever height with a straightedge and machinists rule as shown in Fig. 108. Distance from straightedge to bottom of groove in back side of release lever should be one inch. Use special tool No. 7280 or equivalent, as shown in Fig. 109 to adjust lever to exact dimension. Locknut should be tightened to 60 ft.-lbs. of torque.

Release bearing height can be properly adjusted before reassembly of engine to transmission, using special tool No. 6622. Screw release sleeve (21A – Fig. 105) with locknut (18) into release lever

Fig. 105 – Exploded view of clutch assemblies. Clutch cover assembly (9 through 29) is used prior to tractor serial number 9R6433. Clutch cover assembly (30 through 39) is used on tractor serial number 9R6433 and up.

1. Guide dowel
2. "O" ring
3. Pto coupler
4. Seal
5. Seal shield
6. Flywheel
7. Front clutch disc
7A. Rear clutch disc
8. Floater plate
9. Pressure plate
10. Bearing
11. Lever pivot pin
12. Lever to eyebolt pin
13. Release lever
14. Release lever spring
15. Eyebolt bearing
16. Eyebolt
17. Release spider
18. Spider locknut
19. Eyebolt adjusting nut
20. Grease fitting
21. Release bearing housing
21A. Release sleeve
22. Dark maroon stripe spring
23. Light green stripe spring
24. Clutch cover
25. Locknut
26. Release bearing
27. Bearing spring
28. Retaining ring
29. Bearing cover
30. Pressure plate
30A. Retaining spring
31. Adjusting ring
32. Release lever
33. Retaining ring
34. Lock ring
35. Release sleeve retainer
36. Pressure spring with pivot
37. Clutch cover
38. Lock strap
39. Release bearing assembly

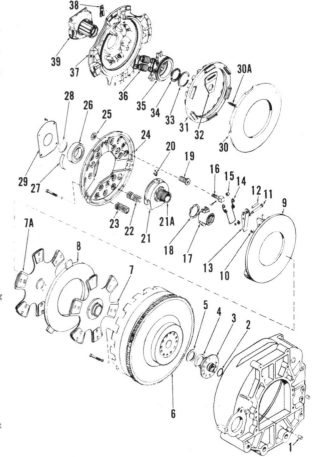

Fig. 104 – Remove pto top link support to remove tension on pto shaft.

spider (17), then adjust release bearing using special tool as shown in Fig. 110. Distance from rear face to flywheel housing to rear face of release bearing should be 5.408-5.418 inch. Tighten locknut after adjustment is correct. Make sure release bearing lube hose is toward left side of tractor.

On late models, remove the two ½-inch blocks of wood, then after reassembly of engine to transmission, adjust transmission brake as outlined in paragraph 96.

OVERHAUL

Early Models

99. Remove clutch cover assembly as outlined in paragraph 98. Compress assembly in a suitable press and remove cap screws (C – Fig. 106) that were installed during removal procedure. Use special tool No. 7280 or equivalent to remove release lever eyebolt locknuts (25 – Fig. 105), then release pressure on cover (24) and springs (22 and 23). Note

spring location on pressure plate (9), then remove springs. Disassemble release levers (13).

Check springs (22 and 23) for overheating, breakage, wear or other damage and against the following pressure specifications:

Light Green Striped Spring:
   Pressure at 1-27/32 inch . . 150-160 lbs.
Dark Maroon Striped Spring:
   Pressure at 1-27/32 inch . . 130-140 lbs.

Inspect remaining components for excessive wear or damage and renew as necessary.

To reassemble, lubricate release lever (13) lightly and assemble so heads of pivot pins (11) are to the side shown in Fig. 111. Place springs (22 and 23 – Fig. 105) onto pressure plate.

NOTE: Three dark maroon stripe springs (22) are installed on center inner pressure plate pins.

Place cover over springs aligning punch marks on cover and pressure plate made before disassembly. Apply pressure to cover until locknuts (25) can just be installed. Install release lever spider (17). Adjust release lever eyebolt nuts (19) until flush with end of bolt threads then tighten locknuts. Reinstall three cap screws (C – Fig. 106) into

Fig. 106 – On early models, install cap screws (C) before removing clutch cover assembly.

Fig. 109 – Use special tool No. 7280 to adjust release lever to proper height.

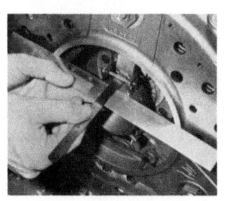

Fig. 107 – On models later than tractor serial number 9R6432. Install two ½-inch blocks of wood (W) before removing clutch cover assembly. Refer to text.

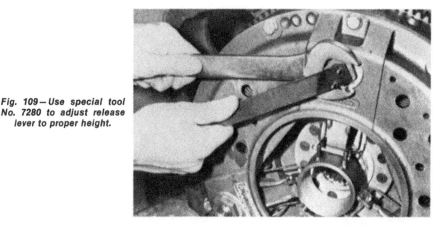

Fig. 110 – Release bearing height can be properly adjusted using tool No. 6622.

Fig. 108 – Use a straightedge and machinists ruler to check release lever height.

clutch assembly, then release pressure of hydraulic press.

Pads of clutch discs (7 and 7A – Fig. 105) are 0.419-0.448 inch thick when new. Renew disc as necessary. Reinstall clutch assembly as outlined in paragraph 98.

Late Models

100. Remove clutch cover assembly as outlined in paragraph 98. Punch mark clutch cover (37 – Fig. 105) and pressure plate (30), then remove the four return springs (30A) and withdraw pressure plate. Remove adjusting ring lockstrap (38), then rotate adjusting ring (31) counterclockwise to remove it with release levers (32). Separate release levers from adjusting ring. Remove retaining ring (33) from release sleeve retainer (35). Insert three 5/16 x 5 inch threaded rods (R – Fig. 112) through mounting holes in release sleeve retainer (35 – Fig. 105) and clutch cover (37) with retaining nuts flush with rods on release bearing side. Use a suitable press as shown in Fig. 112 to compress release sleeve retainer, then tighten the three nuts.

NOTE: Support release bearing on a 2¾ inch diameter plate. Do not apply pressure to the release bearing cover.

Remove clutch cover assembly from press. Remove the wood blocks, then slide release bearing forward to remove lock ring halves (34 – Fig. 105). Withdraw release bearing.

Reinstall clutch cover in press, compress release sleeve retainer and remove the three nuts. Release press, remove retainer (35) and pressure springs with pivots (36).

Inspect all parts for cracks, damage or excessive wear. Check clearance between the contact points of release sleeve retainer lugs with slots in clutch

cover and pressure plate lugs with slots in clutch cover. Wear should not exceed 0.020 inch. Maximum allowable resurface of pressure plate is 0.015 inch. Renew release bearing as required. Reassembly is the reverse of disassembly while noting the following:

Use a high temperature grease on adjusting ring threads. Turn adjusting ring completely in clutch cover, then backout 1½ turns for initial adjustment.

EIGHT-SPEED TRANSMISSION

All models are equipped with a main gear change transmission which has four forward and four reverse speeds and is compounded by a range transmission with two speeds which provides a total of eight forward speeds and six reverse speeds. A lock-out in shift column prevents engaging seventh and eighth speeds when in reverse.

A transmission brake is mounted on front cover of all models. The brake is applied by clutch release bearing after clutch is fully depressed.

Massey-Ferguson Permatran lubricant is used in transmission and center housing. Oil capacity including filter change, for transmission and rear axle is 30 gallons on all models. Fluid should be changed each 750 hours or annually.

Refer to the appropriate following

paragraphs for service to eight speed transmission or to paragraphs 110 through 118 for service to the Multi-Power transmission.

TRANSMISSION BRAKE

101. **R&R AND OVERHAUL.** Split tractor as outlined in paragraph 97. Transmission brake is located in front of transmission front cover and consists of four stationary plates and two rotating plates held in place by snap rings.

Remove snap rings (1 – Fig. 113) and slide transmission brake plates (2 and 3) off input shaft. Renew plates as required. Once installed, contact surface of transmission brake should be 5.790-5.846 inch from front of transmission housing. Reverse removal procedure to reassemble unit.

SHIFT COVER

102. **REMOVE AND REINSTALL.** To remove transmission shift cover, first drain transmission lubricant. Remove forward-reverse shift link (L – Fig. 114) then position gear shift lever (1 – Fig. 116) in bottom of vertical neutral slot on high range side.

Remove plug from side cover and install a cap screw into hole (C – Fig. 114) tight enough to prevent shift shaft in cover from moving while upper portion of shift shaft is removed. Position upper shift shaft out of way, remove shift cover cap screws then withdraw cover. Remove cap screw from hole (C).

Refer to Fig. 115 for exploded view of shift cover and associated parts.

Position shift rails and shift fingers as follows to reinstall shift cover. Move top and bottom shift rails in transmission to neutral position. Slide center shift rail forward engaging high range gear. Move top and bottom shift fingers to center position aligning them with top and bottom shift rails. Move middle shift

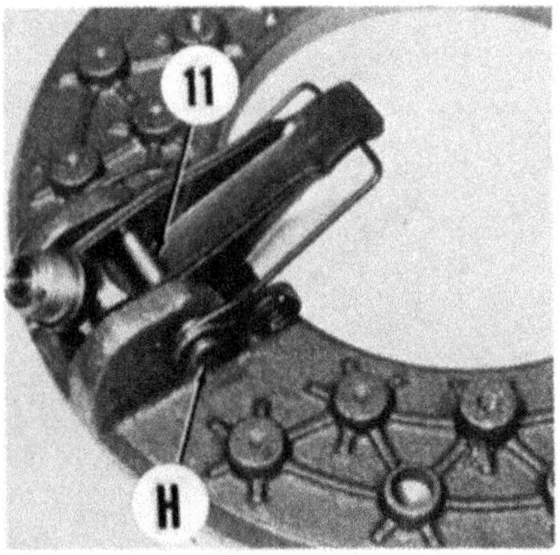

Fig. 112 – Insert threaded rods (R) to aid in overhaul of clutch cover.

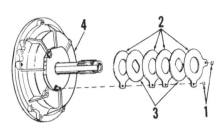

Fig. 113 – Exploded view of transmission brake and associated parts.

1. Snap rings
2. Stationary plates
3. Rotating plates
4. Front cover assembly

Fig. 111 – Install lever pivot pins (11) so pin head (H) is located as shown.

Fig. 114 – To remove shift cover, detach forward-reverse shift link (L). Install cap screw in hole (C) to prevent movement of shift shaft. Refer to text.

finger forward aligning it with middle shift rail. Push shift shaft (3 – Fig. 115) down to hold shift fingers in position. Align forward-reverse shift shaft (12) with uppermost shift rail. Install two guide studs in transmission housing, then install shift cover making sure all shift fingers are engaged. Tighten shift cover cap screws to 25-38 ft.-lbs. torque.

Pull shift shaft (3) up and align hole in cover (C – Fig. 114) with hole in shaft, then install cap screw. Slide upper shift shaft onto lower shift shaft. With shift lever (1 – Fig. 116) centered in horizontal slot between high and low shift patterns, tighten clamp bolt. Reinstall forward-reverse shift link (L – Fig. 114), then check adjustment. Link is properly adjusted when shift lever (1 – Fig. 116) stays in third gear position when moving forward-reverse lever (2) to reverse position and shift lever stays in fourth gear position when moving forward-reverse lever to forward position. Refill transmission with lubricant as required.

FRONT COVER

103. **REMOVE AND REINSTALL.** Separate engine and transmission as outlined in paragraph 97, transmission brake as outlined in paragraph 101 and shift cover as outlined in paragraph 102. Remove pto shaft from input shaft and lubrication line from front cover. Remove front cover retaining cap screws and attach Nuday special tool No. 2801 as shown in Fig. 117. Move front cover forward just enough to clear shift rails and bottom shaft oil tube, then rotate cover approximately ⅛ turn

counterclockwise and withdraw from transmission.

To reinstall, insert two studs, one 2½ inch long and the other 3½ inches long, to aid in installation of front cover. Slide center and bottom shift rails to rear and top shift rail to neutral position. While installing front cover assembly in transmission housing, hold shift rail interlock through shift cover opening to allow shift rails to enter bosses in front cover. Use "Loctite" on all front cover retaining cap screws and tighten to 25-38 ft.-lbs. torque. Remainder of assembly is reverse of disassembly procedure.

104. **OVERHAUL.** Remove forward-reverse shift rail detent (34, 35 and 36 – Fig. 118), set screw (39) and wedge pin (38), then withdraw shift rail (37) and fork (40). Remove shift rail interlock (41 through 46) and bearing support (3).

Stand front cover (1) on input shaft (11) and drive against front gear (15) forcing front bearing cone (9) forward. Slide shift collar (17) forward for access to roll pin (14). Force rubber plug out bottom of roll pin (14) and extract roll pin from coupler hub (18) and shaft (13).

NOTE: Manufacturer recommends using a puller to remove roll pin (14). Puller can be fabricated as shown in Fig. 119 and pin removed as shown in Fig. 120.

Remove rear bearing retaining plate (26 – Fig. 118), then using a suitable puller, push shaft (13) toward front until shaft and parts can be withdrawn from front cover. Keep parts in order to aid in reassembly. Remove roller bearing (25) and snap ring (24) from rear case.

Remove reverse idler set screw (31) and withdraw shaft (30). Remove reverse idler gear (29), needle roller bearings (28) and washers (27). Input shaft (11) and connecting shaft (13) may be separated, if necessary, by using a suitable puller. Snap ring (12) is used to align oil passage holes in shafts. Renew all components that are damaged or worn excessively.

Reassemble reverse idler assembly (27 through 31), securing set screw (31) with lockwire. If oil tube (47) is renewed, apply "Loctite" before installing.

Install thrust washer (10) onto shaft (13) with inside diameter chamfer toward shaft shoulder. Press bearing cone (9) onto shaft (13). Place snap ring (12) inside input shaft (11). Align master splines on shafts, then push the two sections together making sure oil holes are aligned.

Install snap ring (24) in rear case then set housing on end with front up. Place bearing cup (23), bearing cone (22) and thrust washer (21) in housing. Thrust washer (21) should have inside diameter chamfer away from bearing. Set transmission drive gear (20), toothed thrust washer (19), shift coupler hub (18)

Fig. 116 — Shift lever (1) and forward-reverse shift lever (2) are shown in a 24-speed (Multi-Power) model.

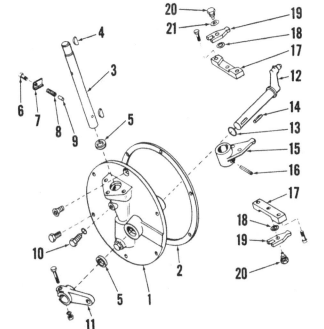

Fig. 115 — Exploded view of shift cover and associated parts. Some models may use a 0.030 inch washer (21) on the top pivot pin (20).

1. Shift cover
2. Gasket
3. Shift shaft
4. Woodruff key
5. Seal
6. Cap screw
7. Bracket
8. Spring
9. Detent
10. Plug
11. Reverse lever
12. Reverse shift shaft
13. "O" ring
14. Woodruff key
15. High-Low shift finger
16. Roll pin
17. Shift finger retainer
18. Washer
19. Shift finger
20. Pivot pin
21. 0.030 inch washer

Fig. 117 — Special tool (T) No. 2801 is used to aid in removal of front cover assembly. Handle for tool (not shown) should be of sufficient length for proper leverage.

and toothed thrust washer (16) on top of thrust washer (21). Hole in hub (18) for roll pin (14) should be accessible. Place shift collar (17) on shift hub (18), then set reverse gear (15) on top of toothed thrust washer (16). Align holes in input shaft and shift coupler hub for roll pin (14), then insert input shaft through gear assembly.

Turn assembly over and rest it on input shaft, then drive bearing cone (22) onto shaft. Install bearing cup (8), shims (7) and bearing support (3) leaving support loose enough to allow access to roll pin hole in shift coupler hub. With rubber plug in bottom side of roll pin (14) and slit in roll pin to rear, drive roll pin in until end is 0.005-0.020 inch below bottom of shift hub splines. Tighten

bearing support cap screws and check input shaft end play. End play should be 0.001-0.006 inch and is adjusted by shims (7). Shims are available in sizes of 0.004, 0.008 and 0.015 inch.

Once end play is correct, remove bearing support and renew seals (4 and 5) as required. Renew "O" ring (6), then reinstall bearing support and tighten cap screws to 25-38 ft.-lbs. torque. Install roller bearing (25) and bearing retaining plate (26). Coat retaining plate cap screws with "Loctite" and tighten to 14-21 ft.-lbs. torque.

Reinstall forward-reverse shift fork, shift rail and detent by reversing disassembly. Use "Loctite" on all threads. Tighten set screw (39) to 21-31 ft.-lbs. torque, then secure with a

lockwire. Tighten plug (34) to 41-62 ft.-lbs. torque. Reassemble shift rail interlock (41 through 46).

TRANSMISSION/CENTER HOUSING SPLIT

105. To separate (split) tractor between transmission and center housing, first drain transmission and center housing. Split engine from transmission as outlined in paragraph 97. Remove front cover, if necessary, as outlined in paragraph 103. Place suitable stands under drawbar and front of center housing to prevent the center housing and axle from tipping when transmission is being removed. Remove batteries, battery tray and step bracket from left side of transmission. Remove cab mounting bolts at front and raise cab approximately one inch, then place stands under each front corner.

On left side of transmission, disconnect hydraulic filters, ground strap, fuel lines and transmission lubrication line. On right side of transmission, disconnect oil sump return line.

Support transmission with a suitable rolling floor jack. Remove two center housing to transmission cap screws and install guide studs. Remove remaining

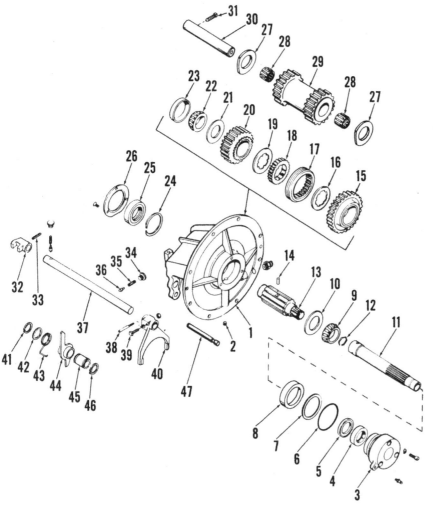

Fig. 118—Exploded view of 8-speed transmission front cover assembly.

Fig. 119—Fabricate puller from stock to remove roll pin (14—Fig. 118). Nut (A) should fit slide hammer. Weld one side of nut (A) to a three-inch long, ¼ inch-28 thread cap screw (B). Grind outside of nut (C) to 13/32 inch diameter.

Fig. 120—After removing rubber plug from input shaft roll pin, pull roll pin using slide hammer and tools as fabricated in Fig. 119.

1. Front cover	24. Snap ring	35. Spring
2. Plug	25. Roller bearing	36. Detent
3. Bearing support	26. Bearing retaining	37. Shift rail
4. Double lip seal	plate	38. Wedge pin
5. Single lip seal	27. Washer	39. Set screw
6. "O" ring	28. Needle roller bearing	40. Shift fork
7. Shim	29. Reverse idler gear	41. Snap ring
8. Bearing cup	30. Idler shaft	42. Washer
9. Bearing cone	31. Set screw	43. Torsion spring
10. Thrust washer	32. Shift finger yoke	44. Shift rail interlock
11. Input shaft	33. Roll pin	45. Bushing
12. Snap ring	34. Plug	46. Snap ring
13. Shaft		47. Oil tube
14. Roll pin		
15. Reverse gear		
16. Toothed thrust washer		
17. Shift collar		
18. Coupler hub		
19. Toothed thrust washer		
20. Drive gear		
21. Thrust washer		
22. Bearing cone		
23. Bearing cup		

cap screws, bolts and nuts, then careful-ly withdraw transmission.

Before installing, manufacturer recommends renewal of the seal and two needle roller bearings located at front of center housing for pto clutch pack. Refer to paragraph 119 for installation procedure. Ensure that output shaft and drive pinion shaft coupler is secured to drive pinion shaft with cotter pin. Use guide studs to aid in alignment during reassembly. Rotate transmission bottom shaft as necessary to align coupler splines.

Transmission to center housing retain-ing cap screws are of varying lengths and should be installed as follows: Top cap screws are 2½ inch, side cap screws are 1½ inch and bottom cap screws are 1¾ inch.

Tighten bottom bolts to 173-260 ft.-lbs. torque, six-point cap screws to 130-200 ft.-lbs. torque and 12-point cap screws to 200-290 ft.-lbs. torque. Coat threads of front cab mounting cap screws with "Loctite" and tighten to 240-350 ft.-lbs. torque. The remainder of reassembly is the reverse of disassembly.

SHIFT RAILS AND FORKS

106. **REMOVE AND REINSTALL.** Split transmission from center housing as outlined in paragraph 105. Remove top and bottom shift rail detent assemblies from right side of transmis-sion housing as shown in Fig. 121. Remove middle shift rail detent assembly (14 through 17 – Fig. 122) from interlock plate (18). Remove interlock plate retaining cap screws and withdraw interlock plate (18) with interlock pins (19 and 20). Loosen set screws in bottom shift fork (5) and shift finger yoke (7), then remove shift rail (6). Loosen set screw in middle shift fork (8) and remove shift rail (9). Loosen set screw in top shift fork (11) and slide shift rail (12)·

to rear out of fork, then slide rail for-ward to remove.

When reassembling shift rails and forks make sure top and middle shift forks have set screw hub toward front. Bottom shift fork should have set screw hub toward rear. Use "Loctite" or equi-valent and lockwires on all set screws.

Install interlock plate (18) on shift rails with short pin (19) on top and long pin (20) on bottom. With special centering cap screw in top, tighten lockplate re-taining cap screws to 63-95 ft.-lbs. torque. Install top and bottom shift rail detent assemblies with ball (2) inserted in top. Tighten retaining plugs to 41-62 ft.-lbs. torque.

NOTE: Forward/reverse shift rail and fork shown in Fig. 118 are serviced in paragraph 104.

TOP SHAFT

107. **R&R AND OVERHAUL.** To remove top shaft, first separate transmission from center housing as outlined in paragraph 105 and remove

front cover as outlined in paragraph 103. Remove shift rails and forks as outlined in paragraph 106. Use Nuday special tool No. 6143 to remove locknut (1 – Fig. 123), then remove rear bearing cone (2). Slide pto shaft or a suitable bar through top shaft (9). Lift assembly to clear gears on bottom shaft, then withdraw top shaft assembly out front of transmission case.

Using Fig. 123 as a guide, remove components from top shaft. Renew any components that are damaged or worn excessively. Install bushings (17 and 22) 0.025 inch below edge of step in coupler side of gears (16 and 23) as required. After installation, bushings must be machined to an inside diameter of 2.5852-2.5860 inch. Install large snap rings (18) in gears (16 and 23). If thrust washer (15) or gear (16) is renewed, en-sure that thrust washer is 0.1485-0.1500 inch wide and hub on gear (16) is 2.740-2.744 inch long. When installing washer (8), cup side fits over bearing cone (7). Reassemble top shaft, then in-sert assembly in transmission.

Install shift rails and forks as outlined

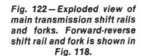

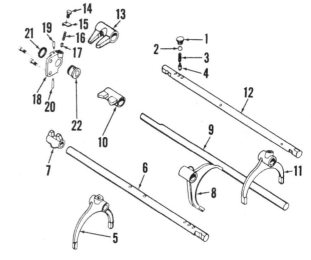

Fig. 122 — Exploded view of main transmission shift rails and forks. Forward-reverse shift rail and fork is shown in Fig. 118.

1. Plug
2. Ball
3. Spring
4. Detent
5. 3-4 shift fork
6. 3-4 shift rail
7. 3-4 shift finger yoke
8. High-Low shift fork
9. High-Low shift rail
10. High-Low shift finger yoke
11. 1-2 shift fork
12. 1-2 shift rail
13. 1-2 shift finger yoke
14. Cap screw
15. Spring retainer
16. Spring
17. Detent
18. Interlock plate
19. Upper interlock pin
20. Lower interlock pin
21. Snap ring
22. Rail bushing

Fig. 121 — Top and bottom shift rail detents. Ball (2) is located in top rail only.

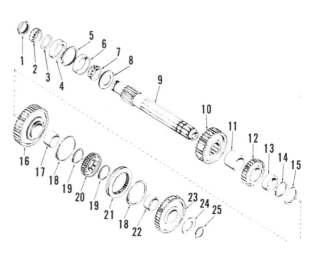

Fig. 123 — Exploded view of top shaft assembly.

1. Locknut
2. Bearing cone
3. Spacer
4. Bearing cup
5. Snap ring
6. Bearing cup
7. Bearing cone
8. Washer
9. Top shaft
10. 3rd driven gear
11. Spacer
12. 4th driven gear
13. Spacer
14. Snap ring
15. Thrust washer
16. 1st driven gear
17. Bushing
18. Snap ring
19. Snap ring
20. Shift hub
21. Shift collar
22. Bushing
23. 2nd driven gear
24. Thrust washer
25. Retaining ring

in paragraph 106 and front cover assembly as outlined in paragraph 103. Install rear bearing cone (2) and nut (1) with "Loctite" 264 or equivalent on nut threads. Tighten nut (using Nuday tool No. 6143) to 45 ft.-lbs. torque while rotating shaft to seat bearings; use Nuday tool No. 6621 or equivalent to prevent bottom shaft from turning.

Top shaft end play should be 0.001 inch loose to 0.003 inch preload. Adjustment is as follows: Back nut (1) off ½ turn. Using a feeler gage, tighten nut until there is 0.010 inch between rear bearing cone and nut. Tighten nut ¼ turn further to provide 0.0025 inch preload. Stake nut in place.

Some models may use spacers (3) between bearing cones (2 and 7) to adjust top shaft end play. Spacers are available in 0.004 inch increments, ranging from 0.106 inch to 0.126 inch. Once end play is correct, tighten nut (1) to 100-150 ft.-lbs. torque. Stake nut in place.

The remainder of reassembly is the reverse of disassembly.

BOTTOM SHAFT

108. R&R AND OVERHAUL REAR SECTION. To remove bottom shaft rear section (1 through 15 – Fig. 124), first remove top shaft as outlined in paragraph 107. Remove locknut (1), using Nuday tool No. 6143, then bearing cone (2). Unbolt and withdraw plate (5) with bearing cups (3 and 6) and snap rings (4). Move shaft assembly to rear, then out through shift cover opening on side of transmission housing. Needle bearing (26) is now accessible and may be removed from rear of front shaft (27).

Remove snap ring (15) and remove components in order from shaft as shown in Fig. 124.

Clean, inspect and renew any components that are damaged or worn excessively.

Reassemble rear section (7 through 15), then check gear (12) end play. End play should be 0.000-0.007 inch and is adjusted by snap ring (15). Snap ring is available in thicknesses of 0.085, 0.090 and 0.0925 inch.

Reinstall shaft and rear plate (5). Tighten small cap screw to 27-40 ft.-lbs. torque and the two large cap screws to 63-95 ft.-lbs. torque. Install rear bearing cone (2) and locknut (1). Tighten locknut to 20 ft.-lbs. torque while rotating shaft to seat bearings.

Shaft position must be checked and adjusted. Mount and zero a dial indicator at end of shaft. Back off locknut (1) and push shaft forward. Shaft position should be within the limits of 0.032-0.055 inch and is adjusted by shim (8). Shims are available in 0.015, 0.030 and 0.060 inch sizes. Once shaft position

is correct, tighten locknut (1) to 20 ft.-lbs. torque. Do not stake locknut at this time.

Shaft end play should be 0.001 inch loose to 0.003 inch preload. Adjustment is as follows: Back locknut (1) off ½ turn. Using a feeler gage, tighten locknut until there is 0.010 inch clearance between rear bearing cone and locknut. Tighten locknut ¼ turn further to provide 0.0025 inch preload. Stake locknut in place.

Some models may use spacers between bearing cones (2 and 7) to adjust end play. Spacers are available in 0.004 inch increments, ranging from 0.106 inch to 0.126 inch. Once end play is correct, tighten locknut to 100-150 ft.-lbs. torque. Stake locknut in place.

The remainder of reassembly is the reverse of disassembly.

109. R&R AND OVERHAUL FRONT SECTION. To remove bottom shaft front section (16 through 33 – Fig. 124), first remove top shaft as outlined in paragraph 107 and bottom shaft rear section as outlined in paragraph 108. At rear of shaft (27), remove snap ring (16), shift collar (18), shift hub (17), gear (20), retaining plate (21) and shims (22). At front of shaft, remove snap ring (33), gear (32) and spacer (31), then withdraw shaft.

Clean, inspect and renew any components that are damaged or worn excessively. Bushing (19) should be installed 0.025 inch below flush at small end of gear hub. After installation, bushing must be machined to an inside diameter of 2.5025-2.5033 inch. When pressing bearing cones (24 and 28) onto shaft (27), an installing force of 8000

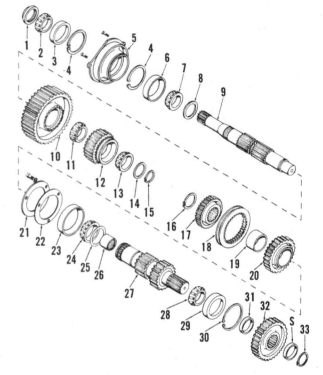

Fig. 124 – Exploded view of bottom shaft assembly. Spacer (S) is used on Multi-Power models only. Refer to text.

1. Locknut
2. Bearing cone
3. Bearing cup
4. Snap ring
5. Bearing plate
6. Bearing cup
7. Bearing cone
8. Shim
9. Rear shaft
10. Output gear
11. Bearing cone
12. Output gear
13. Bearing cone
14. Washer
15. Snap ring
16. Snap ring
17. Shift hub
18. Shift collar
19. Bushing
20. 4th driving gear
21. Retaining plate
22. Shim
23. Bearing cup
24. Bearing cone
25. Shim (0.020 inch)
26. Needle bearing
27. Front shaft
28. Bearing cone
29. Bearing cup
30. Snap ring
31. Spacer
32. Input gear
33. Snap ring

Fig. 125 – Connect hydraulic test gages to shift valve as illustrated. Refer to text.

pounds is recommended to seat cones.

Install snap ring (30), bearing cup (29), shaft assembly (27), bearing cup (23), shims (22) and retaining plate (21). Tighten retaining plate cap screws to 14-21 ft.-lbs. torque, then check shaft end play as follows:

Shaft should have 0.001-0.006 inch preload and is adjusted by shims (22). Shims are available in 0.004, 0.008 and 0.015 inch sizes. Mount a dial indicator at end of shaft, then install or remove shims necessary to obtain the required preload.

The remainder of reassembly is the reverse of disassembly.

NOTE: Spacer (S) is 0.216 inch thick and is used on Multi-Power (24-speed) models only.

MULTI-POWER (24-SPEED) TRANSMISSION

All models are equipped with a basic eight-speed, sliding gear, constant mesh transmission. Available as a factory installed option is a hydraulically actuated, three-range unit located on transmission input shaft which can be shifted when tractor is moving forward. Transmissions with the hydraulically activated three-range unit are referred to as "Multi-Power" transmissions and are capable of 24 forward and 6 reverse speeds.

Since Multi-Power is a factory installed option, service procedures for the main transmission are the same on all models as outlined in the preceding paragraphs. Refer to the appropriate following paragraphs for service of the Multi-Power (24-speed) option.

TROUBLESHOOTING

110. The following are symptoms which may occur during operation of the Multi-Power transmission. By using this information in conjunction with the Test and Adjust information and the R&R and Overhaul information, no trouble should be encountered in servicing the Multi-Power transmission.

EXTERNAL LEAKS. Could be caused by:

a. Loose tube lines or leakage from front cover tube "O" rings.

b. Clutch flywheel seal or "O" ring worn.

c. Input shaft seals in front bearing support worn or damaged.

d. Square cut "O" ring between front cover assembly and transmission housing defective.

e. Low clutch "O" ring between brake support and front cover defective.

NOISE. Noise heard when shifting through ranges or operating in low or intermediate range, could be caused by:

a. Thrust bearings, thrust washers or needle bearings on planetary gear assembly damaged or missing.

b. Planetary gears worn or damaged.

c. Retaining cap screws loose.

d. Snap ring and washer for drive gear in rear cover out of place.

POWER LOSS. Incomplete disengagement of one or more components could be caused by:

a. Low, intermediate or high clutch disc warped.

b. "O" rings on intermediate piston defective.

c. Bleed-off orifice in planetary ring gear plugged.

SLIPPAGE (Pressure OK). If no difficulty in hydraulic system was indicated after testing as outlined in paragraph 111, then slippage during operation could be caused by:

a. Clutch discs worn excessively.

b. Spacers in low and intermediate clutch assembly are incorrect thickness or improperly installed.

c. Locating pin in low and intermediate clutch assembly bent preventing proper engagement of disc.

d. Belleville springs improperly installed.

e. Friction plates and clutch plate in high clutch assembly are incorrect thickness or improperly installed.

f. Too much clearance between high clutch discs and counter plate of high clutch assembly.

g. Belleville spring in high clutch assembly too loose.

h. Planetary gear assembly not fully engaged in high clutch assembly.

SLIPPAGE (Low pressure). If low pressure was indicated after hydraulic tests as outlined in paragraph 111, then slippage during operation could be caused by:

a. Shift control valve improperly adjusted.

b. Improper installation of restrictor pins in shift control valve.

c. Teflon sealing rings on front bearing support worn or damaged.

d. Low piston "O" rings defective.

e. High piston sealing rings defective.

HARSH SHIFTING. Could be caused by:

a. Improper installation of restrictor pins in shift control valve.

b. Intermediate orifice stop in shift control valve base incorrectly adjusted.

c. Intermediate orifices in shift control valve base are incorrect size or improperly installed.

TEST AND ADJUSTMENT

111. Oil to the Multi-Power system is provided by the low volume side of the main hydraulic pump. Normal system pressure should be 290-310 psi and is controlled by the low pressure regulating valve. Test and adjust main hydraulic pump and low pressure regulating valve as outlined in paragraphs 139 and 140.

System pressures below 275 psi can cause rapid wear of clutch friction components and pressures above 400 psi can cause immediate failure of piston "O" rings, sealing rings or retaining rings. If tractor has been operated for extended periods of time with either of above conditions, upon inspection it may be preferred to renew complete front cover assembly.

112. After performing test and adjustment procedures as outlined in paragraph 111, check operation of shift control valve as follows: Attach an individual 400 psi gage to each test port of the shift control valve as shown in Fig. 125. Gage connection (1) is for high range, (2) is for intermediate range and (3) is for low range. Hydraulic fluid should be at operating temperature and engine set at 1500 rpm. All test pressure should fall between 290-310 psi.

Move Multi-Power shift lever to high range. High and intermediate gages should pressurize with high gage pressurizing just before intermediate gage.

Move shift lever from high to intermediate range. Gages should drop to zero together.

Move shift lever from intermediate to low range. Intermediate and low gage should pressurize with low gage pressurizing just before intermediate gage.

Move shift lever back to intermediate. Gages should drop to zero together.

Depress clutch pedal and intermediate gage should pressurize.

Any variation of test results will require overhaul and close examination of shift control valve as outlined in paragraph 113. Pay close attention to the installation of restrictor pins (11 and 12 – Fig. 126), intermediate orifices (6 and 8) and adjustment of orifice stop (5).

SHIFT CONTROL VALVE

113. **R&R AND OVERHAUL.** Refer to Fig. 126 for exploded view of shift control valve and associated parts. To remove control valve, disconnect shift linkage, clutch release link and hydraulic

lines from control valve. Remove retaining cap screws and withdraw control valve. If control valve base (3) requires removal, first remove inspection plate on bottom of transmission and disconnect the three hydraulic lines, then unbolt and withdraw base.

Using Fig. 126 as a guide, disassemble control valve. Clean all parts in a suitable solvent. Visually inspect parts and renew any showing excessive wear or damage. Renew all "O" rings, seals and gaskets. Lubricate all parts with clean oil before reassembly.

When reassembling control valve observe the following: Install "O" ring (29) on pilot spool (27), then insert spool into valve body (1) just far enough to permit installation of "O" ring (30), washer (31) and snap ring (32).

NOTE: Do not turn pilot spool when inserting in valve body or push spool too far. Damage to "O" ring (29) may result.

Plug (19) should be tightened to 41-62 ft.-lbs. torque and lever bracket retaining cap screws 14-21 ft.-lbs. torque. Make certain restrictor pins (11 and 12) are inserted exactly as shown. Check depth of orifice stop (5) in valve base (3). With orifice (8), spring (7) and orifice (6) removed, stop should be 1-5/32 inch below valve base surface. If adjustment is necessary, remove fitting (4) and turn stop (5) as required. The 0.046 inch orifice (6) is inserted first, then spring (7) and the 0.093 inch orifice (8).

Install valve on base and adjust linkage rod (L – Fig. 127) until roll pin (R) is vertical when clutch pedal is up. Cable (C) is properly adjusted when Multi-Power shift lever is in the intermediate position and valve spool is in the intermediate detent position.

FRONT COVER

114. R&R AND OVERHAUL. Separate engine and transmission as outlined in paragraph 97, transmission brake as outlined in paragraph 101 and shift cover as outlined in paragraph 102. Remove pto shaft from input shaft and lubrication line from front cover. Remove front cover retaining cap screws.

NOTE: When removing front cover assembly on Multi-Power models, cover retaining cap screws (C – Fig. 128) are marked. Remaining cap screws retain parts inside cover.

Attach Nuday special tool No. 2801 as shown in Fig. 117, then move front cover forward just enough to clear shift rails and bottom shaft oil tube. Rotate cover approximately ⅛ turn counterclockwise and withdraw from transmission.

To reinstall, insert two studs, one 2½ inches long and the other 3½ inches long, to aid in installation of front cover. Slide center and bottom shift rails to rear and top shift rail to neutral posi-

tion. While installing front cover assembly in transmission housing, hold shift rail interlock through shift cover opening to allow shift rails to enter bosses in front cover. Use "Loctite" on all front cover retaining cap screws and tighten to 25-38 ft.-lbs. torque. Remainder of assembly is reverse of disassembly procedure.

If front cover disassembly is required, match mark all parts during disassembly to aid in reassembly. Unbolt and remove rear cover (3 – Fig. 129) from brake support (33) and refer to the appropriate following paragraphs for servicing. When reinstalling rear cover (3) to brake support (33) tighten retaining cap screws to 25-38 ft.-lbs. torque.

115. INTERMEDIATE AND LOW CLUTCH. Disassemble intermediate clutch assembly as follows: Remove Belleville springs (16), locator plate (17), friction plate (18) and clutch disc (19).

Disassemble low clutch assembly as follows: Remove nine springs (20), nine pins (21), three 9/16-inch spacers (22), six Allen head cap screws (23), friction plate (24), low clutch disc (25), counter plate (26), six round spacers (27) and

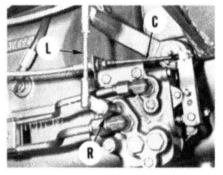

Fig. 127 — Adjust linkage rod (L) and shift cable (C) as outlined in text.

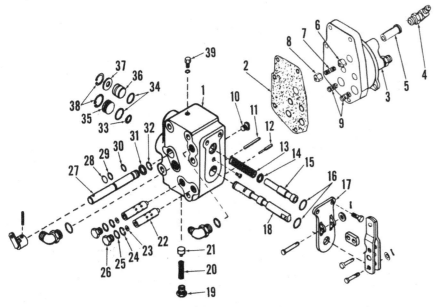

Fig. 126 — Exploded view of shift control valve and associated parts.

1. Valve body	11. Full length restrictor	20. Spring
2. Gasket	pin	21. Detent
3. Valve base	12. Stepped restrictor pin	22. Pressure regulator
4. 90 degree fitting	13. Spring	valve
5. Orifice stop	14. Pilot washer	23. Plug washer
6. 0.046 inch orifice	15. Unloading valve	24. Snap ring
7. Spring	16. "O" ring	25. "O" ring
8. 0.093 inch orifice	17. Lever bracket	26. Plug
9. Spring	18. Spool valve	27. Pilot spool
10. Plug	19. Plug	28. Scraper seal

29. "O" ring
30. "O" ring
31. Washer
32. Snap ring
33. Snap ring
34. "O" ring
35. Spool plug
36. Unloading plug
37. Spool washer
38. Snap ring
39. Plug

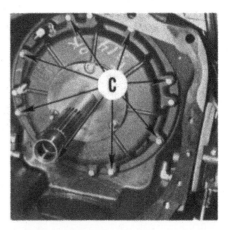

Fig. 128 — Cap screws (C) retain front cover assembly to transmission housing on Multi-Power transmissions.

three 7/16-inch spacers (28).

Unbolt and remove brake support (33) with low piston (30) from front cover (72). Withdraw low piston (30) from brake support. Remove planetary gear assembly (34 through 50) from high clutch assembly, then remove high clutch assembly (51 through 65) from front cover. Remove thrust bearing race (73), thrust bearing (74) and thrust bearing race (75). Detach bearing support (79) from front cover and remove intermediate piston (67) and three springs (69).

Clean, inspect and renew any components that are damaged or worn excessively. Renew "O" rings and seals as required. Check clutch discs for warpage using a flat surface and feeler gage. Discs should not be warped more than 0.030 inch. Lubricate all "O" rings, seals and bearings during reassembly.

Reassemble intermediate and low clutch by first placing springs (69) into front cover (72). Install "O" rings (66 and 68) on intermediate piston (67), then install piston (67) into front cover with small outer diameter of piston facing front of cover.

Place high clutch assembly (51 through 65) into front cover (72), then position assembly with input shaft (64) up. Install thrust bearing race (73), thrust bearing (74) and thrust bearing race (75) into recess in high clutch assembly. Install seal (80), bearing (77) and Teflon sealing rings (78) on bearing support (79), then install bearing support on front cover. Coat threads of bearing support retaining cap screws with "Loctite", then tighten to 25-38 ft.-lbs. torque.

Turn complete assembly over and place on a bucket or equivalent stand to support assembly on the front cover and not the input shaft. Make certain thrust bearing races (73 and 75) and thrust bearing (74) are still in position. Reinstall planetary gear assembly (34 through 50) into high clutch assembly with punch marks on planetary gears (44) to outside. Install "O" rings (29 and 31) on low piston (30), then insert low piston into brake support (33). Install "O" rings (70 and 71) on front cover (72), then install brake support onto front cover aligning marks made during disassembly. Coat threads of retaining cap screws with "Loctite" and tighten to 25-38 ft.-lbs. torque.

Place three 7/16-inch spacers (28) between threaded holes with one over guide dowel and six round spacers (27) over remaining threaded holes as shown in Fig. 130. Install counter plate (26 – Fig. 129), low clutch disc (25) and friction plate (24). Install six Allen head cap screws (23) into same holes as round spacers (27) and tighten to 17-25 ft.-lbs. torque. Install nine pins (21) into holes in counter plate (26) with a spring (20) over each pin. Install intermediate clutch disc (19) and friction plate (18), then place three 9/16-inch spacers (22) in spaces of friction plate (18) as shown in Fig. 131. Place locator plate (17 – Fig. 129) over friction plate (18) with recesses over spacers (22) and narrow relief at rim toward bottom, then install Belleville springs (16), convex side up.

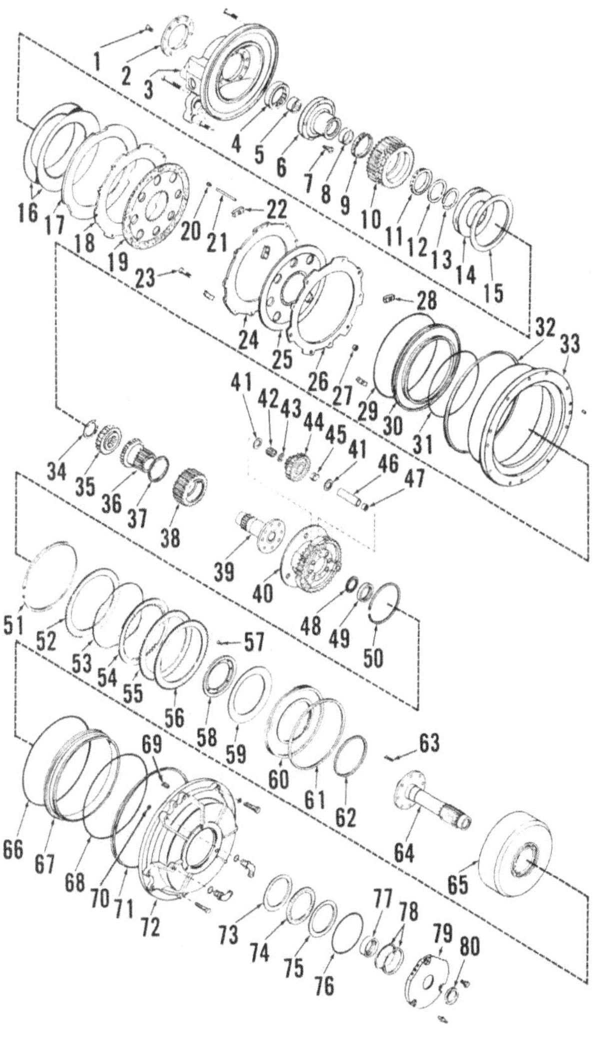

Fig. 129 – Exploded view of front cover assembly used on Multi-Power transmissions.

1. Cap screw
2. Bearing retainer
3. Rear cover
4. Roller bearing
5. Bushing
6. Shaft
7. Cap screw
8. Bushing
9. Bearing cone
10. Drive gear
11. Bearing cone
12. Thrust washer
13. Snap ring
14. Coupler
15. Oil shield
16. Belleville springs
17. Locator plate
18. Friction plate
19. Clutch disc
20. Springs (nine)
21. Pins (nine)
22. 9/16-inch spacer (three)
23. Allen head cap screw (six)
24. Friction plate
25. Clutch disc
26. Counter plate
27. Round spacer (six)
28. 7/16-inch spacer (three)
29. "O" ring
30. Low piston
31. "O" ring
32. Cover seal
33. Brake support
34. Thrust bearing
35. Hub
36. Gear sleeve
37. Thrust washer
38. Sun gear
39. Planetary shaft
40. Carrier
41. Thrust washer
42. Long needle bearings
43. Spacer
44. Planetary gear
45. Short needle bearings
46. Shaft
47. Plug
48. Thrust bearing
49. Needle bearing
50. Snap ring
51. Snap ring
52. Counter plate
53. Wire spring (three)
54. Friction disc (three)
55. Friction plate (two)
56. Clutch plate
57. Cap screw
58. Piston retainer
59. Belleville spring
60. High piston
61. Seal ring
62. Seal ring
63. Cap screw
64. Input shaft

65. Ring gear
66. "O" ring
67. Intermediate piston
68. "O" ring
69. Springs (three)
70. "O" ring
71. "O" ring
72. Front cover

73. Thrust bearing race
74. Thrust bearing
75. Thrust bearing race
76. "O" ring
77. Bearing
78. Teflon sealing rings
79. Bearing support
80. Seal

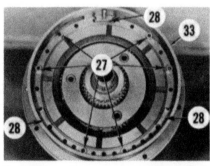

Fig. 130 – Install round spacers (27) and 7/16-inch spacers (28) into support (33).

Fig. 131 – Install 9/16-inch spacers (22) in spaces of friction plate (18). Refer to text.

Install rear cover (3) while aligning match marks and tighten cap screws 25-38 ft.-lbs. torque.

NOTE: The two short cap screws should be installed under shift rail guide.

116. HIGH CLUTCH. To overhaul high clutch assembly (51 through 65 – Fig. 129), first remove intermediate clutch, low clutch and planetary gear assembly as outlined in paragraph 115. Disassemble high clutch assembly by removing snap ring (51), counter plate (52), three friction discs (54) with three wire springs (53) alternating with two friction plates (55) and clutch plate (56).

On models prior to transmission serial number DH 2132, remove flat head cap screws (57), piston retainer (58) and Belleville spring (59), then withdraw piston (60). If required, remove cap screws (63) and separate input shaft (64) and planetary ring gear (65).

Models produced after transmission serial number DH2131 use a heavy duty piston retainer and only eight cap screws to hold piston retainer, Belleville spring, piston, input shaft and planetary ring gear together.

Clean, inspect and renew any components that are damaged or worn excessively. Renew "O" rings and seals as required. Check clutch discs for war-

page using a flat surface and feeler gage. Discs should not be warped more than 0.030 inch. Lubricate all "O" rings, seals and bearings during reassembly.

To reassemble high clutch assembly, first apply "Loctite" to contact surface between input shaft (64) and planetary ring gear (65). Make certain "Loctite" does not cover surfaces for "O" ring seals.

On early production transmissions, renew cap screws (63) and coat threads with "Loctite", then install and tighten to 28 ft.-lbs. torque. Install seal rings (61 and 62) on piston (60) with seal lips toward front. After applying "Permatran" lubricant to seals, piston and piston bore, use Nuday tools No's. 6783 and 6784 to install piston. Install Belleville spring (59), convex side up and retainer plate (58). Coat threads of cap screws (57) and tighten 3-5 ft.-lbs. torque. Check clearance between Belleville spring and retainer plate with a feeler gage. Machine heads of cap screws (63) if required, to provide zero clearance.

On late production transmissions, assemble planetary ring gear, input shaft, piston with seals, Belleville spring and retainer plate. Install and evenly tighten four retaining cap screws, then check compression on Belleville spring as follows: Back retaining cap screws

off, then retighten just until bolts touch retainer plate. Attach a dial indicator as shown in Fig. 132. Tighten cap screws and note reading. Belleville spring should not compress more than 0.020 inch. A 0.020 inch shim is available and installed between retainer plate and input shaft if required.

On all models, reassemble and check clutch pack clearance as follows: Install thick clutch plate (56 – Fig. 129), three friction discs (54) alternating with two thin friction plates (55), counter plate (52) and snap ring (51). Thick ridge on counter plate should be toward rear. Do not install wire springs (53) at this time.

Measure clearance between counter plate and friction disc as shown in Fig. 133. Clearance should be 0.070-0.095 inch and is adjusted by snap ring (51 – Fig. 129). Snap ring is available in thicknesses of 1/8, 5/32 and 7/64-inch. Once clearance is correct, install wire rings (53) around each friction disc (54).

117. PLANETARY GEARS. To overhaul planetary gear assembly (34 through 50 – Fig. 129), first remove intermediate and low clutch as outlined in paragraph 115. Remove thrust bearing (34), hub (35), gear sleeve (36) with thrust washer (37) and sun gear (38), then withdraw planetary carrier assembly from high clutch assembly.

Remove needle bearing (49), thrust bearing (48) and snap ring (50). Press shafts (46) out toward rear thereby releasing planetary gear (44), thrust washers (41), 25 short needle bearings (45), spacer washer (43) and 25 long needle bearings (42).

Clean, inspect and renew any components that are damaged or worn excessively.

When assembling, use Nuday tool No. 6618 and install needle bearings (42 and 45) separated by spacer (43) into gears (44) so longer bearings (42) ride in small end of gear (44). Place thrust washers (41) with flat side inward on each side of gear. Position gear (44) in carrier (40) with small gear toward rear. Use Nuday tool No. 6618 to locate gear, needle bearings and thrust washers, then press shafts (46) into carrier (40) so snap ring groove is towards inside and oil hole points outward. End of shafts (46) should be flush to 0.030 inch below surface of rear of carrier. Reinstall snap ring (50) making sure that snap ring correctly engages groove of shafts (46). Bend tips of snap ring down approximately 30 degrees.

118. FORWARD/REVERSE ASSEMBLY. To overhaul forward/reverse assembly (1 through 15 – Fig. 129), first separate rear cover (3) from brake support (33). Remove shift rail detent and

Fig. 132 – Check compression of Belleville spring with a dial indicator. Refer to text.

Fig. 133 – Measure clearance between counter plate and friction disc of high clutch assembly. Refer to text.

set screw in shift fork, then withdraw shift rail.

NOTE: Shift rail, fork and related parts are not shown in Fig. 129 but are similar to those shown in Fig. 118.

Tilt coupler (14) forward and pull shift fork out, then remove coupler. Remove snap ring (13), thrust washer (12), bearing cone (11), drive gear (10) and bearing cone (9). Unbolt bearing retainer (2) and remove roller bearing (4). Unbolt and withdraw shaft (6).

Reassemble by reversing the disassembly procedure while noting the following: Bearing retaining plate cap screws (1) should be tightened to 21-32 ft.-lbs. torque and shaft retaining cap screws (7) should be tightened to 25-38 ft.-lbs. torque. Bearing end play for drive gear (10) should be 0.000-0.005 inch and is adjusted by snap ring (13). Snap ring is available in 0.117, 0.121 and 0.125 inch sizes. Rear cover (3) to brake support (33) retaining cap screws should be tightened to 25-38 ft.-lbs. torque.

first remove both axle housings as outlined in paragraph 123. Use suitable stands to support rear of cab and rear of transmission. Remove differential as outlined in paragraph 120 and pto clutch pack as outlined in paragraph 132. Disconnect necessary hydraulic lines from rear of hydraulic filters leaving front lines attached. Disconnect filters from mounting bracket and brake lines from center housing. Remove two center housing to transmission cap screws and install guide studs. Remove remaining cap screws, bolts and nuts, then carefully withdraw center housing.

Before installing, manufacturer recommends renewal of the seal and two needle roller bearings located at front of center housing for pto clutch pack. Coat center housing bore with "Loctite", then install bearing (A – Fig. 134) so front edge is 2⅜ inch below flush of housing front surface. Install bearing (B) so front edge is 1¼ inch below flush of

housing front surface. Bearings should be installed with lettered side forward. Coat outside of seal (C) with "Loctite" and install with lip to rear. Ensure that output shaft and drive pinion shaft coupler is secured to drive pinion shaft with cotter pin. Use guide studs to aid in alignment during reassembly. Rotate bottom shaft as necessary to align coupler splines.

Center housing to transmission retaining cap screws are of varying lengths and should be installed as follows: Top cap screws are 2½ inch, side cap screws are 1½ inch and bottom cap screws 1¾ inch.

Tighten bottom bolts to 173-260 ft.-lbs. torque, six point cap screws to 130-200 ft.-lbs. torque and 12 point cap screws to 200-290 ft.-lbs. torque. After reconnecting brake lines, bleed brakes as outlined in paragraph 127. Remainder of procedure is reverse of disassembly.

CENTER HOUSING, MAIN DRIVE BEVEL GEARS AND DIFFERENTIAL

The rear axle center housing contains the main drive bevel gears and differential, power take off gears, pto shaft and pto clutch.

CENTER HOUSING

All Models

119. **REMOVE AND REINSTALL.** To remove center housing assembly,

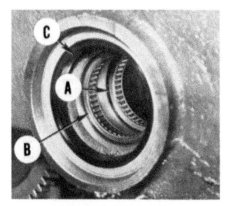

Fig. 134 — Renew bearings (A and B) and seal (C) in center housing as outlined in text.

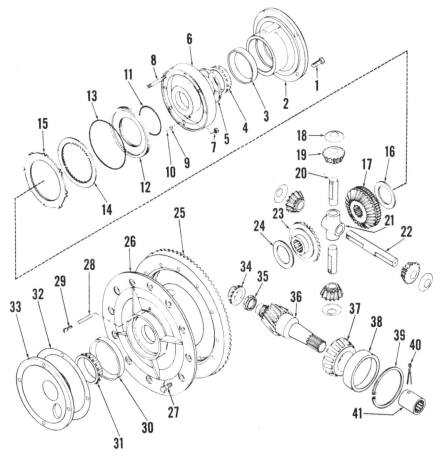

Fig. 135 — Exploded view of differential assembly with differential lock. Models without differential lock option are similar.

1. Cap screw	12. Lock piston	22. Long pin	32. Shim
2. Left carrier plate	13. "O" ring	23. Side gear	33. Right carrier plate
3. Bearing cup	14. Clutch disc	24. Thrust washer	34. Bearing cone
4. Bearing cone	15. Clutch plate	25. Ring gear	35. Adjusting nut
5. Seal ring	16. Thrust washer	26. Case half	36. Pinion gear
6. Case half	17. Side gear	27. Cap screw	37. Bearing cone
7. Case retaining nut	18. Thrust washer	28. Retaining pin	38. Bearing cup
8. Case stud	19. Pinion gears	29. Set screw	39. Snap ring
9. Check valve	20. Half pins	30. Bearing cup	40. Cotter key
10. Ball	21. Differential cross	31. Bearing cone	41. Coupler
11. "O" ring			

DIFFERENTIAL AND
BEVEL GEAR

On models equipped with optional differential lock, refer to paragraphs 145 and 146 for service procedures on differential lock control valve.

All Models

120. **REMOVE AND REINSTALL.** To remove differential and bevel ring gear, first remove axle housings as outlined in paragraph 123 and lift cover as outlined in paragraph 165. Remove pto drive gear assembly as outlined in paragraph 134. Attach "C" clamp to bevel ring gear, lift slightly and remove differential lock standpipe if equipped. Unbolt and withdraw differential carrier plates (2 and 33 – Fig. 135) then remove differential.

To reinstall differential and bevel gear, reverse removal procedure and perform the following additional steps:

Bearing preload and backlash must be checked and adjusted during installation of differential and bevel gear. Check and adjust bearing preload as follows: With differential assembly suspended into center housing, install left side carrier (2) with drain hole down and without shims. Tighten cap screws to 63-95 ft.-lbs. torque. Install right side carrier and tighten cap screws to 36 in.-lbs. torque. Rotate differential several turns then retighten cap screws to 75 in.-lbs. torque. Check clearance between flange of right side carrier and center housing to be sure that clearance is the same all the way around. If clearance is not equal, loosen cap screws, then retighten equally. Rotate differential again and measure gap (next to cap screws) between center housing and flange of carrier. Select proper thickness of shims (32) as determined by chart in Fig. 136. Shims are available in thicknesses of 0.003, 0.006, 0.012 and 0.024 inch. After shim installation, tighten cap screws to 63-95 ft.-lbs. torque.

Backlash between bevel ring gear and pinion gear should be 0.003-0.025 inch and is adjusted as follows: To increase backlash, remove a shim (32 – Fig. 135) from left side and install the same shim on the right side. Tighten carrier plates and check bearing preload. To decrease backlash, move shim from right carrier plate to left carrier plate. If necessary to further decrease backlash after all shims have been moved from right carrier plate to left carrier plate, a 0.010 inch shim can be fabricated from MF part No. 892-120 M1 and installed between bearing cone (31) and carrier plate (33); however, a shim of equal thickness must be installed on left carrier plate to maintain correct bearing preload.

121. **OVERHAUL.** The following overhaul procedure is for models with locking differential; procedure for non-locking differential is basically the same.

To overhaul differential assembly, first scribe alignment marks on major components to aid in reassembly, then remove case retaining nuts (7 – Fig. 135), differential case half (6), clutch discs (14) with plates (15), thrust washer (16) and side gear (17). Unbolt and remove bevel ring gear (25) and three set screws (29) from case half (26). Remove three studs (8) from opposite side of holes for set screws (29). Drive out retaining pins (28), then remove half pins (20) and long pin (22). Withdraw differential cross (21), pinion gears (19), thrust washers (18), side gear (23) and thrust washer (24).

Clean, inspect and renew any components that are damaged or worn excessively. During reassembly lube differential components with "Permatran" fluid; apply a suitable grease to inside bore of differential gears and to outer hubs of side gears.

Install thrust washer (24) and side gear (23). Position differential cross (21) in case half (26) with short ends of cross aligned with bore for long pin (22). Install pinion gears (19) and thrust washers (18). Insert long pin (22) and half pins (20), then insert retaining pins (28) through carrier plate side of case half (26). Install and tighten set screws (29) to 63-95 ft.-lbs. torque. Apply "Loctite" to threads of studs (8) and install in case half (26). Install side gear (17), thrust washer (16) and clutch discs (14) with plates (15). Alternate plates and discs, installing a plate (15) first and ending with a disc (14) last on stack.

Inspect check valve (9) and oil inlet hole in case half (6). Renew "O" rings (11 and 13) on piston (12). Lubricate piston and piston bore with petroleum jelly, then install piston (12) in case half (6)

with raised edge of piston toward clutches.

Assemble case halves and tighten nuts (7) to 27-40 ft.-lbs. torque. Install bevel ring gear (25) using "Loctite" 271 or equivalent on cap screw threads. Tighten cap screws to 90-130 ft.-lbs. torque. Renew seal rings (5) and bearing cone (4) as required.

DRIVE PINION

All Models

122. **R&R AND OVERHAUL.** To remove differential drive pinion, first remove center housing as outlined in paragraph 119 and differential as outlined in paragraph 120.

Install Nuday tool No. 6628 on adjusting nut (35 – Fig. 135), then loosen nut by turning pinion shaft (36) with Nuday tool No. 6621. It will be necessary to remove coupler (41) to install tool. Remove snap ring (39), then attach a suitable puller and extract pinion assembly from front of center housing.

Clean, inspect and renew any components that are damaged or worn excessively.

Install adjusting nut (35) on pinion shaft (36) with smooth edge of nut toward pinion. Install bearing cones (34 and 37) on pinion shaft. Install rear bearing cup for bearing cone (34) into center housing, then install pinion assembly, bearing cup (38) and snap ring (39). Lubricate bearings, then tighten adjusting nut (35) until 5 ft.-lbs. torque is required to turn pinion shaft. Loosen nut, then lightly tap shaft both directions to ensure bearing cups are fully seated in center housing. Gradually retighten nut to obtain a rolling torque of 12-15 in.-lbs. with used bearings and 24-30 in.-lbs. with new bearings. Stake edge of nut into hole in pinion shaft and recheck rolling torque. Reinstall remaining components by reversing removal procedure.

Fig. 136 – Use chart in conjunction with procedures outlined in text to adjust differential bearing preload.

PRE-LOAD SHIM CHART	
Gap Measured	**Use Shims**
.001-.004"	no shims
.005-.013"	.009"
.014-.022"	.018"
.023-.031"	.027"
.032-.040"	.036"
.041-.049"	.045"
.050-.058"	.054"

REAR AXLE AND FINAL DRIVE

AXLE ASSEMBLY

All Models

All models use a planetary, final drive reduction unit located in inner end of axle housing next to center housing. Final drive lubricant reservoir is common with transmission, differential and hydraulic reservoir. Axle stub shafts are free floating, with inner ends supported by differential side gears and outer ends serving as planetary sun gears. Service to any part of axle or final drive requires removal of unit as outlined in paragraph 123.

123. **REMOVE AND REINSTALL.** To remove either final drive unit as an assembly, first drain transmission and hydraulic system fluid, suitably support rear of tractor and remove rear wheel. Remove bottom connection of lift cylinder, then remove bolts and nuts securing rear of cab. Position shift lever in neutral slot or disconnect shift coupler. Remove hood, then raise rear of cab just enough for axle housing clearance. Attach a hoist or place a suitable jack under housing. Unbolt and withdraw complete final drive assembly from center housing.

Thoroughly clean mating surfaces, removing all oil and old sealer. Make sure cap screw retainer (21–Fig. 137) is properly installed. Slide brake disc (26) on stub axle (22), then insert stub axle into differential. Long stub axle goes in left side and short stub axle goes in right side. Using RTV silicone rubber sealer, evenly apply a bead in recess of center housing. Install a guide dowel into center housing to aid alignment during reassembly. Move axle housing into position making sure planetary meshes properly, then slide assembly completely into position. Coat retaining cap screws with "Loctite" or equivalent then install and tighten to 130-200 ft.-lbs. torque.

Tighten cab retaining bolts to 20-24 ft.-lbs. torque. Remainder of procedure is reverse of disassembly.

AXLE AND BEARINGS

All Models

124. **R&R AND ADJUST.** With final drive unit removed as outlined in paragraph 123, unbolt and remove brake plate (25–Fig. 137) and ring gear (24). Remove cap screw retainer (21), cap screw (20) and shims (2), then withdraw planetary assembly (11 through 18), thrust washer (10) and bearing cone (9). Remove outer seal (3), then withdraw axle assembly. Outer bearing cone (4) and both bearing cups (5 and 8) can be removed at this time.

Thoroughly clean mating surfaces, removing all oil and old sealer. Apply "Loctite" to outer edge of inner oil seal (7), then insert seal into housing (6) 0.18 inch below bearing cup seat. Install bearing cups (5 and 8) filling gap behind cup (5) with a suitable multi-purpose Lithium base grease. Install bearing cone (4); a minimum of 20,000 lbs. pressure is required when pressing cone on shaft (1) to fully seat cone. Insert axle shaft (1) into housing (6), then install bearing cone (9), thrust washer (10) and planetary assembly (11 through 18). Assemble shims (2), retainer plate (19) and cap screw (20), then install and tighten cap screw to 200-325 ft.-lbs. torque.

Check axle end play using a dial indicator, then add or remove shims (2) as required to provide 0.001-0.005 inch preload on bearings. Shims are available in 0.002, 0.005, 0.010 and 0.018 inch sizes. Tighten cap screw (20) to 200-325 ft.-lbs. torque and install cap screw retainer (21).

Apply multi-purpose Lithium base grease to area between bearing (4) and bottom of seal recess then install oil seal (3) flush to 0.030 inch below end of housing. Apply a bead of RTV silicone rubber

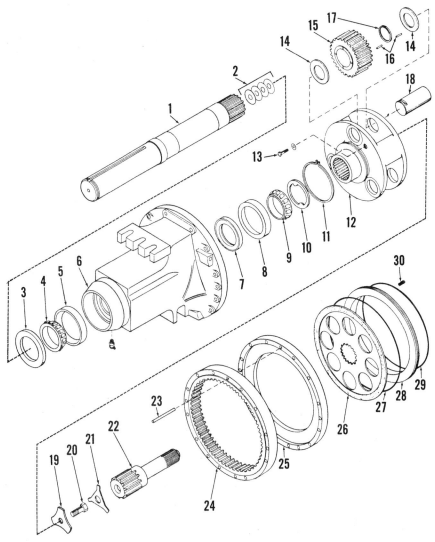

Fig. 137 – Exploded view of right final drive assembly and brake disc.

1. Axle shaft	9. Bearing cone	16. Needle bearings (56 per gear)	23. Dowel pin
2. Shims	10. Thrust washer	17. Spacer	24. Ring gear
3. Seal	11. Retaining ring	18. Planetary shaft	25. Brake plate
4. Bearing cone	12. Planetary carrier	19. Retainer plate	26. Brake disc
5. Bearing cup	13. Cap screw	20. Cap screw	27. "O" ring
6. Axle housing	14. Thrust washer	21. Cap screw retainer	28. Brake piston
7. Seal	15. Planetary gear	22. Stub axle	29. "O" ring
8. Bearing cup			30. Spring

sealer to axle housing side of ring gear (24) and in recess on ring gear side of brake plate (25). Install ring gear and brake plate over dowel pin (23), then install and tighten the two retaining cap screws to 63-95 ft.-lbs. torque. Reinstall final drive unit as outlined in paragraph 123.

PLANETARY ASSEMBLY

All Models

125. **R&R AND OVERHAUL.** With final drive unit removed as outlined in paragraph 123, unbolt and remove brake plate (25 – Fig. 137) and ring gear (24). Remove cap screw retainer (21), cap screw (20) and shims (2), then withdraw planetary assembly (11 through 18).

Planet pinion shafts are a slip fit in carrier (12) and are retained by a snap ring (11) which fits in a slot in all three shafts plus a retaining slot in carrier hub. To disassemble the carrier, place unit on a bench with inner (flat) side down and expand snap ring (11) fully into grooves in shafts (18) as shown in Fig. 138. Work snap ring up out of carrier hub slot, withdrawing all three shafts until snap ring clears shaft grooves.

Slide planetary gear (15 – Fig. 137),

Fig. 138 – To disassemble planet carrier, expand the snap ring fully and lift all three shafts until clear of retaining groove.

Fig. 139 – Loosen locknut (L) and turn adjusting nuts (A) to obtain correct brake pedal travel, then equalize adjustment for proper gap (G) between vertical links and poppets.

together with thrust washers (14) and loose needle bearings (16), out of carrier. There are 56 needle bearings (16) per planetary gear (15).

When assembling, the snap ring and three planet pinion shafts must be installed together. Install planet carrier and adjust axle bearing preload as outlined in paragraph 124.

BRAKES

ADJUSTMENT

All Models

126. Normal brake pedal free play at pedal is 1-1¾ inches. To adjust brakes, loosen locknut (L – Fig. 139), then turn brake rod adjusting nuts (A) until correct pedal travel is obtained. Equalize adjustment so gap (G) does not vary more than 0.015 inch between poppets and vertical links.

FLUID AND BLEEDING

All Models

127. Rear axle housing lubricant is utilized as brake hydraulic fluid. Pressurized oil is supplied to master cylinder from steering control valve. Overflow from master cylinder to pressure relief valve returns to sump at side of transmission.

Whenever servicing brake master cylinder, or if cylinder has been drained, refill with proper lubricant, then bleed brakes as follows: With engine running to keep master cylinder supplied with fluid, have an assistant depress brake pedals while bleed screws (B – Fig. 140) are opened. Bleed screws should remain open until no air appears, then tighten bleed screws. Bleed screws are located on both sides of center housing above rear axle housings.

MASTER CYLINDER

All Models

128. **R&R AND OVERHAUL.** Dis-

Fig. 140 – Brake bleed screws (B) are located at top of rear axle housings on center housing.

connect linkage and lines, plugging all lines and ports to prevent entrance of dirt. Remove master cylinder and mounting bracket assembly, then separate master cylinder from mounting bracket.

Refer to Fig. 141 for exploded view of brake master cylinder. Remove end cap (1) and withdraw plunger assembly (3 through 12). Disassemble plunger assembly noting location and direction of components. Unscrew retainer (14) and using a wire hook, remove check seat (16) from housing (13). Tip inlet port down and remove remaining components. Remove plugs (20) and with-

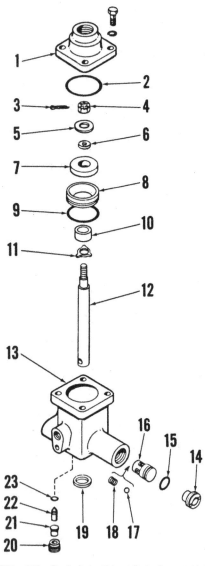

Fig. 141 – Exploded view of brake master cylinder.

1. End cap	
2. "O" ring	13. Housing
3. Cotter key	14. Retainer
4. Locknut	15. "O" ring
5. Washer	16. Check seat
6. Sealing washer	17. Ball
7. Piston washer	18. Spring
8. Piston	19. "O" ring
9. "O" ring	20. Plug
10. Spacer	21. Plunger
11. Washer	22. Metering spool
12. Rod	23. "O" ring

draw plungers (21) and metering spools (22). Remove "O" rings from end cap (1), housing (13), check seat (16), metering spools (22) and piston (8).

Clean, inspect and renew any components that are damaged or worn excessively. Renew all "O" rings. During reassembly, lube components with "Permatran" fluid.

When reassembling, make certain that "O" ring (9) on piston (8) is next to three-prong washer (11). The remainder of reassembly is the reverse of disassembly.

Bleed and adjust brakes as outlined in paragraphs 126 and 127.

PISTONS AND DISCS

All Models

129. **R&R AND OVERHAUL.** Before any service can be performed on brake discs or actuating pistons, final drive unit must be removed as outlined in paragraph 123.

With final drive assembly off, brake disc (26 – Fig. 137) and piston (28) can be removed. Brake plate (25) is bolted to axle housing (6) and can be serviced after removing the two retaining cap screws.

Clean, inspect and renew any components that are damaged or worn excessively. Check clearance of piston (28) in center housing bore, with "O" rings (27 and 29) removed, using three feeler gages as shown in Fig. 142. Piston should have an equal clearance of 0.007-0.014 inch and can be adjusted using oversize pistons. Brake pistons are available in four sizes. Standard size diameter is 16.367-16.370 inches; other sizes are: 16.368-16.369, 16.372-16.373 and 16.373-16.375 inches.

Renew "O" rings (27 and 29 – Fig. 137) after first coating with petroleum jelly. Align holes in back of piston (28) with pins in center housing and push piston into bore. Remainder of assembly is

reverse of disassembly procedure. Bleed brakes as outlined in paragraph 127.

PARKING BRAKE

All Models

130. **ADJUSTMENT.** The parking brake assembly mounts on top of the center housing under the fuel tank. An exploded view is shown in Fig. 143. Shoe (23) engages a groove in bevel ring gear of differential assembly.

Parking brake is properly adjusted when lever (9) is at right angles with cable (3) when in the fully engaged position and brake handle is pulled approximately 2½ inches. Minor adjustment is made with adjusting nut (6) on cable (3). If major adjustment is necessary, remove lever (9) and reposition on shaft (11) as required.

131. **R&R AND OVERHAUL.** To remove parking brake assembly, it is first necessary to drain and remove the fuel tank. Disconnect cable (3 – Fig. 143) and return spring (8) from actuating lever (9), then unbolt and withdraw parking brake assembly.

Overhaul procedure is evident after inspection of parking brake assembly and referral to Fig. 143.

Upon reassembly, tighten parking brake assembly retaining cap screws (13) to 80-95 ft.-lbs. torque and adjust as outlined in paragraph 130.

POWER TAKE-OFF

The pto input shaft is splined into an

adapter attached to engine crankshaft along with flywheel. The front shaft runs inside the hollow clutch shaft and transmission top shaft into the hydraulically operated, multiple-disc clutch located in rear axle center housing. A secondary shaft is located between pto clutch and top link support at rear of tractor and carries pto drive gears. Located directly below pto drive gears are pto driven gears carried by pto output shaft. Information on pto control valve is outlined to paragraphs 145 and 146 under main hydraulic section.

MF 2745 models are equipped with a two-speed (540 and 1000 rpm) pto and MF 2775 and MF 2805 models are equipped with a single-speed (1000 rpm) pto. Pto shaft speed on two-speed models is altered by changing the output shaft. Service procedures are similar on both models as outlined in the following paragraphs.

PTO CLUTCH ASSEMBLY

All Models

132. **REMOVE AND REINSTALL.** To remove pto clutch assembly (34 – Fig. 150), first drain center housing. Remove drive gear assembly as outlined in paragraph 134 and left side cover with auxiliary hydraulic pump as outlined in paragraph 151.

Remove clutch assembly alignment dowel (D – Fig. 144) and clip (C) on standpipe. Completely remove cap screws (1) and loosen cap screw (2) enough to allow fitting (F) to turn down for clearance. Slide sleeve (S – Fig. 145) onto standpipe (P), then remove standpipe. Withdraw pto clutch assembly from housing.

Before reinstallation, check the two needle roller bearings located at front of

Fig. 142 – Measure clearance between piston and housing using three feeler gages (F) as shown.

Fig. 143 – Exploded view of parking brake assembly.

1. Washer
2. "C" clip
3. Cable
4. Pin
5. Spacer
6. Nut
7. Bracket
8. Spring
9. Lever
10. Snap ring
11. Shaft
12. Seal
13. Cap screw
14. Cover
15. Gasket
16. Guide pin
17. Spring pin
18. Spring
19. Wedge block
20. Snap ring
21. Shoe pin
22. Roller
23. Shoe

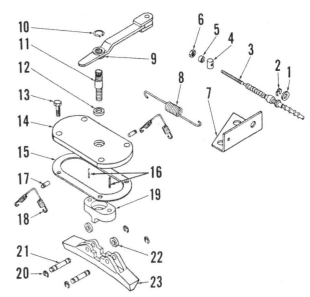

center housing for pto clutch assembly. If renewal is required, center housing will have to be removed and bearings installed as outlined in paragraph 119.

To reinstall pto clutch assembly, renew "O" rings on standpipe (P – Fig. 145), then install standpipe and pto clutch assembly. Renew "O" ring under fitting (F – Fig. 144), then tighten cap screws (1 and 2) to 8-10 ft.-lbs. torque. Install left side cover as outlined in paragraph 151 and drive gear assembly as outlined in paragraph 134. Refill center housing.

133. **OVERHAUL.** To overhaul pto clutch assembly, first remove unit from tractor as outlined in paragraph 132. Refer to Fig. 146 or Fig. 147 for exploded view of pto clutch assembly. Fig. 147 is for clutch assemblies used after rear axle DK 410 and Fig. 146 is used prior to rear axle DK 411. Overhaul procedures will be for early model pto clutch assemblies but will also apply to late model clutch assemblies.

Remove cap screws (27 – Fig. 146) and detach hydraulic pump drive gear (26). Carefully remove springs (24), pins (25), friction discs (21), clutch plates (20),

separating plate (19), thrust bearing (18), bearing race (17), inner hub (16) and brake ring (14). Remove clutch piston (10) from housing (5) using three ¼-inch 20 puller cap screws. Remove clutch hub cap screws (15), plate (13), thrust bearing (12) and thrust bearing race (11), then separate clutch arm (2) from housing (5).

After thoroughly cleaning and air drying all components, perform the following steps before reassembly. Renew all "O" rings and sealing rings. Check oil passage and check valve (6 and 7) in housing (5). Check contact areas of thrust bearings for excessive wear. Check plates (20) and friction discs (21) for warpage. Springs (24) should have a

free length of 1.62 inches and should require 28-34 lbs. pressure to compress to 0.96 inch. Drive gear (26) has a renewable bushing (22) that will require machining after installation. Inside diameter of fitted bushing should be 0.626-0.628 inch. Bushing bore should also be concentric with pitch diameter of splines in cover within 0.002 inch and bushing flange end should be square with axis of splines within 0.001 inch. If bearings (1 and 3) require renewal, rear of front bearing (3) should be 2 inches below rear surface of hub (2) and rear of rear bearing (1) should be 0.12 inch below rear surface. Lubricate all "O" rings, sealing rings and bearings during reassembly.

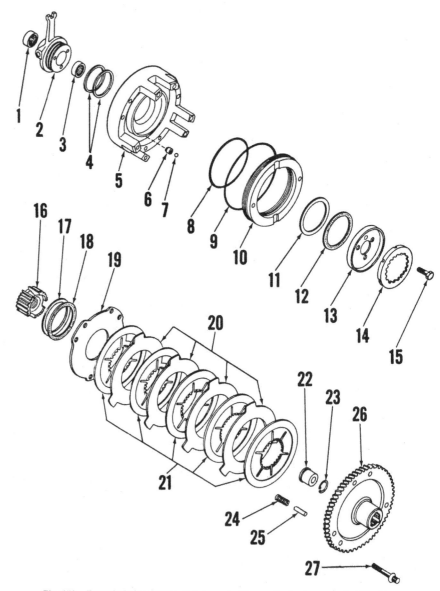

Fig. 144—Pto clutch assembly installed in center housing. Refer to text for removal procedure.

Fig. 145—Slide sleeve (S) onto standpipe (P) after removing retaining clip (C – Fig. 144).

Fig. 146—Exploded view of pto clutch assembly used prior to rear axle DK 411.

1. Bearing	8. "O" ring	15. Cap screw
2. Clutch arm	9. "O" ring	16. Inner hub
3. Bearing	10. Piston	17. Bearing race	22. Bushing
4. Sealing rings	11. Bearing race	18. Thrust bearing	23. Snap ring
5. Housing	12. Thrust bearing	19. Separating plate	24. Springs
6. Check valve	13. Plate	20. Clutch plates	25. Pins
7. Ball	14. Brake ring	21. Friction discs	26. Drive gear
			27. Cap screws

To reassemble pto clutch assembly, first insert clutch arm (2) in housing (5), then install thrust bearing race (11), thrust bearing (12) and plate (13) while aligning lubricating hole in plate (13) with oil slot in clutch arm (2). Coat threads of cap screws (15) with "Loctite" and tighten to 6-10 ft.-lbs. torque. Turn assembly over and check clearance between clutch hub and housing as shown in Fig. 148. Clearance should be 0.004-0.010 inch. Install "O" rings (8 and 9 – Fig. 146) on piston (10), then install piston (10) into housing (5). Install brake ring (14) with notched edge toward plate (13). Insert inner hub (16) into brake ring (14) with raised lip on hub toward plate (13). Temporarily install drive gear (26), tighten cap screws (27) to 25-38 ft.-lbs. torque and check for 0.004 inch minimum clearance between gear and inner hub (16). Remove drive gear (26) and place thrust race (17) and thrust bearing (18) over inner hub (16). Insert

pins (25), then install separating plate (19), clutch plates (20) and friction discs (21). Alternate discs and plates, starting with and ending with a disc. Place springs (24) over pins (25), then install drive gear (26) and tighten cap screws (27) to 25-38 ft.-lbs. torque.

Check clearance between separating plate (19) and housing (5) by inserting a feeler gage under tabs on separating plate. Clearance should be 0.099-0.139 inch for proper brake operation. Check clearance between the top friction disc (21) and drive gear (26). Clearance should be 0.070-0.188 inch for minimum clutch drag.

Check drive gear (26) and housing (5) for concentricity using a magnetic based dial indicator. With dial indicator base on the protruding machined hub of gear (26), position dial indicator so that plunger contacts machined surface of housing (5) as shown in Fig. 149. Slide magnetic base around gear hub and

observe runout. If runout exceeds 0.004 inch, loosen cover retaining cap screws (27 – Fig. 146) and tap edge of gear with a soft faced hammer, then retighten cap screws to 26-38 ft.-lbs. torque. Recheck runout and adjust as required.

Test brake release by applying 9 psi of air pressure to inlet port on clutch arm (2). Inner hub (16) should be turnable. To reinstall pto clutch assembly, refer to paragraph 132.

DRIVE GEAR ASSEMBLY

All Models

134. **R&R AND OVERHAUL.** To remove pto drive gear assembly (23 through 33 – Fig. 150) first remove top link support (23), then withdraw drive gear (28), spring (31), washer (32) and secondary shaft (33).

NOTE: Drive gear (28A – Fig. 150) and spacer (19A) are used in place of drive gear (28) and 540 rpm driven gear (19) on single-speed pto models.

Renew bearings in drive gear (28) and "O" ring seal on top link support (23) as required.

Before reinstalling drive gear assembly, check end play of pto drive

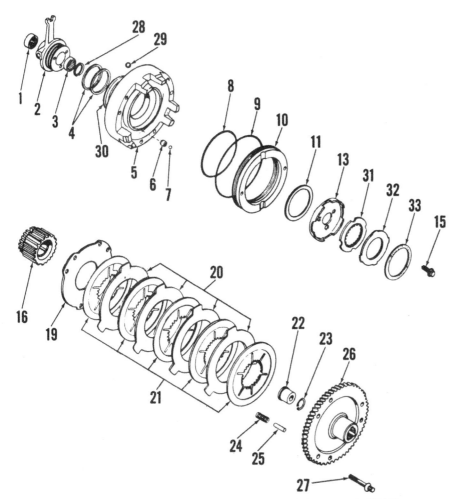

Fig. 147 – Exploded view of pto clutch assembly used after rear axle DK 410.

1. Bearing	9. "O" ring	20. Clutch plates	27. Cap screw
2. Clutch arm	10. Piston	21. Friction discs	28. Snap ring
3. Bearing	11. Thrust washer	22. Bushing	29. Ball
4. Sealing rings	13. Plate	23. Snap ring	30. Thrust washer
5. Housing	15. Cap screw	24. Spring	31. Brake disc
6. Check valve	16. Inner hub	25. Pin	32. Brake plate
7. Ball	19. Separating plate	26. Drive gear	33. Thrust bearing
8. "O" ring			

Fig. 148 – Check clearance between clutch arm (2) and housing (5) using two feeler gages as shown.

Fig. 149 – Mount dial indicator on drive gear (26) and slide plunger along machined surface (S) of housing (5) to check concentricity. Refer to text.

gear (28) as follows: Install secondary shaft (33), pto drive gear (28) and top link support (23). Remove control valve cover on hydraulic lift cover as outlined in paragraph 163 to obtain access to pto drive gear. Attach a dial indicator on side of pto drive gear as shown in Fig. 151. End play of pto drive gear (28 – Fig. 150) should be 0.001-0.009 inch and is adjusted by 0.008 inch shims (25) located behind bearing cup (26) in top link support (23). Remove or install shims as required.

Once end play is correct, reinstall control valve cover as outlined in paragraph 163, washer (32) and spring (31) on secondary shaft (33), then tighten top link support retaining cap screws to 130-200 ft.-lbs. torque.

DRIVEN GEAR ASSEMBLY

All Models

135. **R&R AND OVERHAUL.** To remove pto driven gear assembly (1 through 22-Fig. 150), first drain center housing. Remove lift cover as outlined in paragraph 165 and pto drive gear assembly as outlined in paragraph 134.

NOTE: Drive gear (28A – Fig. 150) and spacer (19A) are used in place of drive gear (28) and 540 rpm driven gear (19) on single speed pto models.

Position a piece of wood under gears (16 and 19) to prevent damage to gear teeth during removal. Remove oil fill tube located at rear of tractor for additional clearance. Remove pto shaft locknut (1) and withdraw pto shaft (2). Remove snap ring (3), seal retainer (4) with seal (5) and "O" ring (6). Withdraw bearing assembly (7 through 12) and thrust washer (14). Insert a suitable hook into a hole in 540 rpm gear (19) and remove through top of housing, then remove 1000 rpm gear (16) and thrust washers (17 and 20). Remove bearing (21) and plug (22) if required.

Clean, inspect and renew any components that are damaged or worn excessively. Renew "O" ring (6 and 9) and seal (5) as required. Driven gears (16 and 19) have renewable bushings (15 and 18) that should not require machining after installation. Fitted bushings should have an inside diameter of 1.8165-1.8175 inch and should be installed 0.020 inch below surface of bushing bore.

To reinstall driven gear assembly, first apply grease to thrust washers (14, 17 and 20) to hold them in position. Install large thrust washer (20) against bearing (21) and thrust washers (14 and 17) in recesses on both sides of 1000 rpm gear (16). Insert bearing assembly (7 through 12), but not completely to allow installation of driven gears. Lower 1000 rpm gear (16) into housing with protruding gear hub towards rear. Partially insert pto shaft (2) to hold gear in position, then install 540 rpm gear (19) with smooth side of gear towards front. Make certain all thrust washers are still in place, then install bearing assembly (7 through 12), "O" ring (6), seal retainer (4) with seal (5), and snap ring (3). Install pto shaft (2) and pto shaft locknut (1).

Check pto shaft end play. End play should be 0.001-0.009 inch and is adjusted by snap ring (3). Snap ring is available in sizes of 0.132, 0.140, 0.150 and 0.156 inch.

Install drive gear assembly as outlined in paragraph 134 and lift cover as outlined in paragraph 165. Refill center housing.

MAIN HYDRAULIC SYSTEM

All models are equipped with basically two separate hydraulic systems: One referred to as the main hydraulic

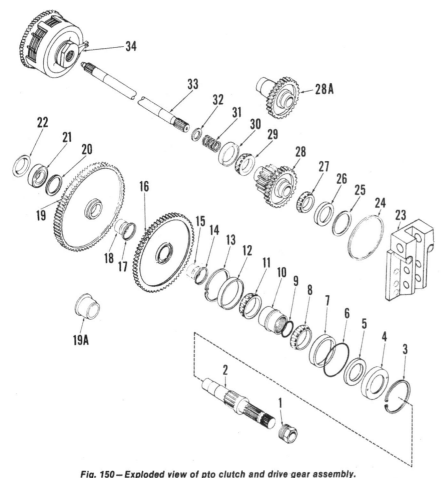

Fig. 151 — Method for checking end play on pto drive gear. Dial indicator (D) has a magnetic base (M).

Fig. 150 — Exploded view of pto clutch and drive gear assembly.

1. Locknut	11. Bearing cone	19A. Spacer
2. Pto shaft	12. Bearing cup	20. Thrust washer
3. Snap ring	13. Snap ring	21. Bearing
4. Seal retainer	14. Thrust washer	22. Plug
5. Seal	15. Bushing	23. Top link support
6. "O" ring	16. 1000 rpm gear	24. "O" ring
7. Bearing cup	17. Thrust washer	25. Shim
8. Bearing cone	18. Bushing	26. Bearing cup
9. "O" ring	19. 540 rpm gear	27. Bearing cone
10. Hub		28. Two-speed drive gear
		28A. Single-speed drive gear
		29. Bearing cone
		30. Bearing cup
		31. Spring
		32. Washer
		33. Secondary shaft
		34. Pto clutch assembly

system, the other referred to as the auxiliary hydraulic system.

The main hydraulic system has a dual-element hydraulic pump mounted on the inside of a cover attached to the right side of the center housing. The low volume side of the pump supplies oil to the pto and differential lock valve, and Multi-Power shift valve if equipped. The high volume side of the pump supplies oil to the three-point lift linkage control valve and optional third-spool auxiliary hydraulic control valve. On late model tractors, high volume pump oil is also supplied to the priority flow control valve for steering and brake circuits.

Information on Multi-Power shift valve is outlined in paragraph 113, lift linkage control valve is outlined in paragraph 163 and steering and brake circuits are outlined in their respective sections. Refer to the appropriate following paragraphs for service to main hydraulic system or to paragraphs 149 through 158 for service to the auxiliary hydraulic system.

LUBRICATION

All Models

136. The hydraulic system fluid serves as a lubricant for the transmission, differential, final drive, power steering and pto gear train. The transmission and rear axle housings serve as reservoir for system.

Recommended fluid is Massey-Ferguson Permatran lubricant. System capacity including filter change on all models is 30 U.S. gallons. To check fluid level, be sure tractor is on level ground and remove dipstick from rear of center housing. Dipstick opening is also filling port.

System filters, located on left side of center housing, should be renewed after first 25 hours of operation and every 750 hours or annually to coincide with lubricant change. Do not attempt to flush the system as a considerable amount of the flushing solution cannot be drained. If fluid is badly contaminated, system must be disassembled for cleaning.

TEST AND ADJUST

All Models

All tests should be conducted with tractor hydraulic fluid at normal operating temperature of 110-120 degrees F. Be sure reservoir is filled to proper level with Permatran fluid at all times during tests.

Always cover openings in lines and ports as soon as possible to prevent entrance of dirt and unnecessary loss of fluid.

CAUTION: To prevent injury, engine should be stopped, all hydraulically actuated equipment should be resting on the ground and all hydraulic pressure within circuits should be relieved of pressure before removing any test port plug or disconnecting any hydraulic line.

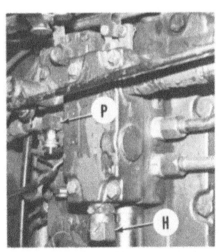

Fig. 153 — View of right side cover showing location of adjustable high pressure relief valve (H) and pto control valve test port (P).

137. **SYSTEM OPERATING PRESSURE.** The dual-element main hydraulic pump system operating pressure is controlled by high and low pressure relief valve.

138. HIGH PRESSURE TEST. Two types of high pressure relief valves may be encountered when servicing. The non-adjustable high pressure relief valve (H – Fig. 152) is located in right side cover and should have a pressure setting of 2050-2300 psi. The adjustable high pressure relief valve (H – Fig. 153) is located in an outside cover plate that is attached directly to right side cover or to the optional third spool auxiliary hydraulic control valve. Pressure setting of adjustable high pressure relief valve should be 1950-2150 psi.

Two different methods for testing high pressure relief valves are possible due to the availability of optional equipment.

On models equipped with optional third-spool auxiliary hydraulic control valve and remote quick coupler, test high pressure relief cartridge as follows: Install a 5000 psi gage in the top right quick coupler (A – Fig. 154), then with engine running, actuate auxiliary valve lever and read relief pressure.

On models without the optional remote quick coupler, it will first be necessary to remove lift control valve cover as outlined in paragraph 163 and test lift control safety relief valve as outlined in paragraph 164 (a false reading may be obtained when testing high pressure relief valve, if lift control safety relief valve is malfunctioning). Reinstall lift control valve cover, then install a "T" fitting on supply line to either lift cylinder and attach a 5000 psi gage. With engine running at 1500 rpm, remove plug and turn transport set screw (T – Fig. 155) counterclockwise until maximum pressure is obtained. Read relief pressure.

NOTE: After test and adjustments are made, adjust draft and transport control as outlined in paragraph 161.

Fig. 152 — View of right side cover showing location of non-adjustable high pressure relief valve (H), adjustable low pressure relief valve (L), low pressure test connection (C) and for models without remote quick coupler, flow test connection (E).

Fig. 154 — View at rear of tractor showing optional right side remote quick coupler (A), left inside coupler (B), left outside coupler (C) and hydraulic fluid level dipstick and filler tube (D).

HIGH PRESSURE RELIEF VALVE ADJUSTMENT. On models equipped with non-adjustable relief valve (H – Fig. 152), correct pressure setting by renewing relief valve. Tighten valve to 30-40 ft.-lbs. torque. Non-adjustable relief valve may be disassembled for cleaning and inspection only. Refer to Fig. 156 and remove snap ring (1), then withdraw internal components of relief valve noting their positions. Thoroughly clean all parts and inspect for damage or wear. Renew back-up washer and all "O" rings prior to reassembly. Make certain snap ring (1) is installed with sharp edge out.

On models equipped with adjustable relief valve cartridge (H – Fig. 153), correct pressure setting by removing cap covering adjusting screw, loosen locknut and turn adjusting screw out to decrease or in to increase pressure. Refer to Fig. 157 to disassemble valve for cleaning and inspection. Renew all "O" rings and back-up washers prior to reassembly. Test and readjust as required.

139. LOW PRESSURE TEST. Low pressure relief valve cartridge (L – Fig. 152) is located in right side cover and should have a pressure setting of 290-310 psi. To test low pressure relief valve, install a "T" fitting at connection (C) and attach a 600 psi gage.

NOTE: On models without Multi-Power transmissions, a plug will be installed at (C). Remove plug and install test gage.

With engine running, actuate pto lever then read relief pressure.

LOW PRESSURE RELIEF VALVE ADJUSTMENT. Refer to Fig. 158 for exploded view of relief valve cartridge. To adjust, remove relief valve from side cover then clean and air dry cartridge. Unscrew cartridge (7) from plug (1). Adjustment can be accomplished by varying thickness of shims (2). Upon reassembly, renew all "O" rings.

140. SYSTEM FLOW TEST. Flow test high volume section of dual element pump as follows:

On models equipped with optional third-spool auxiliary hydraulic control valve and remote quick coupler, connect inlet side of flow meter in top right quick coupler (A – Fig. 154) and outlet to sump (D). Actuate control lever during test.

On models without optional remote quick coupler, it will be necessary to attach flow meter inlet to elbow fitting (E – Fig. 152) on top of right side cover and outlet to sump.

Run engine to heat oil to 110-120 degrees F. Once fluid has reached operating temperature, set engine at 2000 rpm. Slowly close flow meter load valve until a pressure of 2000 psi is obtained and observe rate of flow. If meter reads less than minimum of 11.6 gpm, then repair of pump is necessary.

Flow test low volume section of dual element pump as follows:

Attach flow meter inlet at connection (C – Fig. 152) on right side cover and outlet to sump at rear of tractor.

NOTE: On models without Multi-Power transmissions, a plug will be installed at (C). Remove plug and attach flow meter.

Run engine to heat oil to 110-120 degrees F. Once fluid has reached operating temperature, set engine at 2000 rpm. Slowly close flow meter load valve until a pressure of 200 psi is obtained and observe rate of flow. Tractors prior to serial number 466 should have a minimum reading of 3.75 gpm and tractors after serial number 465 should have a minimum reading of 4.5 gpm. Any reading less than minimum will require repair or renewal of pump.

MAIN PUMP

Two types of main pumps may be encountered when servicing and can be identified by part number stamped in pump housing. Models prior to tractor serial number 9R0466 should be equipped with main pump part number 1610465M2. Models after tractor serial number 9R0465, should be equipped with main pump part number 3038730M1. Refer to the appropriate following paragraphs to service main pump.

Early Models

141. REMOVE AND REINSTALL. To remove main pump, first disconnect hydraulic lines, control linkages and sending unit wires noting their locations, from side cover, then unbolt and carefully withdraw side cover with pump. Remove pto and differential lock control valve to obtain access to external pump retaining cap screw. Remove retaining cap screws and separate pump from side cover.

Before reinstalling main pump, check clearance between main pump drive gear and pump housing. Clearance should be 0.010 inch. Renew "O" rings, seals and gaskets as required.

To reinstall pump and side cover, reverse removal procedure.

142. OVERHAUL. To overhaul main hydraulic pump on early models, first remove pump as outlined in paragraph 141. Clean pump thoroughly, then remove drive gear with Woodruff key and inlet manifold.

Refer to Fig. 159 for exploded view of main pump. Scribe alignment marks on

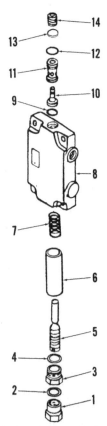

Fig. 157 – Exploded view of adjustable high pressure relief valve cartridge.

1. Cap	8. Plate
2. Copper washer	9. Washer
3. Locknut	10. Relief valve
4. Copper washer	11. Relief valve seat
5. Adjusting screw	12. "O" ring
6. Sleeve	13. Aluminum disc
7. Spring	14. Plug

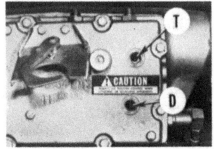

Fig. 155 – Transport (T) and draft (D) adjusting screw locations.

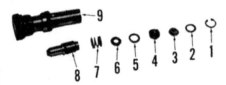

Fig. 156 – Disassembled view of non-adjustable high pressure relief valve cartridge.

1. Snap ring	
2. Washer	
3. Filter screen	6. Washer
4. Orifice plate	7. Spring
5. "O" ring	8. Pilot cartridge
	9. Main cartridge

front cover (3), center housing (12) and rear cover (20) to aid in reassembly. Remove retaining cap screws (1 and 21) then carefully separate rear cover (20) from center housing (12). Remove diaphragm plate (15), back-up gasket (16), protector gasket (17) and diaphragm seal (18) from rear cover. Withdraw idler gear with shaft (13), then slide drive gear (14) off idler shaft (10). Remove drive key (11) from idler shaft.

Carefully separate front cover (3) from center housing and remove diaphragm plate, gaskets and seals (5 through 8) and front seal (2). Withdraw drive gear with shaft (9) and idler gear with shaft (10) from center housing.

Thoroughly clean and air dry all metal parts and check for wear or damage that will require replacement of parts. Minimum diameter of gear shafts (9, 10 and 13) is 0.7494 inch. Maximum inside diameter of bushings in front cover (3), center housing (12) and rear cover (20) is 0.7515 inch. Minimum face width of gears (9 and 10) is 0.9270 inch while minimum width of gear faces (13 and 14) is 0.374 inch. Check for gear pocket wear in center housing. Gear pocket diameter in both sides of center housing (12) should not be greater than 1.7163 inch. Renew front cover seal (2), diaphragm seals (5 and 18), protector gaskets (6 and 17), back-up gaskets (7 and 16) and diaphragm plates (8 and 15).

Lubricate all metal parts with clean hydraulic oil. Insert drive gear with shaft (9) into top bore of center housing and idler gear with shaft (10) into bottom bore. Insert diaphragm seal (5) in front cover with open side "V" facing into groove. Install protector gasket (6), then back-up gasket (7) into diaphragm seal. With bronze side of diaphragm plate (8) away from seals and slot (S – Fig. 160) positioned over hole (H) in front cover, set plate into back-up gasket (7). Make certain diaphragm plate is fully inserted within raised lip on back-up gasket. Use a thin coat of petroleum jelly to seal machined surfaces, then press front cover onto center housing while aligning marks made previously.

Install drive key (11 – Fig. 159) and drive gear (14) on idler shaft (10). Insert idler gear with shaft (13) with short end of shaft toward center housing. Assemble diaphragm seal (18), protector gasket (17), back-up gasket (16) and diaphragm plate (15) into rear cover (20), then install rear cover to center housing following the same procedure for assembling front cover.

Install and alternately tighten cap screws (1 and 21) to 23-26 ft.-lbs. torque. Install front cover seal (2) using a suitable seal installer. Install Woodruff key, drive gear, washer and retaining nut, tightening retaining nut to 35-50 ft.-lbs. torque. Secure with a new cotter key. Install inlet manifold then pour hydraulic fluid into pump and turn pump drive gear by hand. Pump should turn freely after a few revolutions. Reinstall pump as outlined in paragraph 141 and perform break-in procedure as outlined in paragraph 144.

Late Models

143. R&R AND OVERHAUL. Remove pump following same procedure for early models as outlined in paragraph 141.

Clean pump thoroughly, then remove drive gear with Woodruff key and inlet manifold.

Refer to Fig. 161 for exploded view of main pump. Scribe alignment marks on front cover (3), center housing (14) and rear cover (23) to aid in reassembly. Remove retaining cap screws (1 and 24) then carefully separate front cover (3) from center housing (14). Remove diaphragm plate (10), back-up gasket (9), protector gasket (8), diaphragm seal (7), spring (5), ball (4) and front seal (2) from front cover. Withdraw drive gear (11), idler gear (12) and thrust plate (13) from center housing.

Carefully separate rear cover (23) from center housing and remove diaphram plate, gaskets and seals (18 through 21). Withdraw drive gear (15) and idler gear (16) from rear section of center housing.

Thoroughly clean and air dry all metal parts and check for wear or damage that will require replacement of parts. Minimum diameter of gear shafts (11 and 12) is 0.8094 inch while minimum diameter of gear shafts (15 and 16) is 0.6845 inch. Maximum inside diameter of bushings in front cover and front section of center housing is 0.8156 inch. Maximum diameter of bushings in rear cover and

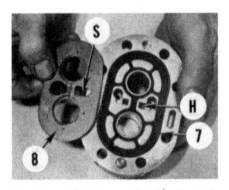

Fig. 160 — Install diaphragm plate (8) with slot (S) positioned over hole (H). Refer to text.

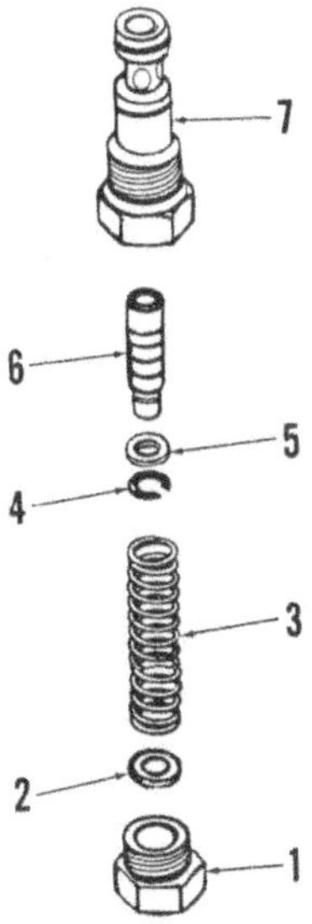

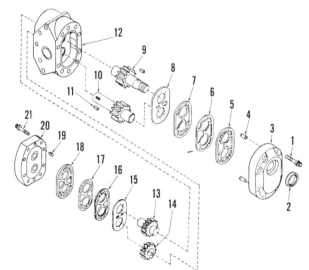

Fig. 159 — Exploded view of main hydraulic pump used on models prior to tractor serial number 9R0466.

1. Cap screw
2. Seal
3. Front cover
4. Dowel
5. Diaphragm seal
6. Protector gasket
7. Back-up gasket
8. Diaphragm plate
9. Drive gear
10. Idler shaft
11. Drive key
12. Center housing
13. Idler gear
14. Drive gear
15. Diaphragm plate
16. Back-up gasket
17. Protector gasket
18. Diaphragm seal
19. Dowel
20. Rear cover
21. Cap screw

Fig. 158 — Exploded view of low pressure relief valve cartridge.

1. Plug
2. Shims
3. Spring
4. Retaining ring
5. Washer
6. Spool
7. Main cartridge

rear section of center housing is 0.6906 inch. Minimum face width of gears (11 and 12) is 0.7051 inch while minimum gear face width of gears (15 and 16) is 0.3716 inch. Check for wear of center housing gear pockets. Gear pocket diameter in front section of center housing should not exceed 2.002 inch and in rear section 1.7185 inch. Renew front cover seal (2), diaphragm seals (7 and 21), protector gaskets (8 and 20), back-up gaskets (9 and 19), diaphram plates (10 and 18) and thrust plate (13).

Lubricate all metal parts with clean hydraulic oil. Insert thrust plate (13) into center housing with bronze side toward front cover and cut-away section toward inlet manifold side of pump. Insert drive gear (11) into top bore of center housing and idler gear (12) into bottom bore with painted end of idler gear shaft toward front cover. Insert diaphragm seal (7) in front cover with open side of "V" facing into groove. Install protector gasket (8), then back-up gasket (9) into diaphragm seal. Insert ball (4) and spring (5) into front cover bore with angle drilling as shown in Fig. 162. With bronze side of diaphragm plate (10 – Fig. 161) away from seals, set plate in place over spring and into back-up gasket. Make certain diaphragm plate is fully inserted within raised lip on back-up gasket. Use a thin coat of petroleum jelly to seal machined surfaces, then press front cover onto center housing while aligning marks made previously. Install and tighten retaining cap screws (1) just enough to hold assembly together.

Insert drive gear (15) into center housing while aligning splines with drive gear (11), then insert idler gear (16) with painted shaft end towards rear. Assemble diaphragm seal (21), protector gasket (20), back-up gasket (19) and diaphragm plate (18) into rear cover, then install rear cover on center housing

following ᴜne same procedure for assembling front cover.

Install and alternately tighten cap screws (1 and 24) to 23-26 ft.-lbs. torque. Install front cover seal (2) using a suitable seal installer. Install Woodruff key, drive gear, washer and retaining nut, tightening retaining nut to 35-50 ft.-lbs. torque. Secure with a new cotter key. Install inlet manifold then pour hydraulic fluid into pump and turn pump drive gear by hand. Pump should turn freely after a few revolutions. Reinstall pump as outlined in paragraph 141 and perform break-in procedure as outlined in paragraph 144.

144. **PUMP BREAK-IN PROCEDURE.** To prevent damage to a new or rebuilt hydraulic pump, the following break-in procedure is recommended. Fill hydraulic system to proper level with Massey-Ferguson Permatran lubricant or equivalent. Start engine and run at approximately 1400 rpm without operating any hydraulic system component for first three minutes. For the next three minutes, intermittently operate pto. For an additional three minutes, run engine at 2600 rpm and intermit-

tently operate pto. Shut off engine and check for leaks.

PTO/DIFFERENTIAL LOCK CONTROL VALVE

All Models

145. **PTO VALVE TEST.** To test pto "feathering" capability, remove sending switch and attach a 600 psi gage to test port (P – Fig. 153).

NOTE: Some models may have a plug in place of sending switch. Remove plug and attach gage.

With engine running, it should be possible to modulate pressure between 0-310 psi by moving pto control lever from disengaged to engaged position.

146. **R&R AND OVERHAUL.** To remove pto/differential lock control valve, disconnect control linkages and sending unit wire, then unbolt and remove valve.

NOTE: Service procedures for models without differential lock will be similar.

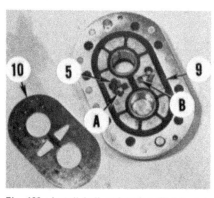

Fig. 162 – Install ball and spring (5) into hole with angle drilling (A). Refer to text.

Fig. 161 – Exploded view of main hydraulic pump used on models after tractor serial number 9R0465.

1. Cap screw
2. Seal
3. Front cover
4. Ball
5. Spring
6. Dowel
7. Diaphragm seal
8. Protector gasket
9. Back-up gasket
10. Diaphragm plate
11. Drive gear
12. Idler gear
13. Thrust plate
14. Center housing
15. Drive gear
16. Idler gear
17. Retaining ring
18. Diaphragm plate
19. Back-up gasket
20. Protector gasket
21. Diaphragm seal
22. Dowel
23. Rear cover
24. Cap screw

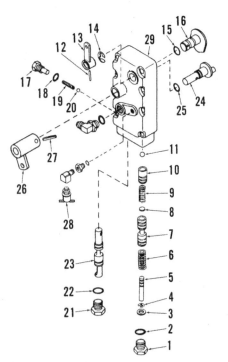

Fig. 163 – Exploded view of pto/differential lock control valve.

1. Plug		16. Camshaft
2. "O" ring		17. Plug
3. Washer		18. "O" ring
4. Circlip		19. Spring
5. Plunger		20. Ball
6. Large spring		21. Plug
7. Double land spool		22. "O" ring
8. Shim		23. Spool
9. Small spring		24. Shaft
10. Spool		25. "O" ring
11. Steel ball		26. Lever
12. Roll pin		27. Roll pin
13. Lever		28. Pressure switch
14. Snap ring		29. Valve body
15. "O" ring		

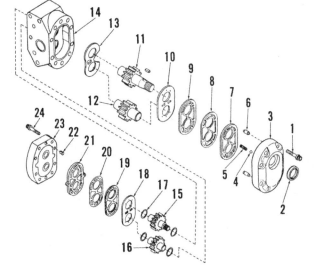

Specifications for pto portion of control valve are the same with or without differential lock.

To overhaul pto portion of control valve, remove plug (1 – Fig. 163) and withdraw washer (3), plunger (5) with circlip (4), large spring (6), double land spool (7), shims (8) if installed, small spring (9), spool (10) and steel ball (11). Drive out roll pin (12) and remove lever (13), snap ring (14) and camshaft (16).

Clean, inspect and renew any components that are damaged or worn excessively. Free length of spring (9) should be 1.540 inch. A 0.006 inch shim (8) can be added if necessary to obtain proper free length. Once free length of spring (9) is correct, 7-8 lbs. of pressure should be required to compress spring to 1.44 inches. Renew all "O" rings as required.

Reassemble pto control valve by reversing disassembly. Make sure recess end in spool (10) is towards steel ball (11) and internal drilling of double land spool (7) is towards spring (9).

To overhaul differential portion of control valve, first remove detent assembly (17 through 20). Remove plug (21), then rotate shaft (24) to release spool (23) and withdraw spool. Drive out roll pin (27) and remove lever (26) and shaft (24).

Clean, inspect and renew any components that are damaged or worn ex-

cessively. Renew all "O" rings as required.

Reassemble differential lock valve by reversing disassembly.

BOSCH REMOTE CONTROL VALVES

All Models So Equipped

Bosch auxiliary remote control valves are the only valves used as an optional right-side third-spool valve with remote quick coupler. Bosch or Husco auxiliary remote control valves may be used on left side.

Three types of Bosch auxiliary valves are used: Spring-centered; detented three-position with hydraulic kickout; and detented four-position with hydraulic kickout. Two variations of the spring-centered control valve were also used during production. All Bosch control valves can be identified by a part number on each valve. Spring-centered valve used prior to tractor serial number 9R0466 is part number 0521602021, spring-centered control valve used after tractor serial number 9R0465 is part number 0521601163, detented three-position control valve is part number 0521601164 and detented four-position is part number 0521601165. System operating pressure is controlled by an adjustable relief valve cartridge located in a separate plate attached to the remote control valve, except for right-side remote control valve used on models prior to tractor serial number 9R0466 which uses the non-adjustable relief cartridge located directly in right-side cover. Test and adjust both types of relief valve cartridges as outlined in paragraph 138.

147. R&R AND OVERHAUL. The following service procedures will cover detented three- and four-position control valves. Service procedures for spring-centered control valves will be similar. Refer to the appropriate exploded view in Fig. 164, Fig. 165 or Fig. 166 when servicing.

To remove right- or left-side remote control valves, disconnect hydraulic lines and linkages, then unbolt and remove valve. Before beginning overhaul, check for parts availability. Valve components should be clearly identified and placed in order to facilitate proper reassembly. Do not interchange components between valves and use care when handling to prevent damage to machined surfaces.

Clean exterior thoroughly with a suitable solvent. Remove end cap (1 – Fig. 166) then unscrew retainer sleeve (3) and withdraw spool assembly through bottom of valve body (29).

Remove seal retainer plate (33), wiper seal (32) and spool seal (31). Separate mounting plate from valve body and remove seals (42) and back-up rings (41).

Thread a suitable machine screw into load check plug (36), press plug in and remove retaining ring (35). Withdraw load check assembly (36 through 40) from valve body. Separate spring (39), "O" ring (38) and back-up ring (37) from plug (36).

If spool assembly requires disassembly, reinsert spool and thread retainer sleeve (3) back into valve body. Push in on end plug (7) and remove retaining ring (6), then remove end plug and detent spring (8). Slide sleeve (5) from detent tube (14), simultaneously removing detent balls (15). Withdraw detent spool (9) from inside of tube (14). Unscrew retainer sleeve (3) and withdraw remainder of spool assembly from valve body. Compress spring (18), then unscrew detent tube (14) from

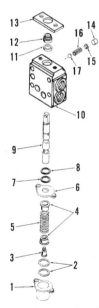

Fig. 164 – Exploded view of spring-centered remote control valve Bosch part number 0521602021 used on models prior to tractor serial number 9R0466.

1. End cap
2. Seals
3. Cap screw
4. Retainers
5. Spring
6. Retainer plate
7. Washer
8. Seal
9. Spool
10. Valve body
11. Spool seal
12. Wiper seal
13. Seal plate
14. Plug
15. Guide
16. Spring
17. Detent ball

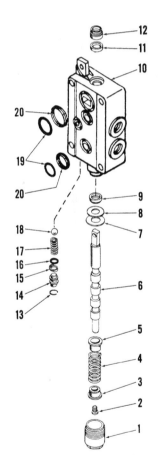

Fig. 165 – Exploded view of spring-centered remote control valve Bosch part number 0521601163 used on models after tractor serial number 9R0465.

1. End cap
2. Cap screw
3. Retainer
4. Spring
5. Retainer
6. Spool
7. Steel washer
8. Teflon washer
9. Spool seal
10. Valve body
11. Spool seal
12. Wiper seal
13. Retaining ring
14. Plug
15. Back-up ring
16. "O" ring
17. Spring
18. Steel ball
19. "O" rings
20. Back-up rings

spool end. Remove spring (18) and retainers (17 and 19). Remove kickout adjustment plug (10) counting number of turns required to remove plug. Remove spring (12) and piston (13) from detent tube. Use a short burst of compressed air in upper crossover hole in spool (30) to remove cone piston (28), steel washers (27 and 24), "O" ring (26), back-up washer (25) and pin (23) from spool bore.

Clean, inspect and renew any components that are damaged or worn excessively. Renew all "O" rings and seals.

Lubricate all components with clean hydraulic oil, then assemble unit by reversing disassembly procedure while noting the following: Install sleeve (5) over detent tube (14) and balls (15) with notch in sleeve positioned away from spool. Install kickout adjustment plug (10) the same number of turns required to remove it. Test and adjust kickout pressure as outlined in paragraph 148.

148. TEST AND ADJUST HYDRAULIC KICKOUT. To test kickout pressure on detented three- and four-position remote control valves, install a 5000 psi gage into top remote coupler of control valve to be tested. Start engine and actuate corresponding auxiliary valve lever forward to first position and read kickout pressure. Adjust valve if kickout does not occur between 1900-2040 psi.

To adjust kickout pressure, first turn off engine, then remove end cap (1 – Fig. 166) from bottom of spool valve requiring adjustment. Push inward on plug (7) and while preventing detent sleeve (5) from moving, remove internal snap ring (6). Withdraw spring (8) and detent spool (9). Turn adjusting screw (10) in to increase or out to decrease kickout pressure. Reinstall components.

NOTE: If detent balls (15) drop out, remove detent sleeve (5) from end of valve. Install detent spool (9) into tube (14), then install detent balls (15) into holes around tube. Slide detent sleeve (5) over balls. Make sure detent balls enter center recess of detent spool (9) and detent sleeve (5) is positioned so its internal detent notch located nearest end of sleeve is away from tube (14).

AUXILIARY HYDRAULIC SYSTEM

All models are equipped with basically two separate hydraulic systems: One referred to as the main hydraulic system, the other referred to as the auxiliary hydraulic system.

The auxiliary hydraulic system has a single-element hydraulic pump mounted on the inside of a cover attached to the left side of the center housing. The auxiliary pump supplies oil to the lubrication control valve, oil cooler relief valve and auxiliary remote control valves.

Refer to the appropriate following paragraphs for service to auxiliary hydraulic system or to paragraphs 136 through 148 for service to main hydraulic system.

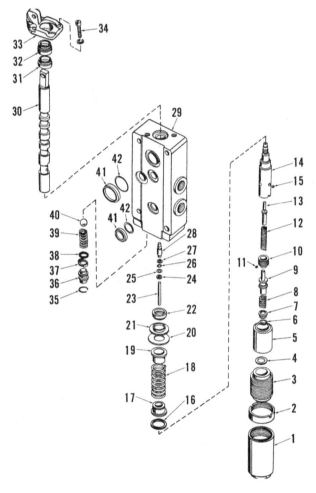

Fig. 166 — Exploded view of detented remote control valve Bosch part number 0521601164 or 0521601165.

1. End cap
2. Locknut
3. Retainer sleeve
4. Shim
5. Sleeve
6. Retaining ring
7. End plug
8. Detent spring
9. Detent spool
10. Adjustment plug
11. Pin
12. Spring
13. Piston
14. Detent tube
15. Detent balls
16. Shim
17. Retainer
18. Spring
19. Retainer
20. Steel washer
21. Teflon washer
22. Spool seal
23. Pin
24. Steel washer
25. Back-up washer
26. "O" ring
27. Steel washer
28. Cone piston
29. Valve body
30. Spool
31. Spool seal
32. Wiper seal
33. Seal retainer plate
34. Cap screw
35. Retaining ring
36. Plug
37. Back-up ring
38. "O" ring
39. Spring
40. Steel ball
41. Back-up rings
42. "O" rings

TEST AND ADJUST

All Models

All tests should be conducted with tractor hydraulic fluid at normal operating temperature of 110-120 degrees F. Be sure reservoir is filled to proper level with Permatran fluid at all times during tests.

Always cover openings in lines and ports as soon as possible to prevent en-

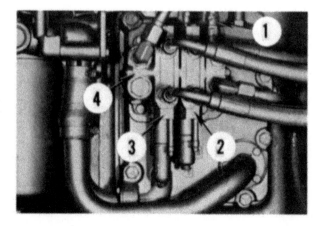

Fig. 167 — View of Husco auxiliary control valve attached to left cover plate showing location of adjustable relief valve (1), valve working sections (2 and 3) and valve outlet plate (4).

trance of dirt and unnecessary loss of fluid.

CAUTION: To prevent injury, engine should be stopped, all hydraulically actuated equipment should be resting on the ground and all hydraulic pressure within circuits should be relieved of pressure before removing any test port plug or disconnecting any hydraulic line.

149. **SYSTEM OPERATING PRESSURE.** Auxiliary hydraulic system operating pressure is controlled by an adjustable high pressure relief valve attached directly to remote control valves.

If equipped with Bosch auxiliary remote control valves, refer to paragraph 137 for test and adjustment procedures. If equipped with Husco auxiliary remote control valves, proceed as follows: Install a 5000 psi gage in the top-left, inside quick coupler, then with engine running, actuate corresponding auxiliary valve lever and read relief pressure. If hydraulic kickout is working properly, it may be necessary to hold lever in position to obtain reading. Relief pressure should be 2050-2300 psi.

To adjust relief pressure, remove acorn nut (1—Fig. 167) on relief valve then loosen jam nut on adjusting screw. Turn screw clockwise to increase or counterclockwise to decrease pressure setting. Refer to Fig. 168 to disassemble

valve for cleaning and inspection. Renew all "O" rings and back-up washers, then prior to reassembly lubricate components with clean hydraulic fluid. Test and readjust as required.

150. **SYSTEM FLOW TEST.** To test auxiliary pump flow, connect inlet side of flow meter in top-left, outer quick coupler and outlet in bottom-left, outer quick coupler. Run engine to heat oil to 110-120 degrees F. Once fluid has reached operating temperature, run engine at 2000 rpm. Actuate auxiliary control valve lever, then slowly close flow meter load valve until a pressure of 2000 psi is obtained and observe rate of flow. If hydraulic kickout is working properly, it may be necessary to hold lever against detent.

If meter reads less than minimum of 11.6 gpm, then repair of pump is necessary.

AUXILIARY PUMP

Two types of single-element, auxiliary pumps may be encountered when servicing and can be identified by part number stamped in pump housing. Models prior to tractor serial number 9R0466 should be equipped with auxiliary pump part number 1610596M3. Models after tractor serial number 9R0465, should be

equipped with auxiliary pump part number 3038732M1. Refer to the appropriate following paragraphs to service auxiliary pump.

Early Models

151. **REMOVE AND REINSTALL.** To remove auxiliary pump, first disconnect hydraulic lines and control linkages to auxiliary control valve and remove valve to obtain access to external pump retaining cap screw. Remove oil supply tubes connected to side cover and hydraulic filters, then unbolt and carefully withdraw side cover with pump. Remove inlet manifold and pump retaining cap screws, then separate pump from side cover.

Before reinstalling auxiliary pump, check clearance between auxiliary pump drive gear and pump housing. Clearance should be 0.010 inch. Renew "O" rings, seals and gaskets as required.

To reinstall pump and side cover, reverse removal procedure.

152. **OVERHAUL.** To overhaul auxiliary pump on early models, first remove pump as outlined in paragraph 151. Clean pump thoroughly, then remove drive gear with Woodruff key.

Refer to Fig. 169 for exploded view of auxiliary pump. Scribe alignment marks on front cover (3) and pump housing (11) to aid in reassembly. Remove retaining cap screws (1) then carefully separate front cover from pump housing. Remove diaphragm plate (8), back-up gasket (7), protector gasket (6) and diaphragm seal (5) from front cover. Withdraw drive gear (9) and idler gear (10) from pump housing.

Thoroughly clean and air dry all metal parts and check for wear or damage that will require replacement of parts. Minimum diameter of gear shafts (9 and 10) is 0.7494 inch. Maximum inside bushing diameter is 0.7515 inch. Minimum face width of gears (9 and 10) is 0.9270 inch. Check for gear pocket wear in pump housing. Gear pocket diameter should not exceed 1.7163 inch.

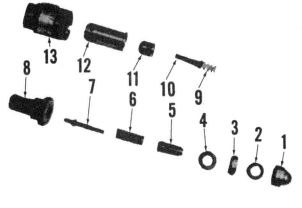

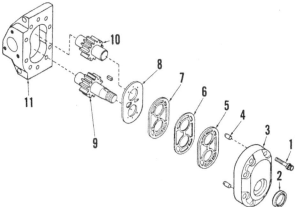

Fig. 168 — Disassembled view of Husco adjustable relief valve.

1. Acorn nut
2. Washer
3. Jam nut
4. Copper washer
5. Adjusting screw
6. Pilot spring
7. Pilot poppet
8. Pilot housing
9. Spring
10. Piston poppet
11. Relief poppet
12. Check poppet
13. Main housing

Fig. 169 — Exploded view of auxiliary hydraulic pump used on models prior to tractor serial number 9R0466.

1. Cap screw
2. Seal
3. Front cover
4. Dowel
5. Diaphragm seal
6. Protector gasket
7. Back-up gasket
8. Diaphragm plate
9. Drive gear
10. Idler gear
11. Pump housing

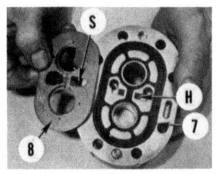

Fig. 170 — Install diaphragm plate (8) with slot (S) positioned over hole (H). Refer to text.

Renew front seal (2), diaphragm seal (5), protector gasket (6), back-up gasket (7) and diaphragm plate (8).

Lubricate all metal parts with clean hydraulic oil. Insert drive gear (9) into bottom bore of pump housing and idler gear (10) into top bore. Insert diaphragm seal (5) in front cover with open side of "V" facing into groove. Install protector gasket (6), then back-up gasket (7) into diaphragm seal. With bronze side of diaphragm plate (8) away from seals and slot (S–Fig. 170) positioned over hole (H) in front cover, set plate into back-up gasket (7). Make certain diaphragm plate is fully inserted within raised lip on back-up gasket. Use a thin coat of petroleum jelly to seal machined surfaces, then press front cover onto pump housing while aligning marks made previously.

Install and alternately tighten cap screws (1–Fig. 169) to 23-26 ft.-lbs. torque. Install front cover seal (2) using a suitable seal installer. Install Woodruff key, drive gear, washer and retaining nut, tightening retaining nut to 35-50 ft.-lbs. torque. Secure with a new cotter key. Pour hydraulic fluid into pump inlet port and turn pump drive gear by hand. Pump should turn freely after a few revolutions. Reinstall pump as outlined in paragraph 151 and perform break-in procedure as outlined in paragraph 154.

Late Models

153. **R&R AND OVERHAUL.** Remove pump following the same procedure for early models as outlined in paragraph 151.

Clean pump thoroughly, then remove drive gear with Woodruff key.

Refer to Fig. 171 for exploded view of auxiliary pump. Scribe alignment marks on front cover (3) and pump housing (14) to aid in reassembly. Remove retaining cap screws (1) then carefully separate front cover from pump housing. Remove

diaphragm plate (10), back-up gasket (9), protector gasket (8), diaphragm seal (7), spring (5), ball (4) and front seal (2) from front cover. Withdraw drive gear (11), idler gear (12) and thrust plate (13) from pump housing.

Thoroughly clean and air dry all metal parts and check for wear or damage that will require replacement of parts. Minimum diameter of gear shafts (11 and 12) is 0.8094 inch. Maximum inside bushing diameter is 0.8156 inch. Minimum face width of gears (11 and 12) is 0.7051 inch. Check for gear pocket wear in pump housing. Gear pocket diameter should not exceed 2.002 inch. Renew front cover seal (2), diaphragm seal (7), protector gasket (8), back-up gasket (9), diaphragm plate (10) and thrust plate (13).

Lubricate all metal parts with clean hydraulic oil. Insert thrust plate (13) into pump housing with bronze side toward front cover and cut away section toward inlet manifold side of pump. Insert drive gear (11) into bottom bore of pump housing and idler gear (12) into top bore. Insert diaphragm seal (7) in front cover with open side of "V" facing into groove. Install protector gasket (8), then back-up gasket (9) into diaphragm seal. Insert ball (4) and spring (5) into bore in front cover with angle drilling as shown in Fig. 172. With bronze side of diaphragm plate (10–Fig. 171) away from seals, install plate over spring and into back-up gasket. Make certain diaphragm plate is fully inserted within raised lip on back-up gasket. Use a thin coat of petroleum jelly to seal machined surfaces, then press front cover onto pump housing while aligning marks made previously.

Install and alternately tighten cap screws (1) to 25-28 ft.-lbs. torque. Install front cover seal (2) using a suitable seal installer. Install Woodruff key, drive gear, washer and retaining nut, tightening retaining nut to 35-50 ft.-lbs. torque. Secure with a new cotter key. Pour hydraulic fluid into pump inlet port and

turn pump drive gear by hand. Pump should turn freely after a few revolutions. Reinstall pump as outlined in paragraph 151 and perform break-in procedure as outlined in paragraph 154.

154. **PUMP BREAK-IN PROCEDURE.** To prevent damage to a new or rebuilt hydraulic pump, the following break-in procedure is recommended. Fill hydraulic system to proper level with Massey-Ferguson Permatran lubricant or equivalent. Start engine and run at approximately 1400 rpm without operating any hydraulic system component for the first three minutes. For the next three minutes, intermittently operate auxiliary controls. For an additional three minutes, run engine at 2600 rpm and intermittently operate auxiliary controls. Shut off engine and check for leaks.

HUSCO REMOTE CONTROL VALVES

All Models So Equipped

Husco auxiliary remote control valves will only be used on left side in conjunction with auxiliary pump, but Bosch valves may be used on either side. Two types of Husco auxiliary valves are used, detented three-position and detented four-position, both having hydraulic kickout mechanisms. Minor differences may be found between valves used on early and late production tractors, but basic service procedures will be the same.

System operating pressure is controlled by an adjustable relief valve cartridge located in a plate that is part of the remote control valve assembly. Test and adjust relief valve cartridge as outlined in paragraph 149.

155. **R&R AND OVERHAUL.** To remove left-side remote control valves, disconnect hydraulic lines and linkages, then unbolt and remove valves. Before

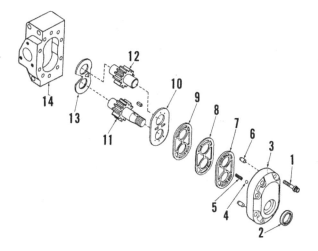

Fig. 171 – Exploded view of auxiliary hydraulic pump used on models after tractor serial number 9R0465.

1. Cap screw
2. Seal
3. Front cover
4. Ball
5. Spring
6. Dowel
7. Diaphragm seal
8. Protector gasket
9. Back-up gasket
10. Diaphragm plate
11. Drive gear
12. Idler gear
13. Thrust plate
14. Pump housing

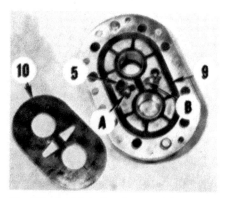

Fig. 172 – Install ball and spring (5) into hole with angle drilling (A). Refer to text.

beginning overhaul, check for parts availability. Valve components should be clearly identified and placed in order to facilitate proper reassembly. Do not interchange components between valves and use care when handling to prevent damage to machined surfaces.

To overhaul, first separate valve sections. Remove retainer plate (10 – Fig. 173), wiper ring (11) and "O" ring (12) from valve spool (13). Remove bolts, spacers (15) and cap (31). Withdraw spool (13) from bore, compress centering spring (24) and remove cap screws (27). Release compression tool and remove centering spring with related components from end of spool.

Tip spool (13) downward and catch detent parts (18, 19, 21, 22 and 23), then carefully slide detent sleeve (16) from spool while catching detent balls (20).

Clean and inspect all components carefully. Renew components that are worn excessively or otherwise damaged. Renew all "O" rings and seals, lubricate components with clean hydraulic fluid, then reassemble by reversing disassembly procedure. Reinstall spool assemblies in their respective bores.

Assemble valve sections as shown in Fig. 173. Make sure machined surfaces are clean and free of burrs. Tighten 3/8-inch retaining nut to 33 ft.-lbs. torque and 5/16-inch nuts to 14 ft.-lbs. torque.

Reinstall unit on tractor, check and add "Permatran" fluid to hydraulic system as required. Test and adjust system pressure as outlined in paragraph 149 and hydraulic kickout as outlined in paragraph 156.

156. TEST AND ADJUST HYDRAULIC KICKOUT. To test kickout pressure on remote control valves, install a 5000 psi gage into top remote coupler of control valve to be tested. Start engine and actuate corresponding auxiliary valve lever forward to first position and read kickout pressure. Adjust valve if kickout does not occur between 1900-2100 psi.

To adjust kickout pressure, first turn off engine, remove rubber plug (32 – Fig. 173), then turn screw (29) clockwise to increase or counterclockwise to decrease spool kickout pressure.

LUBRICATION CONTROL VALVE

The lubrication control valve is attached to oil cooler relief valve and dual hydraulic filters on left side of tractor. Lubrication control valve regulates oil flow to transmission at a minimum of 15 psi-4 gpm to a maximum of 32 psi-12 gpm.

157. R&R AND OVERHAUL. Lubrication control valve and oil cooler relief valve must be removed together, then separated. To remove, disconnect hydraulic lines, then unbolt and remove lube control valve/oil cooler relief valve assembly. Unbolt and separate oil cooler relief valve from lube control valve.

Remove plus (1 – Fig. 174) and withdraw spring (3) and spool (4). Check orifices in elbow fittings (6 and 9). Fitting (6) in center of valve body should have a 0.078 inch diameter orifice and fitting (9) at side of valve body should have a 0.030 inch diameter orifice.

Clean, inspect and renew valve components as required. Renew "O" rings and gaskets.

Reassemble valve tightening plug (1) to 25-35 ft.-lbs. torque. Reinstall valve by reversing removal procedure.

OIL COOLER RELIEF VALVE

The oil cooler relief valve is located between lubrication control valve and dual element hydraulic filters on left side of tractor. The oil cooler relief valve is used to prevent circuit pressure from exceeding 80 psi. Relieved oil is dumped back into hydraulic system filters.

158. R&R AND OVERHAUL. Oil cooler relief valve and lubrication control valve must be removed together, then separated. To remove, disconnect hydraulic lines, then unbolt and remove oil cooler relief/lube control valve assembly. Unbolt and separate oil cooler relief valve from lube control valve.

Overhaul procedure is evident after inspection of unit and referral to Fig. 175.

Lubricate all components with clean hydraulic oil when reassembling.

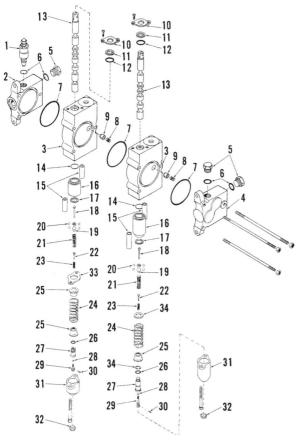

Fig. 173 – Exploded view of Husco auxiliary remote control valves.

1. Relief valve
2. Relief valve plate
3. Valve body
4. Outlet plate
5. Plug
6. "O" ring
7. "O" ring
8. Spring
9. Valve poppet
10. Retainer plate
11. Wiper ring
12. "O" ring
13. Spool
14. "O" ring
15. Spacer
16. Detent sleeve
17. Sleeve seal
18. Poppet
19. Detent cam
20. Detent balls
21. Detent spring
22. Guide
23. Guide spring
24. Centering spring
25. Retainer
26. "O" ring
27. Cap screw
28. "O" ring
29. Adjusting screw
30. Screw insert
31. Caps
32. Plug
33. Seal plate
34. Washer

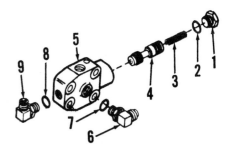

Fig. 174 – Exploded view of lubrication control valve.

1. Plug
2. "O" ring
3. Spring
4. Spool
5. Valve body
6. Elbow fitting (0.078 in. orifice)
7. "O" ring
8. "O" ring
9. Elbow fitting (0.030 in. orifice)

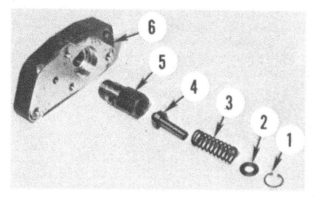

HYDRAULIC LIFT SYSTEM

The high volume/high pressure side of dual-element main hydraulic pump supplies fluid for system operation. Refer to main hydraulic section for testing procedures.

Draft and position controls, as shown in Fig. 176, regulate lift system operation. Draft control operates through a series of sensing devices to maintain a consistent working depth of implement and can be adjusted for smooth operation with intermix control knob (I – Fig. 177). Draft control also has a full up or transport position.

Position control is used for lifting and lowering implement when draft control is set at working depth. Position control can also be used in the same manner as draft control without the benefit of depth sensing devices.

QUICK CHECKS

All Models

159. Attach 1000 lbs. of weight to lift links, warm system fluid temperature to at least 68 degrees F., then raise links to the full transport position. A noticeable change in sound will indicate proper functioning of unloading valve. Measure position of links using scribe marks across lift cover and lift arm, shut off engine, then move control levers to full down position. Links should not move for five minutes (use scribe marks for reference). If links move, check adjustments as outlined in paragraph 161 or for internal leakage within system.

Move draft control lever (D – Fig. 176) to transport position, run engine at 1000 rpm then slowly move drift control to down position while observing lift linkage. Components should show no signs of excessive binding or looseness throughout travel.

Run engine at 2000 rpm and move depth control to full transport position.

Links should raise from full down position to transport position within 2½ seconds, indicating proper oil flow to linkage control valve. Repeat with draft control moved to full down position. Links should fully lower within 2 seconds. Repeat last two tests using position control lever, results should be the same.

ADJUSTMENTS

All Models

160. **INTERMIX LINKAGE.** To adjust intermix linkage, first unbolt and remove lift control valve cover. Lower lift arms to full down position, turn intermix control knob (I – Fig. 177) fully clockwise, then turn knob counterclockwise approximately eight full turns to locate follower arm (F – Fig. 178) at mid-position on intermix cam (C).

Disconnect control links (L – Fig. 179) from draft control arm (D). Loosen bolt (B) on position control arm (P). Install special gage tool number 6204 (G – Fig. 180) to rear surface of lift cover, then attach special tool number 6623 (T) between gage and draft control arm (D).

Disconnect vertical sensing rod (R – Fig. 181) from draft feed-back arm (A), then adjust socket of vertical sensing rod until ball on draft feed-back arm can be connected without binding or lifting vertical sensing rod. Adjustment is correct if end of breakout plunger (E) is

Fig. 176 – View of draft (D) and position (P) control levers for hydraulic lift system.

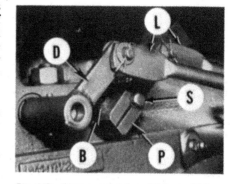

Fig. 179 – Disconnect external linkage (L) from draft control arm (D) and loosen bolt (B) so shaft (S) may turn during intermix linkage adjustment. Refer to text.

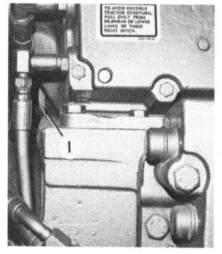

Fig. 177 – View of intermix control knob (I). Refer to text for adjustment procedure.

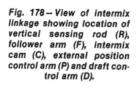

Fig. 178 – View of intermix linkage showing location of vertical sensing rod (R), follower arm (F), intermix cam (C), external position control arm (P) and draft control arm (D).

flush or extends beyond gage plate (G).

Tighten clamp bolt (B – Fig. 179), then remove special tools. Disconnect control links (L) and operate arms (D and P) at the same time in same direction and check components for freedom of movement without binding. Moving both arms in opposite directions should result in same free movement. Reconnect control links and install lift valve cover. Adjust position, draft and transport controls as outlined in paragraphs 161 and 162.

161. DRAFT AND TRANSPORT ADJUSTMENT. To adjust draft and transport control (D – Fig. 176), attach 1000 lbs. of weight to lift links. Start and run engine at 1500 rpm, then actuate draft control lever a few times to remove any air in hydraulic lift system. Remove plugs covering adjusting screws (T and D – Fig. 182).

Move draft control lever (D – Fig. 176) in full up or transport position and position control lever (P) in full down position. Lift cylinders should be fully extended and relief valve releasing as indicated by an audible sound. It may be necessary to turn transport adjusting screw (T – Fig. 182) counterclockwise until relief valve blows.

Scribe reference lines across lift cover casting and lift arm hub. Turn transport adjust screw (T) clockwise until relief valve just stops blowing and note distance between scribe lines. Scribe line separation of more than 1/8-inch may indicate inspection and repair of main

hydraulic pump, high pressure relief valve or lift control valve is required.

Move draft control lever to full down position lowering lift links completely, then move draft control lever 4 11/16 inches back from full down position. Turn draft adjusting screw (D) in the direction required to move lift arms to horizontal position. When adjustment is correct, weighted links will retain their position. Install plugs over adjusting screws.

162. POSITION ADJUSTMENT. To adjust position control (P – Fig. 176), run engine at 1500 rpm and place position control lever in full down position. Loosen clamp bolt (B – Fig. 179) on position control arm and rotate shaft (S) until lift links just start to rise. Tighten clamp bolt to 12-18 ft.-lbs. torque.

LIFT CONTROL VALVE

All Models

163. REMOVE AND REINSTALL. To remove lift control valve, first lower lift links to relieve all hydraulic system pressure. Disconnect hydraulic lines at control valve cover. Unbolt and withdraw cover with lift control valve. Remove elbows (1, 2 and 3 – Fig. 183) and standpipes (4, 5 and 6), then unbolt and separate control valve from cover. Make sure spool (8) does not fall out.

Check lift control valve actuating linkage for excessive wear or damage.

Renew linkage components as required. Renew all "O" rings and back-up washers. Reinstall control valve by reversing removal procedure. Perform control adjustments as outlined in paragraphs 160, 161 and 162.

164. OVERHAUL. Remove valve as outlined in paragraph 163. Disassembly is evident after inspection of valve and reference to Fig. 184. Using a suitable hand pump, check pressure setting of safety relief cartridge (9). Renew cartridge if relief pressure is not 2550 psi. Remove plug (A – Fig. 185) to obtain access to 0.030-0.032 inch dampening orifice (B). Pilot orifice (C – Fig. 186) should be 0.060 inch.

Clean, inspect and renew any components that are damaged or worn excessively. Renew all "O" rings and reassemble unit by reversing disassembly procedure.

LIFT COVER

All Models

165. REMOVE AND REINSTALL. To remove lift cover, it is first necessary to drain and remove the fuel tank. Disconnect draft and position control linkages at cover. Disconnect parking brake cable (C – Fig. 187), return spring (S), actuating lever (L) and bracket (B). Remove lift control valve cover as outlined in paragraph 163. Remove snap

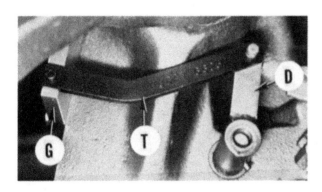

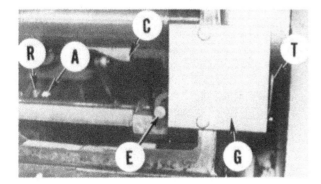

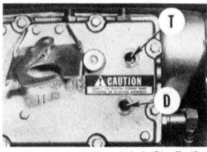

Fig. 180 — Attach special tool number 6623 (T) between gage plate (G) and draft control arm (D). Refer to text for details.

Fig. 181 — Intermix linkage adjustment is correct if breakout plunger (E) is flush or extends beyond gage plate (G).

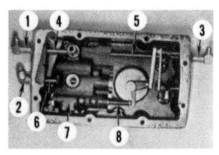

Fig. 182 — Transport (T) and draft (D) adjusting screw locations.

Fig. 183 — Remove elbow fittings (1, 2 and 3) and standpipes (4, 5 and 6) before separating lift control valve from cover.

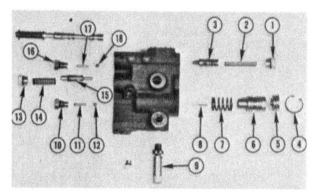

Fig. 184 – Disassembled view of lift control valve.
1. Plug
2. Spring
3. Inlet throttle valve
4. Snap ring
5. Plug
6. Servo piston
7. Large spring
8. Discharge valve pin
9. Safety relief valve
10. Plug
11. Spring
12. Steel ball
13. Plug
14. Spring
15. Unloading valve spool
16. Plug
17. Spring
18. Steel ball

guide studs to aid alignment and install new gasket onto top of center housing. Renew all standpipe "O" rings as required. Make sure differential lock standpipe, if equipped, is properly installed in center housing. Check adjustment of vertical sensing rod (R – Fig. 188) as outlined in paragraph 160.

Carefully lower cover into position making sure that binding does not occur. Remainder of installation is reverse of removal procedure. Adjust park brake as outlined in paragraph 130 and draft and position controls as outlined in paragraphs 161 and 162.

LIFT ARMS AND CROSS SHAFT

All Models

ring (S – Fig. 188) and disconnect position control link (L) from intermix cam (C). Remove standpipe (P) and disconnect vertical sensing rod (R). Disconnect hydraulic lines from top left side of lift cover. Slightly raise lift arms and secure by placing a bar between arms and cover to prevent damage to internal linkage. Disconnect lift rods and cylinders from lift arms. Remove lift cover retaining cap screws and nuts, then carefully remove cover using a suitable hoist. Make sure components do not bind or catch.

To reinstall lift cover, first install

166. **REMOVE AND REINSTALL.** To remove lift arms and cross shaft, first remove lift cover as outlined in paragraph 165. Remove retaining cap screw (1 – Fig. 189), lock (2) and washer (3), then slide arms (4) off of shaft (5). Unbolt and remove intermix cam (6), then withdraw shaft (5) from either side of cover (13). Reinstall by reversing removal procedure.

167. **OVERHAUL.** Refer to Fig. 189 for exploded view of lift arm, cross shaft and related components. Bushing (11A) and "O" ring seal (10A) are used in place of components (7 through 12) on models prior to rear axle serial number DG 1554. Overhaul procedures will be for early model cross shaft assemblies. Overhaul procedures for late models will

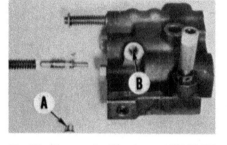

Fig. 185 – Remove plug (A) to inspect 0.030-0.032 inch dampening orifice (B).

Fig. 188 – Remove snap ring (S) and disconnect link (L) from cam (C), then remove standpipe (P) and disconnect rod (R) before removing lift cover.

Fig. 186 – Pilot orifice (C) should be 0.060 inch.

Fig. 189 – Exploded view of lift arms and cross shaft with related components.
1. Cap screw
2. Lock plate
3. Washer
4. Lift arm
5. Cross shaft
6. Intermix cam
7. Dust seal
8. Seal
9. Sleeve
10. "O" ring
10A. "O" ring
11. Bushing
11A. Bushing
12. Snap ring
13. Lift cover

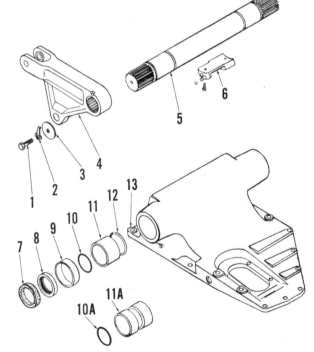

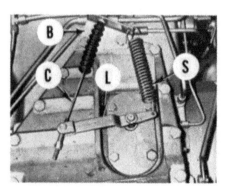

Fig. 187 – Disconnect park brake cable (C) and spring (S) from lever (L), then remove lever (L) and bracket (B) from lift cover.

be evident after inspection of unit and referral to Fig. 189.

After removing cross shaft as outlined in paragraph 166, remove "O" ring seals (10A) from shaft. Drive out bushings (11A) from lift cover (13). Coat outside of new bushings with "Loctite" and install in lift cover slightly below chamfered edge at each end. Insert cross shaft with bolt holes for intermix cam (6) toward right side. Install special tool No. 7072 (T – Fig. 190) over right side of shaft with flared end of tool squarely into bushing chamfer. Lubricate seal (10A), then install over shaft into seal groove. Push shaft into

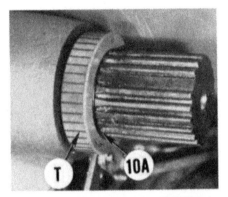

Fig. 190 — Use special tool number 7072 (T) to install seal (10A). Refer to text.

lift cover while holding special tool. Remove tool and repeat procedure for opposite side.

Remainder of reassembly is the reverse of disassembly.

LIFT CYLINDERS

All Models

168. **R&R AND OVERHAUL.** Before disconnecting lift cylinders, raise and block lift arms to avoid damage to internal controls of three-point linkage. Disconnect hydraulic line at cylinder then disconnect cylinder from lift arm. Remove bolt securing lower collar to lower pin. Use a puller bolt, nut, plate and a piece of pipe to pull pin enough to permit collar removal. Bolt hole is slightly offset toward lift cylinder. Pull pin and remove lift cylinder.

Refer to Fig. 191 for exploded view of lift cylinder. To disassemble the cylinder, refer to Fig. 192. Remove hose fitting and work through open port to move retainer ring into deep groove in piston rod. Piston rod can now be withdrawn. Assemble by reversing disassembly sequence.

SENSING SHAFT

All Models

169. **R&R AND OVERHAUL.** The following service procedures will cover sensing shaft assemblies used prior to rear axle serial number DH 1334. Ser-

vice procedures for assemblies used after rear axle serial number DH 1333 will be similar. Refer to the appropriate exploded view Fig. 193 or Fig. 194 when servicing.

To remove sensing shaft and linkage, first block lift arms to prevent damage to internal linkage when lift arms are removed. Drain center housing then remove lower links from sensing shaft. Remove lift cylinders following procedure outlined in paragraph 168 and lift cover as outlined in paragraph 165.

Remove cap screws (1 and 17 – Fig. 193) with washers and square keys, plates (2 and 16), seals (3 and 15) and snap rings (4 and 14). Remove spring (23) and loosen sensing arm clamp bolt (19). Slide sensing shaft (7) towards right side of tractor and push right bushing (5) and thrust washer (6) out of case. Continue withdrawing until guide plate (11) is free of pin (24). Slide guide plate (11), draft signal cam (10), with vertical sensing rod (25) still attached, and washer (9) from left end of shaft. Slide sensing shaft towards left side of tractor and push left bushing (13) and thrust washer (12) out of case. Continue withdrawing shaft out of case leaving sensing arm assembly in housing, then withdraw sensing arm assembly.

Clean, inspect and renew any components that are damaged or worn excessively. Renew snap rings (4 and 14) and seals (3 and 15). If guide pin (24) was removed, apply "Loctite" to threads when reinstalling.

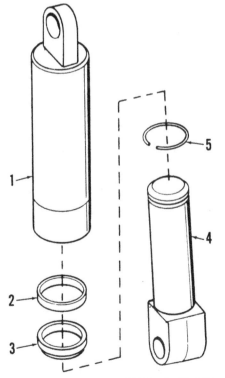

Fig. 191 — Exploded view of hydraulic lift cylinder.

1. Barrel	4. Rod assembly
2. Barrel seal	5. Retainer ring
3. Wiper seal	

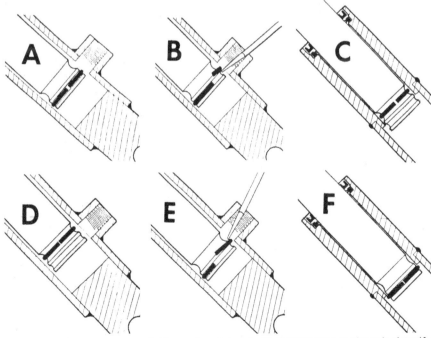

Fig. 192 — Cylinder is disassembled by moving retainer ring to deeper groove as shown in views (A, B and C). Reassemble with retainer ring in deep groove, then move ring to shallow groove at end as shown in views (E, E and F). Retainer can be moved from one groove to the other using screwdriver or similar tool through port opening as shown.

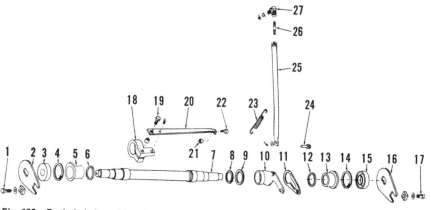

Fig. 193 — Exploded view of sensing shaft assembly used prior to rear axle serial number DH 1334.

1. Cap screw	8. Snap ring	15. Seal
2. Plate	9. Washer	16. Plate
3. Seal	10. Draft signal cam	17. Cap screw
4. Snap ring	11. Guide plate	18. Clamp
5. Bushing	12. Thrust washer	19. Clamp bolt
6. Thrust washer	13. Bushing	20. Sensing arm
7. Sensing shaft	14. Snap ring	21. Spacer

22. Roller
23. Spring
24. Pin
25. Sensing rod
26. Rod stud
27. Ball joint link

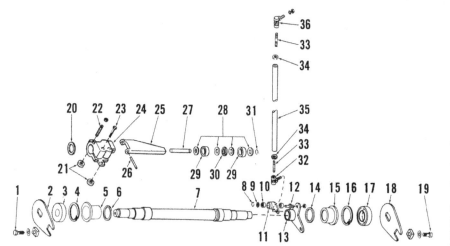

Fig. 194 — Exploded view of sensing shaft assembly used after rear axle serial number DH 1333.

1. Cap screw	10. Bearing	19. Cap screw	28. Thrust race
2. Plate	11. Draft signal cam	20. Thrust washer	29. Roller assembly
3. Seal	12. Pivot shaft	21. Distributor pad	30. Thrust bearing
4. Snap ring	13. Guide plate	22. Adjusting screw	31. Retaining ring
5. Bushing	14. Thrust washer	23. Clamp bolt	32. Ball joint link
6. Thrust washer	15. Bushing	24. Clamp	33. Rod stud
7. Sensing shaft	16. Snap ring	25. Sensing arm	34. Locknut
8. Retaining ring	17. Seal	26. Pin	35. Sensing rod
9. Washer	'18. Plate	27. Shaft	36. Ball joint link

To reassemble sensing shaft unit, first install sensing arm assembly (18 through 22) with clamp (18) towards right side and head of bolt (19) up and towards front. Install snap ring (8) on shaft (7), then insert shaft through left side of center housing and into clamp (18). Slide shaft far enough through right side of center housing to permit installation of sensing cam components. Slide the 0.120 inch thick washer (9) over left end of shaft with chamfered side towards snap ring (8). Install signal cam (10) and vertical rod (25) onto shaft with cam arm away from washer (9). Install guide plate (11) with narrowest side

Fig. 195 — Check end play on sensing shaft (7) using a dial indicator as illustrated. Refer to text for details.

of slot up. Slide sensing shaft (7) left through bore in housing and install hole in guide plate (11) over pin (24). Roller (22) on sensing arm must enter slot in guide plate (11).

Install a 0.156 inch thrust washer (12), bushing (13) and snap ring (14) on left side of shaft. Push sensing shaft toward left fully seating components. Install bushing (5) and snap ring (4) on right end of shaft without thrust washer (6) then check and adjust shaft end play. Attach a dial indicator as shown in Fig. 195 to check end play. Shaft end play should be 0.020-0.050 inch and is adjusted by thrust washer (6 – Fig. 193). Thrust washers are available in 0.125, 0.156 and 0.188 inch sizes. Remove snap ring (4) and bushing (5), then install thrust washer with chamfered side away from bushing. Reinstall bushing and snap ring.

Install seals (3 and 15), lightly coating outside of seal with "Loctite", then press seals into housing 0.010 inch below flush. Special tool number 6620 is recommended to prevent damage to seal when installing. Install plates (2 and 16), square keys, washers and cap screws (1 and 17).

Rotate sensing arm assembly (18 through 22) as required to center roller (22) in slot of guide plate (11), then move arm assembly left or right on shaft (7) as required until 0.010 inch clearance is obtained between signal cam (10) and end of sensing arm (20). Tighten clamp bolt (19) to 80-90 ft.-lbs. torque. Connect spring (23) between vertical sensing rod (25) and draft signal cam (10).

Remainder of procedure is reverse of disassembly. Adjust linkage as outlined in paragraphs 160, 161 and 162.

NOTES

NOTES

NOTES

NOTES

NOTES

NOTES